Springer Optimization and Its Applications

VOLUME 119

Aims and Scope
Optimization has been expanding in all directions at an astonishing rate during the last few decades. New algorithmic and theoretical techniques have been developed, the diffusion into other disciplines has proceeded at a rapid pace, and our knowledge of all aspects of the field has grown even more profound. At the same time, one of the most striking trends in optimization is the constantly increasing emphasis on the interdisciplinary nature of the field. Optimization has been a basic tool in all areas of applied mathematics, engineering, medicine, economics, and other sciences.

The series *Springer Optimization and Its Applications* publishes undergraduate and graduate textbooks, monographs and state-of-the-art expository work that focus on algorithms for solving optimization problems and also study applications involving such problems. Some of the topics covered include nonlinear optimization (convex and nonconvex), network flow problems, stochastic optimization, optimal control, discrete optimization, multi-objective programming, description of software packages, approximation techniques and heuristic approaches.

More information about this series at http://www.springer.com/series/7393

Alexander J. Zaslavski

Discrete-Time Optimal Control and Games on Large Intervals

Springer

Alexander J. Zaslavski
Department of Mathematics
The Technion – Israel Institute
of Technology
Haifa, Israel

ISSN 1931-6828 ISSN 1931-6836 (electronic)
Springer Optimization and Its Applications
ISBN 978-3-319-52931-8 ISBN 978-3-319-52932-5 (eBook)
DOI 10.1007/978-3-319-52932-5

Library of Congress Control Number: 2017933842

Mathematics Subject Classification (2010): 49J27, 49K40, 90C26, 90C31, 90C48, 93C25

Printed on acid-free paper

This Springer imprint is published by Springer Nature
The registered company is Springer International Publishing AG
The registered company address is: Gewerbestrasse 11, 6330 Cham, Switzerland

Preface

The book is devoted to the study of the structure of approximate solutions of discrete-time optimal control problems and approximate solutions of dynamic discrete-time two-player zero-sum games. We present a number of results on properties of approximate solutions which are independent of the length of the interval, for all sufficiently large intervals. These results were obtained by the author during the last five years. Two types of results on the structure of solutions on large intervals are considered in the book. The results of the first type deal with the so-called turnpike property of optimal control problems, a term which was first coined by P. Samuelson in 1948 when he showed that an efficient expanding economy would spend most of the time in the vicinity of a balanced equilibrium path (also called a von Neumann path). To have the turnpike property means, roughly speaking, that the approximate solutions of the problems are determined mainly by the objective function (integrand) and are essentially independent of the choice of interval and endpoint conditions, except in regions close to the endpoints. Now, it is known that the turnpike property is a general phenomenon which holds for large classes of variational problems. For these classes of problems, using the Baire category (generic) approach, it was shown that the turnpike property holds for a generic (typical) variational problem. In this book, we are interested in individual turnpike results for discrete-time optimal control problems and for dynamic discrete-time two-player zero-sum games.

The results of the second type deal with the structure of approximate solutions of optimal control problems and zero-sum games in the regions close to the endpoints of the time intervals. It is shown that in these regions, solutions are close to solutions of the corresponding infinite horizon problems. This is a new direction in infinite horizon optimal control developed in the recent research of the author.

The monograph contains eight chapters. Chapter 1 is an introduction. In Chap. 2, we study the structure of approximate solutions of autonomous nonconcave discrete-time optimal control Lagrange problems. In this chapter, our main goal is to analyze the structure of approximate solutions in regions close to the endpoints of the time intervals. In Chap. 3, we study the structure of approximate solutions of autonomous discrete-time optimal control problems which are discrete-time analogs of Bolza problems in the calculus of variations. They are described by a pair of objective functions which determines an optimality criterion. We consider two classes of Bolza problems and obtain for each of them the full description of approximate solutions of these problems on large intervals.

In Chap. 4, for discrete-time optimal control Bolza problems, it is shown that the turnpike phenomenon and the structure of solutions on finite intervals in the regions close to the endpoints are stable under small perturbations of the objective functions and the constraint sets.

In Chaps. 5–8, we analyze the structure of approximate solutions of two-player zero-sum games. In Chap. 5, we consider unconstrained dynamic games without using standard convexity–concavity assumptions and prove two turnpike results. Chapter 6 is devoted to the turnpike theory for a class of dynamic constrained games without using convexity–concavity assumptions. In Chap. 7, we study turnpike properties of approximate solutions for a class of nonautonomic dynamic discrete-time games with convexity–concavity assumptions. In Chap. 8, we study the existence and turnpike properties of approximate solutions for a class of dynamic constrained discrete-time two-player zero-sum games which satisfy convexity–concavity assumptions.

The results of Chaps. 2, 3, and 5–7 were obtained in our recent research, while the results of Chaps. 4 and 8 are new.

In the turnpike theory, there are many open problems, and it can be developed in many various directions. One of them is a generalization of the results presented in this book for problems with nonsingleton turnpikes. In principle, one can study turnpike properties of any variational or optimal control problem.

Rishon LeZion, Israel
August 19, 2016

Alexander J. Zaslavski

Contents

1

Introduction

The study of optimal control problems and dynamic games defined on infinite intervals and on sufficiently large intervals has been a rapidly growing area of research [1, 4, 5, 9–16, 21, 23, 24, 26, 27, 29–33, 35, 36, 40, 41, 43, 44, 49–52, 54–56, 61] which has various applications in engineering [2, 19, 38, 70], in models of economic growth [3, 6, 17–20, 25, 28, 34, 37, 42, 48, 53, 59, 60, 62, 70, 79, 85, 94, 95], in infinite discrete models of solid-state physics related to dislocations in one-dimensional crystals [8, 57, 63], and in the theory of thermodynamical equilibrium for materials [22, 39, 45–47].

In this chapter we discuss turnpike properties and optimality criterions over infinite horizon for a class of convex dynamic optimization problems.

1.1 Convex Discrete-Time Problems

Let R^n be the n-dimensional Euclidean space with the inner product $\langle \cdot, \cdot \rangle$ which induces the norm

$$|x| = (\sum_{i=1}^{n} x_i^2)^{1/2}, \; x = (x_1, \dots, x_n) \in R^n.$$

Let K be a nonempty convex subset of R^n. A function $f : K \to R^1$ is called convex (strictly convex respectively) if for all $x, y \in K$ such that $x \neq y$ and all $\alpha \in (0, 1)$,

$$f(\alpha x + (1-\alpha)y) \leq \alpha f(x) + (1-\alpha)f(y)$$

$$(f(\alpha x + (1-\alpha)y) < \alpha f(x) + (1-\alpha)f(y)$$

respectively) [58].

A.J. Zaslavski, *Discrete-Time Optimal Control and Games on Large Intervals*, Springer Optimization and Its Applications 119,
DOI 10.1007/978-3-319-52932-5_1

Let $v : R^n \times R^n \to R^1$ be a bounded from below function. We consider the minimization problem

$$\sum_{i=0}^{T-1} v(x_i, x_{i+1}) \to \min, \tag{P_0}$$

$$\text{such that } \{x_i\}_{i=0}^{T} \subset R^n \text{ and } x_0 = z,\ x_T = y,$$

where T is a natural number and the points $y, z \in R^n$.

The interest in discrete-time optimal problems of type (P_0) stems from the study of various optimization problems which can be reduced to it, e.g., continuous-time control systems which are represented by ordinary differential equations whose cost integrand contains a discounting factor [37], tracking problems in engineering [2, 38], the study of Frenkel–Kontorova model [8, 63], and the analysis of a long slender bar of a polymeric material under tension in [22, 39, 45–47]. Optimization problems of the type (P_0) were considered in [64–66, 68, 69, 74].

In this section we suppose that the function $v : R^n \times R^n \to R^1$ is strictly convex and differentiable and satisfies the growth condition

$$v(y, z)/(|y| + |z|) \to \infty \text{ as } |y| + |z| \to \infty. \tag{1.1}$$

We intend to analyze the structure of solutions of the problem (P_0) when the points y, z and the real number T vary and T is sufficiently large. More precisely, we are interested to study a turnpike property of solutions of (P_0) which is independent of the length of the interval T, for all sufficiently large intervals. To have this property means, roughly speaking, that solutions of the optimal control problems are determined mainly by the objective function v, and are essentially independent of T, y, and z. Turnpike properties are well known in mathematical economics (see, for example, [42, 48, 59, 60] and the references mentioned there). Many turnpike results are collected in [70, 81, 84, 93, 95].

In order to meet our goal we consider the auxiliary optimization problem

$$v(x, x) \to \min,\ x \in R^n. \tag{P_1}$$

In view of the strict convexity of v and (1.1), problem (P_1) has a unique solution $\bar{x}$. Let

$$\nabla v(\bar{x}, \bar{x}) = (l_1, l_2), \tag{1.2}$$

where $l_1,\ l_2 \in R^n$. Since $\bar{x}$ is a solution of (P_1) it follows from (1.2) that for every $h \in R^n$, we have

$$\langle l_1, h\rangle + \langle l_2, h\rangle = \langle (l_1, l_2), (h, h)\rangle$$
$$= \lim_{t \to 0^+} t^{-1}[v(\bar{x} + th, \bar{x} + th) - v(\bar{x}, \bar{x})] \geq 0.$$

Therefore

$$\langle l_1 + l_2, h \rangle \geq 0 \text{ for all } h \in R^n,$$

$l_2 = -l_1$ and

$$\nabla v(\bar{x}, \bar{x}) = (l_1, -l_1), \tag{1.3}$$

For every $(y, z) \in R^n \times R^n$ define

$$\begin{aligned} L(y, z) &= v(y, z) - v(\bar{x}, \bar{x}) - \langle \nabla v(\bar{x}, \bar{x}), (y - \bar{x}, z - \bar{x}) \rangle \\ &= v(y, z) - v(\bar{x}, \bar{x}) - \langle l_1, y - z \rangle. \end{aligned} \tag{1.4}$$

It is easy to see that the function $L : R^n \times R^n \to R^1$ is differentiable and strictly convex. By (1.1) and (1.4), we have

$$L(y, z)/(|y| + |z|) \to \infty \text{ as } |y| + |z| \to \infty. \tag{1.5}$$

Since the functions v and L are both strictly convex [58] it follows from (1.4) that

$$L(y, z) \geq 0 \text{ for all } (y, z) \in R^n \times R^n \tag{1.6}$$

and

$$L(y, z) = 0 \text{ if and only if } y = \bar{x}, \ z = \bar{x}. \tag{1.7}$$

We show that the function $L : R^n \times R^n \to R^1$ possesses the following property:

(C) If a sequence $\{(y_i, z_i)\}_{i=1}^{\infty} \subset R^n \times R^n$ satisfies the equality

$$\lim_{i \to \infty} L(y_i, z_i) = 0,$$

then

$$\lim_{i \to \infty} (y_i, z_i) = (\bar{x}, \bar{x}).$$

Assume that a sequence $\{(y_i, z_i)\}_{i=1}^{\infty} \subset R^n \times R^n$ and that $\lim_{i \to \infty} L(y_i, z_i) = 0$. It follows from (1.5) that the sequence $\{(y_i, z_i)\}_{i=1}^{\infty}$ is bounded. Let (y, z) be its limit point. Then it is not difficult to see that the equality

$$L(y, z) = \lim_{i \to \infty} L(y_i, z_i) = 0$$

is true and in view of (1.7), we have $(y, z) = (\bar{x}, \bar{x})$. This implies that $(\bar{x}, \bar{x}) = \lim_{i \to \infty} (y_i, z_i)$.

Therefore property (C) holds.

Consider an auxiliary minimization problem

$$\sum_{i=0}^{T-1} L(x_i, x_{i+1}) \to \min, \tag{P_2}$$

$$\text{such that } \{x_i\}_{i=0}^{T} \subset R^n \text{ and } x_0 = z, \ x_T = y,$$

where $T \geq 1$ is an integer and $y, z \in R^n$.

By (1.4), for every natural number T and every sequence $\{x_i\}_{i=0}^T \subset R^n$,

$$\sum_{i=0}^{T-1} L(x_i, x_{i+1}) = \sum_{i=0}^{T-1} v(x_i, x_{i+1}) - Tv(\bar{x}, \bar{x}) - \sum_{i=0}^{T-1} \langle l_1, x_i - x_{i+1} \rangle$$

$$= \sum_{i=0}^{T-1} v(x_i, x_{i+1}) - Tv(\bar{x}, \bar{x}) - \langle l_1, x_0 - x_T \rangle. \tag{1.8}$$

It follows from (1.8) that problems (P_0) and (P_2) are equivalent. More precisely, $\{x_i\}_{i=0}^T \subset R^n$ is a solution of problem (P_0) if and only if it is a solution of problem (P_2).

Let $T \geq 1$ be an integer and Δ be a positive number. A sequence $\{x_i\}_{i=0}^T \subset R^n$ is called (Δ)-optimal if for every sequence $\{x_i'\}_{i=0}^T \subset R^n$ which satisfies $x_i = x_i'$, $i = 0, T$ the inequality

$$\sum_{i=0}^{T-1} v(x_i, x_{i+1}) \leq \sum_{i=0}^{T-1} v(x_i', x_{i+1}') + \Delta$$

is valid. It is clear that if a sequence $\{x_i\}_{i=0}^T \subset R^n$ is (0)-optimal, then it is a solution of problems (P_0) and (P_2) with $z = x_0$ and $y = x_T$.

We prove the following existence result.

Proposition 1.1. *Let $T > 1$ be an integer and $y, z \in R^n$. Then problem (P_0) possesses a solution.*

Proof. In order to prove the proposition it is sufficient to show that problem (P_2) possesses a solution. Consider a sequence $\{x_i'\}_{i=0}^T \subset R^n$ which satisfies $x_0' = z$, $x_T' = y$. Put

$$M_1 = \sum_{i=0}^{T-1} L(x_i', x_{i+1}')$$

and

$$M_2 = \inf\{\sum_{i=0}^{T-1} L(x_i, x_{i+1}) : \ \{x_i\}_{i=0}^T \subset R^n, \ x_0 = z, \ x_T = y\}. \tag{1.9}$$

Evidently,

$$0 \leq M_2 \leq M_1.$$

We may assume without loss of generality that

$$M_2 < M_1. \tag{1.10}$$

There exists a sequence $\{x_i^{(k)}\}_{i=0}^T \subset R^n$, $k = 1, 2, \ldots$ such that for every integer $k \geq 1$, we have

$$x_0^{(k)} = z, \ x_T^{(k)} = y \tag{1.11}$$

and

$$\lim_{k\to\infty} \sum_{i=0}^{T-1} L(x_i^{(k)}, x_{i+1}^{(k)}) = M_2. \tag{1.12}$$

By (1.10), (1.11), and (1.12), we may assume that

$$\sum_{i=0}^{T-1} L(x_i^{(k)}, x_{i+1}^{(k)}) < M_1 \text{ for all integers } k \geq 1. \tag{1.13}$$

It follows from (1.13) and (1.5) that there exists a positive number M_3 such that

$$|x_i^{(k)}| \leq M_3 \text{ for all } i = 0, \dots, T \text{ and all integers } k \geq 1. \tag{1.14}$$

By (1.14), extracting subsequences, using diagonalization process and re-indexing, if necessary, we may assume without loss of generality that for every integer $i \in \{0, \dots, T\}$ there exists

$$\widehat{x}_i = \lim_{k\to\infty} x_i^{(k)}. \tag{1.15}$$

In view of (1.15) and (1.11), we have

$$\widehat{x}_0 = z, \ \widehat{x}_T = y. \tag{1.16}$$

Relations (1.15) and (1.12) imply that

$$\sum_{i=0}^{T-1} L(\widehat{x}_i, \widehat{x}_{i+1}) = M_2.$$

Combined with (1.16) and (1.9) this implies that the finite sequence $\{\widehat{x}_i\}_{i=0}^T$ is a solution of problem (P_2). This completes the proof of Proposition 1.1.

Denote by Card(A) the cardinality of a set A.

The following result establishes a turnpike property for approximate solutions of the problem (P_0).

Proposition 1.2. *Let $M_1, M_2, \epsilon > 0$. Then there exists an integer $k_0 \geq 1$ such that for every natural number $T > 1$ and every (M_1)-optimal sequence $\{x_i\}_{i=0}^T \subset R^n$ which satisfies*

$$|x_0| \leq M_2, \ |x_T| \leq M_2 \tag{1.17}$$

the inequality

$$Card(\{i \in \{0, \dots, T-1\} : \ |x_i - \bar{x}| + |x_{i+1} - \bar{x}| > \epsilon\}) \leq k_0$$

is valid.

Proof. Condition (C) implies that there exists a positive number δ such that for every point $(y, z) \in R^n \times R^n$ which satisfies

$$L(y, z) \le \delta \tag{1.18}$$

the inequality

$$|y - \bar{x}| + |z - \bar{x}| \le \epsilon \tag{1.19}$$

is true. Set

$$M_3 = \sup\{L(y, z) : \; y, z \in R^n \text{ and } |y| + |z| \le |\bar{x}| + M_2\} \tag{1.20}$$

and fix a natural number

$$k_0 > \delta^{-1}(M_1 + 2M_3). \tag{1.21}$$

Assume that an integer $T > 1$ and that an (M_1)-optimal sequence $\{x_i\}_{i=0}^T \subset R^n$ satisfies (1.17). Define

$$y_0 = x_0, \; y_T = x_T, \; y_i = \bar{x}, \; i = 1, \dots, T - 1. \tag{1.22}$$

Since the sequence $\{x_i\}_{i=0}^T$ is (M_1)-optimal it follows from (1.22) that

$$\sum_{i=0}^{T-1} v(x_i, x_{i+1}) \le \sum_{i=0}^{T-1} v(y_i, y_{i+1}) + M_1.$$

Combined with (1.7), (1.8), and (1.22) this implies that

$$\sum_{i=0}^{T-1} L(x_i, x_{i+1}) \le \sum_{i=0}^{T-1} L(y_i, y_{i+1}) + M_1 = L(x_0, \bar{x}) + L(\bar{x}, x_T) + M_1.$$

Together with (1.17) and (1.20) this implies that

$$\sum_{i=0}^{T-1} L(x_i, x_{i+1}) \le M_1 + 2M_3.$$

In view of the choice of δ (see (1.18) and (1.19)), (1.21) and the inequality above, we have

$$\begin{aligned} &\mathrm{Card}(\{i \in \{0, \dots, T-1\} : \; |x_i - \bar{x}| + |x_{i+1} - \bar{x}| > \epsilon\}) \\ &\le \mathrm{Card}(\{i \in \{0, \dots, T-1\} : \; L(x_i, x_{i+1}) > \delta\}) \\ &\le \delta^{-1} \sum_{i=0}^{T-1} L(x_i, x_{i+1}) \le \delta^{-1}(M_1 + 2M_3) \le k_0. \end{aligned}$$

Proposition 1.2 is proved.

Proposition 1.2 implies the following turnpike result for exact solutions of the problem (P_0).

Proposition 1.3. *Let $M, \epsilon > 0$. Then there exists an integer $k_0 \geq 1$ such that for every natural number $T > 1$, every pair of points $y, z \in R^n$ which satisfies $|y|,\ |z| \leq M$ and every optimal solution $\{x_i\}_{i=0}^T \subset R^n$ of problem (P_0) the inequality*

$$Card(\{i \in \{0, \dots, T-1\} :\ |x_i - \bar{x}| + |x_{i+1} - \bar{x}| > \epsilon\}) \leq k_0$$

is valid.

Now it is clear that the optimal solution $\{x_i\}_{i=0}^T$ of problem (P_0) spends most of the time in an ϵ-neighborhood of the point $\bar{x}$. In view of Proposition 1.3, the number of all integers $i \in \{0, \dots, T-1\}$ for which that x_i does not belong to this ϵ-neighborhood, does not exceed the constant k_0 which depends only on M, ϵ and does not depend on T. Following the tradition, the point $\bar{x}$ is called the turnpike. Moreover we can show that the set

$$\{i \in \{0 \dots, T\} : |x_i - \bar{x}| > \epsilon\}$$

is contained in the union of two intervals $[0, k_1] \cup [T - k_1, T]$, where k_1 is a constant depending only on M, ϵ.

We also study the infinite horizon problem associated with problem (P_0).

By (1.1) there is $M_* > 0$ such that

$$v(y, z) > |v(\bar{x}, \bar{x})| + 1 \tag{1.23}$$

for any $(y, z) \in R^n \times R^n$ satisfying $|y| + |z| \geq M_*$.

We suppose that the sum over empty set is zero.

Proposition 1.4. *Let M_0 be a positive number. Then there exists a positive number M_1 such that for every natural number T and every finite sequence $\{x_i\}_{i=0}^T \subset R^n$ which satisfies $|x_0| \leq M_0$ the inequality*

$$\sum_{i=0}^{T-1} v(x_i, x_{i+1}) \geq Tv(\bar{x}, \bar{x}) - M_1 \tag{1.24}$$

holds.

Proof. Set

$$M_1 = |l_1|(M_0 + M_*).$$

Assume that T is a natural number and a that a finite sequence $\{x_i\}_{i=0}^T \subset R^n$ satisfies

$$|x_0| \leq M_0. \tag{1.25}$$

If $|x_i| > M_*$, $i = 1, \dots, T$, then in view of (1.23), we have

$$\sum_{i=0}^{T-1} v(x_i, x_{i+1}) \geq T v(\bar{x}, \bar{x})$$

and inequality (1.24) is true. Therefore we may assume that there exists an integer $q \geq 1$ for which

$$q \leq T, \ |x_q| \leq M_*. \tag{1.26}$$

We may assume without loss of generality that

$$|x_i| > M_* \text{ for all integers } i \text{ satisfying } q < i \leq T. \tag{1.27}$$

In view of (1.23) and (1.27), we have

$$\sum_{i=0}^{T-1} (v(x_i, x_{i+1}) - v(\bar{x}, \bar{x})) = \sum_{i=0}^{q-1} (v(x_i, x_{i+1}) - v(\bar{x}, \bar{x}))$$

$$+ \sum \{v(x_i, x_{i+1}) - v(\bar{x}, \bar{x})) : \text{ an integer } i \text{ satisfies } q \leq i < T\}$$

$$\geq \sum_{i=0}^{q-1} (v(x_i, x_{i+1}) - v(\bar{x}, \bar{x})).$$

By the equation above, (1.8), (1.6), (1.25), (1.26) and the choice of M_1, we have

$$\sum_{i=0}^{T-1} (v(x_i, x_{i+1}) - v(\bar{x}, \bar{x})) \geq \sum_{i=0}^{q-1} (v(x_i, x_{i+1}) - v(\bar{x}, \bar{x}))$$

$$= \sum_{i=0}^{q-1} L(x_i, x_{i+1}) + \langle l_1, x_0 - x_q \rangle \geq -|l_1|(|x_0| + |x_q|)$$

$$\geq -|l_1|(M_0 + M_*) = -M_1.$$

This completes the proof of Proposition 1.4.

Choose a positive number $\tilde{M}$ for which

$$\text{Proposition 1.4 holds with } M_0 = M_* \text{ and } M_1 = \tilde{M}. \tag{1.28}$$

Proposition 1.5. *Let $\{x_i\}_{i=0}^{\infty} \subset R^n$. Then either the sequence*

$$\{\sum_{i=0}^{T-1} (v(x_i, x_{i+1}) - v(\bar{x}, \bar{x}))\}_{T=1}^{\infty}$$

is bounded or

$$\lim_{T \to \infty} \sum_{i=0}^{T-1} (v(x_i, x_{i+1}) - v(\bar{x}, \bar{x})) = \infty. \tag{1.29}$$

Proof. In view of (1.23), if for all sufficiently large natural numbers i, we have $|x_i| \geq M_*$, then equality (1.29) is valid. Therefore we may assume without loss of generality that there exists a strictly increasing sequence of natural numbers $\{t_k\}_{k=1}^{\infty}$ such that

$$|x_{t_k}| < M_* \text{ for all integers } k \geq 1. \tag{1.30}$$

Proposition 1.4 implies that the sequence $\{\sum_{i=0}^{T-1}(v(x_i, x_{i+1}) - v(\bar{x}, \bar{x}))\}_{T=1}^{\infty}$ is bounded from below.

Assume that this sequence is not bounded from above. In order to complete the proof it is sufficient to show that equality (1.29) is valid.

Let $Q > 0$ be given. Then there exists an integer $T_0 \geq 1$ for which

$$\sum_{i=0}^{T_0-1}(v(x_i, x_{i+1}) - v(\bar{x}, \bar{x})) > Q + \tilde{M}. \tag{1.31}$$

Fix an integer $k \geq 1$ such that

$$t_k > T_0 + 4. \tag{1.32}$$

Let an integer

$$T > t_k. \tag{1.33}$$

In view of (1.30), (1.32), and (1.33), there exists an integer S such that

$$T > S \geq T_0, \tag{1.34}$$

$$|x_S| \leq M_*, \tag{1.35}$$

$$|x_t| > M_* \text{ for all integers } t \text{ satisfying} \tag{1.36}$$
$$S > t \geq T_0.$$

By (1.31), (1.34), (1.36), (1.23), (1.35), (1.28), and Proposition 1.4, we have

$$\sum_{i=0}^{T-1}(v(x_i, x_{i+1}) - v(\bar{x}, \bar{x})) = \sum_{i=0}^{T_0-1}(v(x_i, x_{i+1}) - v(\bar{x}, \bar{x}))$$
$$+ \sum\{v(x_i, x_{i+1}) - v(\bar{x}, \bar{x}) : \ i \text{ is an integer and } T_0 \leq i < S\}$$
$$+ \sum_{i=S}^{T-1}(v(x_i, x_{i+1}) - v(\bar{x}, \bar{x}))$$
$$> Q + \tilde{M} + \sum_{i=S}^{T-1}(v(x_i, x_{i+1}) - v(\bar{x}, \bar{x})) > Q.$$

Thus for any integer $T > t_k$,

$$\sum_{i=0}^{T-1}(v(x_i, x_{i+1}) - v(\bar{x}, \bar{x})) > Q.$$

Since Q is any positive number (1.29) is valid. This completes the proof of Proposition 1.5.

A sequence $\{x_i\}_{i=0}^{\infty} \subset R^n$ is called good [28, 70, 81] if the sequence $\{\sum_{i=0}^{T-1}(v(x_i, x_{i+1}) - v(\bar{x}, \bar{x}))\}_{T=1}^{\infty}$ is bounded.

Proposition 1.6. *1. A sequence $\{x_i\}_{i=0}^{\infty} \subset R^n$ is good if and only if*

$$\sum_{i=0}^{\infty} L(x_i, x_{i+1}) < \infty.$$

2. If a sequence $\{x_i\}_{i=0}^{\infty} \subset R^n$ is good, then it converges to $\bar{x}$.

Proof. Assume that a sequence $\{x_i\}_{i=0}^{\infty} \subset R^n$ is good. Then there exists a positive number M_0 such that

$$\sum_{i=0}^{T-1}(v(x_i, x_{i+1}) - v(\bar{x}, \bar{x})) < M_0 \text{ for all integers } T \geq 1. \tag{1.37}$$

In view of (1.37) and (1.23), there exists a strictly increasing sequence of natural numbers $\{t_k\}_{k=1}^{\infty}$ such that

$$|x_{t_k}| < M_* \text{ for all natural numbers } k. \tag{1.38}$$

Let $k \geq 1$ be an integer. It follows from (1.8), (1.37), and (1.38) that

$$M_0 > \sum_{i=0}^{t_k-1}(v(x_i, x_{i+1}) - v(\bar{x}, \bar{x})) = \sum_{i=0}^{t_k-1} L(x_i, x_{i+1}) + \langle l_1, x_0 - x_{t_k}\rangle$$

$$\geq \sum_{i=0}^{t_k-1} L(x_i, x_{i+1}) - |l_1|(|x_0| + |x_{t_k}|)$$

$$\geq \sum_{i=0}^{t_k-1} L(x_i, x_{i+1}) - |l_1|(|x_0| + M_*)$$

and

$$\sum_{i=0}^{t_k-1} L(x_i, x_{i+1}) \leq M_0 + |l_1|(|x_0| + M_*).$$

Since the inequality above is valid for all natural numbers k we conclude that

$$\sum_{i=0}^{\infty} L(x_i, x_{i+1}) \leq M_0 + |l_1|(|x_0| + M_*).$$

Property (C) implies that the sequence $\{x_i\}_{i=0}^{\infty}$ converges to $\bar{x}$ and Assertion 2 is proved.

Assume that

$$M_1 := \sum_{i=0}^{\infty} L(x_i, x_{i+1}) < \infty. \tag{1.39}$$

In view of (1.5), there exists a positive number M_2 such that

$$|x_i| < M_2 \text{ for all integers } i \geq 0. \tag{1.40}$$

By (1.8), (1.39), and (1.40), for all integers $T \geq 1$, we have

$$\sum_{i=0}^{T-1}(v(x_i, x_{i+1}) - v(\bar{x}, \bar{x})) = \sum_{i=0}^{T-1} L(x_i, x_{i+1}) + \langle l_1, x_0 - x_T \rangle$$

$$\leq M_1 + 2|l_1|M_2.$$

Combined with Proposition 1.5 this implies that the sequence $\{x_i\}_{i=0}^{\infty}$ is good and completes the proof of Proposition 1.6.

Proposition 1.7. *Let $x \in R^n$. Then there exists a sequence $\{x_i\}_{i=0}^{\infty} \subset R^n$ such that $x_0 = x$ and for each sequence $\{y_i\}_{i=0}^{\infty} \subset R^n$ satisfying $y_0 = x$ the inequality*

$$\sum_{i=0}^{\infty} L(x_i, x_{i+1}) \leq \sum_{i=0}^{\infty} L(y_i, y_{i+1})$$

holds.

Proof. Set

$$M_0 = \inf\{\sum_{i=0}^{\infty} L(y_i, y_{i+1}) : \{y_i\}_{i=0}^{\infty} \subset R^n \text{ and } y_0 = x\}. \tag{1.41}$$

It is clear that M_0 is well defined and nonnegative. There exists a sequence $\{x_i^{(k)}\}_{i=0}^{\infty} \subset R^n$, $k = 1, 2, \ldots$ such that

$$x_0^{(k)} = x, \ k = 1, 2, \ldots, \tag{1.42}$$

$$\lim_{k \to \infty} \sum_{i=0}^{\infty} L(x_i^{(k)}, x_{i+1}^{(k)}) = M_0. \tag{1.43}$$

In view of (1.43) and (1.5), there exists a positive number M_1 such that

$$|x_i^{(k)}| < M_1 \text{ for all integers } i \geq 0 \text{ for all integers } k \geq 1. \tag{1.44}$$

By (1.44) using diagonalization process, extracting subsequences and re-indexing we may assume without loss of generality that for every nonnegative integer i there exists

$$x_i = \lim_{k \to \infty} x_i^{(k)}. \tag{1.45}$$

It follows from (1.42) and (1.45) that

$$x_0 = x. \tag{1.46}$$

In view of (1.6), (1.43), and (1.45), for every integer $T \geq 1$, we have

$$\sum_{i=0}^{T-1} L(x_i, x_{i+1}) = \lim_{k\to\infty} \sum_{i=0}^{T-1} L(x_i^{(k)}, x_{i+1}^{(k)}) \leq \lim_{k\to\infty} \sum_{i=0}^{\infty} L(x_i^{(k)}, x_{i+1}^{(k)}) = M_0.$$

Since T is an arbitrary natural number we conclude that

$$\sum_{i=0}^{\infty} L(x_i, x_{i+1}) \leq M_0.$$

Combined with (1.41) and (1.46) this implies that

$$\sum_{i=0}^{\infty} L(x_i, x_{i+1}) = M_0.$$

Proposition 1.7 is proved.

In our study we use the following optimality criterion introduced in the economic literature [6, 28, 62] and used in the optimal control [19, 70, 81, 84].

A sequence $\{x_i\}_{i=0}^{\infty} \subset R^n$ is called overtaking optimal if

$$\limsup_{T\to\infty} [\sum_{i=0}^{T-1} v(x_i, x_{i+1}) - \sum_{i=0}^{T-1} v(y_i, y_{i+1})] \leq 0$$

for any sequence $\{y_i\}_{i=0}^{\infty} \subset R^n$ satisfying $y_0 = x_0$.

Proposition 1.8. *Let $\{x_i\}_{i=0}^{\infty} \subset R^n$. Then the following assertions are equivalent:*

1. the sequence $\{x_i\}_{i=0}^{\infty}$ is overtaking optimal;

2.

$$\sum_{i=0}^{\infty} L(x_i, x_{i+1}) \leq \sum_{i=0}^{\infty} L(y_i, y_{i+1})$$

for every sequence $\{y_i\}_{i=0}^{\infty} \subset R^n$ which satisfies $y_0 = x_0$.

Proof. Assume that the sequence $\{x_i\}_{i=0}^{\infty}$ is overtaking optimal. Evidently, it is good. Proposition 1.6 implies that

$$\sum_{i=0}^{\infty} L(x_i, x_{i+1}) < \infty.$$

Let a sequence $\{y_i\}_{i=0}^{\infty} \subset R^n$ satisfy

$$y_0 = x_0. \tag{1.47}$$

We claim that

$$\sum_{i=0}^{\infty} L(x_i, x_{i+1}) \le \sum_{i=0}^{\infty} L(y_i, y_{i+1}).$$

We may assume that

$$\sum_{i=0}^{\infty} L(y_i, y_{i+1}) < \infty.$$

Property (C) implies that

$$\lim_{i \to \infty} y_i = \bar{x}, \ \lim_{i \to \infty} x_i = \bar{x}. \tag{1.48}$$

Since the sequence $\{x_i\}_{i=0}^{\infty}$ is overtaking optimal it follows from (1.47), (1.8), and (1.48) that

$$0 \ge \limsup_{T \to \infty} [\sum_{i=0}^{T-1} v(x_i, x_{i+1}) - \sum_{i=0}^{T-1} v(y_i, y_{i+1})]$$

$$= \limsup_{T \to \infty} [\sum_{i=0}^{T-1} L(x_i, x_{i+1}) + \langle l_1, x_0 - x_T \rangle - \sum_{i=0}^{T-1} L(y_i, y_{i+1}) - \langle l_1, y_0 - y_T \rangle]$$

$$= \limsup_{T \to \infty} [\sum_{i=0}^{T-1} L(x_i, x_{i+1}) - \sum_{i=0}^{T-1} L(y_i, y_{i+1}) + \langle l_1, y_T - x_T \rangle]$$

$$= \sum_{i=0}^{\infty} L(x_i, x_{i+1}) - \sum_{i=0}^{\infty} L(y_i, y_{i+1}).$$

Thus assertion 2 holds.

Assume that assertion 2 holds. We claim that the sequence $\{x_i\}_{i=0}^{\infty}$ is overtaking optimal. It is clear that

$$\sum_{i=0}^{\infty} L(x_i, x_{i+1}) < \infty.$$

Proposition 1.6 implies that the sequence $\{x_i\}_{i=0}^{\infty}$ is good and that

$$\lim_{i \to \infty} x_i = \bar{x}. \tag{1.49}$$

Assume that a sequence $\{y_i\}_{i=0}^{\infty} \subset R^n$ satisfies

$$y_0 = x_0. \tag{1.50}$$

We show that

$$\limsup_{T \to \infty} [\sum_{i=0}^{T-1} v(x_i, x_{i+1}) - \sum_{i=0}^{T-1} v(y_i, y_{i+1})] \le 0.$$

We may assume without loss of generality that the sequence $\{y_i\}_{i=0}^{\infty}$ is good. In view of Proposition 1.6, we have

$$\lim_{i\to\infty} y_i = \bar{x}, \ \sum_{i=0}^{\infty} L(y_i, y_{i+1}) < \infty. \tag{1.51}$$

In view of (1.8), (1.49), (1.50), (1.51), and assertion 2, we have

$$\limsup_{T\to\infty}[\sum_{i=0}^{T-1} v(x_i, x_{i+1}) - \sum_{i=0}^{T-1} v(y_i, y_{i+1})]$$

$$= \limsup_{T\to\infty}[\sum_{i=0}^{T-1} L(x_i, x_{i+1}) + \langle l_1, x_0 - x_T\rangle - \sum_{i=0}^{T-1} L(y_i, y_{i+1}) - \langle l_1, y_0 - y_T\rangle]$$

$$= \sum_{i=0}^{\infty} L(x_i, x_{i+1}) - \sum_{i=0}^{\infty} L(y_i, y_{i+1}) + \langle l_1, \lim_{T\to\infty} y_T - \lim_{T\to\infty} x_T\rangle]$$

$$= \sum_{i=0}^{\infty} L(x_i, x_{i+1}) - \sum_{i=0}^{\infty} L(y_i, y_{i+1}) \le 0.$$

Thus assertion 1 holds and Proposition 1.8 is proved.

Propositions 1.7 and 1.8 imply the following existence result.

Proposition 1.9. *For every point $x \in R^n$ there exists an overtaking optimal sequence $\{x_i\}_{i=0}^{\infty} \subset R^n$ such that $x_0 = x$.*

1.2 The Turnpike Phenomenon

In the previous section we proved the turnpike result and the existence of overtaking optimal solutions for rather simple class of discrete-time problems. The problems of this class are unconstrained and their objective functions are convex and differentiable. In this book our goal is to study the structure of approximate solutions over large intervals for a class of discrete-time constrained optimal control problems without convexity (concavity) assumptions. In particular, in Chaps. 2–4 we study the structure of approximate solutions of an autonomous discrete-time control system with a compact metric space of states X. This control system is described by a bounded upper semicontinuous function $v : X \times X \to R^1$ which determines an optimality criterion and by a nonempty closed set $\Omega \subset X \times X$ which determines a class of admissible trajectories (programs). We study the problems

$$\sum_{i=0}^{T-1} v(x_i, x_{i+1}) \to \max, \ \{(x_i, x_{i+1})\}_{i=0}^{T-1} \subset \Omega, \ x_0 = z, \ x_T = y, \tag{P1}$$

$$\sum_{i=0}^{T-1} v(x_i, x_{i+1}) \to \max, \ \{(x_i, x_{i+1})\}_{i=0}^{T-1} \subset \Omega, \ x_0 = z \tag{P2}$$

and

$$\sum_{i=0}^{T-1} v(x_i, x_{i+1}) \to \max, \ \{(x_i, x_{i+1})\}_{i=0}^{T-1} \subset \Omega, \tag{P3}$$

where $T \geq 1$ is an integer and the points $y, z \in X$.

In the classical turnpike theory the objective function v possesses the turnpike property (TP) if there exists a point $\bar{x} \in X$ (a turnpike) such that the following condition holds:

For each positive number ϵ there exists an integer $L \geq 1$ such that for each integer $T \geq 2L$ and each solution $\{x_i\}_{i=0}^{T} \subset X$ of the problem (P1) the inequality $\rho(x_i, \bar{x}) \leq \epsilon$ is true for all $i = L, \ldots, T - L$.

It should be mentioned that the constant L depends neither on T nor on y, z.

The turnpike phenomenon has the following interpretation. If one wishes to reach a point A from a point B by a car in an optimal way, then one should turn to a turnpike, spend most of time on it, and then leave the turnpike to reach the required point.

In the classical turnpike theory [28, 48, 59, 62] the space X is a compact convex subset of a finite-dimensional Euclidean space, the set Ω is convex, and the function v is strictly concave. Under these assumptions the turnpike property can be established and the turnpike $\bar{x}$ is a unique solution of the maximization problem $v(x, x) \to \max$, $(x, x) \in \Omega$. In this situation it is shown that for each program $\{x_t\}_{t=0}^{\infty}$ either the sequence $\{\sum_{t=0}^{T-1} v(x_t, x_{t+1}) - Tv(\bar{x}, \bar{x})\}_{T=1}^{\infty}$ is bounded (in this case the program $\{x_t\}_{t=0}^{\infty}$ is called (v)-good) or it diverges to $-\infty$. Moreover, it is also established that any (v)-good program converges to the turnpike $\bar{x}$. In the sequel this property is called as the asymptotic turnpike property.

Recently it was shown that the turnpike property is a general phenomenon which holds for large classes of variational and optimal control problems without convexity assumptions. (See, for example, [70] and the references mentioned therein). For these classes of problems a turnpike is not necessarily a singleton but may instead be a nonstationary trajectory (in the discrete time nonautonomous case) or an absolutely continuous function on the interval $[0, \infty)$ (in the continuous time nonautonomous case) or a compact subset of the space X (in the autonomous case). Note that all of these results were obtained for unconstrained problems. In particular, the turnpike results for the problems of the type (P1) were obtained in the case $\Omega = X \times X$.

For classes of problems considered in [70], using the Baire category approach, it was shown that the turnpike property holds for a generic (typical) problem. In this book we are interested in individual (non-generic) results describing the structure of approximate solutions. We study the problems (P1)–(P3) with the constraint $\{(x_i, x_{i+1})\}_{i=0}^{T-1} \subset \Omega$ where Ω is an arbitrary nonempty closed subset of $X \times X$. Clearly, these constrained problems are more difficult and less understood than their unconstrained prototypes in the previous section and in [64–66, 68]. They are also more realistic from the point of view of mathematical economics. As we have mentioned before in general a turnpike is not necessarily a singleton. Nevertheless problems of the type (P1)–(P3) for which the turnpike is a singleton are of great importance because of the following reasons: there are many models of economic growth for which a turnpike is a singleton; if a turnpike is a singleton, then approximate solutions have very simple structure and this is very important for applications; if a turnpike is a singleton, then it can be easily calculated as a solution of the problem $v(x, x) \to \max$, $(x, x) \in \Omega$.

The turnpike property is very important for applications. Suppose that our objective function v has the turnpike property and we know a finite number of "approximate" solutions of the problem (P1). Then we know the turnpike $\bar{x}$, or at least its approximation, and the constant L (see the definition of (TP)) which is an estimate for the time period required to reach the turnpike. This information can be useful if we need to find an "approximate" solution of the problem (P1) with a new time interval $[m_1, m_2]$ and the new values $z, y \in X$ at the end points m_1 and m_2. Namely instead of solving this new problem on the "large" interval $[m_1, m_2]$ we can find an "approximate" solution of the problem (P1) on the "small" interval $[m_1, m_1+L]$ with the values $z, \bar{x}$ at the end points and an "approximate" solution of the problem (P1) on the "small" interval $[m_2 - L, m_2]$ with the values $\bar{x}, y$ at the end points. Then the concatenation of the first solution, the constant sequence $x_i = \bar{x}$, $i = m_1 + L, \dots, m_2 - L$ and the second solution is an "approximate" solution of the problem (P1) on the interval $[m_1, m_2]$ with the values z, y at the end points. Sometimes as an "approximate" solution of the problem (P1) we can choose any admissible sequence $\{x_i\}_{i=m_1}^{m_2}$ satisfying

$$x_{m_1} = z,\ x_{m_2} = y \text{ and } x_i = \bar{x} \text{ for all } i = m_1 + L, \dots, m_2 - L.$$

1.3 Nonconcave (Nonconvex) Problems

In Chap. 2 we study the structure of approximate solutions of discrete-time optimal control problems introduced in Sect. 1.2.

Let (X, ρ) be a compact metric space, Ω be a nonempty closed subset of $X \times X$, and let $v : X \times X \to R^1$ be a bounded upper semicontinuous function.

A sequence $\{x_t\}_{t=0}^{\infty} \subset X$ is called an (Ω)-program (or just a program if the set Ω is understood) if $(x_t, x_{t+1}) \in \Omega$ for all nonnegative integers t. A sequence $\{x_t\}_{t=0}^{T}$ where $T \geq 1$ is an integer is called an (Ω)-program (or just a program if the set Ω is understood) if $(x_t, x_{t+1}) \in \Omega$ for all integers $t \in [0, T - 1]$.

We consider the problems

$$\sum_{i=0}^{T-1} v(x_i, x_{i+1}) \to \max, \ \{(x_i, x_{i+1})\}_{i=0}^{T-1} \subset \Omega, \ x_0 = y, \ x_T = z,$$

$$\sum_{i=0}^{T-1} v(x_i, x_{i+1}) \to \max, \ \{(x_i, x_{i+1})\}_{i=0}^{T-1} \subset \Omega, \ x_0 = z$$

and

$$\sum_{i=0}^{T-1} v(x_i, x_{i+1}) \to \max, \ \{(x_i, x_{i+1})\}_{i=0}^{T-1} \subset \Omega,$$

where $T \geq 1$ is an integer and the points $y, z \in X$.

We suppose that there exist a point $\bar{x} \in X$ and a positive number $\bar{c}$ such that the following assumptions hold:

(i) $(\bar{x}, \bar{x})$ is an interior point of Ω;

(ii) $\sum_{t=0}^{T-1} v(x_t, x_{t+1}) \leq Tv(\bar{x}, \bar{x}) + \bar{c}$ for any natural number T and any program $\{x_t\}_{t=0}^{T}$.

The property (ii) implies that for each program $\{x_t\}_{t=0}^{\infty}$ either the sequence

$$\{\sum_{t=0}^{T-1} v(x_t, x_{t+1}) - Tv(\bar{x}, \bar{x})\}_{T=1}^{\infty}$$

is bounded or $\lim_{T\to\infty}[\sum_{t=0}^{T-1} v(x_t, x_{t+1}) - Tv(\bar{x}, \bar{x})] = -\infty$.

A program $\{x_t\}_{t=0}^{\infty}$ is called (v)-good if the sequence

$$\{\sum_{t=0}^{T-1} v(x_t, x_{t+1}) - Tv(\bar{x}, \bar{x})\}_{T=1}^{\infty}$$

is bounded.

We suppose that the following assumption holds.

(iii) (the asymptotic turnpike property) For any (v)-good program $\{x_t\}_{t=0}^{\infty}$, $\lim_{t\to\infty} \rho(x_t, \bar{x}) = 0$.

Note that the properties (i)–(iii) hold for models of economic dynamics considered in the classical turnpike theory.

For each positive number M denote by X_M the set of all points $x \in X$ for which there exists a program $\{x_t\}_{t=0}^{\infty}$ such that $x_0 = x$ and that for all natural numbers T the following inequality holds:

$$\sum_{t=0}^{T-1} v(x_t, x_{t+1}) - Tv(\bar{x}, \bar{x}) \geq -M.$$

It is not difficult to see that $\cup\{X_M : M \in (0, \infty)\}$ is the set of all points $x \in X$ for which there exists a (v)-good program $\{x_t\}_{t=0}^{\infty}$ satisfying $x_0 = x$.

Let $T \geq 1$ be an integer and $\Delta \geq 0$. A program $\{x_i\}_{i=0}^T \subset X$ is called (Δ)-optimal if for any program $\{x'_i\}_{i=0}^T$ satisfying $x_0 = x'_0$, the inequality

$$\sum_{i=0}^{T-1} v(x_i, x_{i+1}) \geq \sum_{i=0}^{T-1} v(x'_i, x'_{i+1}) - \Delta$$

holds.

The turnpike theory for problems (P1) and (P2) is presented in [84] which summaries our research [71–73, 75–77].

In particular, in Chap. 2 of [84] we prove the following turnpike result for approximate solutions of our second optimization problem stated above.

Theorem 1.10. *Let ϵ, M be positive numbers. Then there exist a natural number L and a positive number δ such that for each integer $T > 2L$ and each (δ)-optimal program $\{x_t\}_{t=0}^T$ which satisfies $x_0 \in X_M$ there exist non-negative integers $\tau_1, \tau_2 \leq L$ such that $\rho(x_t, \bar{x}) \leq \epsilon$ for all $t = \tau_1, \dots, T - \tau_2$ and if $\rho(x_0, \bar{x}) \leq \delta$, then $\tau_1 = 0$.*

An analogous turnpike result for approximate solutions of our first optimization problem is also proved in Chap. 2 of [84].

A program $\{x_t\}_{t=0}^\infty$ is called (v)-overtaking optimal if for each program $\{y_t\}_{t=0}^\infty$ satisfying $y_0 = x_0$ the inequality $\limsup_{T\to\infty} \sum_{t=0}^{T-1} [v(y_t, y_{t+1}) - v(x_t, x_{t+1})] \leq 0$ holds.

In Chap. 2 of [84] we prove the following result which establishes the existence of an overtaking optimal program.

Theorem 1.11. *Assume that $x \in X$ and that there exists a (v)-good program $\{x_t\}_{t=0}^\infty$ such that $x_0 = x$. Then there exists a (v)-overtaking optimal program $\{x_t^*\}_{t=0}^\infty$ such that $x_0^* = x$.*

In Chap. 2 of [84] for problems which satisfy concavity assumption common in the literature we study the structure of approximate solutions in the regions containing end points and obtain a full description of the structure of approximate solutions. More precisely, we study the structure of approximate solutions of our second optimization problem stated above in the regions $[0, L]$ and $[T - L, T]$ (see the definition of the turnpike property). We show that if $\{x_i\}_{i=0}^T \subset X$ is an approximate solution of our problem, then for all integers $t = 0, \dots, L$ the state x_t is closed enough to z_t where $\{z_t\}_{t=0}^\infty \subset X$ is a unique solution of a certain infinite horizon optimal control problem satisfying $z_0 = z$. We also show that if $\{x_i\}_{i=0}^T \subset X$ is an approximate solution of our second optimization problem, then for all integers $t = 0, \dots, L$ the state x_{T-t} is closed enough to Λ_t where $\{\Lambda_t\}_{t=0}^\infty \subset X$ is a unique solution of a certain infinite horizon optimal control problem which does not depend on z. These results are established when the set X is a convex subset of the Euclidean space R^n, the set Ω is convex, and the function v is strictly concave. In this case we obtain the full description of the structure of approximate solutions of our second optimization problem. Note that the structure of approximate solutions in the region $[0, L]$ depends on z while their structure in the region $[T-L, T]$ does not depend on z. Actually it depends only on v and Ω.

In Chap. 2 of the present book we prove the generalizations of the results of Chap. 2 of [84] on the structure of approximate solutions in the regions containing end points. These generalizations are established for problems (P1)–(P3) without concavity assumptions. The results of this chapter were obtained in [83, 86, 90].

In Chap. 3 we consider optimal control systems which are discrete-time analogs of Bolza problems in the calculus of variations. They are described by a pair of objective functions which determines an optimality criterion. We consider two classes of Bolza problems and obtain for each of them the full description of approximate solutions of these problems on large intervals. This description shows that on large intervals the approximate solutions are determined mainly by our optimality criterion and are essentially independent of the choice of time intervals and data. The results of Chap. 4 were obtained in [91, 92].

In Chap. 4 we continue to study the discrete-time analogs of Bolza problems in the calculus of variations. We show that the turnpike phenomenon and the structure of solutions on finite intervals in the regions close to the endpoints are stable under small perturbations of the objective functions and the set Ω. The results of this chapter are new.

1.4 Examples

Example 1.12. Let (X, ρ) be a compact metric space, Ω be a nonempty closed subset of $X \times X$, $\bar{x} \in X$, $(\bar{x}, \bar{x})$ be an interior point of Ω, $\pi : X \to R^1$ be a continuous function, α be a real number, and $L : X \times X \to [0, \infty)$ be a continuous function such that for each $(x, y) \in X \times X$ the equality $L(x, y) = 0$ holds if and only if $(x, y) = (\bar{x}, \bar{x})$. Set

$$v(x, y) = \alpha - L(x, y) + \pi(x) - \pi(y)$$

for all $x, y \in X$. It is not difficult to see that assumptions (i), (ii), and (iii) hold.

Example 1.13. Let X be a compact convex subset of the Euclidean space R^n with the norm $|\cdot|$ induced by the scalar product $\langle \cdot, \cdot \rangle$, let $\rho(x, y) = |x - y|$, $x, y \in R^n$, Ω be a nonempty closed subset of $X \times X$, a point $\bar{x} \in X$, $(\bar{x}, \bar{x})$ be an interior point of Ω, and let $v : X \times X \to R^1$ be a strictly concave continuous function such that

$$v(\bar{x}, \bar{x}) = \sup\{v(z, z) : \; z \in X \text{ and } (z, z) \in \Omega\}.$$

We assume that there exists a positive constant $\bar{r}$ such that

$$\{(x, y) \in R^n \times R^n : \; |x - \bar{x}|, \; |y - \bar{x}| \le \bar{r}\} \subset \Omega.$$

It is a well-known fact of convex analysis [58] that there exists a point $l \in R^n$ such that

$$v(x, y) \leq v(\bar{x}, \bar{x}) + \langle l, x - y \rangle$$

for any point $(x, y) \in X \times X$. Set

$$L(x, y) = v(\bar{x}, \bar{x}) + \langle l, x - y \rangle - v(x, y)$$

for all $(x, y) \in X \times X$. It is not difficult to see that this example is a particular case of Example 1.12. Therefore assumptions (i), (ii), and (iii) hold.

Example 1.14. Let $X = [0, 1]$, $\Omega = \{(x, y) \in [0, 1] \times [0, 1] : \ y \leq x^{1/2}\}$, $v(x, y) = x^{1/2} - y^2$, $x, y \in X$. It is not difficult to see that the set Ω is convex, the function v is strictly concave, the optimization problem $v(z, z) \to \max$, $z \in X$, and $(z, z) \in \Omega$ has a unique solution $16^{-1/3}$ and $(16^{-1/3}, 16^{-1/3})$ is an interior point of Ω. Therefore this example is a particular case of Example 1.13 and assumptions (i), (ii), and (iii) hold.

Example 1.15. Let $X = [0, 1]$, $\Omega = \{(x, y) \in [0, 1] \times [0, 1] : \ y \leq x^{1/2}\}$, $v(x, y) = x^{1/2} - y$, $x, y \in X$. It is not difficult to see that the set Ω is convex, the function v is concave but not strictly concave, the optimization problem $v(z, z) \to \max$, $z \in X$, and $(z, z) \in \Omega$ has a unique solution 4^{-1} and $(4^{-1}, 4^{-1})$ is an interior point of Ω. Since the function v is concave for all $x, y \in X$,

$$v(x, y) \leq v(4^{-1}, 4^{-1}) + x - y = 4^{-1} + x - y$$

and

$$4^{-1} + x - y - v(x, y) = (x^{1/2} - 2^{-1})^2$$

is equal zero if and only if $x = 4^{-1}$. Now it is not difficult to see that assumptions (i), (ii), and (iii) hold.

Example 1.16. Consider the sets X, Ω, and the function v defined in Example 1.15 and set $u(x, y) = x^{1/2} - x^2 - y + y^2$, $x, y \in X$. The function u is strictly convex with respect to the variable y. Nevertheless assumptions (i), (ii), and (iii) hold for the function u because for any integer T and any program $\{x_t\}_{t=0}^T$,

$$\sum_{t=0}^{T-1} u(x_t, x_{t+1}) = \sum_{t=0}^{T-1} v(x_t, x_{t+1}) + x_T^2 - x_0^2.$$

1.5 Two-Player Zero-Sum Games

In Chaps. 5–8 we prove turnpike results for classes of dynamic discrete-time two-player zero-sum games. These results describe the structure of approximate solutions, for all sufficiently large intervals. We also study the structure of approximate solutions on large intervals in the regions close to the endpoints of the intervals and examine the existence of a pair of overtaking equilibria strategies over an infinite horizon.

Let $X \subset R^{m_1}$ and $Y \subset R^{m_2}$ be nonempty convex compact sets. Denote by $\mathcal{M}$ the set of all continuous functions $f : X \times X \times Y \times Y \to R^1$ such that:

for each point $(y_1, y_2) \in Y \times Y$ the function $(x_1, x_2) \to f(x_1, x_2, y_1, y_2)$, $(x_1, x_2) \in X \times X$ is convex;

for each point $(x_1, x_2) \in X \times X$ the function $(y_1, y_2) \to f(x_1, x_2, y_1, y_2)$, $(y_1, y_2) \in Y \times Y$ is concave.

The set $\mathcal{M}$ is equipped with a metric $\rho : \mathcal{M} \times \mathcal{M} \to R^1$ defined by

$$\rho(f, g) = \sup\{|f(x_1, x_2, y_1, y_2) - g(x_1, x_2, y_1, y_2)| :$$

$$x_1, x_2 \in X, \quad y_1, y_2 \in Y\}, \quad f, g \in \mathcal{M}.$$

It is clearly that $(\mathcal{M}, \rho)$ is a complete metric space.

Given $f \in \mathcal{M}$ and a natural number n we consider a discrete-time two-player zero-sum game over the interval $[0, n]$. For this game $\{\{x_i\}_{i=0}^n : x_i \in X, i = 0, \dots n\}$ is the set of strategies for the first player, $\{\{y_i\}_{i=0}^n : y_i \in Y, i = 0, \dots n\}$ is the set of strategies for the second player, and the objective function for the first player associated with the strategies $\{x_i\}_{i=0}^n$, $\{y_i\}_{i=0}^n$ is given by $\sum_{i=0}^{n-1} f(x_i, x_{i+1}, y_i, y_{i+1})$.

Let $f \in \mathcal{M}$, n be a natural number and let $M \in [0, \infty)$. A pair of sequences $\{\bar{x}_i\}_{i=0}^n \subset X$, $\{\bar{y}_i\}_{i=0}^n \subset Y$ is called (f, M)-good if the following properties hold:

(i) for each sequence $\{x_i\}_{i=0}^n \subset X$ satisfying $x_0 = \bar{x}_0$, $x_n = \bar{x}_n$ the inequality

$$M + \sum_{i=0}^{n-1} f(x_i, x_{i+1}, \bar{y}_i, \bar{y}_{i+1}) \geq \sum_{i=0}^{n-1} f(\bar{x}_i, \bar{x}_{i+1}, \bar{y}_i, \bar{y}_{i+1})$$

holds;

(ii) for each sequence $\{y_i\}_{i=0}^n \subset Y$ satisfying $y_0 = \bar{y}_0$, $y_n = \bar{y}_n$ the inequality

$$M + \sum_{i=0}^{n-1} f(\bar{x}_i, \bar{x}_{i+1}, \bar{y}_i, \bar{y}_{i+1}) \geq \sum_{i=0}^{n-1} f(\bar{x}_i, \bar{x}_{i+1}, y_i, y_{i+1})$$

holds.

If a pair of sequences $\{x_i\}_{i=0}^n \subset X$, $\{y_i\}_{i=0}^n \subset Y$ is $(f, 0)$-good then it is called (f)-optimal.

Let $f \in \mathcal{M}$. We say that the function f possesses the *turnpike property* if there exists a unique pair $(x_f, y_f) \in X \times Y$ for which the following assertion holds:

For each positive number ϵ there exist an integer $n_0 \geq 2$ and a positive number δ such that for each integer $n \geq 2n_0$ and each (f, δ)-good pair of sequences $\{x_i\}_{i=0}^n \subset X$, $\{y_i\}_{i=0}^n \subset Y$ the inequalities $|x_i - x_f|$, $|y_i - y_f| \leq \epsilon$ holds for all integers $i \in [n_0, n - n_0]$.

In [67] we showed that the turnpike property holds for a generic $f \in \mathcal{M}$. Namely, in [67] we proved the existence of a set $\mathcal{F} \subset \mathcal{M}$ which is a countable intersection of open everywhere dense sets in $\mathcal{M}$ such that each $f \in \mathcal{F}$ has the turnpike property. Thus for most functions $f \in \mathcal{M}$ the turnpike property holds. Nevertheless it is very important to have conditions on $f \in \mathcal{M}$ which imply the turnpike property. These conditions are discussed in [78] and in Chap. 5 of [84].

In Chaps. 5–8 we study the structure of solutions of more general and complicated classes of dynamic discrete-time two-player zero-sum games. In Chap. 5 we study a class of unconstrained dynamic discrete-time two-player zero-sum games without using standard convexity–concavity assumptions and prove two turnpike results of [82]. Chapter 6 contains the study of the existence and turnpike properties of approximate solutions for a class of dynamic constrained discrete-time two-player zero-sum games without convexity–concavity assumptions. Its results were obtained in [88]. In Chap. 7 we study turnpike properties of approximate solutions for a class of dynamic discrete-time two-player zero-sum nonautonomous games with convexity–concavity assumptions which were established in [87]. In Chap. 8 we analyze the existence and turnpike properties of approximate solutions for a class of dynamic constrained discrete-time two-player zero-sum games which satisfy convexity–concavity assumptions. We describe the structure of approximate solutions which is independent of the length of the interval, for all sufficiently large intervals and show that approximate solutions are determined mainly by the objective function, and are essentially independent of the choice of intervals and endpoint conditions. We show that the turnpike phenomenon is stable under small perturbations of objective functions and analyze the structure of approximate solutions in regions closed to the endpoints of domains.

2

Lagrange Problems

In this chapter we study the structure of approximate solutions of an autonomous nonconcave discrete-time optimal control system with a compact metric space of states. This control system is described by a bounded upper semicontinuous objective function which determines an optimality criterion. In the turnpike theory, it is known that approximate solutions are determined mainly by the objective function, and are essentially independent of the choice of time intervals and data, except in regions close to the endpoints of the time intervals. In this chapter our main goal is to analyze the structure of approximate solutions in regions close to the endpoints of the time intervals.

2.1 Discrete-Time Optimal Control Systems

Let (X, ρ) be a compact metric space and Ω be a nonempty closed subset of $X \times X$.

A sequence $\{x_t\}_{t=0}^{\infty} \subset X$ is called an (Ω)-program if $(x_t, x_{t+1}) \in \Omega$ for all integers $t \geq 0$. A sequence $\{x_t\}_{t=T_1}^{T_2} \subset X$, where integers T_1, T_2 satisfy $0 \leq T_1 < T_2$, is called an (Ω)-program if $(x_t, x_{t+1}) \in \Omega$ for all integers $t \in [T_1, T_2 - 1]$.

We consider the problem

$$\sum_{t=0}^{T-1} v(x_t, x_{t+1}) \to \max, \ \{(x_t, x_{t+1})\}_{t=0}^{T-1} \subset \Omega, \ x_0 = z_1, \ x_T = z_2 \tag{P1}$$

studied in [72], the problem

$$\sum_{t=0}^{T-1} v(x_t, x_{t+1}) \to \max, \ \{(x_t, x_{t+1})\}_{t=0}^{T-1} \subset \Omega, \ x_0 = z_1 \tag{P2}$$

A.J. Zaslavski, *Discrete-Time Optimal Control and Games on Large Intervals*, Springer Optimization and Its Applications 119,
DOI 10.1007/978-3-319-52932-5_2

studied in [71], and the problem

$$\sum_{t=0}^{T-1} v(x_t, x_{t+1}) \to \max, \ \{(x_t, x_{t+1})\}_{t=0}^{T-1} \subset \Omega, \tag{P3}$$

considered in [83], where T is a natural number, $z_1, z_2 \in X$, and $v : \Omega \to R^1$ is a bounded upper semicontinuous objective function.

In models of economic growth the set X is the space of states, v is a utility function, and $v(x_t, x_{t+1})$ evaluates consumption at moment t. The interest in discrete-time optimal problems of types (P1)–(P3) also stems from the study of various optimization problems which can be reduced to it [8, 37–39, 46, 70].

In [71–73, 75–77, 83, 84] we were interested in a turnpike property of the approximate solutions of problems (P1)–(P3) which is independent of the length of the interval T, for all sufficiently large intervals. In [71, 72, 83, 84] we studied the problems (P1), (P2), and (P3) and showed under certain assumptions that the turnpike property holds and that the turnpike $\bar{x}$ is a unique solution of the maximization problem $v(x, x) \to \max$, $(x, x) \in \Omega$. More precisely, we considered a class of (v)-good programs which are approximate solutions of the corresponding infinite horizon optimal control problem associated with the objective function v. It was shown that the turnpike property holds and $\bar{x}$ is the turnpike if the following asymptotic turnpike property holds: the all (v)-good programs converge to $\bar{x}$.

In this chapter we show that the asymptotic turnpike property holds for most objective functions in the sense of Baire category. In other words, the asymptotic turnpike property holds for a generic (typical) objective function. This result was obtained in [83].

In this chapter we also study the structure of approximate solutions of the problems $(P2)$ and $(P3)$ in regions close to the endpoints of the time intervals. We show that in regions close to the right endpoint T of the time interval these approximate solutions are determined only by the objective function, and are essentially independent of the choice of interval and endpoint value z_1. For the problems $(P3)$, approximate solutions are determined only by the objective function also in regions close to the left endpoint 0 of the time interval.

More precisely, we define $\bar{\Omega} = \{(y, x) \in X \times X : \ (x, y) \in \Omega\}$ and $\bar{v}(y, x) = v(x, y)$ for all $(x, y) \in \Omega$ and consider the set $\mathcal{P}(\bar{v})$ of all solutions of a corresponding infinite horizon optimal control problem associated with the pair $(\bar{v}, \bar{\Omega})$. For given a real positive number ϵ and a natural number τ, we show that if T is large enough and $\{x_t\}_{t=0}^{T}$ is an approximate solution of the problem $(P2)$, then $\rho(x_{T-t}, y_t) \le \epsilon$ for all integers $t \in [0, \tau]$, where $\{y_t\}_{t=0}^{\infty} \in \mathcal{P}(\bar{v})$.

Moreover, using the Baire category approach, we show that for most objective functions v the set $\mathcal{P}(\bar{v})$ is a singleton.

These results were obtained in [86, 90].

2.2 The Turnpike Results

Let (X, ρ) be a compact metric space and Ω be a nonempty closed subset of $X \times X$. Denote by $\mathcal{M}(\Omega)$ the set of all bounded functions $u : \Omega \to R^1$. For each $w \in \mathcal{M}(\Omega)$ set

$$\|w\| = \sup\{|w(x,y)| : (x,y) \in \Omega\}. \tag{2.1}$$

For each $x, y \in X$, each integer $T \geq 1$, and each $u \in \mathcal{M}(\Omega)$ set

$$\sigma(u, T, x) = \sup\{\sum_{i=0}^{T-1} u(x_i, x_{i+1}) : \{x_i\}_{i=0}^{T} \text{ is an } (\Omega) - \text{program and } x_0 = x\}, \tag{2.2}$$

$$\sigma(u, T, x, y) = \sup\{\sum_{i=0}^{T-1} u(x_i, x_{i+1}) : \tag{2.3}$$

$$\{x_i\}_{i=0}^{T} \text{ is an } (\Omega) - \text{program and } x_0 = x,\ x_T = y\},$$

$$\sigma(u, T) = \sup\{\sum_{i=0}^{T-1} u(x_i, x_{i+1}) : \{x_i\}_{i=0}^{T} \text{ is an } (\Omega) - \text{program}\}. \tag{2.4}$$

(Here we use the convention that the supremum of an empty set is $-\infty$).

For each $x, y \in X$, each pair of integers T_1, T_2 satisfying $0 \leq T_1 < T_2$, and each sequence $\{u_t\}_{t=T_1}^{T_2-1} \subset \mathcal{M}(\Omega)$ set

$$\sigma(\{u_t\}_{t=T_1}^{T_2-1}, T_1, T_2, x) = \sup\{\sum_{t=T_1}^{T_2-1} u_t(x_t, x_{t+1}) :$$

$$\{x_t\}_{t=T_1}^{T_2} \text{ is an } (\Omega) - \text{program and } x_{T_1} = x\}, \tag{2.5}$$

$$\sigma(\{u_t\}_{t=T_1}^{T_2-1}, T_1, T_2, x, y) = \sup\{\sum_{t=T_1}^{T_2-1} u_t(x_t, x_{t+1}) :$$

$$\{x_t\}_{t=T_1}^{T_2} \text{ is an } (\Omega) - \text{program and } x_{T_1} = x,\ x_{T_2} = y\}, \tag{2.6}$$

$$\sigma(\{u_t\}_{t=T_1}^{T_2-1}, T_1, T_2) = \sup\{\sum_{t=T_1}^{T_2-1} u_t(x_t, x_{t+1}) : \{x_t\}_{t=T_1}^{T_2} \text{ is an } (\Omega) - \text{program}\}, \tag{2.7}$$

$$\widehat{\sigma}(\{u_t\}_{t=T_1}^{T_2-1}, T_1, T_2, y) = \sup\{\sum_{t=T_1}^{T_2-1} u_t(x_t, x_{t+1}) :$$

$$\{x_t\}_{t=T_1}^{T_2} \text{ is an } (\Omega) - \text{program and } x_{T_2} = y\}. \tag{2.8}$$

Assume that $v \in \mathcal{M}(\Omega)$ is an upper semicontinuous function. We suppose that there exist $\bar{x}_v \in X$ and constants $\bar{c}_v > 0$ and $\bar{r}_v > 0$ such that the following assumptions hold.

(A1) $\{(x,y) \in X \times X : \ \rho(x,\bar{x}_v),\ \rho(y,\bar{x}_v) \le \bar{r}_v\} \subset \Omega$ and v is continuous at $(\bar{x}_v,\bar{x}_v)$.

(A2) $\sigma(v,T) \le Tv(\bar{x}_v,\bar{x}_v) + \bar{c}_v$ for all integers $T \ge 1$.

It is easy to see that for each natural number T and each (Ω)-program $\{x_t\}_{t=0}^T$,

$$\sum_{t=0}^{T-1} v(x_t,x_{t+1}) \le \sigma(v,T) \le Tv(\bar{x}_v,\bar{x}_v) + \bar{c}_v. \tag{2.9}$$

Inequality (2.9) implies the following result.

Proposition 2.1. *For each (Ω)-program $\{x_t\}_{t=0}^\infty$ either the sequence*

$$\{\sum_{t=0}^{T-1} v(x_t,x_{t+1}) - Tv(\bar{x}_v,\bar{x}_v)\}_{T=1}^\infty$$

is bounded or $\lim_{T\to\infty}[\sum_{t=0}^{T-1} v(x_t,x_{t+1}) - Tv(\bar{x}_v,\bar{x}_v)] = -\infty$.

An (Ω)-program $\{x_t\}_{t=0}^\infty$ is called (v,Ω)-good if the sequence

$$\{\sum_{t=0}^{T-1} v(x_t,x_{t+1}) - Tv(\bar{x}_v,\bar{x}_v)\}_{T=1}^\infty$$

is bounded.

We suppose that the following assumption holds:

(A3) (the asymptotic turnpike property or, briefly, (ATP)) For any (v,Ω)-good program $\{x_t\}_{t=0}^\infty$, $\lim_{t\to\infty} \rho(x_t,\bar{x}_v) = 0$.

In view of (A1) and (A3), if $(\bar{x}_v,\bar{x}_v)$ is not an isolated point of $X \times X$, then $\|v\| > 0$.

Examples of optimal control problems satisfying (A1)–(A3) are given in [71, 72, 84].

Denote by $\mathrm{Card}(A)$ the cardinality of a set A and suppose that the sum over empty set is zero.

It is clear that for each pair of integers T_1, T_2 satisfying $0 \le T_1 < T_2$, each sequence $\{w_t\}_{t=T_1}^{T_2-1} \subset \mathcal{M}(\Omega)$, and each pair of points $x, y \in X$ satisfying $\rho(x,\bar{x}_v), \rho(y,\bar{x}_v) \le \bar{r}_v$ the value $\sigma(\{w_t\}_{t=T_1}^{T_2-1}, T_1, T_2, x, y)$ is finite.

Definition 2.2. *Let $T \ge 1$ be an integer. We denote by Y_T the set of all points $x \in X$ for which there exists an (Ω)-program $\{x_t\}_{t=0}^T$ satisfying $x_0 = \bar{x}_v$ and $x_T = x$ and denote by $\bar{Y}_T$ the set of all points $x \in X$ for which there exists an (Ω)-program $\{x_t\}_{t=0}^T$ such that $x_0 = x$ and $x_T = \bar{x}_v$.*

The following two theorems obtained in [76] and presented in [84] establish the turnpike property for approximate solutions of the optimal control problems of the types (P1) and (P2) with objective functions u_t, $t = 0, \dots, T-1$ which belong to a small neighborhood of v.

Theorem 2.3. *Let a positive number $\epsilon < \bar{r}_v$, $L_0 \geq 1$ be an integer and $M_0 > 0$. Then there exist an integer $L \geq 1$ and $\delta \in (0, \epsilon)$ such that for each integer $T > 2L$, each $\{u_t\}_{t=0}^{T-1} \subset \mathcal{M}(\Omega)$ satisfying*

$$\|u_t - v\| \leq \delta, \ t = 0 \ldots T-1,$$

and each (Ω)-program $\{x_t\}_{t=0}^{T}$ which satisfies

$$x_0 \in \bar{Y}_{L_0}, \ x_T \in Y_{L_0},$$

$$\sum_{t=0}^{T-1} u_t(x_t, x_{t+1}) \geq \sigma(\{u_t\}_{t=0}^{T-1}, 0, T, x_0, x_T) - M_0$$

and

$$\sum_{t=\tau}^{\tau+L-1} u_t(x_t, x_{t+1}) \geq \sigma(\{u_t\}_{t=\tau}^{\tau+L-1}, \tau, \tau+L, x_\tau, x_{\tau+L}) - \delta$$

for each integer $\tau \in [0, T-L]$ there exist integers $\tau_1 \in [0, L]$, $\tau_2 \in [T-L, T]$ such that

$$\rho(x_t, \bar{x}_v) \leq \epsilon, \ t = \tau_1, \ldots, \tau_2.$$

Moreover, if $\rho(x_0, \bar{x}_v) \leq \delta$, then $\tau_1 = 0$ and if $\rho(x_T, \bar{x}_v) \leq \delta$, then $\tau_2 = T$.

Theorem 2.4. *Let a positive number $\epsilon < \bar{r}_v$, $L_0 \geq 1$ be an integer and $M_0 > 0$. Then there exist an integer $L \geq 1$ and a number $\delta \in (0, \epsilon)$ such that for each integer $T > 2L$, each $\{u_t\}_{t=0}^{T-1} \subset \mathcal{M}(\Omega)$ satisfying*

$$\|u_t - v\| \leq \delta, \ t = 0 \ldots, T-1$$

and each (Ω)-program $\{x_t\}_{t=0}^{T}$ which satisfies

$$x_0 \in \bar{Y}_{L_0}, \ \sum_{t=0}^{T-1} u_t(x_t, x_{t+1}) \geq \sigma(\{u_t\}_{t=0}^{T-1}, 0, T, x_0) - M_0$$

and

$$\sum_{t=\tau}^{\tau+L-1} u_t(x_t, x_{t+1}) \geq \sigma(\{u_t\}_{t=\tau}^{\tau+L-1}, \tau, \tau+L, x_\tau, x_{\tau+L}) - \delta$$

for each integer $\tau \in [0, T-L]$ there exist integers $\tau_1 \in [0, L]$, $\tau_2 \in [T-L, T]$ such that

$$\rho(x_t, \bar{x}_v) \leq \epsilon, \ t = \tau_1, \ldots, \tau_2.$$

Moreover if $\rho(x_0, \bar{x}_v) \leq \delta$, then $\tau_1 = 0$.

In this chapter we prove the following theorem obtained in [83] which establishes the turnpike property for approximate solutions of the optimal control problems of the type (P3).

Theorem 2.5. *Let $\epsilon \in (0, \bar{r}_v)$ and $M > 0$. Then there exist a positive number $\delta < \min\{1, M\}$ and a natural number L such that the following assertions hold.*

1. Assume that an integer $T \geq L$, $\{u_t\}_{t=0}^{T-1} \subset \mathcal{M}(\Omega)$ and an (Ω)-program $\{x_t\}_{t=0}^{T}$ satisfy

$$\|u_t - v\| \leq \delta,\ t = 0, \dots, T-1, \tag{2.10}$$

$$\sum_{t=0}^{T-1} u_t(x_t, x_{t+1}) \geq \sigma(\{u_t\}_{t=0}^{T-1}, 0, T) - M. \tag{2.11}$$

Then the inequality

$$Card(\{t \in \{0, \dots, T\} :\ \rho(x_t, \bar{x}_v) > \epsilon\}) < L$$

holds.

2. Assume that an integer $T \geq 2L$, $\{u_t\}_{t=0}^{T-1} \subset \mathcal{M}(\Omega)$ and an (Ω)-program $\{x_t\}_{t=0}^{T}$ satisfy (2.10), (2.11), and

$$\sum_{t=0}^{T-1} u_t(x_t, x_{t+1}) \geq \sigma(\{u_t\}_{t=0}^{T-1}, 0, T, x_0, x_T) - \delta. \tag{2.12}$$

Then there exist integers $\tau_1 \in [0, L]$ and $\tau_2 \in [T-L, T]$ such that

$$\rho(x_t, \bar{x}_v) \leq \epsilon,\ t = \tau_1, \dots, \tau_2.$$

Moreover, if $\rho(x_0, \bar{x}_v) \leq \delta$, then $\tau_1 = 0$ and if $\rho(x_T, \bar{x}_v) \leq \delta$, then $\tau_2 = T$.

3. Assume that $\{u_t\}_{t=0}^{\infty} \subset \mathcal{M}(\Omega)$ and an (Ω)-program $\{x_t\}_{t=0}^{\infty}$ satisfy

$$\|u_t - v\| \leq \delta \text{ for all integers } t \geq 0,$$

$$\limsup_{T\to\infty}[\sum_{t=0}^{T-1} u_t(x_t, x_{t+1}) - \sigma(\{u_t\}_{t=0}^{T-1}, 0, T)] > -M.$$

Then

$$Card(\{t \text{ is a nonnegative integer such that } \rho(x_t, \bar{x}_v) > \epsilon\}) < L.$$

4. Assume that an integer $T > 0$, $\{u_t\}_{t=0}^{T-1} \subset \mathcal{M}(\Omega)$, an (Ω)-program $\{x_t\}_{t=0}^{T}$ satisfy (2.10), (2.11), (2.12), and integers T_1, T_2 satisfy $0 \leq T_1 < T_2 \leq T$. Then

$$\sum_{t=T_1}^{T_2-1} u_t(x_t, x_{t+1}) \geq \sigma(\{u_t\}_{t=T_1}^{T_2-1}, T_1, T_2) - (4L+2)(2\|v\| + 2) - M - 1.$$

Assertions 1 and 2 establish turnpike properties for approximate solutions of the problem (P3) while Assertion 3 establish the turnpike property for approximate solutions of the corresponding infinite horizon problem. Moreover, they also show the stability of the turnpike phenomenon under small perturbations of the objective function v.

It is clear that $(\mathcal{M}(\Omega), \|\cdot\|)$ is a Banach space. Denote by $\mathcal{M}_0(\Omega)$ the set of all upper semicontinuous functions $v \in \mathcal{M}(\Omega)$ for which there exist $\bar{x}_v \in X$, $\bar{c}_v > 0$ and $\bar{r}_v \in (0, 1)$ such that

$$\{(x, y) \in X \times X : \rho(x, \bar{x}_v), \ \rho(y, \bar{x}_v) \le \bar{r}_v\} \subset \Omega, \tag{2.13}$$

$$v \text{ is continuous at } (\bar{x}_v, \bar{x}_v), \tag{2.14}$$

$$\sigma(v, T) \le Tv(\bar{x}_v, \bar{x}_v) + \bar{c}_v \text{ for all integers } T \ge 1. \tag{2.15}$$

In other words, $\mathcal{M}_0(\Omega)$ is the set of all upper semicontinuous functions $v \in \mathcal{M}(\Omega)$ which satisfy assumptions (A1) and (A2) with some $\bar{x}_v \in X$, $\bar{r}_v \in (0, 1)$, $\bar{c}_v > 0$.

Denote by $\mathcal{M}_{c,0}(\Omega)$ the set of all continuous functions $v \in \mathcal{M}_0(\Omega)$. Denote by $\bar{\mathcal{M}}_{c,0}(\Omega)$ and $\bar{\mathcal{M}}_0(\Omega)$ the closure of subspaces $\mathcal{M}_{c,0}(\Omega)$ and $\mathcal{M}_0(\Omega)$ in $\mathcal{M}(\Omega)$, respectively.

We equip the sets $\bar{\mathcal{M}}_{c,0}(\Omega)$ and $\bar{\mathcal{M}}_0(\Omega)$ with the metric d induced by the norm $\|\cdot\|$: $d(u_1, u_2) = \|u_1 - u_2\|$, $u_1, u_2 \in \bar{\mathcal{M}}_0(\Omega)$.

For each $u \in \bar{\mathcal{M}}_0(\Omega)$ and each $r > 0$ set

$$B_d(u, r) = \{w \in \bar{\mathcal{M}}_0(\Omega) : \ \|u - w\| < r\}.$$

We associate with any $v \in \mathcal{M}_0(\Omega)$ the triplet $(\bar{x}_v, \bar{c}_v, \bar{r}_v)$ satisfying (2.13), (2.14), and (2.15).

Denote by $\mathcal{M}_*(\Omega)$ the set of all $v \in \mathcal{M}_0(\Omega)$ such that for any (v, Ω)-good program $\{x_i\}_{i=0}^{\infty}$,

$$\lim_{i \to \infty} \rho(x_i, \bar{x}_v) = 0.$$

Set

$$\mathcal{M}_{c*}(\Omega) = \mathcal{M}_*(\Omega) \cap \mathcal{M}_c(\Omega).$$

In this chapter we prove the following theorem established in [83].

Theorem 2.6. *$\mathcal{M}_*(\Omega)$ contains a set which is a countable intersection of open everywhere dense subsets of $\bar{\mathcal{M}}_0(\Omega)$ and $\mathcal{M}_{c*}(\Omega)$ contains a set which is a countable intersection of open everywhere dense subsets of $\bar{\mathcal{M}}_{c,0}(\Omega)$.*

2.3 Auxiliary Results for Theorem 2.5

Lemma 2.7. *Let ϵ and M_0 be positive numbers. Then there exists an integer $T \ge 1$ such that for each (Ω)-program $\{x_t\}_{t=0}^{T}$ which satisfies*

$$\sum_{t=0}^{T-1} v(x_t, x_{t+1}) \ge Tv(\bar{x}_v, \bar{x}_v) - M_0$$

the inequality $\min\{\rho(x_i, \bar{x}_v) : \ i = 1, \ldots, T\} \le \epsilon$ holds.

Proof. Assume the contrary. Then for each integer $k \geq 1$ there exists an (Ω)-program $\{x_t^{(k)}\}_{t=0}^k$ such that

$$\sum_{t=0}^{k-1} v(x_t^{(k)}, x_{t+1}^{(k)}) \geq kv(\bar{x}_v, \bar{x}_v) - M_0, \tag{2.16}$$

$$\rho(x_t^{(k)}, \bar{x}_v) > \epsilon \text{ for all integers } t = 1, \dots, k. \tag{2.17}$$

Let k be a natural number. Relations (2.16) and (2.4) and assumption (A2) imply that for each natural number $j < k$,

$$\sum_{t=0}^{j-1} v(x_t^{(k)}, x_{t+1}^{(k)}) = \sum_{t=0}^{k-1} v(x_t^{(k)}, x_{t+1}^{(k)}) - \sum_{t=j}^{k-1} v(x_t^{(k)}, x_{t+1}^{(k)})$$

$$\geq kv(\bar{x}_v, \bar{x}_v) - M_0 - \sum_{t=j}^{k-1} v(x_t^{(k)}, x_{t+1}^{(k)})$$

$$\geq kv(\bar{x}_v, \bar{x}_v) - M_0 - \sigma(v, k-j) \geq kv(\bar{x}_v, \bar{x}_v) - M_0 - (k-j)v(\bar{x}_v, \bar{x}_v) - \bar{c}_v.$$

Combined with (2.16) this inequality implies that for each natural number k and each integer $j \in \{1, \dots, k\}$ we have

$$\sum_{t=0}^{j-1} v(x_t^{(k)}, x_{t+1}^{(k)}) \geq jv(\bar{x}_v, \bar{x}_v) - \bar{c}_v - M_0. \tag{2.18}$$

There exists a strictly increasing sequence of natural numbers $\{k_i\}_{i=1}^{\infty}$ such that for each nonnegative integer t there exists a limit

$$x_t = \lim_{i \to \infty} x_t^{(k_i)}. \tag{2.19}$$

It is easy to see that the sequence $\{x_t\}_{t=0}^{\infty}$ is an (Ω)-program. In view of relations (2.19) and (2.17), we have

$$\rho(x_t, \bar{x}_v) \geq \epsilon \text{ for all natural numbers } t. \tag{2.20}$$

By relations (2.19) and (2.18), for each natural number T,

$$\sum_{t=0}^{T-1} v(x_t, x_{t+1}) \geq Tv(\bar{x}_v, \bar{x}_v) - M_0 - \bar{c}_v.$$

This inequality implies that the sequence $\{x_t\}_{t=0}^{\infty}$ is a (v, Ω)-good program. It follows from assumption (A3) that the equality $\lim_{t \to \infty} \rho(x_t, \bar{x}_v) = 0$ holds. This equality contradicts (2.20). The contradiction we have reached proves the lemma. □

Lemma 2.8 (Lemma 3.3 of [72]). *Let M_0, ϵ be positive numbers. Then there exists a natural number T_0 such that for each integer $T \geq T_0$, each (Ω)-program $\{x_t\}_{t=0}^T$ which satisfies*

$$\sum_{t=0}^{T-1} v(x_t, x_{t+1}) \geq Tv(\bar{x}_v, \bar{x}_v) - M_0 \tag{2.21}$$

and each integer $S \in [0, T - T_0]$ the inequality

$$\min\{\rho(x_t, \bar{x}_v) : t = S + 1, \dots, S + T_0\} \leq \epsilon$$

holds.

Proof. Lemma 2.7 implies that there exists an integer $T_0 \geq 1$ such that the following property holds:

(i) For each (Ω)-program $\{x_t\}_{t=0}^{T_0}$ which satisfies

$$\sum_{t=0}^{T_0-1} v(x_t, x_{t+1}) \geq T_0 v(\bar{x}_v, \bar{x}_v) - M_0 - 2\bar{c}_v$$

the inequality

$$\min\{\rho(x_i, \bar{x}_v) : i = 1, \dots, T_0\} \leq \epsilon$$

holds.

Let an integer $T \geq T_0$, let an (Ω)-program $\{x_t\}_{t=0}^T$ satisfy (2.21), and let an integer $S \in [0, T - T_0]$. By relations (2.21) and (2.9), we have

$$\sum_{t=S}^{S+T_0-1} v(x_t, x_{t+1}) - T_0 v(\bar{x}_v, \bar{x}_v) \geq -M_0 - 2\bar{c}_v.$$

It follows from this inequality and property (i) that

$$\min\{\rho(x_i.\bar{x}_v) : i = S + 1, \dots, S + T_0\} \leq \epsilon.$$

Lemma 2.8 is proved. □

Lemma 2.9. *Let $\epsilon > 0$. Then there exists $\delta \in (0, \bar{r}_v)$ such that for each integer $T \geq 1$ and each (Ω)-program $\{x_t\}_{t=0}^T$ satisfying*

$$\rho(x_0, \bar{x}_v),\ \rho(x_T, \bar{x}_v) \leq \delta, \tag{2.22}$$

$$\sum_{t=0}^{T-1} v(x_t, x_{t+1}) \geq \sigma(v, T, x_0, x_T) - \delta \tag{2.23}$$

the inequality $\rho(x_t, \bar{x}_v) \leq \epsilon$ holds for all $t = 0, \dots, T$.

Proof. Since the function v is continuous at the point $(\bar{x}_v, \bar{x}_v)$ for any integer $k \geq 1$ there exists a number

$$\delta_k \in (0, 2^{-k}\bar{r}_v) \tag{2.24}$$

such that

$$|v(x,y) - v(\bar{x}_v, \bar{x}_v)| \leq 2^{-k} \tag{2.25}$$

for each pair of points $x, y \in X$ satisfying

$$\rho(x, \bar{x}_v),\ \rho(y, \bar{x}_v) \leq \delta_k. \tag{2.26}$$

Assume that the lemma is wrong. Then for each integer $k \geq 1$ there exist a natural number T_k and an (Ω)-program $\{x_t^{(k)}\}_{t=0}^{T_k}$ such that

$$\rho(x_0^{(k)}, \bar{x}_v),\ \rho(x_{T_k}^{(k)}, \bar{x}_v) \leq \delta_k, \tag{2.27}$$

$$\sum_{t=0}^{T_k-1} v(x_t^{(k)}, x_{t+1}^{(k)}) \geq \sigma(v, x_0^{(k)}, x_{T_k}^{(k)}, T_k) - \delta_k, \tag{2.28}$$

$$\max\{\rho(x_t^{(k)}, \bar{x}_v) :\ t = 0, \ldots, T_k\} > \epsilon. \tag{2.29}$$

Let a natural number k be given. Define a sequence $\{z_t\}_{t=0}^{T_k} \subset X$ as follows:

$$z_0 = x_0^{(k)},\ z_{T_k} = x_{T_k}^{(k)},\ z_t = \bar{x}_v,\ t \in \{0, \ldots, T_k\} \setminus \{0, T_k\}. \tag{2.30}$$

In view of relations (2.30), (2.27), (2.24), and (A1), the sequence $\{z_t\}_{t=0}^{T_k}$ is a program. By (2.28) and (2.30), we have

$$\sum_{t=0}^{T_k-1} v(x_t^{(k)}, x_{t+1}^{(k)}) \geq \sigma(v, x_0^{(k)}, x_{T_k}^{(k)}, T_k) - \delta_k \geq \sum_{t=0}^{T_k-1} v(z_t, z_{t+1}) - \delta_k. \tag{2.31}$$

It follows from relations (2.27), (2.30), and the choice of δ_k (see (2.24)–(2.26)) that

$$|v(z_0, z_1) - v(\bar{x}_v, \bar{x}_v)| \leq 2^{-k},\ |v(z_{T_k-1}, z_{T_k}) - v(\bar{x}_v, \bar{x}_v)| \leq 2^{-k},$$

$$v(z_t, z_{t+1}) = v(\bar{x}_v, \bar{x}_v),\ t \in \{0, \ldots, T_k - 1\} \setminus \{0, T_k - 1\}. \tag{2.32}$$

In view of (2.32) and (2.31),

$$\sum_{t=0}^{T_k-1} v(x_t^{(k)}, x_{t+1}^{(k)}) \geq T_k v(\bar{x}_v, \bar{x}_v) - 2 \cdot 2^{-k} - \delta_k. \tag{2.33}$$

Put

$$S_0 = 0,\ S_k = \sum_{i=1}^{k} (T_i + 1) - 1 \text{ for all natural numbers } k. \tag{2.34}$$

Define a sequence $\{x_t\}_{t=0}^{\infty} \subset X$ as follows:

$$x_t = x_t^{(1)},\ t = 0, \dots, T_1,\ x_t = x_i^{(k+1)} \tag{2.35}$$

for each natural number k, each $i \in \{0, \dots, T_{k+1}\}$ and $t = S_k + i + 1$.

By relations (2.35), (2.27), (2.24), and (A1), the sequence $\{x_t\}_{t=0}^{\infty}$ is an (Ω)-program. It follows from (2.34), (2.35), (2.27), and (2.24) that for each natural number k we have

$$|v(x_{S_k}, x_{S_k+1}) - v(\bar{x}_v, \bar{x}_v)| \le 2 \cdot 2^{-k}. \tag{2.36}$$

In view of relations (2.34), (2.35), (2.33), (2.30), and the choice of δ_j, $j = 1, 2, \dots$ (see (2.24)–(2.27)) for any natural number $k \ge 2$, we have

$$\sum_{t=0}^{S_k-1} v(x_t, x_{t+1}) - S_k v(\bar{x}_v, \bar{x}_v) = \sum_{j=1}^{k} \left(\sum_{t=0}^{T_j-1} [v(x_t^{(j)}, x_{t+1}^{(j)}) - v(\bar{x}_v, \bar{x}_v)] \right)$$

$$+ \sum_{j=1}^{k-1} [v(x_{T_j}^{(j)}, x_0^{(j+1)}) - v(\bar{x}_v, \bar{x}_v)] \ge - \sum_{j=1}^{k} (2 \cdot 2^{-j} + \delta_j) - 2 \sum_{j=1}^{k-1} 2^{-j}.$$

Together with inclusion (2.24) this relation implies that for any natural number $k \ge 2$,

$$\sum_{t=0}^{S_k-1} v(x_t, x_{t+1}) - S_k v(\bar{x}_v, \bar{x}_v) \ge -5 \sum_{j=1}^{k} 2^{-j} \ge -10.$$

It follows from this inequality and Proposition 2.1 that the sequence $\{x_t\}_{t=0}^{\infty}$ is a (v, Ω)-good program. By assumption (A3), we have

$$\lim_{t \to \infty} \rho(x_t, \bar{x}_v) = 0.$$

On the other hand it follows from relations (2.29), (2.34), and (2.35) that $\limsup_{t \to \infty} \rho(x_t, \bar{x}_v) \ge \epsilon$. The contradiction we have reached proves Lemma 2.9. □

Lemma 2.10. *Let $\epsilon \in (0, \bar{r}_v)$ and $M_0 > 0$. Then there exist a number $\delta_0 \in (0, 1)$ and an integer $L_0 > 4$ such that for each integer $T \ge L_0$, each finite sequence of functions $\{u_t\}_{t=0}^{T-1} \subset \mathcal{M}(\Omega)$, each (Ω)-program $\{x_t\}_{t=0}^{T}$ which satisfy*

$$\|u_t - v\| \le \delta_0,\ t = 0, \dots, T - 1, \tag{2.37}$$

$$\min\{\rho(x_t, \bar{x}_v) : t = 1, \dots, T\} > \epsilon \tag{2.38}$$

and each pair of points $z_0, z_1 \in X$ satisfying

$$\rho(\bar{x}_v, z_i) \le \bar{r}_v,\ i = 0, 1, \tag{2.39}$$

the following relation is valid:

$$\sum_{t=0}^{T-1} u_t(x_t, x_{t+1}) \le u_0(z_0, \bar{x}_v) + \sum_{t=1}^{T-2} u_t(\bar{x}_v, \bar{x}_v) + u_{T-1}(\bar{x}_v, z_1) - M_0.$$

Proof. Fix a real number

$$M_1 > 8 + 4\|v\| + M_0. \tag{2.40}$$

In view of Lemma 2.8 there exists an integer $L_0 > 4$ such that the following property holds:

(i) for each integer $T \ge L_0$, each (Ω)-program $\{x_t\}_{t=0}^T$ which satisfies

$$\sum_{t=0}^{T-1} v(x_t, x_{t+1}) \ge Tv(\bar{x}_v, \bar{x}_v) - M_1$$

and each integer $S \in [0, T - L_0]$ the inequality

$$\min\{\rho(x_t, \bar{x}_v) : t = S+1, \dots, S+L_0\} \le \epsilon$$

is valid.

Fix a positive number

$$\delta_0 < (4L_0)^{-1}. \tag{2.41}$$

Assume that an integer $T \ge L_0$, $\{u_t\}_{t=0}^{T-1} \subset \mathcal{M}$, an (Ω)-program $\{x_t\}_{t=0}^T$ satisfy (2.37) and (2.38) and that a pair of points $z_0, z_1 \in X$ satisfy (2.39). There exists an integer $k \ge 1$ such that

$$kL_0 \le T < (k+1)L_0. \tag{2.42}$$

In view of (2.39) the sequence $z_0, \bar{x}, \dots, \bar{x}$ $(T-2$ times$), z_1$ is an (Ω)-program. Property (i) and (2.38) imply that for every integer i satisfying $0 < i \le k-1$,

$$\sum_{t=(i-1)L_0}^{iL_0-1} v(x_t, x_{t+1}) < L_0 v(\bar{x}_v, \bar{x}_v) - M_1, \tag{2.43}$$

$$\sum_{t=(k-1)L_0}^{T-1} v(x_t, x_{t+1}) < (T-(k-1)L_0)v(\bar{x}_v, \bar{x}_v) - M_1. \tag{2.44}$$

It follows from (2.37), (2.41), (2.43), and (2.44) that for every integer i satisfying $0 < i \le k-1$, we have

$$\sum_{t=(i-1)L_0}^{iL_0-1} u_t(x_t, x_{t+1}) \le \sum_{t=(i-1)L_0}^{iL_0-1} v(x_t, x_{t+1}) + \delta_0 L_0$$

$$\le L_0 v(\bar{x}_v, \bar{x}_v) - M_1 + \delta_0 L_0$$

$$\leq \sum_{t=(i-1)L_0}^{iL_0-1} u_t(\bar{x}_v, \bar{x}_v) + \delta_0 L_0 - M_1 + \delta_0 L_0$$

$$\leq \sum_{t=(i-1)L_0}^{iL_0-1} u_t(\bar{x}_v, \bar{x}_v) - M_1 + 1 \tag{2.45}$$

and

$$\sum_{t=(k-1)L_0}^{T-1} u_t(x_t, x_{t+1}) \leq \sum_{t=(k-1)L_0}^{T-1} v(x_t, x_{t+1}) + 2\delta_0 L_0$$

$$\leq (T-(k-1)L_0)v(\bar{x}_v, \bar{x}_v) - M_1 + 2\delta_0 L_0$$

$$\leq \sum_{t=(k-1)L_0}^{T-1} u_t(\bar{x}_v, \bar{x}_v) + 2\delta_0 L_0 - M_1 + 2\delta_0 L_0$$

$$\leq \sum_{t=(k-1)L_0}^{T-1} u_t(\bar{x}_v, \bar{x}_v) - M_1 + 1. \tag{2.46}$$

It follows from (2.37), (2.40), (2.42), (2.45), and (2.46) that

$$\sum_{t=0}^{T-1} u_t(x_t, x_{t+1}) \leq \sum_{t=0}^{T-1} u_t(\bar{x}_v, \bar{x}_v) - M_1 + 1$$

$$\leq u_0(z_0, \bar{x}_v) + \sum_{t=1}^{T-2} u_t(\bar{x}_v, \bar{x}_v) + u_{T-1}(\bar{x}_v, z_1) + 2\|u_0\| + 2\|u_T\| - M_1 + 1$$

$$\leq u_0(z_0, \bar{x}_v) + \sum_{t=1}^{T-2} u_t(\bar{x}_v, \bar{x}_v) + u_{T-1}(\bar{x}_v, z_1) - M_0.$$

Lemma 2.10 is proved. □

Lemma 2.11. *Let $\epsilon \in (0, \bar{r}_v)$, M_0 be a positive number and let $\delta_0 \in (0,1)$ and an integer $L_0 > 4$ be as guaranteed by Lemma 2.10.*

Assume that an integer $T \geq L_0$, $\{u_t\}_{t=0}^{T-1} \subset \mathcal{M}$ and that an (Ω)-program $\{x_t\}_{t=0}^{T}$ satisfy

$$\|u_t - v\| \leq \delta_0, \ t = 0, \dots, T-1 \tag{2.47}$$

and

$$\sum_{t=0}^{T-1} u_t(x_t, x_{t+1}) > \sigma(\{u_t\}_{t=0}^{T-1}, 0, T) - M_0 + 2\|v\| + 2. \tag{2.48}$$

Then there exists a finite strictly increasing sequence of natural numbers $\{T_j\}_{j=1}^q$ *where* q *is a natural number such that*

$$T_1 \in [1, L_0], \; T_q \in [T - L_0 + 1, T], \tag{2.49}$$

$$T_{j+1} - T_j \le L_0 \text{ for all natural numbers } j < q, \tag{2.50}$$

$$\rho(\bar{x}_v, x_{T_j}) \le \epsilon, \; j =, 1, \dots, q. \tag{2.51}$$

Proof. We claim that there exists a natural number $T_1 \in [1, L_0]$ such that $\rho(\bar{x}_v, x_{T_1}) \le \epsilon$.

Assume the contrary. Then

$$\rho(x_t, \bar{x}_v) > \epsilon, \; t = 1, \dots, L_0. \tag{2.52}$$

There are two cases:

$$\rho(x_t, \bar{x}_v) > \epsilon, \; t = 1, \dots, T; \tag{2.53}$$

$$\min\{\rho(x_t, \bar{x}_v) : \; t = 1, \dots, T\} \le \epsilon. \tag{2.54}$$

Assume that (2.53) is valid. It follows from (2.47), (2.53), the inequality $T \ge L_0$, the choice of δ_0, L_0, and Lemma 2.10 that

$$\sum_{t=0}^{T-1} u_t(x_t, x_{t+1}) \le \sum_{t=0}^{T-1} u_t(\bar{x}_v, \bar{x}_v) - M_0 \le \sigma(\{u_t\}_{t=0}^{T-1}, 0, T) - M_0.$$

This contradicts (2.48). The contradiction we have reached proves (2.54). In view of (2.52) and (2.54) there exists an integer

$$S \in (L_0, T] \tag{2.55}$$

for which

$$\rho(x_S, \bar{x}_v) \le \epsilon, \; \rho(x_t, \bar{x}_v) > \epsilon, \; t = 1, \dots, S - 1. \tag{2.56}$$

Set

$$z_t = \bar{x}_v, \; t = 0, \dots, S - 1, \; z_t = x_t, \; t = S, \dots, T. \tag{2.57}$$

Relations (2.56) and (2.57) imply that $\{z_t\}_{t=0}^T$ is an (Ω)-program. By (2.47), (2.55), (2.56), (2.57), the choice of δ_0, and L_0 and Lemma 2.10 (with $T = S - 1$), we have

$$\sum_{t=0}^{S-2} u_t(x_t, x_{t+1}) \le \sum_{t=0}^{S-2} u_t(\bar{x}_v, \bar{x}_v) - M_0 = \sum_{t=0}^{S-2} u_t(z_t, z_{t+1}) - M_0. \tag{2.58}$$

It follows from (2.57) and (2.58) that

$$\sum_{t=0}^{T-1} u_t(x_t, x_{t+1}) - \sum_{t=0}^{T-1} u_t(z_t, z_{t+1})$$

$$= \sum_{t=0}^{S-2} u_t(x_t, x_{t+1}) + u_{S-1}(x_{S-1}, x_S) - \sum_{t=0}^{S-2} u_t(\bar{x}_v, \bar{x}_v) - u_{S-1}(z_{S-1}, z_S)$$

$$\leq -M_0 + 2\|u_{S-1}\| \leq -M_0 + 2\|v\| + 2.$$

This implies that

$$\sum_{t=0}^{T-1} u_t(x_t, x_{t+1}) \leq \sigma(\{u_t\}_{t=0}^{T-1}, 0, T) - M_0 + 2\|v\| + 2.$$

The inequality above contradicts (2.48). The contradiction we have reached proves that there exists a natural number

$$T_1 \in [1, L_0] \tag{2.59}$$

for which

$$\rho(\bar{x}_v, x_{T_1}) \leq \epsilon. \tag{2.60}$$

Assume now that k is a natural number and we defined a strictly increasing sequence of positive integers $\{T_i\}_{i=1}^k$ such that

$$T_1 \in [1, L_0],\ T_k \leq T, \tag{2.61}$$

$$\rho(\bar{x}_v, x_{T_i}) \leq \epsilon,\ i =, 1, \dots, k, \tag{2.62}$$

for every natural number $i < k$, we have

$$T_{i+1} - T_i \leq L_0. \tag{2.63}$$

(It is clear that for $k = 1$ our assumption is true.)

If $T - T_k < L_0$, then the construction is completed. Assume that

$$T - T_k \geq L_0. \tag{2.64}$$

We claim that there exists an integer T_{k+1} such that

$$T_{k+1} - T_k \leq [1, L_0],\ \rho(\bar{x}_v, x_{T_{k+1}}) \leq \epsilon. \tag{2.65}$$

Assume the contrary. Then

$$\rho(\bar{x}_v, x_t) > \epsilon,\ t = T_k + 1, \dots, T_k + L_0. \tag{2.66}$$

There are two cases:

$$\rho(\bar{x}_v, x_t) > \epsilon,\ t = T_k + 1, \dots, T; \tag{2.67}$$

$$\min\{\rho(\bar{x}_v, x_t) :\ t = T_k + 1, \dots, T\} \leq \epsilon. \tag{2.68}$$

Assume that (2.67) is true. It follows from (2.47), (2.62), (2.64), (2.67), the choice of δ_0, L_0, and Lemma 2.10 that

$$\sum_{t=T_k}^{T-1} u_t(x_t, x_{t+1}) \leq u_{T_k}(x_{T_k}, \bar{x}_v) + \sum_{t=T_k+1}^{T-1} u_t(\bar{x}_v, \bar{x}_v) - M_0. \tag{2.69}$$

Define

$$z_t = x_t,\ t = 0, \dots, T_k,\ z_t = \bar{x}_v,\ t = T_k + 1, \dots, T. \tag{2.70}$$

In view of (2.70) and (2.62), $\{z_t\}_{t=0}^T$ is an (Ω)-program. By (2.70) and (2.69), we have

$$\sum_{t=0}^{T-1} u_t(x_t, x_{t+1}) - \sigma(\{u_t\}_{t=0}^{T-1}, 0, T)$$

$$\leq \sum_{t=0}^{T-1} u_t(x_t, x_{t+1}) - \sum_{t=0}^{T-1} u_t(z_t, z_{t+1})$$

$$= \sum_{t=T_k}^{T-1} u_t(x_t, x_{t+1}) - u_{T_k}(x_{T_k}, \bar{x}_v) - \sum_{t=T_k+1}^{T-1} u_t(\bar{x}_v, \bar{x}_v) \leq -M_0.$$

The relation above contradicts (2.48). The contradiction we have reached proves (2.68). In view of (2.66) and (2.68), there exists an integer S such that

$$T_k + L_0 < S \leq T,\ \rho(x_S, \bar{x}_v) \leq \epsilon, \tag{2.71}$$

$$\rho(x_t, \bar{x}_v) > \epsilon,\ t = T_k + 1, \dots, S - 1. \tag{2.72}$$

Set

$$z_t = x_t,\ t = 0, \dots, T_k,\ z_t = \bar{x}_v,\ t = T_k + 1, \dots, S - 1,\ z_t = x_t,\ t = S, \dots, T. \tag{2.73}$$

It follows from (2.62), (2.71), and (2.73) that $\{z_t\}_{t=0}^T$ is an (Ω)-program. By (2.37), (2.71), (2.72), the choice of δ_0 and L_0 and Lemma 2.10, we have

$$\sum_{t=T_k}^{S-2} u_t(x_t, x_{t+1})$$

$$\leq u_{T_k}(x_{T_k}, \bar{x}_v) + \sum_{t=T_k+1}^{S-2} u_t(\bar{x}_v, \bar{x}_v) - M_0$$

$$= \sum_{t=T_k}^{S-2} u_t(z_t, z_{t+1}) - M_0. \tag{2.74}$$

In view of (2.73) and (2.74), we have

$$\sum_{t=0}^{T-1} u_t(x_t, x_{t+1}) - \sigma(\{u_t\}_{t=0}^{T-1}, 0, T) \leq \sum_{t=0}^{T-1} u_t(x_t, x_{t+1}) - \sum_{t=0}^{T-1} u_t(z_t, z_{t+1})$$

$$= \sum_{t=T_k}^{T-1} u_t(x_t, x_{t+1}) - \sum_{t=T_k}^{T-1} u_t(z_t, z_{t+1})$$

$$= \sum_{t=T_k}^{S-1} u_t(x_t, x_{t+1}) - u_{T_k}(x_{T_k}, \bar{x}_v) - \sum_{t=T_k+1}^{S-2} u_t(\bar{x}_v, \bar{x}_v) - u_{S-1}(\bar{x}_v, x_S)$$

$$\leq -M_0 + 2\|u_{S-1}\| \leq -M_0 + 2\|v\| + 2.$$

The relation above contradicts (2.48). The contradiction we have reached proves that there is an integer T_{k+1} satisfying (2.65).

It is clear that the assumption made for k also holds for $k+1$. Thus by induction we construct the sequence T_i, $i = 1, \dots, q$ which is obtained by a finite number of steps and T_q is its last element. It follows from the construction that (2.49)–(2.51) are true. This completes the proof of Lemma 2.11. □

2.4 Proof of Theorem 2.5

Lemma 2.9 implies that there exists a positive number

$$\delta_0 < \min\{\bar{r}_v, \epsilon, 1\}$$

such that the following property holds:

(i) for every natural number T and every (Ω)-program $\{x_t\}_{t=0}^T$ which satisfy

$$\rho(x_0, \bar{x}_v),\ \rho(x_T, \bar{x}_v) \leq \delta_0$$

and

$$\sum_{t=0}^{T-1} v(x_t, x_{t+1}) \geq \sigma(v, T, x_0, x_T) - \delta_0$$

the inequality $\rho(x_t, \bar{x}_v) \leq \epsilon$ is true for all integers $t = 0, \dots, T$.

Lemma 2.11 implies that there exist $\delta_1 \in (0, 1)$ and an integer $L_1 > 4$ such that the following property holds:

(ii) for every natural number $T \geq L_1$, every finite sequence of functions $\{u_t\}_{t=0}^{T-1} \subset \mathcal{M}(\Omega)$ and every (Ω)-program $\{x_t\}_{t=0}^T$ which satisfy

$$\|u_t - v\| \leq \delta_1,\ t = 0, \dots, T-1,$$

and

$$\sum_{t=0}^{T-1} u_t(x_t, x_{t+1}) > \sigma(\{u_t\}_{t=0}^{T-1}, 0, T) - M - 4$$

there exists a strictly increasing sequence of natural numbers $\{T_j\}_{j=1}^q$ where q is a natural number such that

$$T_1 \in [1, L_1],\ T_q \in [T - L_1 + 1, T], \tag{2.75}$$

$$T_{j+1} - T_j \leq L_1 \text{ for all natural numbers } j < q, \tag{2.76}$$

$$\rho(\bar{x}_v, x_{T_j}) < \delta_0,\ j =, 1, \dots, q. \tag{2.77}$$

Choose a positive number δ and an integer $L \geq 1$ such that

$$\delta < \min\{\delta_1, M, 8^{-1}L_1^{-1}\delta_0\}, \tag{2.78}$$

$$L > 8L_1 + 2L_1(M+1)\delta_0^{-1}.$$

Let us prove Assertion 1. Property (ii), (2.10), (2.11), and (2.78) imply that there exists a strictly increasing sequence of natural numbers $\{T_j\}_{j=1}^q$, where q is a natural number, such that (2.75), (2.76), and (2.77) are true. In view of (2.75), (2.76), and (2.78),

$$q > 2. \tag{2.79}$$

Let

$$E = \{j \in \{1, \dots, q-1\} :$$

$$\sum_{t=T_j}^{T_{j+1}-1} u_t(x_t, x_{t+1}) \geq \sigma(\{u_t\}_{t=T_j}^{T_{j+1}-1}, T_j, T_{j+1}, x_{T_j}, x_{T_{j+1}}) - \delta_0/2. \tag{2.80}$$

By (2.11) and (2.80), we have

$$-M \leq \sum_{t=0}^{T-1} u_t(x_t, x_{t+1}) - \sigma(\{u_t\}_{t=0}^{T-1}, 0, T)$$

$$\leq \sum_{t=0}^{T-1} u_t(x_t, x_{t+1}) - \sigma(\{u_t\}_{t=0}^{T-1}, 0, T, x_0, x_T)$$

$$\leq \sum\{ \sum_{t=T_j}^{T_{j+1}-1} u_t(x_t, x_{t+1}) - \sigma(\{u_t\}_{t=T_j}^{T_{j+1}-1}, T_j, T_{j+1}, x_{T_j}, x_{T_{j+1}}) :$$

$$j \in \{1, \dots, q-1\} \setminus E\}$$

$$\leq (-\delta_0/2)\mathrm{Card}(\{1, \dots, q-1\} \setminus E\}$$

and

$$\mathrm{Card}(\{1, \dots, q-1\} \setminus E\}) \leq 2M\delta_0^{-1}. \tag{2.81}$$

Let

$$j \in E. \tag{2.82}$$

It follows from (2.10), (2.80), and (2.82) that

$$\sum_{t=T_j}^{T_{j+1}-1} v(x_t, x_{t+1}) \geq \sum_{t=T_j}^{T_{j+1}-1} u_t(x_t, x_{t+1}) - \delta L_1$$

$$\geq \sigma(\{u_t\}_{t=T_j}^{T_{j+1}-1}, T_j, T_{j+1}, x_{T_j}, x_{T_{j+1}}) - \delta_0/2 - \delta L_1$$

$$\geq \sigma(v, T_{j+1} - T_j, x_{T_j}, x_{T_{j+1}}) - \delta L_1 - \delta_0/2 - \delta L_1$$

$$\geq \sigma(v, T_{j+1} - T_j, x_{T_j}, x_{T_{j+1}}) - \delta_0. \tag{2.83}$$

In view of (2.77), (2.83), and property (i),

$$\rho(x_t, \bar{x}_v) \le \epsilon,\ t = T_j, \dots, T_{j+1} \text{ and all } j \in E.$$

Combined with (2.77) the relation above implies that

$$\{t \in \{0, \dots, T\} :\ \rho(x_t, \bar{x}_v) > \epsilon\}$$

$$\subset \{0, \dots, T_1 - 1\} \cup \{T - L_1 + 1, \dots, T\}$$

$$\cup \{\{t \text{ is an integer and } T_j < t < T_{j+1}\} :\ j \in \{1, \dots, q-1\} \setminus E\}.$$

Combined with (2.75), (2.76), (2.78), and (2.81) this implies that

$$\mathrm{Card}(\{t \in \{0, \dots, T\} :\ \rho(x_t, \bar{x}_v) > \epsilon\})$$

$$\le 2L_1 + L_1 \mathrm{Card}(\{1, \dots, q-1\} \setminus E\}) \le 2L_1 + 2ML_1\delta_0^{-1} < L.$$

This completes the proof of Assertion 1.

Let us prove Assertion 2. It follows from (2.10), (2.11), (2.78), and property (ii) that there exists a strictly increasing sequence of natural numbers $\{T_j\}_{j=1}^{q}$, where $q \ge 1$ is an integer, such that (2.76), (2.77) are true and that

$$T_q \in [T - L_1 + 1, T],\ T_1 \in [0, L_1].$$

Evidently, if $\rho(x_0, \bar{x}_v) \le \delta$, then we may assume that $T_1 = 0$ and if $\rho(x_T, \bar{x}_v) \le \delta$, then we may assume that $T_q = T$.

Let an integer $j \in \{1, \dots, q-1\}$. In view of (2.10), (2.12), (2.76), and (2.78), we have

$$\sum_{t=T_j}^{T_{j+1}-1} v(x_t, x_{t+1}) \ge \sum_{t=T_j}^{T_{j+1}-1} u_t(x_t, x_{t+1}) - \delta L_1$$

$$\ge \sigma(\{u_t\}_{t=T_j}^{T_{j+1}-1}, T_j, T_{j+1}, x_{T_j}, x_{T_{j+1}}) - \delta(L_1 + 1)$$

$$\ge \sigma(v, T_{j+1} - T_j, x_{T_j}, x_{T_{j+1}}) - \delta L_1 - \delta(L_1 + 1)$$

$$\ge \sigma(v, T_{j+1} - T_j, x_{T_j}, x_{T_{j+1}}) - \delta_0.$$

Combined with (2.77) and property (i) this implies that

$$\rho(x_t, \bar{x}_v) \le \epsilon \text{ for all integers } t \in [T_j, T_{j+1}] \text{ and all } j = 1, \dots, q-1.$$

This implies that

$$\rho(x_t, \bar{x}_v) \le \epsilon \text{ for all integers } t \in [T_1, T_q].$$

This completes the proof of Assertion 2.

Assertion 3 easily follows from Assertion 1.

Let us prove Assertion 4. There are two cases: $T_2 - T_1 \le 4L$; $T_2 - T_1 > 4L$.

Evidently, if $T_2 - T_1 \le 4L$, then by (2.10), Assertion 4 holds. Assume that

$$T_2 - T_1 > 4L. \tag{2.84}$$

There exists an (Ω)-program $\{y_t\}_{t=T_1}^{T_2}$ such that

$$\sum_{t=T_1}^{T_2-1} u_t(y_t, y_{t+1}) \ge \sigma(\{u_t\}_{t=T_1}^{T_2-1}, T_1, T_2) - \delta. \tag{2.85}$$

It follows from (2.10), (2.11), (2.12), (2.84), (2.85), and Assertion 2 that

$$\rho(x_t, \bar{x}_v) \le \epsilon \text{ for all integers } t = L, \dots, T - L, \tag{2.86}$$

$$\rho(y_t, \bar{x}_v) \le \epsilon \text{ for all integers } t = T_1 + L, \dots, T_2 - L. \tag{2.87}$$

Define

$$z_t = x_t,\ t = 0, \dots, T_1 + L,\ z_t = y_t,\ t = T_1 + L + 1, \dots, T_2 - L - 1,$$

$$z_t = x_t,\ t = T_2 - L, \dots, T. \tag{2.88}$$

It follows from (2.88), (2.86), (2.87), and (A1) that $\{z_t\}_{t=0}^{T}$ is an (Ω)-program. In view of (2.10), (2.11), and (2.88), we have

$$-M \le \sum_{t=0}^{T-1} u_t(x_t, x_{t+1}) - \sigma(\{u_t\}_{t=0}^{T-1}, 0, T) \le \sum_{t=0}^{T-1} u_t(x_t, x_{t+1}) - \sum_{t=0}^{T-1} u_t(z_t, z_{t+1})$$

$$= \sum_{t=T_1+L}^{T_2-L-1} u_t(x_t, x_{t+1}) - \sum_{t=T_1+L}^{T_2-L-1} u_t(z_t, z_{t+1})$$

$$\le \sum_{t=T_1}^{T_2-1} u_t(x_t, x_{t+1}) + 2L(\|v\| + 1) - \sum_{t=T_1+L}^{T_2-L-1} u_t(y_t, y_{t+1}) + 4(\|v\| + 1)$$

$$\le \sum_{t=T_1}^{T_2-1} u_t(x_t, x_{t+1}) + 2(\|v\| + 1)(L + 2) - \sum_{t=T_1}^{T_2-1} u_t(y_t, y_{t+1}) + 2L(\|v\| + 1).$$

Together with (2.85) this implies that

$$\sum_{t=T_1}^{T_2-1} u_t(x_t, x_{t+1}) \ge -M - 2(\|v\| + 1)(2L + 2) + \sum_{t=T_1}^{T_2-1} u_t(y_t, y_{t+1})$$

$$\ge -M - 2(\|v\| + 1)(2L + 2) + \sigma(\{u_t\}_{t=T_1}^{T_2-1}, T_1, T_2) - 1.$$

This implies Assertion 4. This completes the proof of Theorem 2.5. □

2.5 Proof of Theorem 2.6

Lemma 2.12. *The set $\mathcal{M}_*(\Omega)$ is an everywhere dense subset of $\bar{\mathcal{M}}_0(\Omega)$ and $\mathcal{M}_{c*}(\Omega)$ is an everywhere dense subset of $\bar{\mathcal{M}}_{c,0}(\Omega)$.*

Proof. Let $v \in \mathcal{M}_0(\Omega)$ and $\gamma \in (0,1)$. Define

$$v_\gamma(x,y) = v(x,y) - \gamma\rho(x,\bar{x}_v) - \gamma\rho(y,\bar{x}_v).$$

It is not difficult to see that $v_\gamma \in \mathcal{M}_*(\Omega)$, if $v \in \mathcal{M}_{c,0}(\Omega)$, then $v_\gamma \in \mathcal{M}_{c*}(\Omega)$ and $\lim_{\gamma\to 0+} v_\gamma = v$. Lemma 2.12 is proved.

Let a function $v \in \mathcal{M}(\Omega)_*$ and $n \geq 1$ be an integer. In view of (A1) and (2.14), there exists a number

$$r_1(v,n) \in (0, 4^{-1}\bar{r}_v) \tag{2.89}$$

such that

$$|v(\bar{x}_v,\bar{x}_v) - v(z_1,z_2)| < 8^{-1}\bar{r}_v n^{-1} \tag{2.90}$$

for every pair of points $z_1, z_2 \in X$ which satisfy $\rho(z_i,\bar{x}_v) \leq r_1(v,n)$, $i = 1,2$.

Theorem 2.5 implies that there exist a positive number

$$\delta(v,n) < (4n)^{-1}$$

and an integer $L(v,n) \geq 1$ such that for every function

$$u \in B_d(v,\delta(v,n)) \tag{2.91}$$

the following properties hold:

(i) for every integer $T \geq L(v,n)$ and every (Ω)-program $\{x_t\}_{t=0}^T$ which satisfies

$$\sum_{t=0}^{T-1} u(x_t,x_{t+1}) \geq \sigma(u,T) - 2n \tag{2.92}$$

the relation

$$\mathrm{Card}(\{t \in \{0,\dots,T\} :\ \rho(x_t,\bar{x}_v) > (4n)^{-1}r_1(v,n)\}) < L(v,n)$$

is true;

(ii) for every (Ω)-program $\{x_t\}_{t=0}^\infty$ which satisfies

$$\limsup_{T\to\infty}[\sum_{t=0}^{T-1} u(x_t,x_{t+1}) - \sigma(u,T)] > -2n$$

the inequality

$$\mathrm{Card}(\{t \text{ is a nonnegative integer such that}$$

$$\rho(x_t,\bar{x}_v) > (4n)^{-1}r_1(v,n)\}) < L(v,n)$$

is true;

(iii) for every integer $T \geq 2L(v,n)$ and every (Ω)-program $\{x_t\}_{t=0}^T$ which satisfies (2.92) and

$$\sum_{t=0}^{T-1} u(x_t, x_{t+1}) \geq \sigma(u, T, x_0, x_T) - \delta(v,n) \tag{2.93}$$

the inequality

$$\rho(x_t, \bar{x}_v) \leq (4n)^{-1} r_1(v,n)$$

is true for all $t = L(v,n), \dots, T - L(v,n)$;

(iv) for every natural number T, every (Ω)-program $\{x_t\}_{t=0}^T$ which satisfies (2.92) and (2.93), and every pair of integers T_1, T_2 satisfying $0 \leq T_1 < T_2 \leq T$ the inequality

$$\sum_{t=T_1}^{T_2-1} u(x_t, x_{t+1}) \geq \sigma(u, T_2 - T_1) - (4L(v,n) + 2)(2\|v\| + 2) - 2n - 1$$

is true.

Set

$$\mathcal{F} = \cap_{n=1}^{\infty} \cup \{B_d(v, \delta(v,n)) : \ v \in \mathcal{M}_*(\Omega), \ n \text{ is a natural number}\}, \tag{2.94}$$

$$\mathcal{F}_c = \cap_{n=1}^{\infty} \cup \{B_d(v, \delta(v,n)) \cap \bar{\mathcal{M}}_{c,0}(\Omega) : \ v \in \mathcal{M}_{c*}(\Omega), n \text{ is a natural number}\}. \tag{2.95}$$

Lemma 2.12 implies that $\mathcal{F}$ is a countable intersection of open everywhere dense subsets of $\bar{\mathcal{M}}_0(\Omega)$ and $\mathcal{F}_c$ is a countable intersection of open everywhere dense subsets of $\bar{\mathcal{M}}_{c,0}(\Omega)$. Let

$$u \in \mathcal{F}. \tag{2.96}$$

In order to complete the proof it is sufficient to show that $u \in \mathcal{M}_*(\Omega)$. It is clear that for every integer $T \geq 1$ there exists an (Ω)-program $\{x_i^{(T)}\}_{i=0}^T$ for which

$$\sum_{t=0}^{T-1} u(x_t^{(T)}, x_{t+1}^{(T)}) = \sigma(u, T). \tag{2.97}$$

Extracting subsequences and using diagonalization process we obtain that there exists a strictly increasing sequence of natural numbers $\{T_k\}_{k=1}^{\infty}$ such that for every nonnegative integer t there exists

$$x_t^* = \lim_{k \to \infty} x_t^{(T_k)}. \tag{2.98}$$

In view of (2.94) and (2.96), for every integer $n \geq 1$ there exists

$$v_n \in \mathcal{M}_*(\Omega) \tag{2.99}$$

such that

$$u \in B_d(v_n, \delta(v_n, n)). \tag{2.100}$$

Let $n \geq 1$ be an integer. It follows from (2.97), (2.99), (2.100), and property (iii) that for every integer $T \geq 2L(v_n, n)$ we have

$$\rho(x_t^{(T)}, \bar{x}_{v_n}) \leq (4n)^{-1} r_1(v_n, n), \; t = L(v_n, n), \dots, T - L(v_n, n). \tag{2.101}$$

In view of (2.98) and (2.101), for every integer $T \geq L(v_n, n)$, we have

$$\rho(x_t^*, \bar{x}_{v_n}) \leq (4n)^{-1} r_1(v_n, n). \tag{2.102}$$

Property (iv), (2.97), (2.99), and (2.100) imply that for every natural number T and every pair of integers T_1, T_2 satisfying $0 \leq T_1 < T_2 \leq T$ the inequality

$$\sum_{t=T_1}^{T_2-1} u(x_t^{(T)}, x_{t+1}^{(T)}) \geq \sigma(u, T_2 - T_1) - (4L(v_n, n) + 2)(2\|v_n\| + 2) - 2n - 1 \tag{2.103}$$

holds. In view of (2.98) and (2.103), for every pair of integers $T_2 > T_1 \geq 0$, we have

$$\sum_{t=T_1}^{T_2-1} u(x_t^*, x_{t+1}^*) \geq \sigma(u, T_2 - T_1) - (4L(v_n, n) + 2)(2\|v_n\| + 2) - 2n - 1. \tag{2.104}$$

Since n is any natural number it follows from (2.102) that $\{x_t^*\}_{t=0}^{\infty}$ is a Cauchy sequence and there exists

$$x^* = \lim_{t \to \infty} x_t^*. \tag{2.105}$$

Relations (2.102) and (2.105) imply that for every natural number n, we have

$$\rho(x^*, \bar{x}_{v_n}) \leq (4n)^{-1} r_1(v_n, n). \tag{2.106}$$

In view of (2.104) and (2.105), for every natural number n and every natural number T, we have

$$Tu(x^*, x^*) \geq \sigma(u, T) - (4L(v_n, n) + 2)(2\|v_n\| + 4) - 2n - 1. \tag{2.107}$$

Let $n \geq 1$ be an integer and points $y, z \in X$ satisfy

$$\rho(y, x^*), \; \rho(z, x^*) \leq (4n)^{-1} r_1(v_n, n). \tag{2.108}$$

By (2.106) and (2.108),

$$\rho(y, \bar{x}_{v_n}) \leq \rho(y, x^*) + \rho(x^*, \bar{x}_{v_n}) \leq (2n)^{-1} r_1(v_n, n), \tag{2.109}$$

$$\rho(z, \bar{x}_{v_n}) \leq \rho(z, x^*) + \rho(x^*, \bar{x}_{v_n}) \leq (2n)^{-1} r_1(v_n, n). \tag{2.110}$$

Relations (2.89), (2.90), (2.99), (2.109), and (2.110) imply that

$$(y, z) \in \Omega, \; (x^*, x^*) \in \Omega. \tag{2.111}$$

It follows from (2.89), (2.90), (2.100), (2.106), (2.109), and (2.110) that

$$|u(y,z) - u(x^*,x^*)| \le |v_n(y,z) - v_n(x^*,x^*)| + 2\delta(v_n,n)$$
$$\le |v_n(y,z) - v_n(\bar{x}_{v_n},\bar{x}_{v_n})| + |v_n(\bar{x}_{v_n},\bar{x}_{v_n}) - v_n(x^*,x^*)| + 2\delta(v_n,n)$$
$$\le (4n)^{-1}\bar{r}_{v_n} + 2n^{-1} < 1/n.$$

Thus for every pair of points $y, z \in X$ satisfying (2.108), we have $(y,z) \in \Omega$ and

$$|u(y,z) - u(x^*,x^*)| < n^{-1}.$$

Since n is any natural number we conclude that the function u is continuous at the point (x^*,x^*) and together with (2.107) this implies that $u \in \mathcal{M}_0(\Omega)$.

In order to complete the proof it is sufficient to show that for every (u,Ω)-good program $\{x_i\}_{i=0}^{\infty}$, we have

$$\lim_{i\to\infty} \rho(x_i, x^*) = 0.$$

Assume that a program $\{x_i\}_{i=0}^{\infty}$ is (u,Ω)-good. Proposition 2.1 implies that there exists a natural number n_0 such that for all natural numbers T,

$$\sum_{t=0}^{T-1} u(x_t, x_{t+1}) - \sigma(u,T) > -n_0. \tag{2.112}$$

Let an integer $n > n_0$. Property (ii), (2.99), (2.100), and (2.112) imply that there exists a natural number $j(n)$ such that for every integer $t \ge j(n)$, we have

$$\rho(x_t, \bar{x}_{v_n}) \le (4n)^{-1} r_1(v_n, n).$$

Combined with (2.116) the inequality above implies that for all integers $t \ge j(n)$,

$$\rho(x_t, x^*) \le \rho(x_t, \bar{x}_{v_n}) + \rho(\bar{x}_{v_n}, x^*) \le (2n)^{-1} r_1(v_n, n) \le (2n)^{-1}.$$

Since n is any natural number we conclude that

$$\lim_{i\to\infty} \rho(x_t, x^*) = 0$$

and $u \in \mathcal{M}_*(\Omega)$. This completes the proof of Theorem 2.6. □

2.6 Overtaking Optimal Programs

In the sequel we use a notion of an overtaking optimal program [70, 84].

An (Ω)-program $\{x_t\}_{t=0}^{\infty}$ is called (v,Ω)-overtaking optimal if for each (Ω)-program $\{y_t\}_{t=0}^{\infty}$ satisfying $y_0 = x_0$ the inequality

$$\limsup_{T\to\infty} \sum_{t=0}^{T-1} [v(y_t, y_{t+1}) - v(x_t, x_{t+1})] \le 0$$

holds.

The following result obtained in [71] and presented in Chap. 2 of [84] establishes the existence of an overtaking optimal program.

Theorem 2.13. *Assume that $x \in X$ and that there exists a (v, Ω)-good program $\{x_t\}_{t=0}^{\infty}$ such that $x_0 = x$. Then there exists a (v, Ω)-overtaking optimal program $\{x_t^*\}_{t=0}^{\infty}$ such that $x_0^* = x$.*

The next result obtained in [71] and presented in Chap. 2 of [84] provides necessary and sufficient conditions for overtaking optimality.

Theorem 2.14. *Assume that $\{x_t\}_{t=0}^{\infty}$ is an (Ω)-program and that there exists a (v, Ω)-good program $\{y_t\}_{t=0}^{\infty}$ such that $y_0 = x_0$. Then the program $\{x_t\}_{t=0}^{\infty}$ is (v, Ω)-overtaking optimal if and only if the following conditions hold:*

(i) $\lim_{t \to \infty} \rho(x_t, \bar{x}_v) = 0$;

(ii) for each natural number T and each (Ω)-program $\{y_t\}_{t=0}^{T}$ satisfying $y_0 = x_0$, $y_T = x_T$ the inequality $\sum_{t=0}^{T-1} v(y_t, y_{t+1}) \leq \sum_{t=0}^{T-1} v(x_t, x_{t+1})$ holds.

The following two theorems which establish the uniform convergence of overtaking optimal programs to $\bar{x}_v$, are proved in Sect. 2.9. They were obtained in [86].

Theorem 2.15. *Let L_0 be a natural number and ϵ be a positive number. Then there exists an integer $T_0 \geq 1$ such that for each (v, Ω)-overtaking optimal program $\{x_t\}_{t=0}^{\infty}$ satisfying $x_0 \in \bar{Y}_{L_0}$ the inequality $\rho(x_t, \bar{x}_v) \leq \epsilon$ holds for all integers $t \geq T_0$.*

Theorem 2.16. *Let ϵ be a positive number. Then there exists a positive number δ such that for each (v, Ω)-overtaking optimal program $\{x_t\}_{t=0}^{\infty}$ satisfying $\rho(x_0, \bar{x}_v) \leq \delta$ the inequality $\rho(x_t, \bar{x}_v) \leq \epsilon$ holds for all nonnegative integers t.*

2.7 Auxiliary Results

Let $v \in \mathcal{M}(\Omega)$ be an upper semicontinuous function. Suppose that $\bar{x}_v \in X$, $\bar{r}_v \in (0, 1)$, $\bar{c}_v > 0$ and that assumptions (A1), (A2), and (A3) hold.

For each positive number M denote by X_M the set of all points $x \in X$ for which there exists a (Ω)-program $\{x_t\}_{t=0}^{\infty}$ such that $x_0 = x$ and that for all natural numbers T the following inequality holds:

$$\sum_{t=0}^{T-1} v(x_t, x_{t+1}) - Tv(\bar{x}_v, \bar{x}_v) \geq -M.$$

It is not difficult to see that $\cup\{X_M : M \in (0, \infty)\}$ is the set of all points $x \in X$ such that there exists a (v, Ω)-good program $\{x_t\}_{t=0}^{\infty}$ satisfying $x_0 = x$.

The boundedness of v implies the following result.

Proposition 2.17. *Let T be a natural number. Then there exists a positive number M such that $\bar{Y}_T \subset X_M$.*

The next result follows from Lemma 2.8.

Proposition 2.18. *Let M be positive number. Then there exists an integer $T \geq 1$ such that the inclusion $X_M \subset \bar{Y}_T$ holds.*

We define a function $\pi^v(x)$, $x \in X$ which plays an important role in our study. For all $x \in X \setminus \cup\{X_M : M \in (0, \infty)\}$ set

$$\pi^v(x) = -\infty.$$

Let

$$x \in \cup\{X_M : M \in (0, \infty)\}. \tag{2.113}$$

Set

$$\pi^v(x) = \sup\{\limsup_{T \to \infty} \sum_{t=0}^{T-1} (v(x_t, x_{t+1}) - v(\bar{x}_v, \bar{x}_v)) :$$

$$\{x_t\}_{t=0}^{\infty} \text{ is an } (\Omega) - \text{program such that } x_0 = x\}. \tag{2.114}$$

In view of (A2), (2.113), and (2.114),

$$-\infty < \pi^v(x) \leq \bar{c}_v. \tag{2.115}$$

By (2.113), (2.114), and Proposition 2.1,

$$\pi^v(x) = \sup\{\limsup_{T \to \infty} \sum_{t=0}^{T-1} (v(x_t, x_{t+1}) - v(\bar{x}_v, \bar{x}_v)) :$$

$$\{x_t\}_{t=0}^{\infty} \text{ is a } (v, \Omega) - \text{program such that } x_0 = x\}. \tag{2.116}$$

Denote by $\mathcal{P}(v, x)$ the set of all (v, Ω)-overtaking optimal programs $\{x_t\}_{t=0}^{\infty}$ satisfying $x_0 = x$. By Theorem 2.13, the set $\mathcal{P}(v, x)$ is nonempty. Definition (2.114) implies the following result.

Proposition 2.19. *1. Let $\{x_t\}_{t=0}^{\infty}$ be a (v, Ω)-good program. Then for each integer $t \geq 0$,*

$$\pi^v(x_t) \geq v(x_t, x_{t+1}) - v(\bar{x}_v, \bar{x}_v) + \pi^v(x_{t+1}). \tag{2.117}$$

2. Let $T \geq 1$ be an integer and $\{x_t\}_{t=0}^{T}$ be an (Ω)-program such that $\pi^v(x_T) > -\infty$. Then (2.117) holds for all integers $t = 0, \ldots, T-1$.

The next result follows from the definition of (v, Ω)-overtaking optimal programs.

Proposition 2.20. *Let $x \in \cup\{X_M : M \in (0, \infty)\}$ and $\{x_t\}_{t=0}^{\infty}$ be a (v, Ω)-overtaking optimal program satisfying $x_0 = x$. Then*

$$\pi^v(x) = \limsup_{T \to \infty} \sum_{t=0}^{T-1} (v(x_t, x_{t+1}) - v(\bar{x}_v, \bar{x}_v)).$$

Corollary 2.21. *Let $\{x_t\}_{t=0}^{\infty}$ be a (v, Ω)-overtaking optimal and (v, Ω)-good program. Then for any integer $t \geq 0$,*

$$\pi^v(x_t) = v(x_t, x_{t+1}) - v(\bar{x}_v, \bar{x}_v) + \pi^v(x_{t+1}).$$

Set

$$\sup(\pi^v) = \sup\{\pi^v(z) : z \in \cup\{X_M : M \in (0, \infty)\}\}, \tag{2.118}$$

$$X_v = \{x \in \cup\{X_M : M \in (0, \infty)\} : \pi^v(x) \geq \sup(\pi^v) - 1\}. \tag{2.119}$$

Proposition 2.22. $\pi^v(\bar{x}_v) = 0$.

Proof. Set $x_t = \bar{x}_v$ for all integers $t \geq 0$. By Theorem 2.14 and (A2), the program $\{x_t\}_{t=0}^{\infty}$ is a (v, Ω)-overtaking optimal. In view of Proposition 2.20, $\pi^v(\bar{x}_v) = 0$. □

Theorem 2.16, Proposition 2.20, and (A1) imply the following result.

Proposition 2.23. *The function π^v is finite in a neighborhood of $\bar{x}_v$ and continuous at $\bar{x}_v$.*

Proposition 2.24. *Assume that $x_0 \in \cup\{X_M : M \in (0, \infty)\}$ and $\{x_t\}_{t=0}^{\infty} \in \mathcal{P}(v, x_0)$. Then*

$$\pi^v(x_0) = \lim_{T \to \infty} \sum_{t=0}^{T-1} (v(x_t, x_{t+1}) - v(\bar{x}_v, \bar{x}_v)).$$

Proof. It follows from (A3), Proposition 2.23, and Corollary 2.21 that

$$\pi^v(x_0) = \lim_{T \to \infty} (\pi^v(x_0) - \pi^v(x_T)) = \lim_{T \to \infty} \sum_{t=0}^{T-1} (v(x_t, x_{t+1}) - v(\bar{x}_v, \bar{x}_v)).$$

Proposition 2.24 is proved. □

The next result is proved in Sect. 2.10.

Proposition 2.25. *There exists an integer $L_v \geq 1$ such that $X_v \subset \bar{Y}_{L_v}$.*

The following result is proved in Sect. 2.11.

Proposition 2.26. *The function $\pi^v : X \to R^1 \cup \{-\infty\}$ is upper semicontinuous.*

Set

$$\mathcal{D}(v) = \{x \in X : \pi^v(x) = \sup(\pi^v)\}. \tag{2.120}$$

By Proposition 2.26 and (2.115) the set $\mathcal{D}(v)$ is nonempty and closed subset of X. The following proposition is proved in Sect. 2.11.

Proposition 2.27. *Let $\{x_t\}_{t=0}^{\infty}$ be a (v, Ω)-good program such that for all integers $t \geq 0$,*

$$v(x_t, x_{t+1}) - v(\bar{x}_v, \bar{x}_v) = \pi^v(x_t) - \pi^v(x_{t+1}).$$

Then $\{x_t\}_{t=0}^{\infty}$ is a (v, Ω)-overtaking optimal program.

The next result easily follows from Proposition 2.25, Theorem 2.15, (2.119), and (2.120).

Proposition 2.28. *For each $\epsilon > 0$ there exists a natural number T_ϵ such that for each $z \in \mathcal{D}(v)$ and each (Ω)-program $\{x_t\}_{t=0}^{\infty} \in \mathcal{P}(v, z)$ the inequality $\rho(x_t, \bar{x}_v) \leq \epsilon$ holds for all integers $t \geq T_\epsilon$.*

In order to study the structure of solutions of the problems (P2) and (P3) we introduce the following notation and definitions.

Set

$$\bar{\Omega} = \{(x, y) \in X \times X : (y, x) \in \Omega\}. \tag{2.121}$$

Clearly, $\bar{\Omega}$ is a nonempty closed subset of $X \times X$ and

$$\{(x, y) \in X \times X : \rho(x, \bar{x}_v), \rho(y, \bar{x}_v) \leq \bar{r}_v\} \subset \bar{\Omega}. \tag{2.122}$$

Then $\mathcal{M}(\bar{\Omega})$ is the set of all bounded functions $u : \bar{\Omega} \to R^1$ with $\|u\| = \sup\{|u(z)| : z \in \bar{\Omega}\}$.

For each $u \in \mathcal{M}(\Omega)$ define $\bar{u} \in \mathcal{M}(\bar{\Omega})$ by

$$\bar{u}(x, y) = u(y, x), \ (x, y) \in \bar{\Omega}. \tag{2.123}$$

Clearly, $u \to \bar{u}$, $u \in \mathcal{M}(\Omega)$ is a linear invertible isometry operator.

Let $0 \leq T_1 < T_2$ be integers and let $\{x_t\}_{t=T_1}^{T_2}$ be an (Ω)-program. Define $\{\bar{x}_t\}_{t=T_1}^{T_2} \subset X$ by

$$\bar{x}_t = x_{T_2 - t + T_1}, \ t = T_1, \dots, T_2. \tag{2.124}$$

Clearly, $\{\bar{x}_t\}_{t=T_1}^{T_2}$ is an $(\bar{\Omega})$-program.

Assume that $\{u_t\}_{t=T_1}^{T_2 - 1} \subset \mathcal{M}(\Omega)$. It is easy to see that

$$\sum_{t=T_1}^{T_2-1} \bar{u}_{T_2-t+T_1-1}(\bar{x}_t, \bar{x}_{t+1}) = \sum_{t=T_1}^{T_2-1} u_{T_2-t+T_1-1}(x_{T_2-t+T_1-1}, x_{T_2-t+T_1})$$

$$= \sum_{t=T_1}^{T_2-1} u_t(x_t, x_{t+1}). \tag{2.125}$$

The next result easily follows from (2.125).

Proposition 2.29. *Let $T \geq 1$ be an integer, $M \geq 0$, $\{u_t\}_{t=0}^{T-1} \subset \mathcal{M}(\Omega)$ and $\{x_t^{(i)}\}_{t=0}^{T}$, $i = 1, 2$ are (Ω)-programs. Then*

$$\sum_{t=0}^{T-1} u_t(x_t^{(1)}, x_{t+1}^{(1)}) \geq \sum_{t=0}^{T-1} u_t(x_t^{(2)}, x_{t+1}^{(2)}) - M$$

if and only if

$$\sum_{t=0}^{T-1} \bar{u}_{T-t-1}(\bar{x}_t^{(1)}, \bar{x}_{t+1}^{(1)}) \geq \sum_{t=0}^{T-1} \bar{u}_{T-t-1}(x_t^{(2)}, x_{t+1}^{(2)}) - M.$$

Proposition 2.29 implies the following result.

Proposition 2.30. *Let $T \geq 1$ be an integer, $M \geq 0$, $\{u_t\}_{t=0}^{T-1} \subset \mathcal{M}(\Omega)$ and $\{x_t\}_{t=0}^{T}$ be an (Ω)-program. Then $\{\bar{x}_t\}_{t=0}^{T}$ is an $(\bar{\Omega})$-program and the following assertions hold:*

if $\sum_{t=0}^{T-1} u_t(x_t, x_{t+1}) \geq \sigma(\{u_t\}_{t=0}^{T-1}, 0, T) - M$, then

$$\sum_{t=0}^{T-1} \bar{u}_{T-t-1}(\bar{x}_t, \bar{x}_{t+1}) \geq \sigma(\{\bar{u}_{T-t-1}\}_{t=0}^{T-1}, 0, T) - M;$$

if $\sum_{t=0}^{T-1} u_t(x_t, x_{t+1}) \geq \sigma(\{u_t\}_{t=0}^{T-1}, 0, T, x_0, x_T) - M$, then

$$\sum_{t=0}^{T-1} \bar{u}_{T-t-1}(\bar{x}_t, \bar{x}_{t+1}) \geq \sigma(\{\bar{u}_{T-t-1}\}_{t=0}^{T-1}, 0, T, \bar{x}_0, \bar{x}_T) - M;$$

if $\sum_{t=0}^{T-1} u_t(x_t, x_{t+1}) \geq \widehat{\sigma}(\{u_t\}_{t=0}^{T-1}, 0, T, x_T) - M$, then

$$\sum_{t=0}^{T-1} \bar{u}_{T-t-1}(\bar{x}_t, \bar{x}_{t+1}) \geq \sigma(\{\bar{u}_{T-t-1}\}_{t=0}^{T-1}, 0, T, \bar{x}_0) - M;$$

if $\sum_{t=0}^{T-1} u_t(x_t, x_{t+1}) \geq \sigma(\{u_t\}_{t=0}^{T-1}, 0, T, x_0) - M$, then

$$\sum_{t=0}^{T-1} \bar{u}_{T-t-1}(\bar{x}_t, \bar{x}_{t+1}) \geq \widehat{\sigma}(\{\bar{u}_{T-t-1}\}_{t=0}^{T-1}, 0, T, \bar{x}_T) - M.$$

The next result is proved in Sect. 2.11.

Proposition 2.31. *Let $v \in \mathcal{M}(\Omega)$ be an upper semicontinuous function. Suppose that $\bar{x}_v \in X$, $\bar{r}_v > 0$, $\bar{c}_v > 0$ and that assumptions (A1), (A2), and (A3) hold. Then the function $\bar{v}$ is upper semicontinuous,*

$$\{(x, y) \in X \times X : \ \rho(x, \bar{x}_v), \ \rho(y, \bar{x}_v) \leq \bar{r}_v\} \subset \bar{\Omega}, \tag{2.126}$$

the function $\bar{v}$ is continuous at $(\bar{x}_v, \bar{x}_v)$,

$$\sigma(\bar{v}, T) \leq T\bar{v}(\bar{x}_v, \bar{x}_v) + \bar{c}_v \text{ for all integers } T \geq 1 \tag{2.127}$$

and for all $(\bar{v}, \bar{\Omega})$-good programs $\{x_t\}_{t=0}^{\infty}$,

$$\lim_{t\to\infty} \rho(x_t, \bar{x}_v) = 0.$$

In view of Proposition 2.31, if $v \in \mathcal{M}(\Omega)$ is upper semicontinuous and satisfies (A1)–(A3), then $\bar{v}$ is also upper semicontinuous and satisfies (A1)–(A3). Therefore all the results presented above for the pair (v, Ω) are also true for the pair $(\bar{v}, \bar{\Omega})$.

2.8 Structure of Solutions in the Regions Close to the Endpoints

In Sect. 2.13 we prove the following result which describes the structure of approximate solutions of the problems of the type (P2) in the regions close to the right endpoints.

Theorem 2.32. *Suppose that $v \in \mathcal{M}(\Omega)$ is an upper semicontinuous function, $\bar{x}_v \in X$, $\bar{r}_v > 0$, $\bar{c}_v > 0$ and that assumptions (A1), (A2), and (A3) hold. Let $L_0 \geq 1$, $\tau_0 \geq 1$ be integers and $\epsilon > 0$. Then there exist $\delta > 0$ and an integer $T_0 \geq \tau_0$ such that for each integer $T \geq T_0$, each $\{u_t\}_{t=0}^{T-1} \subset \mathcal{M}(\Omega)$ satisfying*

$$\|u_t - v\| \leq \delta, \ t = 0\dots, T-1$$

and each (Ω)-program $\{x_t\}_{t=0}^{T}$ which satisfies

$$x_0 \in \bar{Y}_{L_0}, \ \sum_{t=0}^{T-1} u_t(x_t, x_{t+1}) \geq \sigma(\{u_t\}_{t=0}^{T-1}, 0, T, x_0) - \delta$$

there exists an $(\bar{\Omega})$-program

$$\{x_t^*\}_{t=0}^{\infty} \in \cup\{\mathcal{P}(\bar{v}, z) : \ z \in \mathcal{D}(\bar{v})\}$$

such that

$$\rho(x_{T-t}, x_t^*) \leq \epsilon, \ t = 0, \dots, \tau_0.$$

Recall that $\cup\{\mathcal{P}(\bar{v}, z) : \ z \in \mathcal{D}(\bar{v})\}$ is the set of all $(\bar{v}, \bar{\Omega})$-overtaking optimal programs $\{x_t^*\}_{t=0}^{\infty}$ such that x_0^* is the point of maximum of the function $\pi^{\bar{v}}$.

In Sect. 2.14 we prove the following result which describes the structure of approximate solutions of the problems of the type (P3) in the regions close to the endpoints.

Theorem 2.33. *Suppose that $v \in \mathcal{M}(\Omega)$ is an upper semicontinuous function, $\bar{x}_v \in X$, $\bar{r}_v > 0$, $\bar{c}_v > 0$ and that assumptions (A1), (A2), and (A3) hold. Let $\tau_0 \geq 1$ be an integer and $\epsilon > 0$. Then there exist $\delta > 0$ and an integer $T_0 \geq \tau_0$ such that for each integer $T \geq T_0$, each $\{u_t\}_{t=0}^{T-1} \subset \mathcal{M}(\Omega)$ satisfying*

$$\|u_t - v\| \leq \delta,\ t = 0\dots, T-1$$

and each (Ω)-program $\{x_t\}_{t=0}^{T}$ which satisfies

$$\sum_{t=0}^{T-1} u_t(x_t, x_{t+1}) \geq \sigma(\{u_t\}_{t=0}^{T-1}, 0, T) - \delta$$

there exist an (Ω)-program

$$\{y_t^*\}_{t=0}^{\infty} \in \cup\{\mathcal{P}(v, z) :\ z \in \mathcal{D}(v)\}$$

and an $(\bar{\Omega})$-program

$$\{x_t^*\}_{t=0}^{\infty} \in \cup\{\mathcal{P}(\bar{v}, z) :\ z \in \mathcal{D}(\bar{v})\}$$

such that for all integers $t = 0, \dots, \tau_0$,

$$\rho(x_{T-t}, x_t^*) \leq \epsilon,\ \rho(x_t, y_t^*) \leq \epsilon.$$

The next result is proved in Sect. 2.15.

Proposition 2.34. *Suppose that $v \in \mathcal{M}(\Omega)$ is an upper semicontinuous function, $\bar{x}_v \in X$, $\bar{r}_v > 0$, $\bar{c}_v > 0$ and that assumptions (A1), (A2), and (A3) hold. Let $\tau_0 \geq 1$ be an integer and $\epsilon > 0$. Then there exist $\delta > 0$ and an integer $T_0 \geq \tau_0$ such that for each $u \in B_d(v, \delta) \cap \mathcal{M}_*(\Omega)$ the following properties hold:*

for each $\{x_t\}_{t=0}^{\infty} \in \cup\{\mathcal{P}(u, z) :\ z \in \mathcal{D}(u)\}$ there exists

$$\{y_t\}_{t=0}^{\infty} \in \cup\{\mathcal{P}(v, z) :\ z \in \mathcal{D}(v)\}$$

such that $\rho(x_t, y_t) \leq \epsilon$ for all integers $t = 0, \dots, \tau_0$;

for each $\{x_t\}_{t=0}^{\infty} \in \cup\{\mathcal{P}(\bar{u}, z) :\ z \in \mathcal{D}(\bar{u})\}$ there exists

$$\{y_t\}_{t=0}^{\infty} \in \cup\{\mathcal{P}(\bar{v}, z) :\ z \in \mathcal{D}(\bar{v})\}$$

such that $\rho(x_t, y_t) \leq \epsilon$ for all integers $t = 0, \dots, \tau_0$.

We have already mentioned that the mapping $v \to \bar{v}$, $v \in \mathcal{M}(\Omega)$ is a linear isometry which has the inverse. It is not difficult to see that

$$\bar{v} \in \mathcal{M}_0(\bar{\Omega}) \text{ for all } v \in \mathcal{M}_0(\Omega),$$

$$\bar{v} \in \mathcal{M}_{c,0}(\bar{\Omega}) \text{ for all } v \in \mathcal{M}_{c,0}(\Omega),$$

$$\bar{v} \in \mathcal{M}_*(\bar{\Omega}) \text{ for all } v \in \mathcal{M}_*(\Omega),$$

$$\bar{v} \in \mathcal{M}_{c*}(\bar{\Omega}) \text{ for all } v \in \mathcal{M}_{c*}(\Omega),$$
$$\bar{v} \in \bar{\mathcal{M}}_0(\bar{\Omega}) \text{ if and only if } v \in \bar{\mathcal{M}}_0(\Omega),$$
$$\bar{v} \in \bar{\mathcal{M}}_{c,0}(\bar{\Omega}) \text{ if and only if } v \in \bar{\mathcal{M}}_{c,0}(\Omega).$$

The next theorem which is proved in Sect. 2.16 shows that for most objective functions v (in the sense of the Baire category) the sets $\cup\{\mathcal{P}(v,z) : z \in \mathcal{D}(v)\}$ and $\cup\{\mathcal{P}(\bar{v},z) : z \in \mathcal{D}(\bar{v})\}$ are singletons. In this case approximate solutions of the problems of the types (P2) and (P3) in the regions close to the endpoints have a simple structure.

Theorem 2.35. *Let $\mathfrak{M}$ be either $\bar{\mathcal{M}}_0(\Omega)$ or $\bar{\mathcal{M}}_{c,0}(\Omega)$. Then there exists a set $\mathcal{F} \subset \mathfrak{M} \cap \mathcal{M}_*(\Omega)$ which is a countable intersection of open everywhere dense subsets of $\mathfrak{M}$ such that for each $v \in \mathcal{F}$ there exists a unique pair of points $z, \bar{z} \in X$ such that*

$$\pi^v(z) = \sup(\pi^v),\ \pi^{\bar{v}}(\bar{z}) = \sup(\pi^{\bar{v}})$$

and there exist a unique (v,Ω)-overtaking optimal program $\{z_t\}_{t=0}^{\infty}$ satisfying $z_0 = z$ and a unique $(\bar{v},\bar{\Omega})$-overtaking optimal program $\{\widehat{z}_t\}_{t=0}^{\infty}$ satisfying $\widehat{z}_0 = \bar{z}$.

The results of this section were obtained in [86].

2.9 Proof of Theorems 2.15 and 2.16

We prove Theorems 2.15 and 2.16 simultaneously. In the case of Theorem 2.16 set $L_0 = 1$. By Theorem 2.3, there exist an integer $L \geq 1$ and $\delta \in (0, \min\{\epsilon, \bar{r}_v/2\})$ such that for the following property holds:

(i) for each integer $T > 2L$ and each (Ω)-program $\{x_t\}_{t=0}^{T}$ which satisfies

$$x_0 \in \bar{Y}_{L_0},\ x_T \in Y_{L_0},$$

$$\sum_{t=0}^{T-1} v(x_t, x_{t+1}) \geq \sigma(v, T, x_0, x_T) - \delta$$

there exist integers $\tau_1 \in [0, L]$, $\tau_2 \in [T-L, T]$ such that $\rho(x_t, \bar{x}_v) \leq \epsilon$, $t = \tau_1, \dots, \tau_2$; moreover, if $\rho(x_0, \bar{x}_v) \leq \delta$, then $\tau_1 = 0$ and if $\rho(x_T, \bar{x}_v) \leq \delta$, then $\tau_2 = T$.

Assume that $\{x_t\}_{t=0}^{\infty}$ is an (v,Ω)-overtaking optimal program such that

$$x_0 \in \bar{Y}_{L_0}. \tag{2.128}$$

This implies that it is (v,Ω)-good and in view of (A3), $\lim_{t\to\infty} x_t = \bar{x}_v$. Then for all sufficiently large natural numbers T, $\rho(x_T, \bar{x}_v) \leq \delta$ and $x_T \in Y_{L_0}$. Since the program $\{x_t\}_{t=0}^{\infty}$ is (v,Ω)-overtaking optimal it follows from the inclusion above, (2.128) and the property (i) that for all sufficiently large natural numbers T, $\rho(x_t, \bar{x}_v) \le \epsilon$, $t = L, \dots, T$ and if $\rho(x_0, \bar{x}_v) \le \delta$, then

$$\rho(x_t, \bar{x}_v) \le \epsilon, \ t = 0, \dots T$$

This completes the proof of Theorems 2.15 and 2.16. □

2.10 Proof of Proposition 2.25

By Assertion 1 of Theorem 2.5 and (A2) there exists a natural number L_v such that for each integer $T \ge L_v$ and each (Ω)-program $\{x_t\}_{t=0}^T$ satisfying

$$\sum_{t=0}^{T-1} [v(x_t, x_{t+1}) - v(\bar{x}_v, \bar{x}_v)] > -2 + \sup(\pi^v) \tag{2.129}$$

we have

$$\mathrm{Card}(\{t \in \{0, \dots, T\} : \ \rho(x_t, \bar{x}_v) > \bar{r}_v/8\}) < L_v. \tag{2.130}$$

Assume that

$$x \in X_v. \tag{2.131}$$

By (2.119), (2.131), and Proposition 2.24, there exists a (v, Ω)-overtaking optimal program $\{x_t\}_{t=0}^\infty$ such

$$x_0 = x,$$

$$\sup(\pi^v) - 1 \le \pi^v(x) = \lim_{T \to \infty} \sum_{t=0}^{T-1} (v(x_t, x_{t+1}) - v(\bar{x}_v, \bar{x}_v)).$$

There exists a natural number T_0 such that for all integers $T \ge T_0$,

$$\sum_{t=0}^{T-1} (v(x_t, x_{t+1}) - v(\bar{x}_v, \bar{x}_v)) \ge \sup(\pi^v) - 2.$$

Thus

$$\sum_{t=0}^{T_0+L_v} (v(x_t, x_{t+1}) - v(\bar{x}_v, \bar{x}_v)) \ge \sup(\pi^v) - 2.$$

By the inequality above and the choice of L_v (see (2.129) and (2.130)),

$$\mathrm{Card}(\{t \in \{0, \dots, T_0 + L_v + 1\} : \ \rho(x_t, \bar{x}_v) > \bar{r}_v/8\}) < L_v.$$

Thus there exists an integer $\tau \in [0, L_v - 1]$ such that $\rho(x_\tau, \bar{x}_v) \le \bar{r}_v/8$. This implies that $x_0 \in \bar{Y}_{\tau+1} \subset \bar{Y}_{L_v}$. Proposition 2.25 is proved. □

2.11 Proofs of Propositions 2.26, 2.27, and 2.31

Proof of Proposition 2.26. Assume that $\{x^{(i)}\}_{i=1}^{\infty} \subset X$, $x \in X$ and

$$\lim_{i\to\infty} x^{(i)} = x.$$

We show that

$$\pi^v(x) \geq \limsup_{i\to\infty} \pi^v(x^{(i)}).$$

We may assume without loss of generality that

$$-\infty < \limsup_{i\to\infty} \pi^v(x^{(i)}) = \lim_{i\to\infty} \pi^v(x^{(i)}), \tag{2.132}$$

$$\pi^v(x^{(i)}) > -\infty \text{ for all integers } i \geq 1. \tag{2.133}$$

By (2.133), Theorem 2.13, and Proposition 2.24, for each integer $i \geq 1$, there exists a (v, Ω)-overtaking optimal program $\{x_t^{(i)}\}_{t=0}^{\infty}$ such

$$x_0^{(i)} = x^{(i)},$$

$$\pi^v(x^{(i)}) = \lim_{T\to\infty} \sum_{t=0}^{T-1} (v(x_t^{(i)}, x_{t+1}^{(i)}) - v(\bar{x}_v, \bar{x}_v)). \tag{2.134}$$

Extracting a subsequence and re-indexing, if necessary, we may assume without loss of generality that for each integer $t \geq 0$ there exists

$$x_t = \lim_{i\to\infty} x_t^{(i)}. \tag{2.135}$$

In view of (2.135), $\{x_t\}_{t=0}^{\infty}$ is an (Ω)-program.

Let $\epsilon > 0$. By Propositions 2.22 and 2.23, there exists $\delta \in (0, \min\{\epsilon, \bar{r}_v/4\})$ such that for each $x \in X$ satisfying $\rho(x, \bar{x}_v) \leq \delta$,

$$|\pi^v(x)| \leq \epsilon/2. \tag{2.136}$$

It follows from (2.132), (2.134), and (A2) that for all sufficiently large natural numbers i there exists an integer $S_i \geq 1$ such that for all integers $T \geq S_i$,

$$\sum_{t=0}^{T-1} v(x_t^{(i)}, x_{t+1}^{(i)}) \geq Tv(\bar{x}_v, \bar{x}_v) + \lim_{j\to\infty} \pi^v(x^{(j)}) - 4$$

$$\geq \sigma(v, T) - \bar{c}_v + \lim_{j\to\infty} \pi^v(x^{(j)}) - 4.$$

Together with Assertion 1 of Theorem 2.5 this implies the existence of a natural number L_1 such that for all sufficiently large natural numbers i,

$$\min\{\rho(x_t^{(i)}, \bar{x}_v) : t = 0, \dots, L_1 - 1\} \leq \delta.$$

This implies that for all sufficiently large natural numbers i,

$$x_0^{(i)} \in \bar{Y}_{L_1}. \tag{2.137}$$

Since $\{x_t^{(i)}\}_{t=0}^{\infty}$, $i = 1, 2, \dots$ are (v, Ω)-overtaking optimal programs it follows from (2.137) and Theorem 2.15 that there exists a natural number L_2 such that for all sufficiently large natural numbers i,

$$\rho(x_t^{(i)}, \bar{x}_v) \leq \delta \text{ for all integers } t \geq L_2. \tag{2.138}$$

We may assume without loss of generality that (2.138) holds for all integers $i \geq 1$. Let $T \geq L_2$ be an integer. By Corollary 2.21, (2.138), and the choice of δ (see (2.136)), for all integers $i \geq 1$,

$$\sum_{t=0}^{T-1}(v(x_t^{(i)}, x_{t+1}^{(i)}) - v(\bar{x}_v, \bar{x}_v)) = \pi^v(x_0^{(i)}) - \pi^v(x_T^{(i)}) \geq \pi^v(x_0^{(i)}) - \epsilon/2.$$

In view of the relation above and the upper semicontinuity of v, for all integers $T \geq L_2$,

$$\sum_{t=0}^{T-1}(v(x_t, x_{t+1}) - v(\bar{x}_v, \bar{x}_v)) \geq \limsup_{i\to\infty} \sum_{t=0}^{T-1}(v(x_t^{(i)}, x_{t+1}^{(i)}) - v(\bar{x}_v, \bar{x}_v))$$

$$\geq \limsup_{i\to\infty} \pi^v(x_0^{(i)}) - \epsilon/2.$$

By (2.134), (2.135), and the relation above,

$$\pi^v(x) \geq \limsup_{T\to\infty} \sum_{t=0}^{T-1}(v(x_t, x_{t+1}) - v(\bar{x}_v, \bar{x}_v)) \geq \limsup_{i\to\infty} \pi^v(x^{(i)}) - \epsilon/2.$$

Since ϵ is any positive number we conclude that $\pi^v(x) \geq \limsup_{i\to\infty} \pi^v(x^{(i)})$. Proposition 2.26 is proved. □

Proof of Proposition 2.27. In view of (A3),

$$\lim_{t\to\infty} x_t = \bar{x}_v. \tag{2.139}$$

By Theorem 2.13 there exists an (v, Ω)-overtaking optimal program $\{y_i\}_{i=0}^{\infty}$ such that

$$y_0 = x_0. \tag{2.140}$$

Proposition 2.24 and (2.140) imply that

$$\pi^v(x_0) = \lim_{T\to\infty} \sum_{t=0}^{T-1}(v(y_t, y_{t+1}) - v(\bar{x}_v, \bar{x}_v)). \tag{2.141}$$

On the other hand it follows from (2.139) and Propositions 2.22 and 2.23 that for any integer $T \geq 1$,

$$\sum_{t=0}^{T-1}(v(x_t, x_{t+1}) - v(\bar{x}_v, \bar{x}_v)) = \pi^v(x_0) - \pi^v(x_T) \to \pi^v(x_0) \text{ as } T \to \infty. \quad (2.142)$$

By (2.141) and (2.142),

$$\lim_{T\to\infty} \sum_{t=0}^{T-1}(v(x_t, x_{t+1}) - v(y_t, y_t)) = 0.$$

This implies that $\{x_i\}_{i=0}^{\infty}$ is an (v, Ω)-overtaking optimal program. Proposition 2.27 is proved. □

Proof of Proposition 2.31. It is clear that the function $\bar{v}$ is upper semicontinuous and continuous at the point $(\bar{x}_v, \bar{x}_v)$ and that (2.126) and (2.127) hold. Let $\{x_i\}_{i=0}^{\infty}$ be a $(\bar{v}, \bar{\Omega})$-good program. Then there exists a number $M_1 > 0$ such that for each pair of integers $T_2 > T_1 \geq 0$,

$$|\sum_{t=T_1}^{T_2-1}(\bar{v}(x_t, x_{t+1}) - v(\bar{x}_v, \bar{x}_v))| \leq M_1. \quad (2.143)$$

In order to complete the proof of the proposition it is sufficient to show that

$$\lim_{t\to\infty} x_t = \bar{x}_v.$$

Let $\epsilon > 0$. By Lemma 2.9 and Proposition 2.30, there exists $\delta \in (0, \bar{r}_v)$ such that the following property holds:

(ii) for each natural number T and each $(\bar{\Omega})$-program $\{z_t\}_{t=0}^{T}$ which satisfies $\rho(z_0, \bar{x}_v),\ \rho(z_T, \bar{x}_v) \leq \delta$,

$$\sum_{t=0}^{T-1} \bar{v}(z_t, z_{t+1}) \geq \sigma(\bar{v}, z_0, z_T, T) - \delta$$

the inequality $\rho(z_t, \bar{x}_v) \leq \epsilon$ holds for all integers $t = 0, \ldots, T$.

By Lemma 2.8 and Proposition 2.30, there exists an integer $T_0 \geq 1$ such that the following property holds:

(iii) for each $(\bar{\Omega})$-program $\{z_t\}_{t=0}^{T_0}$ which satisfies

$$\sum_{t=0}^{T_0-1} \bar{v}(z_t, z_{t+1}) \geq T_0\bar{v}(\bar{x}_v, \bar{x}_v) - M_1$$

we have

$$\min\{\rho(z_t, \bar{x}_v) :\ t = 0, \ldots, T_0\} \leq \delta.$$

Since $\{x_i\}_{i=0}^{\infty}$ is a $(\bar{v}, \bar{\Omega})$-good program there exists a natural number τ_0 such that for each pair of integers $T_2 > T_1 \geq \tau_0$,

$$\sum_{t=T_1}^{T_2-1} \bar{v}(x_t, x_{t+1}) \geq \sigma(\bar{v}, x_{T_1}, x_{T_2}, T_2 - T_1) - \delta. \tag{2.144}$$

It follows from the choice of M_1 (see (2.143)) and the property (iii) that there exists a strictly increasing sequence of natural numbers $\{S_k\}_{k=1}^{\infty}$ such that $S_1 \geq \tau_0$ and $\rho(x_{S_k}, \bar{x}_v) \leq \delta$ for all integers $k \geq 1$. By the relations above, (2.144) and the property (ii), for every natural number k, $\rho(x_t, \bar{x}_v) \leq \epsilon$, $t = S_k, \dots, S_{k+1}$. Therefore $\rho(x_t, \bar{x}_v) \leq \epsilon$ for all integers $t \geq S_1$ and $\lim_{t\to\infty} x_t = \bar{x}_v$. Proposition 2.31 is proved. □

2.12 A Basic Lemma for Theorem 2.32

Lemma 2.36. *Suppose that $v \in \mathcal{M}(\Omega)$ is an upper semicontinuous function, $\bar{x}_v \in X$, $\bar{r}_v > 0$, $\bar{c}_v > 0$ and that assumptions (A1), (A2), and (A3) hold. Let $T_0 \geq 1$ be an integer and $\epsilon \in (0, 1)$. Then there exists $\delta \in (0, \epsilon)$ such that for each (Ω)-program $\{x_t\}_{t=0}^{T_0}$ which satisfies*

$$\pi^v(x_0) \geq \sup(\pi^v) - \delta, \tag{2.145}$$

$$\sum_{t=0}^{T_0-1} (v(x_t, x_{t+1}) - v(\bar{x}_v, \bar{x}_v)) - \pi^v(x_0) + \pi^v(x_T) \geq -\delta \tag{2.146}$$

there exists an (Ω)-program

$$\{z_t\}_{t=0}^{\infty} \in \cup\{\mathcal{P}(v, z) :\ z \in \mathcal{D}(v)\} \tag{2.147}$$

such that

$$\rho(z_t, x_t) \leq \epsilon,\ t = 0, \dots, T_0. \tag{2.148}$$

Proof. Assume that the lemma does not hold. Then there exist a sequence $\{\delta_k\}_{k=1}^{\infty} \subset (0, 1]$ and a sequence of (Ω)-programs $\{x_t^{(k)}\}_{t=0}^{T_0}$, $k = 1, 2, \dots$ such that

$$\lim_{k\to\infty} \delta_k = 0 \tag{2.149}$$

and that for each integer $k \geq 1$ and each (Ω)-program $\{z_t\}_{t=0}^{\infty}$ satisfying (2.147),

$$\pi^v(x_0^{(k)}) \geq \sup(\pi^v) - \delta_k, \tag{2.150}$$

$$\sum_{t=0}^{T_0-1} (v(x_t^{(k)}, x_{t+1}^{(k)}) - v(\bar{x}_v, \bar{x}_v)) - \pi^v(x_0^{(k)}) + \pi^v(x_{T_0}^{(k)}) \geq -\delta_k \tag{2.151}$$

we have

$$\max\{\rho(z_t, x_t^{(k)}) :\ t = 0, \dots, T_0\} > \epsilon. \tag{2.152}$$

In view of (2.150) and (2.151), for each integer $k \geq 1$, $\pi^v(x_0^{(k)})$ and $\pi^v(x_{T_0}^{(k)})$ are finite. Extracting a subsequence and re-indexing, if necessary, we may assume without loss of generality that for each integer $t \in [0, T_0]$ there exists

$$x_t = \lim_{k \to \infty} x_t^{(k)}. \tag{2.153}$$

Proposition 2.26, (2.150), and (2.153) imply that

$$\sup(\pi^v) \geq \pi^v(x_0) \geq \limsup_{k \to \infty} \pi^v(x_0^{(k)}) \geq \sup(\pi^v). \tag{2.154}$$

By upper semicontinuity of v and π^v (see Proposition 2.26), (2.149)–(2.151), (2.153), and (2.154),

$$\sum_{t=0}^{T_0-1} (v(x_t, x_{t+1}) - v(\bar{x}_v, \bar{x}_v)) - \pi^v(x_0) + \pi^v(x_{T_0})$$

$$\geq \limsup_{k \to \infty} (\sum_{t=0}^{T_0-1} (v(x_t^{(k)}, x_{t+1}^{(k)}) - v(\bar{x}_v, \bar{x}_v)) - \pi^v(x_0^{(k)}) + \pi^v(x_{T_0}^{(k)}))$$

$$\geq \limsup_{k \to \infty} (-\delta_k) = 0.$$

Thus

$$\sum_{t=0}^{T_0-1} (v(x_t, x_{t+1}) - v(\bar{x}_v, \bar{x}_v)) - \pi^v(x_0) + \pi^v(x_{T_0}) \geq 0. \tag{2.155}$$

In view of (2.154) and (2.155), $\pi^v(x_{T_0})$ is finite. Together with Proposition 2.19 and (2.155) this implies that for all integers $t = 0, \dots, T_0 - 1$,

$$v(x_t, x_{t+1}) - v(\bar{x}_v, \bar{x}_v) = \pi^v(x_t) - \pi^v(x_{t+1}). \tag{2.156}$$

Since $\pi^v(x_{T_0})$ is finite Theorem 2.13 implies that there is a (v, Ω)-overtaking optimal and (v, Ω)-good program $\{\tilde{x}_t\}_{t=0}^{\infty}$ satisfying

$$\tilde{x}_0 = x_{T_0}. \tag{2.157}$$

For all integers $t > T_0$ set

$$x_t = \tilde{x}_{t-T_0}. \tag{2.158}$$

It is clear that $\{x_t\}_{t=0}^{\infty}$ is an (Ω)-program. Since $\{\tilde{x}_t\}_{t=0}^{\infty}$ is (v, Ω)-overtaking optimal program Corollary 2.21 and (2.158) imply that (2.156) holds for all integers $t \geq 0$. Since $\{\tilde{x}_t\}_{t=0}^{\infty}$ is (v, Ω)-good program it follows from (2.156), (2.158), and Proposition 2.27 that $\{x_t\}_{t=0}^{\infty}$ is (v, Ω)-overtaking optimal program. In view of (2.120) and (2.154), $x_0 \in \mathcal{D}(v)$ and

$$\{x_t\}_{t=0}^{\infty} \in \cup\{\mathcal{P}(v, z) : \; z \in \mathcal{D}(v)\}.$$

In view of (2.153), for all sufficiently large natural numbers k, $\rho(x_t, x_t^{(k)}) \leq \epsilon/4$, $t = 0, \dots, T_0$. This contradicts (2.147) and (2.152). The contradiction we have reached proves Lemma 2.36. □

2.13 Proof of Theorem 2.32

We may assume that $\bar{r}_v \in (0,1)$. Recall that

$$\{(x,y) \in X \times X : \ \rho(x,\bar{x}_v),\ \rho(y,\bar{x}_v) \le \bar{r}_v\} \subset \Omega. \tag{2.159}$$

By Lemma 2.36 applied to the function $\bar{v}$ there exists $\delta_1 \in (0, \min\{\epsilon, \bar{r}_v/2\})$ such that the following property holds:

(Pi) for each $(\bar{\Omega})$-program $\{y_t\}_{t=0}^{\tau_0}$ which satisfies

$$\pi^{\bar{v}}(y_0) \ge \sup(\pi^{\bar{v}}) - \delta_1, \tag{2.160}$$

$$\sum_{t=0}^{\tau_0 - 1} (\bar{v}(y_t, y_{t+1}) - \bar{v}(\bar{x}_v, \bar{x}_v)) - \pi^{\bar{v}}(y_0) + \pi^{\bar{v}}(y_T) \ge -\delta_1 \tag{2.161}$$

there exists an $(\bar{\Omega})$-program

$$\{z_t\}_{t=0}^{\infty} \in \cup\{\mathcal{P}(\bar{v}, z) : \ z \in \mathcal{D}(\bar{v})\} \tag{2.162}$$

such that

$$\rho(z_t, y_t) \le \epsilon,\ t = 0, \dots, \tau_0. \tag{2.163}$$

By Propositions 2.22 and 2.23 and (A2), there exists $\delta_2 \in (0, \delta_1)$ such that for each $z \in X$ satisfying $\rho(z, \bar{x}_v) \le 2\delta_2$,

$$|\pi^{\bar{v}}(z)| = |\pi^{\bar{v}}(z) - \pi^{\bar{v}}(\bar{x}_v)| \le \delta_1/8 \tag{2.164}$$

and for each $(x,y) \in X \times X$ satisfying $\rho(x,\bar{x}_v) \le 2\delta_2$, $\rho(y,\bar{x}_v) \le 2\delta_2$,

$$|v(x,y) - \bar{v}(\bar{x}_v, \bar{x}_v)| \le \delta_1/8. \tag{2.165}$$

By Theorem 2.4, there exist an integer $L \ge 1$ and a number $\delta_3 > 0$ such that the following property holds:

(Pii) for each integer $T > 2L$, each $\{u_t\}_{t=0}^{T-1} \subset \mathcal{M}(\Omega)$ satisfying $\|u_t - v\| \le \delta_3$, $t = 0 \dots, T-1$ and each (Ω)-program $\{z_t\}_{t=0}^{T}$ which satisfies

$$z_0 \in \bar{Y}_{L_0},\ \sum_{t=0}^{T-1} u_t(z_t, z_{t+1}) \ge \sigma(\{u_t\}_{t=0}^{T-1}, 0, T, z_0) - \delta_3$$

we have $\rho(z_t, \bar{x}_v) \le \delta_2$, $t = L, \dots, T-L$.

By Proposition 2.25, (2.119), (2.120), and Theorem 2.15 applied to the function $\bar{v}$ there exists a natural number τ_1 such that for each $(\bar{\Omega})$-program

$$\{z_t\}_{t=0}^{\infty} \in \cup\{\mathcal{P}(\bar{v}, z) : \ z \in \mathcal{D}(\bar{v})\} \tag{2.166}$$

we have

$$\rho(z_t, \bar{x}_v) \le \delta_2 \text{ for all integers } t \ge \tau_1. \tag{2.167}$$

Choose a positive number δ and an integer T_0 such that

$$\delta < (16(L + \tau_1 + \tau_0 + 6))^{-1} \min\{\delta_1, \delta_2, \delta_3\}, \tag{2.168}$$

$$T_0 > 2L + 2\tau_0 + 2\tau_1 + 4. \tag{2.169}$$

Assume that an integer $T \geq T_0$, $\{u_t\}_{t=0}^{T-1} \subset \mathcal{M}(\Omega)$ satisfies

$$\|u_t - v\| \leq \delta,\ t = 0 \ldots, T - 1 \tag{2.170}$$

and $\{x_t\}_{t=0}^{T}$ is an (Ω)-program which satisfies

$$x_0 \in \bar{Y}_{L_0},\ \sum_{t=0}^{T-1} u_t(x_t, x_{t+1}) \geq \sigma(\{u_t\}_{t=0}^{T-1}, 0, T, x_0) - \delta. \tag{2.171}$$

By (2.168)–(2.171) and the property (Pii),

$$\rho(x_t, \bar{x}_v) \leq \delta_2,\ t = L, \ldots, T - L. \tag{2.172}$$

In view of (2.169),

$$[T - L - \tau_0 - \tau_1 - 4, T - L - \tau_0 - \tau_1] \subset [L, T - L - \tau_0 - \tau_1]. \tag{2.173}$$

Relations (2.172) and (2.173) imply that

$$\rho(x_t, \bar{x}_v) \leq \delta_2,\ t \in \{T - L - \tau_0 - \tau_1 - 4, T - L - \tau_0 - \tau_1\}. \tag{2.174}$$

Fix an $(\bar{\Omega})$-program $\{z_t\}_{t=0}^{\infty}$ satisfying (2.166). Then (2.167) is true. Define a sequence $\{\tilde{x}_t\}_{t=0}^{T}$ by

$$\tilde{x}_t = x_t,\ t = 0, \ldots, T - L - \tau_0 - \tau_1 - 4,$$

$$\tilde{x}_t = z_{T-t},\ t = T - L - \tau_0 - \tau_1 - 3, \ldots, T. \tag{2.175}$$

In view of (2.167) and (2.175),

$$\rho(\tilde{x}_{T-L-\tau_0-\tau_1-3}, \bar{x}_v) = \rho(z_{L+\tau_0+\tau_1+3}, \bar{x}_v) \leq \delta_2. \tag{2.176}$$

By (2.159) and (2.174)–(2.176), $\{\tilde{x}_t\}_{t=0}^{T}$ is an (Ω)-program. It follows from (2.170), (2.171), and (2.175) that

$$\delta \geq \sum_{t=0}^{T-1} (u_t(\tilde{x}_t, \tilde{x}_{t+1}) - u_t(x_t, x_{t+1}))$$

$$= \sum_{t=T-L-\tau_0-\tau_1-4}^{T-1} (u_t(\tilde{x}_t, \tilde{x}_{t+1}) - u_t(x_t, x_{t+1}))$$

$$\geq \sum_{t=T-L-\tau_0-\tau_1-4}^{T-1} v(\tilde{x}_t, \tilde{x}_{t+1}) - \sum_{t=T-L-\tau_0-\tau_1-4}^{T-1} v(x_t, x_{t+1}) - 2\delta(L + \tau_0 + \tau_1 + 4).$$

Together with (2.175) this implies that

$$\sum_{t=T-L-\tau_0-\tau_1-4}^{T-1} v(x_t, x_{t+1}) \geq \sum_{t=T-L-\tau_0-\tau_1-4}^{T-1} v(\tilde{x}_t, \tilde{x}_{t+1}) - 2\delta(L+\tau_0+\tau_1+5)$$

$$= v(x_{T-L-\tau_0-\tau_1-4}, z_{L+\tau_0+\tau_1+3}) + \sum_{t=0}^{L+\tau_0+\tau_1+2} \bar{v}(z_t, z_{t+1}) - 2\delta(L+\tau_0+\tau_1+5). \tag{2.177}$$

By (2.165), (2.167), and (2.174),

$$|\bar{v}(z_{L+\tau_0+\tau_1+3}, x_{T-L-\tau_0-\tau_1-4}) - \bar{v}(z_{L+\tau_0+\tau_1+3}, z_{L+\tau_0+\tau_1+4})| \leq \delta_1/4.$$

Together with (2.177) this implies that

$$\sum_{t=T-L-\tau_0-\tau_1-4}^{T-1} v(x_t, x_{t+1}) \geq \sum_{t=0}^{L+\tau_0+\tau_1+3} \bar{v}(z_t, z_{t+1}) - \delta_1/4 - 2\delta(L+\tau_0+\tau_1+5). \tag{2.178}$$

Set

$$y_t = x_{T-t},\ t = 0, \dots, L+\tau_0+\tau_1+4. \tag{2.179}$$

It follows from (2.125), (2.168), (2.178), and (2.179) that

$$\sum_{t=0}^{L+\tau_0+\tau_1+3} \bar{v}(y_t, y_{t+1}) \geq \sum_{t=0}^{L+\tau_0+\tau_1+3} \bar{v}(z_t, z_{t+1}) - \delta_1/4 - \delta_1/8. \tag{2.180}$$

By (2.166), (2.174), (2.179), (2.180), Proposition 2.19, and Corollary 2.21,

$$\pi^{\bar{v}}(y_0) - \sup(\pi^{\bar{v}}) + \sum_{t=0}^{\tau_0-1} (\bar{v}(y_t, y_{t+1}) - \bar{v}(\bar{x}_v, \bar{x}_v)) - \pi^{\bar{v}}(y_0) + \pi^{\bar{v}}(y_{\tau_0})$$

$$\geq \pi^{\bar{v}}(y_0) - \pi^{\bar{v}}(z_0)$$

$$+ \sum_{t=0}^{L+\tau_0+\tau_1+3} (\bar{v}(y_t, y_{t+1}) - \bar{v}(\bar{x}_v, \bar{x}_v)) - \pi^{\bar{v}}(y_0) + \pi^{\bar{v}}(y_{L+\tau_0+\tau_1+4})$$

$$\geq \pi^{\bar{v}}(y_0) - \pi^{\bar{v}}(z_0) + \sum_{t=0}^{L+\tau_0+\tau_1+3} (\bar{v}(z_t, z_{t+1}) - \bar{v}(\bar{x}_v, \bar{x}_v))$$

$$-\pi^{\bar{v}}(y_0) + \pi^{\bar{v}}(y_{L+\tau_0+\tau_1+4}) - 3\delta_1/8$$

$$\geq \pi^{\bar{v}}(y_0) - \pi^{\bar{v}}(z_0) + \pi^{\bar{v}}(z_0) - \pi^{\bar{v}}(z_{L+\tau_0+\tau_1+3}) - \pi^{\bar{v}}(y_0) - \pi^{\bar{v}}(y_{L+\tau_0+\tau_1+4}) - 3\delta_1/8$$

$$= \pi^{\bar{v}}(y_{L+\tau_0+\tau_1+4}) - \pi^{\bar{v}}(z_{L+\tau_0+\tau_1+3}) - 3\delta_1/8. \tag{2.181}$$

By (2.164) and (2.167),

$$\pi^{\bar{v}}(z_{L+\tau_0+\tau_1+3}) \leq \delta_1/8. \tag{2.182}$$

In view of (2.164), (2.179), and (2.174),

$$|\pi^{\bar{v}}(y_{L+\tau_0+\tau_1+4})| = |\pi^{\bar{v}}(x_{T-L-\tau_0-\tau_1-4})| \leq \delta_1/8. \tag{2.183}$$

It follows from (2.181) to (2.183),

$$\pi^{\bar{v}}(y_0) - \sup(\pi^{\bar{v}}) + \sum_{t=0}^{\tau_0-1}(\bar{v}(y_t, y_{t+1}) - \bar{v}(\bar{x}_v, \bar{x}_v)) - \pi^{\bar{v}}(y_0) + \pi^{\bar{v}}(y_{\tau_0}) \geq -\delta_1.$$

Together with (2.174), (2.179), and Proposition 2.19 this implies that

$$\pi^{\bar{v}}(y_0) - \sup(\pi^{\bar{v}}) \geq -\delta_1, \tag{2.184}$$

$$\sum_{t=0}^{\tau_0-1}(\bar{v}(y_t, y_{t+1}) - \bar{v}(\bar{x}_v, \bar{x}_v)) - \pi^{\bar{v}}(y_0) + \pi^{\bar{v}}(y_{\tau_0}) \geq -\delta_1. \tag{2.185}$$

By (2.184), (2.185), and the property (Pi), there exists

$$\{\xi_t\}_{t=0}^{\infty} \in \cup\{\mathcal{P}(\bar{v}, z) : \ z \in \mathcal{D}(\bar{v})\}$$

such that $\rho(\xi_t, x_{T-t}) = \rho(\xi_t, y_t) \leq \epsilon$, $t = 0, \ldots, \tau_0$. Theorem 2.32 is proved. □

2.14 Proof of Theorem 2.33

Theorem 2.32 and Assertion 1 of Theorem 2.5 imply the following result.

Proposition 2.37. *Suppose that $v \in \mathcal{M}(\Omega)$ is an upper semicontinuous function, $\bar{x}_v \in X$, $\bar{r}_v > 0$, $\bar{c}_v > 0$ and that assumptions (A1), (A2), and (A3) hold. Let $\tau_0 \geq 1$ be an integer and $\epsilon > 0$. Then there exist $\delta > 0$ and an integer $T_0 \geq \tau_0$ such that for each integer $T \geq T_0$, each $\{u_t\}_{t=0}^{T-1} \subset \mathcal{M}(\Omega)$ satisfying $\|u_t - v\| \leq \delta$, $t = 0 \ldots, T-1$ and each (Ω)-program $\{x_t\}_{t=0}^{T}$ which satisfies*

$$\sum_{t=0}^{T-1} u_t(x_t, x_{t+1}) \geq \sigma(\{u_t\}_{t=0}^{T-1}, 0, T) - \delta$$

there exists an $(\bar{\Omega})$-program $\{x_t^*\}_{t=0}^{\infty} \in \cup\{\mathcal{P}(\bar{v}, z) : \ z \in \mathcal{D}(\bar{v})\}$ such that for all integers $t = 0, \ldots, \tau_0$, $\rho(x_{T-t}, x_t^*) \leq \epsilon$.

Since the mapping $v \to \bar{v}$, $v \in \mathcal{M}(\Omega)$ is an isometry Theorem 2.33 follows from Propositions 2.30, 2.31, and 2.37.

2.15 Proof of Proposition 2.34

Since the mapping $v \to \bar{v}$, $v \in \mathcal{M}(\Omega)$ is an isometry Proposition 2.34 follows from Propositions 2.30, 2.31 and the following result.

Proposition 2.38. *Suppose that $v \in \mathcal{M}_*(\Omega)$, $\tau_0 \geq 1$ is an integer and $\epsilon > 0$. Then there exist $\delta > 0$ and an integer $T_0 \geq \tau_0$ such that for each $u \in B_d(v, \delta) \cap \mathcal{M}_*(\Omega)$ and each $\{x_t\}_{t=0}^{\infty} \in \cup\{\mathcal{P}(\bar{u}, z) : z \in \mathcal{D}(\bar{u})\}$ there exists*

$$\{y_t\}_{t=0}^{\infty} \in \cup\{\mathcal{P}(\bar{v}, z) : z \in \mathcal{D}(\bar{v})\}$$

such that $\rho(x_t, y_t) \leq \epsilon$ for all integers $t = 0, \dots, \tau_0$.

Proof. By (A2), for each $u \in \mathcal{M}_0(\Omega)$ and each integer $T \geq 1$,

$$\sigma(u, T, \bar{x}_u, \bar{x}_u) = Tu(\bar{x}_u, \bar{x}_u). \tag{2.186}$$

Together with Assertion 1 of Theorem 2.5 this implies that there exists $\delta_0 \in (0, \epsilon)$ such that for each $u \in B_d(v, \delta_0) \cap \mathcal{M}_0(\Omega)$,

$$\rho(\bar{x}_u, \bar{x}_v) \leq \bar{r}_v/4. \tag{2.187}$$

By Theorem 2.32, there exist $\delta \in (0, \delta_0)$ and a natural number $T_0 \geq \tau_0$ such that the following property holds:

(Piii) for each integer $T \geq T_0$, each $u \in \mathcal{M}_0(\Omega)$ satisfying $\|u - v\| \leq \delta$ and each (Ω)-program $\{z_t\}_{t=0}^{T}$ which satisfies

$$\rho(z_0, \bar{x}_v) \leq \bar{r}_v/2, \ \sum_{t=0}^{T-1} u(z_t, z_{t+1}) \geq \sigma(u, T, z_0) - \delta$$

there exists an $(\bar{\Omega})$-program

$$\{x_t^*\}_{t=0}^{\infty} \in \cup\{\mathcal{P}(\bar{v}, z) : z \in \mathcal{D}(\bar{v})\} \tag{2.188}$$

such that

$$\rho(z_{T-t}, x_t^*) \leq \epsilon, \ t = 0, \dots, \tau_0. \tag{2.189}$$

Assume that

$$u \in \mathcal{M}_*(\Omega), \ \|u - v\| \leq \delta, \tag{2.190}$$

$$\{x_t\}_{t=0}^{\infty} \in \cup\{\mathcal{P}(\bar{u}, z) : z \in \mathcal{D}(\bar{u})\}. \tag{2.191}$$

In view of (2.190),

$$\lim_{t \to \infty} x_t = \bar{x}_u. \tag{2.192}$$

By (2.192), there exists an integer $S_0 > T_0$ such that

$$\rho(x_{S_0}, \bar{x}_u) \leq \bar{r}_v/4. \tag{2.193}$$

By (2.190) and the choice of δ_1, (2.187) holds. In view of (2.187) and (2.193),

$$\rho(x_{S_0}, \bar{x}_v) \leq \bar{r}_v/2. \tag{2.194}$$

Set

$$z_t = x_{S_0 - t},\ t = 0, \dots, S_0. \tag{2.195}$$

By (2.194) and (2.195),

$$\rho(z_0, \bar{x}_v) \leq \bar{r}_v/2. \tag{2.196}$$

It is clear that $\{z_t\}_{t=0}^{S_0}$ is an (Ω)-program. We show that

$$\sum_{t=0}^{S_0-1} u(z_t, z_{t+1}) = \sigma(u, S_0, z_0). \tag{2.197}$$

Let $\{y_t\}_{t=0}^{S_0}$ be an (Ω)-program satisfying

$$y_0 = z_0. \tag{2.198}$$

In order to prove (2.197) it is sufficient to show that

$$\sum_{t=0}^{S_0-1} u(z_t, z_{t+1}) \geq \sum_{t=0}^{S_0-1} u(y_t, y_{t+1}).$$

It follows from (2.195) that

$$\sum_{t=0}^{S_0-1} u(z_t, z_{t+1}) = \sum_{t=0}^{S_0-1} \bar{u}(x_{S_0-t-1}, x_{S_0-t}). \tag{2.199}$$

Set

$$\bar{y}_t = y_{S_0 - t},\ t = 0, \dots, S_0. \tag{2.200}$$

In view of (2.200),

$$\sum_{t=0}^{S_0-1} u(y_t, y_{t+1}) = \sum_{t=0}^{S_0-1} \bar{u}(\bar{y}_t, \bar{y}_{t+1}). \tag{2.201}$$

By (2.190), (2.191), (2.193), (2.196), (2.198), (2.199), (2.201), and Corollary 2.21,

$$\sum_{t=0}^{S_0-1} u(z_t, z_{t+1}) - \sum_{t=0}^{S_0-1} u(y_t, y_{t+1}) = \sum_{t=0}^{S_0-1} \bar{u}(x_t, x_{t+1}) - \sum_{t=0}^{S_0-1} \bar{u}(\bar{y}_t, \bar{y}_{t+1})$$

$$= \sum_{t=0}^{S_0-1} [\bar{u}(x_t, x_{t+1}) - u(\bar{x}_u, \bar{x}_u) - \pi^{\bar{u}}(x_t) + \pi^{\bar{u}}(x_{t+1})] + \pi^{\bar{u}}(x_0) - \pi^{\bar{u}}(x_{S_0})$$

$$-\sum_{t=0}^{S_0-1}[\bar{u}(\bar{y}_t,\bar{y}_{t+1})-u(\bar{x}_u,\bar{x}_u)-\pi^{\bar{u}}(\bar{y}_t)+\pi^{\bar{u}}(\bar{y}_{t+1})]-\pi^{\bar{u}}(\bar{y}_0)+\pi^{\bar{u}}(\bar{y}_{S_0})$$

$$=\pi^{\bar{u}}(x_0)-\pi^{\bar{u}}(\bar{y}_0)-\sum_{t=0}^{S_0-1}[\bar{u}(\bar{y}_t,\bar{y}_{t+1})-u(\bar{x}_u,\bar{x}_u)-\pi^{\bar{u}}(\bar{y}_t)+\pi^{\bar{u}}(\bar{y}_{t+1})]$$

$$\geq \pi^{\bar{u}}(x_0)-\pi^{\bar{u}}(\bar{y}_0)\geq 0.$$

Thus (2.197) holds. By (2.190), (2.195)–(2.197), and the property (Pii) applied to $\{z_t\}_{t=0}^{S_0}$ there exists an $(\bar{\Omega})$-program $\{x_t^*\}_{t=0}^{\infty}$ satisfying (2.188) such that for all $t=0,\dots,\tau_0$,

$$\epsilon\geq\rho(z_{S_0-t},x_t^*)=\rho(x_t^*,x_t).$$

Proposition 2.38 is proved. □

2.16 Proof of Theorem 2.35

In the sequel we use the following auxiliary results.

Proposition 2.39. *Let $v\in\mathcal{M}_*(\Omega)$, $\{x_t\}_{t=0}^{\infty}$ be a (v,Ω)-overtaking optimal and (v,Ω)-good program and $t_2>t_1$ be nonnegative integers such that $x_{t_1}=x_{t_2}$. Then $x_t=\bar{x}_v$ for all integers $t=t_1,\dots,t_2$.*

Proof. By Corollary 2.21, for all integers $t\geq 0$,

$$v(x_t,x_{t+1})-v(\bar{x}_v,\bar{x}_v)=\pi^v(x_t)-\pi^v(x_{t+1}). \tag{2.202}$$

We may assume without loss of generality that $t_1=0$. There exists a sequence $\{\tilde{x}_t\}_{t=0}^{\infty}$ such that

$$\tilde{x}_t=x_t,\ t=0,\dots,t_2,\ \tilde{x}_{t+t_2}=\tilde{x}_t,\ t=0,1,\dots. \tag{2.203}$$

In view of (2.202) and (2.203), for all integers $t\geq 0$,

$$v(\tilde{x}_t,\tilde{x}_{t+1})-v(\bar{x}_v,\bar{x}_v)=\pi^v(\tilde{x}_t)-\pi^v(\tilde{x}_{t+1}). \tag{2.204}$$

Together with (2.203) this implies that $\{\tilde{x}_t\}_{t=0}^{\infty}$ is a (v,Ω)-good program. By (A3), $\lim_{t\to\infty}\tilde{x}_t=\bar{x}_v$. Combined with (2.203) this implies that $x_t=\bar{x}_v$ for all integers $t=0,\dots,t_2$. Proposition 2.39 is proved. □

The next result follows from Theorem 2.16.

Proposition 2.40. *Let $v\in\mathcal{M}_*(\Omega)$, $\{x_t\}_{t=0}^{\infty}$ be a (v,Ω)-overtaking optimal program such that $x_0=\bar{x}_v$. Then $x_t=\bar{x}_v$ for all integers $t\geq 0$.*

For any $(x,y)\in X\times X$ and any nonempty set $D\subset X\times X$ put

$$\rho((x,y),D)=\inf\{\rho(x,z_1)+\rho(y,z_2):\ (z_1,z_2)\in D\}.$$

Since the mapping $v\to\bar{v}$, $v\in\mathcal{M}(\Omega)$ is an isometry Theorem 2.35 follows from Propositions 2.30, 2.31 and the following result.

Proposition 2.41. *Let $\mathfrak{M}$ be either $\bar{\mathcal{M}}_0(\Omega)$ or $\bar{\mathcal{M}}_{c,0}(\Omega)$. Then there exists a set $\mathcal{F} \subset \mathfrak{M} \cap \mathcal{M}_*(\Omega)$ which is a countable intersection of open everywhere dense subsets of $\mathfrak{M}$ such that for each $v \in \mathcal{F}$ there exists a unique point $z_v \in X$ such that $\pi^v(z_v) = \sup(\pi^v)$ and there exists a unique (v, Ω)-overtaking optimal program $\{z_t^v\}_{t=0}^\infty$ satisfying $z_0^v = z_v$.*

Proof. By Theorem 2.6, there exists a set $\mathcal{F}_0 \subset \mathfrak{M} \cap \mathcal{M}_*(\Omega)$ which is a countable intersection of open everywhere dense subsets of $\mathfrak{M}$. Denote by E the set of all $v \in \mathfrak{M} \cap \mathcal{M}_*(\Omega)$ for which the following property holds:

(Piv) there exists a unique point $z_v \in X$ such that $\pi^v(z_v) = \sup(\pi^v)$ and there exists a unique (v, Ω)-overtaking optimal program $\{z_t^v\}_{t=0}^\infty$ satisfying $z_0^v = z_v$.

We show that E is an everywhere dense subset of $\mathfrak{M}$. Let $v \in \mathfrak{M} \cap \mathcal{M}_*(\Omega)$. It is sufficient to show that for any neighborhood $\mathcal{U}$ of v in $\mathfrak{M}$, $\mathcal{U} \cap E \neq \emptyset$. There are two cases:

$$\pi^v(\bar{x}_v) = \sup(\pi^v); \tag{2.205}$$

$$\pi^v(\bar{x}_v) < \sup(\pi^v); \tag{2.206}$$

Assume that (2.205) holds. Let $\gamma \in (0, 1)$. Define

$$v_\gamma(x, y) = v(x, y) - \gamma(\rho(x, \bar{x}_v) + \rho(y, \bar{x}_v)),\ (x, y) \in \Omega. \tag{2.207}$$

It is not difficult to see that

$$v_\gamma \in \mathfrak{M} \cap \mathcal{M}_0(\Omega) \text{ with } \bar{x}_{v_\gamma} = \bar{x}_v. \tag{2.208}$$

By (2.207), any (v_γ, Ω)-good program $\{x_t\}_{t=0}^\infty$ is (v, Ω)-good and $\lim_{t\to\infty} x_t = \bar{x}_v$. Thus

$$v_\gamma \in \mathcal{M}_*(\Omega). \tag{2.209}$$

It follows from (2.207) to (2.209) and Proposition 2.22 that

$$v_\gamma(\bar{x}_v, \bar{x}_v) = v(\bar{x}_v, \bar{x}_v),\ \pi^{v_\gamma}(y) \le \pi^v(y),\ y \in X,\ \pi^{v_\gamma}(\bar{x}_v) = \pi^v(\bar{x}_v) = 0. \tag{2.210}$$

Assume that $z \in X$ satisfies

$$\pi^{v_\gamma}(z) = \sup(\pi^{v_\gamma}) \tag{2.211}$$

and that $\{z_t\}_{t=0}^\infty$ is a (v_γ, Ω)-overtaking optimal program satisfying

$$z_0 = z. \tag{2.212}$$

By (2.205), (2.207)–(2.212), and Proposition 2.24,

$$0 = \pi^{v_\gamma}(z) = \lim_{T\to\infty} \sum_{t=0}^{T-1} [v_\gamma(z_t, z_{t+1}) - v_\gamma(\bar{x}_v, \bar{x}_v)]$$

$$= \lim_{T\to\infty} [\sum_{t=0}^{T-1} v(z_t, z_{t+1}) - Tv(\bar{x}_v, \bar{x}_v) - \gamma \sum_{t=0}^{T-1} (\rho(z_t, \bar{x}_v) + \rho(z_{t+1}, \bar{x}_v))]$$

$$\leq \limsup_{T\to\infty} \sum_{t=0}^{T-1} (v(z_t, z_{t+1}) - v(\bar{x}_v, \bar{x}_v)) - \gamma \sum_{t=0}^{\infty} (\rho(z_t, \bar{x}_v) + \rho(z_{t+1}, \bar{x}_v))$$

$$\leq \sup(\pi^v) - \gamma \sum_{t=0}^{\infty} (\rho(z_t, \bar{x}_v) + \rho(z_{t+1}, \bar{x}_v)) = -\gamma \sum_{t=0}^{\infty} (\rho(z_t, \bar{x}_v) + \rho(z_{t+1}, \bar{x}_v)).$$

This implies that $z_t = \bar{x}_v$ for all integers $t \geq 0$ and $v_\gamma \in E$.

Assume that (2.206) holds. There exist $z_* \in X$ such that

$$\pi^v(z_*) = \sup(\pi^v) \tag{2.213}$$

and a (v, Ω)-overtaking optimal program $\{z_t^*\}_{t=0}^\infty$ such that

$$z_0^* = z_*.$$

Proposition 2.24 implies that

$$\pi^v(z_*) = \lim_{T\to\infty} \sum_{t=0}^{T-1} [v(z_t^*, z_{t+1}^*) - v(\bar{x}_v, \bar{x}_v)]. \tag{2.214}$$

In view of (A3),

$$\lim_{t\to\infty} z_t^* = \bar{x}_v. \tag{2.215}$$

It follows from (2.206), (2.213), (2.215), and Proposition 2.23 that

$$\pi^v(z_t^*) < \pi^v(z_0^*) \text{ for all large enough integers } t \geq 1. \tag{2.216}$$

By (2.216), there exists an integer $\tau_0 \geq 0$ such that $\pi^v(z_{\tau_0}^*) = \pi^v(z^*)$ and that $\pi^v(z_t^*) < \pi^v(z^*)$ for all integers $t > \tau_0$. We may assume without loss of generality $\tau_0 = 0$. Thus

$$\pi^v(z_0^*) = \sup(\pi^v),\ \pi^v(z_t^*) < \pi^v(z_0^*) \text{ for all integers } t \geq 1. \tag{2.217}$$

Let $\gamma \in (0, 1)$. For all $(x, y) \in \Omega$ define

$$v_\gamma(x, y) = v(x, y) - \gamma\rho((x, y), (\{(z_t^*, z_{t+1}^*) :\ t = 0, 1, \dots\} \cup \{(\bar{x}_v, \bar{x}_v)\})). \tag{2.218}$$

In follows from (2.218) that

$$v_\gamma \in \mathfrak{M} \cap \mathcal{M}_0(\Omega),\ v_\gamma(\bar{x}_v, \bar{x}_v) = v(\bar{x}_v, \bar{x}_v),$$

any (v_γ, Ω)-good program is (v, Ω)-good and converges to $\bar{x}_v$. Thus

$$v_\gamma \in \mathcal{M}_*(\Omega),\ \bar{x}_{v_\gamma} = \bar{x}_v. \tag{2.219}$$

By (2.213), (2.214), (2.218), (2.219), and the equality $z_0^* = z_*$,

$$\pi^{v_\gamma}(y) \le \pi^v(y),\ y \in X,\ \pi^{v_\gamma}(z_*) = \pi^v(z_*). \tag{2.220}$$

In view of (2.219), for all integers $T \ge 1$,

$$\sigma(v_\gamma, T, \bar{x}_v, \bar{x}_v) = Tv(\bar{x}_v, \bar{x}_v). \tag{2.221}$$

Proposition 2.22 and (2.219) imply that

$$\pi^{v_\gamma}(\bar{x}_v) = 0. \tag{2.222}$$

Set

$$K = \{(z_t^*, z_{t+1}^*) :\ t = 0, 1, \dots\} \cup \{(\bar{x}_v, \bar{x}_v)\}. \tag{2.223}$$

Assume that $y \in X$ satisfies

$$\pi^{v_\gamma}(y) = \sup(\pi^{v_\gamma}) \tag{2.224}$$

and that $\{y_t\}_{t=0}^\infty$ is a (v_γ, Ω)-overtaking optimal program satisfying

$$y_0 = y. \tag{2.225}$$

By (2.213), (2.218)–(2.220), (2.223)–(2.225), Proposition 2.24 and the inclusion $\{y_t\}_{t=0}^\infty \in \mathcal{P}(v_\gamma, y)$,

$$\pi^v(z_*) = \pi^{v_\gamma}(z_*) = \pi^{v_\gamma}(y) = \lim_{T\to\infty} \sum_{t=0}^{T-1} [v_\gamma(y_t, y_{t+1}) - v_\gamma(\bar{x}_v, \bar{x}_v)]$$

$$= \lim_{T\to\infty} [\sum_{t=0}^{T-1} (v(y_t, y_{t+1}) - v(\bar{x}_v, \bar{x}_v) - \gamma\rho((y_t, y_{t+1}), K))]$$

$$\le \pi^v(y) - \gamma \sum_{t=0}^{\infty} \rho((y_t, y_{t+1}), K).$$

Together with (2.221) this implies that

$$\pi^v(y) = \pi^v(z_*), \tag{2.226}$$

$$(y_t, y_{t+1}) \in K \text{ for all integers } t \ge 0. \tag{2.227}$$

In view of (2.206), (2.214), (2.217), (2.223), (2.225)–(2.227), and the equality $z_0^* = z_*$,

$$y = y_0 = z_*. \tag{2.228}$$

We show by induction that $y_t = z_t^*$ for all integers $t \ge 0$. There are two cases:

$$z_t^* \ne \bar{x}_v \text{ for all integers } t \ge 0; \tag{2.229}$$

$$\bar{x}_v \in \{z_t^* :\ t = 0, 1, \dots\}. \tag{2.230}$$

Assume that (2.229) holds. By Proposition 2.39, (2.213), (2.214), and (2.229),

$$z^*_{t_1} \neq z^*_{t_2} \text{ for all integers } t_2 > t_1 \geq 0. \tag{2.231}$$

Assume that $T \geq 0$ is an integer and that

$$y_t = z^*_t, \ t = 0, \dots, T. \tag{2.232}$$

(Note that in view of (2.228) and the equality $z^*_0 = z_*$ our assumption holds for $T = 0$.) By (2.223), (2.227), (2.229), (2.231), (2.232),

$$(z^*_T, y_{T+1}) = (y_T, y_{T+1}) \in K = \{(z^*_t, z^*_{t+1}) : \ t = 0, 1, \dots\} \cup \{(\bar{x}_v, \bar{x}_v)\}$$

and $y_{T+1} = z^*_{T+1}$. Thus $y_t = z^*_t$ for all integers $t \geq 0$.

Assume that (2.230) holds. By (2.206), (2.213), and the equality $z^*_0 = z_*$, there is a natural number S such that

$$z^*_S = \bar{x}_v, \ z^*_t \neq \bar{x}_v \text{ for all integers } t \in [0, S). \tag{2.233}$$

Propositions 2.39 and 2.40 imply that,

$$z^*_t = \bar{x}_v \ \text{ for all integers } t \geq S, \tag{2.234}$$

$$z^*_{t_2} \neq z^*_{t_1} \text{ for all integers } t_1, t_2 \in [0, S] \text{ such that } t_1 < t_2. \tag{2.235}$$

Assume that $T \geq 0$ is an integer and that

$$y_t = z^*_t, \ t = 0, \dots, T. \tag{2.236}$$

(Note that in view of (2.228), our assumption holds for $T = 0$.) If $T < S$, then by (2.223), (2.227), (2.233)–(2.236),

$$(z^*_T, y_{T+1}) = (y_T, y_{T+1}) \in \{(z^*_t, z^*_{t+1}) : \ t = 0, 1, \dots\} \cup \{(\bar{x}_v, \bar{x}_v)\}$$

and $y_{T+1} = z^*_{T+1}$. If $T \geq S$, then by (2.223), (2.227), (2.232), (2.233), (2.234),

$$(\bar{x}_v, y_{T+1}) = (y_T, y_{T+1}) \in \{(z^*_t, z^*_{t+1}) : \ t = 0, 1, \dots\} \cup \{(\bar{x}_v, \bar{x}_v)\}$$

and $y_{T+1} = \bar{x}_v = z^*_{T+1}$. Thus $y_t = z^*_t$ for all integers $t \geq 0$ in both the cases (see (2.229), (2.230)). This implies that $v_\gamma \in E$. Therefore the inclusion above holds in both the cases (see (2.205), (2.206)). Since $v_\gamma \to v$ as $\gamma \to 0^+$ in $\mathfrak{M}$ we conclude that for any neighborhood $\mathcal{U}$ of v in $\mathfrak{M}$, $\mathcal{U} \cap E \neq \emptyset$. Thus E is an everywhere dense subset of $\mathfrak{M}$.

By definition, for every $v \in E$, there exists a unique (v, Ω)-overtaking optimal program $\{z^v_t\}_{t=0}^\infty$ satisfying $\pi^v(z^v_0) = \sup(\pi^v)$. Let $v \in E$ and $k \geq 1$ be an integer. By Proposition 2.34, there exist an open neighborhood $\mathcal{U}(v, k)$ of v in $\mathfrak{M}$ and an integer $T(v, k) \geq k$ such that the following property holds:

(Pv) for each $u \in \mathcal{U}(v, k) \cap \mathcal{M}_*(\Omega)$ and each $\{x_t\}_{t=0}^\infty \in \cup\{\mathcal{P}(u, z) : \ z \in \mathcal{D}(u)\}$ we have $\rho(x_t, z^v_t) \leq k^{-1}$, $t = 0, \dots, k$.

Set

$$\mathcal{F}_1 = \cap_{p=1}^{\infty} \cup \{\mathcal{U}(v,k) :\ v \in E,\ k \geq p\},\ \mathcal{F} = \mathcal{F}_1 \cap \mathcal{F}_0. \tag{2.237}$$

Clearly, $\mathcal{F}$ is a countable intersection of open everywhere dense subsets of $\mathfrak{M}$ and $\mathcal{F} \subset \mathcal{F}_0 \subset \mathcal{M}_*(\Omega)$.

Let $u \in \mathcal{F}$, $p \geq 1$ be an integer and $\{x_t^{(i)}\}_{t=0}^{\infty}$, $i = 1,2$ be (u,Ω)-overtaking optimal programs such that

$$\pi^v(x_0^{(i)}) = \sup(\pi^v),\ i = 1,2. \tag{2.238}$$

By (2.237), there exist $v_p \in E$ and an integer $k_p \geq p$ such that

$$u \in \mathcal{U}(v_p, k_p). \tag{2.239}$$

In view of (2.239) and (Pv), $\rho(x_t^{(i)}, z_t^{v_p}) \leq k_p^{-1} \leq p^{-1}$, $t = 0,\dots,p$, $i = 1,2$. This implies that $\rho(x_t^{(1)}, x_t^{(2)}) \leq 2p^{-1}$, $t = 0,\dots,p$. Since p is any natural number we conclude that $x_t^{(1)} = x_t^{(2)}$ for all integers $t \geq 0$. Proposition 2.41 is proved. □

2.17 Structure of Solutions of the Problem (*P*1)

Suppose that $v \in \mathcal{M}_*(\Omega)$, $\bar{x}_v \in X$, $\bar{r}_v > 0$, $\bar{c}_v > 0$ and that assumptions (A1), (A2), and (A3) hold.

We prove the following result obtained in [86] which describes the structure of approximate solutions of the problems of the type (P1) in the regions close to the right and the left endpoints.

Theorem 2.42. *Let $L_0 \geq 1$, $\tau_0 \geq 1$ be integers, $\epsilon > 0$, $x \in \bar{Y}_{L_0}$ and $y \in Y_{L_0}$. Then there exist $\delta \in (0,\epsilon)$ and an integer $T_0 \geq \tau_0$ such that for each integer $T \geq T_0$, each $\{u_t\}_{t=0}^{T-1} \subset \mathcal{M}(\Omega)$ satisfying $\|u_t - v\| \leq \delta$, $t = 0\dots,T-1$ and each (Ω)-program $\{x_t\}_{t=0}^{T}$ which satisfies*

$$x_0 = x,\ x_T = y, \sum_{t=0}^{T-1} u_t(x_t, x_{t+1}) \geq \sigma(\{u_t\}_{t=0}^{T-1}, 0, T, x_0, x_T) - \delta$$

there exist a (v,Ω)-overtaking optimal program $\{y_t\}_{t=0}^{\infty}$ and a $(\bar{v},\bar{\Omega})$-overtaking optimal program $\{z_t\}_{t=0}^{\infty}$ such that $y_0 = x$, $z_0 = y$ and for all $t = 0,\dots,\tau_0$,

$$\rho(x_t, y_t) \leq \epsilon,\ \rho(x_{T-t}, z_t) \leq \epsilon.$$

Since the mapping $v \to \bar{v}$, $v \in \mathcal{M}(\Omega)$ is an isometry Theorem 2.42 follows from Propositions 2.30, 2.31 and the following result.

Theorem 2.43. *Let $L_0 \geq 1$, $\tau_0 \geq 1$ be integers, $\epsilon > 0$ and $x \in \bar{Y}_{L_0}$. Then there exist $\delta \in (0, \epsilon)$ and an integer $T_0 \geq \tau_0$ such that for each integer $T \geq T_0$, each $\{u_t\}_{t=0}^{T-1} \subset \mathcal{M}(\Omega)$ satisfying $\|u_t - v\| \leq \delta$, $t = 0 \dots, T-1$ and each (Ω)-program $\{x_t\}_{t=0}^{T}$ which satisfies*

$$x_0 = x,\ x_T \in Y_{L_0},\ \sum_{t=0}^{T-1} u_t(x_t, x_{t+1}) \geq \sigma(\{u_t\}_{t=0}^{T-1}, 0, T, x_0, x_T) - \delta$$

there exists a (v, Ω)-overtaking optimal program $\{y_t\}_{t=0}^{\infty}$ such that $y_0 = x$ and for all $t = 0, \dots, \tau_0$, $\rho(x_t, y_t) \leq \epsilon$.

Theorems 2.43 and 2.4 imply the following result.

Theorem 2.44. *Let $L_0 \geq 1$, $\tau_0 \geq 1$ be integers, $\epsilon > 0$ and $x \in \bar{Y}_{L_0}$. Then there exist $\delta \in (0, \epsilon)$ and an integer $T_0 \geq \tau_0$ such that for each integer $T \geq T_0$, each $\{u_t\}_{t=0}^{T-1} \subset \mathcal{M}(\Omega)$ satisfying $\|u_t - v\| < \delta$, $t = 0 \dots, T-1$ and each (Ω)-program $\{x_t\}_{t=0}^{T}$ which satisfies*

$$x_0 = x,\ \sum_{t=0}^{T-1} u_t(x_t, x_{t+1}) \geq \sigma(\{u_t\}_{t=0}^{T-1}, 0, T, x_0) - \delta$$

there exists a (v, Ω)-overtaking optimal program $\{y_t\}_{t=0}^{\infty}$ such that $y_0 = x$ and for all $t = 0, \dots, \tau_0$, $\rho(x_t, y_t) \leq \epsilon$.

2.18 Proof of Theorem 2.43

Assume that Theorem 2.43 does not hold. Then for each natural number k there exist an integer

$$T_k \geq \tau_0 + k, \tag{2.240}$$

$\{u_t^{(k)}\}_{t=0}^{T_k-1} \subset \mathcal{M}(\Omega)$ satisfying

$$\|u_t^{(k)} - v\| \leq k^{-1},\ t = 0 \dots, T_k - 1, \tag{2.241}$$

an (Ω)-program $\{x_t^{(k)}\}_{t=0}^{T_k}$ such that

$$x_0^{(k)} = x,\ x_{T_k}^{(k)} \in Y_{L_0}, \tag{2.242}$$

$$\sum_{t=0}^{T_k-1} u_t^{(k)}(x_t^{(k)}, x_{t+1}^{(k)}) \geq \sigma(\{u_t^{(k)}\}_{t=0}^{T_k-1}, 0, T_k, x_0^{(k)}, x_{T_k}^{(k)}) - k^{-1} \tag{2.243}$$

and that for each (Ω)-program

$$\{y_t\}_{t=0}^{\infty} \in \mathcal{P}(v, x) \tag{2.244}$$

we have

$$\max\{\rho(x_t^{(k)}, y_t) : t = 0 \ldots, \tau_0\} > \epsilon. \quad (2.245)$$

Extracting a subsequence and re-indexing, if necessary, we may assume without loss of generality that for any integer $t \geq 0$ there exists

$$x_t = \lim_{k \to \infty} x_t^{(k)}. \quad (2.246)$$

Clearly, $\{x_t\}_{t=0}^{\infty}$ is an (Ω)-program. We show that for each integer $T \geq 1$,

$$\sum_{t=0}^{T-1} v(x_t, x_{t+1}) = \sigma(v, 0, T, x_0, x_T).$$

Assume the contrary. Then there exist a natural number S and $\Delta > 0$ such that

$$\sum_{t=0}^{S-1} v(x_t, x_{t+1}) < \sigma(v, 0, S, x_0, x_S) - 7\Delta.$$

Clearly, there exists an (Ω)-program $\{\tilde{x}_t\}_{t=0}^{S}$ such that

$$\tilde{x}_0 = x, \ \tilde{x}_S = x_S, \ \sum_{t=0}^{S-1} v(x_t, x_{t+1}) < \sum_{t=0}^{S-1} v(\tilde{x}_t, \tilde{x}_{t+1}) - 6\Delta. \quad (2.247)$$

By (A1), there exists $\delta \in (0, \bar{r}_v/4)$ such that for each $x, y \in X$ satisfying $\rho(x, \bar{x}_v)$, $\rho(y, \bar{x}_v) \leq \delta$,

$$|v(x, y) - v(\bar{x}_v, \bar{x}_v)| \leq \Delta/2. \quad (2.248)$$

In view of (2.240)–(2.243), the inclusion $x \in \bar{Y}_{L_0}$ and Theorem 2.3, there exist natural numbers $k_1 > L_1$ such that for each integer $k \geq k_1$,

$$\rho(x_t^{(k)}, \bar{x}_v) \leq \delta, \ t = L_1, \ldots, T_k - L_1. \quad (2.249)$$

It follows from (2.246) and (2.249) that

$$\rho(x_t, \bar{x}_v) \leq \delta \text{ for all integers } t \geq L_1. \quad (2.250)$$

Choose a natural number

$$k_2 > k_1 + S + 2L_1 + 8. \quad (2.251)$$

Relations (2.240) and (2.251) imply that for each integer $k \geq k_2$,

$$T_k - L_1 \geq k - L_1 > k_1 + S + L_1 + 8. \quad (2.252)$$

By (2.241), (2.246), and the upper semicontinuity of v,

$$\limsup_{k\to\infty}\sum_{t=0}^{k_1+S+L_1+7} u_t^{(k)}(x_t^{(k)},x_{t+1}^{(k)}) = \limsup_{k\to\infty}\sum_{t=0}^{k_1+S+L_1+7} v(x_t^{(k)},x_{t+1}^{(k)})$$
$$\le \sum_{t=0}^{k_1+S+L_1+7} v(x_t,x_{t+1}). \tag{2.253}$$

For each integer $k\ge k_2$ set

$$\widehat{x}_t^{(k)}=\tilde{x}_t,\ t=0,\dots,S,\ \widehat{x}_t^{(k)}=x_t,\ t=S+1,\dots,k_1+S+L_1+8,$$
$$\widehat{x}_{k_1+S+L_1+9}^{(k)}=x_{k_1+S+L_1+9}^{(k)}. \tag{2.254}$$

In view of (2.247), (2.249), (2.250), (2.252), and (2.254), for each integer $k\ge k_2$, $\{\widehat{x}_t^{(k)}\}_{t=0}^{k_1+S+L_1+9}$ is an (Ω)-program. By (2.242), (2.247), and (2.254), for each integer $k\ge k_2$,

$$\widehat{x}_0^{(k)}=x=x_0^{(k)},\ \widehat{x}_{k_1+S+L_1+9}^{(k)}=x_{k_1+S+L_1+9}^{(k)}. \tag{2.255}$$

By (2.253), there exists an integer $k\ge k_2$ such that

$$\sum_{t=0}^{k_1+S+L_1+7} u_t^{(k)}(x_t^{(k)},x_{t+1}^{(k)}) \le \sum_{t=0}^{k_1+S+L_1+7} v(x_t,x_{t+1})+\Delta, \tag{2.256}$$

$$k^{-1}(k_1+S+L_1+10)<\Delta. \tag{2.257}$$

Relations (2.241), (2.247)–(2.250), (2.254), (2.256), and (2.257) imply that

$$\sum_{t=0}^{k_1+S+L_1+8} u_t^{(k)}(x_t^{(k)},x_{t+1}^{(k)}) - \sum_{t=0}^{k_1+S+L_1+8} u_t^{(k)}(\widehat{x}_t^{(k)},\widehat{x}_{t+1}^{(k)})$$
$$= \sum_{t=0}^{k_1+S+L_1+7} u_t^{(k)}(x_t^{(k)},x_{t+1}^{(k)}) - \sum_{t=0}^{k_1+S+L_1+7} u_t^{(k)}(\widehat{x}_t^{(k)},\widehat{x}_{t+1}^{(k)})$$
$$+u_{k_1+S+L_1+8}^{(k)}(x_{k_1+S+L_1+8}^{(k)},x_{k_1+S+L_1+9}^{(k)})$$
$$-u_{k_1+S+L_1+8}^{(k)}(\widehat{x}_{k_1+S+L_1+8}^{(k)},\widehat{x}_{k_1+S+L_1+9}^{(k)})$$
$$\le \sum_{t=0}^{k_1+S+L_1+7} v(x_t,x_{t+1})+\Delta-\sum_{t=0}^{k_1+S+L_1+7} v(\widehat{x}_t^{(k)},\widehat{x}_{t+1}^{(k)})$$
$$+(k_1+S+L_1+8)k^{-1}+2k^{-1}$$
$$+v(x_{k_1+S+L_1+8}^{(k)},x_{k_1+S+L_1+9}^{(k)})-v(x_{k_1+S+L_1+8},x_{k_1+S+L_1+9}^{(k)})$$

$$\leq \Delta + \sum_{t=0}^{S-1} v(x_t, x_{t+1}) - \sum_{t=0}^{S-1} v(\tilde{x}_t, \tilde{x}_{t+1}) + k^{-1}(k_1 + S + L_1 + 10) + \Delta$$

$$\leq -6\Delta + 2\Delta + k^{-1}(k_1 + S + L_1 + 10) \leq -3\Delta_1 \leq -3k^{-1}.$$

Together with (2.255) this contradicts (2.243). The contradiction we have reached proves that for each integer $T > 0$,

$$\sum_{t=0}^{T-1} v(x_t, x_{t+1}) = \sigma(v, 0, T, x, x_T). \tag{2.258}$$

Let $\epsilon > 0$. By (2.240)–(2.243), the inclusion $x \in \bar{Y}_{L_0}$ and Theorem 2.3, there exist natural numbers $k(\epsilon) > L(\epsilon)$ such that for each integer $k \geq k(\epsilon)$, the inequality $\rho(x_t^{(k)}, \bar{x}_v) \leq \epsilon$ holds for all $t = L(\epsilon), \dots, T_k - L(\epsilon)$. Together with (2.256) this implies that $\rho(x_t, \bar{x}_v) \leq \epsilon$ for all integers $t \geq L(\epsilon)$. Since ϵ is any positive number we conclude that $\lim_{t\to\infty} x_t = \bar{x}_v$. Together with (2.242), (2.246), (2.258), and Theorem 2.14 this implies that $\{x_t\}_{t=0}^{\infty}$ is a (v, Ω)-overtaking optimal program and $\{x_t\}_{t=0}^{\infty} \in \mathcal{P}(v, x)$. By (2.246), for all sufficiently large natural numbers k, $\rho(x_t, x_t^{(k)}) \leq \epsilon/2$ for all $t = 0, \dots, L_0$, a contradiction. The contradiction we have reached proves Theorem 2.43. □

3

Bolza Problems

In this chapter we study the structure of approximate solutions of an autonomous discrete-time optimal control system with a compact metric space of states. These optimal control systems are discrete-time analogs of Bolza problems in the calculus of variations. They are described by a pair of objective functions which determines an optimality criterion. We consider two classes of Bolza problems and obtain for each of them the full description of approximate solutions of these problems on large intervals. This description shows that on large intervals the approximate solutions are determined mainly by our optimality criterion and are essentially independent of the choice of time intervals and data.

3.1 The First Class of Bolza Problems

Let (X, ρ) be a compact metric space and Ω be a nonempty closed subset of $X \times X$. In this chapter we use the notation, definitions, and assumptions introduced in Chap. 2.

Assume that $v \in \mathcal{M}(\Omega)$ is an upper semicontinuous function. We suppose that there exist $\bar{x}_v \in X$ and constants $\bar{c}_v > 0$ and $\bar{r}_v > 0$ such that assumptions (A1), (A2), and (A3) hold.

For each nonempty set Y and each function $h : Y \to R^1 \cup \{-\infty\}$ put

$$\sup(h) = \sup\{h(y) : \; y \in Y\}.$$

Denote by $\mathcal{M}(X)$ the set of all bounded functions $h : X \to R^1$. For each $h \in \mathcal{M}(X)$ set

$$\|h\| = \sup\{|h(x)| : \; x \in X\}.$$

A.J. Zaslavski, *Discrete-Time Optimal Control and Games on Large Intervals*, Springer Optimization and Its Applications 119,
DOI 10.1007/978-3-319-52932-5_3

Clearly, $(\mathcal{M}(X), \|\cdot\|)$ is a Banach space. For each $h_1, h_2 \in \mathcal{M}(X)$ set

$$d_X(h_1, h_2) = \|h_1 - h_2\|.$$

For each $x \in X$, each pair of integers T_1, T_2 satisfying $0 \leq T_1 < T_2$, each sequence $\{u_t\}_{t=T_1}^{T_2-1} \subset \mathcal{M}(\Omega)$ and each $h \in \mathcal{M}(X)$ we consider the problem

$$\sum_{t=T_1}^{T_2-1} u_t(x_t, x_{t+1}) + h(x_{T_2}) \to \max, \; \{(x_t, x_{t+1})\}_{t=0}^{T-1} \subset \Omega, \; x_{T_1} = x \qquad (P4)$$

and set

$$\sigma(h, \{u_t\}_{t=T_1}^{T_2-1}, T_1, T_2, x) = \sup\{\sum_{t=T_1}^{T_2-1} u_t(x_t, x_{t+1}) + h(x_{T_2}) :$$

$$\{x_t\}_{t=T_1}^{T_2} \text{ is an } (\Omega) - \text{program and } x_{T_1} = x\}. \qquad (3.1)$$

For each $x \in X$, each pair of integers T_1, T_2 satisfying $0 \leq T_1 < T_2$, each $u \in \mathcal{M}(\Omega)$ and each $h \in \mathcal{M}(X)$ set

$$\sigma(h, u, T_1, T_2, x) = \sigma(h, \{u_t\}_{t=T_1}^{T_2-1}, T_1, T_2, x) \text{ where } u_t = u, \; t = T_1, \dots, T_2 - 1. \qquad (3.2)$$

In this section we study turnpike properties of approximate solutions of problems of the type (P4) and establish the following three results.

Theorem 3.1. *Let $\epsilon \in (0, \bar{r}_v)$ and $M > 0$. Then there exist a positive number $\delta < \min\{1, M\}$ and a natural number L such that for each integer $T \geq L$, each $\{u_t\}_{t=0}^{T-1} \subset \mathcal{M}(\Omega)$, each $h \in \mathcal{M}(X)$ and each (Ω)-program $\{x_t\}_{t=0}^{T}$ which satisfy*

$$\|h\| \leq M, \; \|u_t - v\| \leq \delta, \; t = 0, \dots, T - 1, \qquad (3.3)$$

$$\sum_{t=0}^{T-1} u_t(x_t, x_{t+1}) + h(x_T) \geq \sigma(h, \{u_t\}_{t=0}^{T-1}, 0, T) - M \qquad (3.4)$$

the inequality

$$Card(\{t \in \{0, \dots, T\} : \; \rho(x_t, \bar{x}_v) > \epsilon\}) < L \qquad (3.5)$$

holds.

Proof. By Assertion 1 of Theorem 2.5, there exist a positive number

$$\delta < \min\{1, M\}$$

and a natural number L such that the following property holds:

(i) for each integer $T \geq L$, each $\{u_t\}_{t=0}^{T-1} \subset \mathcal{M}(\Omega)$ and each (Ω)-program $\{x_t\}_{t=0}^{T}$ which satisfy

$$\|u_t - v\| \leq \delta,\ t = 0, \dots, T-1,$$

$$\sum_{t=0}^{T-1} u_t(x_t, x_{t+1}) \geq \sigma(\{u_t\}_{t=0}^{T-1}, 0, T) - 8M \tag{3.6}$$

inequality (3.5) holds.

Assume that an integer $T \geq L$, $\{u_t\}_{t=0}^{T-1} \subset \mathcal{M}(\Omega)$, $h \in \mathcal{M}(X)$, $\|h\| \leq M$, $\{x_t\}_{t=0}^{T}$ is an (Ω)-program and (3.3) and (3.4) hold. In view of (3.3) and (3.4), inequality (3.6) holds. By (3.3), (3.6), and property (i), inequality (3.5) holds. Theorem 3.1 is proved. □

Theorem 3.2. *Let a positive number $\epsilon < \bar{r}_v$, $L_0 \geq 1$ be an integer and $M_0 > 0$. Then there exist an integer $L \geq 1$ and a number $\delta \in (0, \epsilon)$ such that for each integer $T > 2L$, each $\{u_t\}_{t=0}^{T-1} \subset \mathcal{M}(\Omega)$ satisfying*

$$\|u_t - v\| \leq \delta,\ t = 0 \dots, T-1, \tag{3.7}$$

each $h \in \mathcal{M}(X)$ satisfying

$$\|h\| \leq M_0 \tag{3.8}$$

and each (Ω)-program $\{x_t\}_{t=0}^{T}$ which satisfies

$$x_0 \in \bar{Y}_{L_0}, \tag{3.9}$$

$$h(x_T) + \sum_{t=0}^{T-1} u_t(x_t, x_{t+1}) \geq \sigma(h, \{u_t\}_{t=0}^{T-1}, 0, T, x_0) - M_0 \tag{3.10}$$

and

$$\sum_{t=\tau}^{\tau+L-1} u_t(x_t, x_{t+1}) \geq \sigma(\{u_t\}_{t=\tau}^{\tau+L-1}, \tau, \tau+L, x_\tau, x_{\tau+L}) - \delta \tag{3.11}$$

for each integer $\tau \in [0, T-L]$ there exist integers $\tau_1 \in [0, L]$, $\tau_2 \in [T-L, T]$ such that

$$\rho(x_t, \bar{x}_v) \leq \epsilon,\ t = \tau_1, \dots, \tau_2. \tag{3.12}$$

Moreover if $\rho(x_0, \bar{x}_v) \leq \delta$, then $\tau_1 = 0$.

Proof. By Theorem 2.4, there exist an integer $L \geq 1$ and a number $\delta \in (0, \epsilon)$ such that the following property holds:

(ii) for each integer $T > 2L$, each $\{u_t\}_{t=0}^{T-1} \subset \mathcal{M}(\Omega)$ satisfying (3.7) and each (Ω)-program $\{x_t\}_{t=0}^{T}$ which satisfies (3.9), the inequality

$$\sum_{t=0}^{T-1} u_t(x_t, x_{t+1}) \geq \sigma(\{u_t\}_{t=0}^{T-1}, 0, T, x_0) - 4M_0 \tag{3.13}$$

and (3.11) for each integer $\tau \in [0, T-L]$, there exist integers $\tau_1 \in [0, L]$, $\tau_2 \in [T-L, T]$ such that (3.12) holds; moreover if $\rho(x_0, \bar{x}_v) \le \delta$, then $\tau_1 = 0$.

Assume that an integer $T > 2L$, $\{u_t\}_{t=0}^{T-1} \subset \mathcal{M}(\Omega)$ satisfies (3.7), $h \in \mathcal{M}(X)$ satisfies (3.8) and that an (Ω)-program $\{x_t\}_{t=0}^{T}$ satisfies (3.9), (3.10), and (3.11) for each integer $\tau \in [0, T-L]$. In view of (3.8) and (3.10), relation (3.13) holds. By (3.7), (3.9), (3.11), (3.13) and property (ii), there exist integers $\tau_1 \in [0, L]$ and $\tau_2 \in [T-L, T]$ such that (3.12) holds and if $\rho(x_0, \bar{x}_v) \le \delta$, then $\tau_1 = 0$. Theorem 3.2 is proved. □

Theorem 3.2 implies the following turnpike result.

Theorem 3.3. *Let a positive number $\epsilon < \bar{r}_v$, $L_0 \ge 1$ be an integer and $M_0 > 0$. Then there exist an integer $L \ge 1$ and a number $\delta \in (0, \epsilon)$ such that for each integer $T > 2L$, each $\{u_t\}_{t=0}^{T-1} \subset \mathcal{M}(\Omega)$ satisfying*

$$\|u_t - v\| \le \delta,\ t = 0 \dots, T-1,$$

each $h \in \mathcal{M}(X)$ satisfying

$$\|h\| \le M_0$$

and each (Ω)-program $\{x_t\}_{t=0}^{T}$ which satisfies

$$x_0 \in \bar{Y}_{L_0},$$

$$h(x_T) + \sum_{t=0}^{T-1} u_t(x_t, x_{t+1}) \ge \sigma(h, \{u_t\}_{t=0}^{T-1}, 0, T, x_0) - \delta$$

there exist integers $\tau_1 \in [0, L]$, $\tau_2 \in [T-L, T]$ such that

$$\rho(x_t, \bar{x}_v) \le \epsilon,\ t = \tau_1, \dots, \tau_2.$$

Moreover if $\rho(x_0, \bar{x}_v) \le \delta$, then $\tau_1 = 0$.

The results of this section were obtained in [92].

3.2 Structure of Solutions of the Problem (P4) in the Regions Close to the Endpoints

In Sect. 3.4 we prove the following result which describes the structure of approximate solutions of the problems of the type (P4) in the regions close to the right endpoints.

Theorem 3.4. *Suppose that $g \in \mathcal{M}(X)$ and $v \in \mathcal{M}(\Omega)$ are upper semicontinuous functions, $\bar{x}_v \in X$, $\bar{r}_v > 0$, $\bar{c}_v > 0$ and that assumptions (A1), (A2) and (A3) hold. Let $L_0 \ge 1$, $\tau_0 \ge 1$ be integers and $\epsilon > 0$, $M_0 > 1$. Then there exist $\delta > 0$ and an integer $T_0 \ge \tau_0$ such that for each integer $T \ge T_0$, each $h \in \mathcal{M}(X)$ satisfying*

$$\|h - g\| \le \delta,$$

each $\{u_t\}_{t=0}^{T-1} \subset \mathcal{M}(\Omega)$ *satisfying*

$$\|u_t - v\| \le \delta,\ t = 0\dots, T-1$$

and each (Ω)*-program* $\{x_t\}_{t=0}^{T}$ *which satisfies*

$$x_0 \in \bar{Y}_{L_0},$$

$$h(x_T) + \sum_{t=0}^{T-1} u_t(x_t, x_{t+1}) \ge \sigma(h, \{u_t\}_{t=0}^{T-1}, 0, T, x_0) - M_0,$$

$$\sum_{t=\tau}^{\tau+T_0-1} u_t(x_t, x_{t+1}) \ge \sigma(\{u_t\}_{t=\tau}^{\tau+T_0-1}, \tau, \tau + T_0, x_\tau, x_{\tau+T_0}) - \delta$$

for each $\tau \in \{0, \dots, T - T_0\}$,

$$h(x_T) + \sum_{t=T-T_0}^{T-1} u_t(x_t, x_{t+1}) \ge \sigma(h, \{u_t\}_{t=T-T_0}^{T-1}, T - T_0, T, x_{T-T_0}) - \delta$$

there exists an $(\bar{v}, \bar{\Omega})$*-overtaking optimal program* $\{x_t^*\}_{t=0}^{\infty}$ *such that*

$$(\pi^{\bar{v}} + g)(x_0^*) = \sup(\pi^{\bar{v}} + g),$$

$$\rho(x_{T-t}, x_t^*) \le \epsilon,\ t = 0, \dots, \tau_0.$$

Let $g \in \mathcal{M}(X)$ and $v \in \mathcal{M}(\Omega)$ be as in the statement of Theorem 3.4 and satisfy all the assumptions posed there. Then $\pi^v + g : X \to R^1 \cup \{-\infty\}$ is an upper semicontinuous, bounded from above function such that $(\pi^v + g)(\bar{x}_v) = g(\bar{x}_v)$ is finite. Therefore there exists $x \in X$ such that

$$(\pi^v + g)(x) = \sup(\pi^v + g).$$

Denote by $\mathcal{M}_u(X)$ the set of all upper semicontinuous functions belonging to the space $\mathcal{M}(X)$ and denote by $\mathcal{M}_c(X)$ the set of all continuous functions belonging to the space $\mathcal{M}(X)$. Clearly, $\mathcal{M}_u(X)$ and $\mathcal{M}_c(X)$ are closed subsets of $\mathcal{M}(X)$. We consider the complete metric spaces $\mathcal{M}_u(X)$ and $\mathcal{M}_c(X)$ equipped with the metric d_X.

In order to state our next result we need the following notion of porosity [81].

Let (Y, d) be a complete metric space. We denote by $B_Y(y, r)$ the closed ball of center $y \in Y$ and radius $r > 0$. A subset $E \subset Y$ is called porous (with respect to d) if there exist $\alpha \in (0, 1]$ and $r_0 > 0$ such that for each $r \in (0, r_0]$ and each $y \in Y$ there exists $z \in Y$ for which

$$B_Y(z, \alpha r) \subset B_Y(y, r) \setminus E.$$

A subset of the space Y is called σ-porous (with respect to d) if it is a countable union of porous (with respect to d) subsets of Y.

Since porous sets are nowhere dense, all σ-porous sets are of the first category. If Y is a finite dimensional Euclidean space, then σ-porous sets are of Lebesgue measure 0. In fact, the class of σ-porous sets in such a space is much smaller than the class of sets which have measure 0 and are of the first category. To point out the difference between porous and nowhere dense sets note that if $E \subset Y$ is nowhere dense, $y \in Y$ and $r > 0$, then there is a point $z \in Y$ and a number $s > 0$ such that $B_Y(z,s) \subset B_Y(y,r) \setminus E$. If, however, E is also porous, then for small enough r we can choose $s = \alpha r$, where $\alpha \in (0,1)$ is a constant which depends only on E.

The discussion of the porosity notion and the corresponding references can be found in [81]. Theorem 5.9 of [81] and Theorem 3.4 imply the following result.

Theorem 3.5. *Suppose that $v \in \mathcal{M}(\Omega)$ is an upper semicontinuous function, $\bar{x}_v \in X$, $\bar{r}_v > 0$, $\bar{c}_v > 0$ and that assumptions (A1), (A2), and (A3) hold. Then there exists a set $\mathcal{F} \subset \mathcal{M}_c(\Omega)$ such that the set $\mathcal{M}_c(\Omega) \setminus \mathcal{F}$ is σ-porous in $\mathcal{M}_c(\Omega)$ and that for each $g \in \mathcal{F}$ the following assertions hold.*

1. There exists a unique point $x_g \in X$ such that

$$\{x \in X : (\pi^{\bar{v}} + g)(x) = \sup(\pi^{\bar{v}} + g)\} = \{x_g\}.$$

2. Let $L_0 \geq 1$, $\tau_0 \geq 1$ be integers and $\epsilon > 0$, $M_0 > 1$. Then there exist $\delta > 0$ and an integer $T_0 \geq \tau_0$ such that for each integer $T \geq T_0$, each $h \in \mathcal{M}(X)$ satisfying

$$\|h - g\| \leq \delta,$$

each $\{u_t\}_{t=0}^{T-1} \subset \mathcal{M}(\Omega)$ satisfying

$$\|u_t - v\| \leq \delta,\ t = 0 \dots, T-1$$

and each (Ω)-program $\{x_t\}_{t=0}^{T}$ which satisfies

$$x_0 \in \bar{Y}_{L_0},$$

$$h(x_T) + \sum_{t=0}^{T-1} u_t(x_t, x_{t+1}) \geq \sigma(h, \{u_t\}_{t=0}^{T-1}, 0, T, x_0) - M_0,$$

$$\sum_{t=\tau}^{\tau+T_0-1} u_t(x_t, x_{t+1}) \geq \sigma(\{u_t\}_{t=\tau}^{\tau+T_0-1}, \tau, \tau + T_0, x_\tau, x_{\tau+T_0}) - \delta$$

for each $\tau \in \{0, \dots, T - T_0\}$,

$$h(x_T) + \sum_{t=T-T_0}^{T-1} u_t(x_t, x_{t+1}) \geq \sigma(h, \{u_t\}_{t=T-T_0}^{T-1}, T - T_0, T, x_{T-T_0}) - \delta$$

there exists a $(\bar{v}, \bar{\Omega})$-overtaking optimal program $\{x_t^*\}_{t=0}^{\infty}$ such that

$$x_0^* = x_g,$$

$$\rho(x_{T-t}, x_t^*) \leq \epsilon,\ t = 0, \dots, \tau_0.$$

The next theorem which is proved in Sect. 3.6 shows that given a function $g \in \mathcal{M}_u(X)$, for most objective functions v (in the sense of the Baire category) there exists a unique pair of a (v, Ω)-overtaking optimal program $\{z_t\}_{t=0}^{\infty}$ and a $(\bar{v}, \bar{\Omega})$-overtaking optimal program $\{\widehat{z}_t\}_{t=0}^{\infty}$ such that

$$(\pi^v + g)(z_0) = \sup(\pi^v + g),$$

$$(\pi^{\bar{v}} + g)(\widehat{z}_0) = \sup(\pi^{\bar{v}} + g).$$

In this case approximate solutions of the problems of the types (P4) in the regions close to the right endpoints have a simple structure.

Theorem 3.6. *Let $\mathfrak{M}$ be either $\bar{\mathcal{M}}_0(\Omega)$ or $\bar{\mathcal{M}}_{c,0}(\Omega)$ and let $g \in \mathcal{M}_u(X)$. Then there exists a set $\mathcal{F} \subset \mathfrak{M} \cap \mathcal{M}_*(\Omega)$ which is a countable intersection of open everywhere dense subsets of $\mathfrak{M}$ such that for each $v \in \mathcal{F}$ there exists a unique pair of points $z, \bar{z} \in X$ such that*

$$(g + \pi^v)(z) = \sup(g + \pi^v), \ (g + \pi^{\bar{v}})(\bar{z}) = \sup(g + \pi^{\bar{v}})$$

and there exist a unique (v, Ω)-overtaking optimal program $\{z_t\}_{t=0}^{\infty}$ satisfying $z_0 = z$ and a unique $(\bar{v}, \bar{\Omega})$-overtaking optimal program $\{\widehat{z}_t\}_{t=0}^{\infty}$ satisfying $\widehat{z}_0 = \bar{z}$.

Let $v \in \mathcal{M}_*(\Omega)$ and $g \in \mathcal{M}_u(X)$. Set

$$\mathcal{D}(g, v) = \{z \in X : \ (g + \pi^v)(z) = \sup(g + \pi^v)\}$$

and denote by $\tilde{\mathcal{P}}(g, v)$ the set of all (v, Ω)-overtaking optimal programs $\{z_t\}_{t=0}^{\infty}$ satisfying $z_0 \in \mathcal{D}(g, v)$.

In Sect. 3.5 we prove the following result.

Proposition 3.7. *Suppose that $g \in \mathcal{M}_u(X)$ and $v \in \mathcal{M}(\Omega)$ are upper semicontinuous functions, $\bar{x}_v \in X$, $\bar{r}_v > 0$, $\bar{c}_v > 0$ and that assumptions (A1), (A2) and (A3) hold. Let $\tau_0 \geq 1$ be an integer and $\epsilon > 0$. Then there exist $\delta > 0$ and an integer $T_0 \geq \tau_0$ such that for each $u \in B_d(v, \delta) \cap \mathcal{M}_*(\Omega)$ and each $h \in \mathcal{M}_u(X)$ satisfying $\|h - g\| \leq \delta$ the following properties hold:*

for each $\{x_t\}_{t=0}^{\infty} \in \tilde{\mathcal{P}}(h, u)$ there exists $\{y_t\}_{t=0}^{\infty} \in \tilde{\mathcal{P}}(g, v)$ such that

$$\rho(x_t, y_t) \leq \epsilon$$

for all integers $t = 0, \ldots, \tau_0$;

for each $\{x_t\}_{t=0}^{\infty} \in \tilde{\mathcal{P}}(h, \bar{u})$ there exists $\{y_t\}_{t=0}^{\infty} \in \tilde{\mathcal{P}}(g, \bar{v})$ such that

$$\rho(x_t, y_t) \leq \epsilon$$

for all integers $t = 0, \ldots, \tau_0$.

The next result is an extension of Theorem 3.6.

Theorem 3.8. *Let $\mathfrak{M}$ be either $\bar{\mathcal{M}}_0(\Omega)$ or $\bar{\mathcal{M}}_{c,0}(\Omega)$ and $\mathfrak{A}$ be either $\mathcal{M}_u(X)$ or $\mathcal{M}_c(X)$. Then there exists a set $\mathcal{F} \subset (\mathfrak{M} \cap \mathcal{M}_*(\Omega)) \times \mathfrak{A}$ which is a countable intersection of open everywhere dense subsets of $\mathfrak{M} \times \mathfrak{A}$ such that for each $(v, g) \in \mathcal{F}$ there exists a unique pair of programs $\{x_t\}_{t=0}^{\infty} \in \tilde{\mathcal{P}}(g, v)$ and $\{\bar{x}_t\}_{t=0}^{\infty} \in \tilde{\mathcal{P}}(\bar{g}, v)$.*

Since the mapping $v \to \bar{v}$, $v \in \mathcal{M}(\Omega)$ is a linear isometry which has the inverse, Theorem 3.8 follows from the next proposition which is proved in Sect. 3.7.

Proposition 3.9. *Let $\mathfrak{M}$ be either $\bar{\mathcal{M}}_0(\Omega)$ or $\bar{\mathcal{M}}_{c,0}(\Omega)$ and $\mathfrak{A}$ be either $\mathcal{M}_u(X)$ or $\mathcal{M}_c(X)$. Then there exists a set $\mathcal{F} \subset (\mathfrak{M} \cap \mathcal{M}_*(\Omega)) \times \mathfrak{A}$ which is a countable intersection of open everywhere dense subsets of $\mathfrak{M} \times \mathfrak{A}$ such that for each $(v, g) \in \mathcal{F}$ there exists a unique program $\{x_t\}_{t=0}^{\infty} \in \tilde{\mathcal{P}}(g, v)$.*

The results of this section were obtained in [92].

3.3 A Basic Lemma for Theorem 3.4

Lemma 3.10. *Suppose that $g \in \mathcal{M}(X)$ and $v \in \mathcal{M}(\Omega)$ are upper semicontinuous functions, $\bar{x}_v \in X$, $\bar{r}_v > 0$, $\bar{c}_v > 0$ and that assumptions (A1), (A2), and (A3) hold. Let $T_0 \geq 1$ be an integer and $\epsilon \in (0, 1)$. Then there exists $\delta \in (0, \epsilon)$ such that for each (Ω)-program $\{x_t\}_{t=0}^{T_0}$ which satisfies*

$$(\pi^v + g)(x_0) \geq \sup(\pi^v + g) - \delta, \tag{3.14}$$

$$\sum_{t=0}^{T_0-1} (v(x_t, x_{t+1}) - v(\bar{x}_v, \bar{x}_v)) - \pi^v(x_0) + \pi^v(x_{T_0}) \geq -\delta \tag{3.15}$$

there exists a (v, Ω)-overtaking optimal program $\{z_t\}_{t=0}^{\infty}$ such that

$$(\pi^v + g)(z_0) = \sup(\pi^v + g), \tag{3.16}$$

$$\rho(z_t, x_t) \leq \epsilon, \ t = 0, \dots, T_0. \tag{3.17}$$

Proof. Assume that the lemma does not hold. Then there exist a sequence of real numbers $\{\delta_k\}_{k=1}^{\infty} \subset (0, 1]$ and a sequence of (Ω)-programs $\{x_t^{(k)}\}_{t=0}^{T_0}$, $k = 1, 2, \dots$ such that

$$\lim_{k \to \infty} \delta_k = 0 \tag{3.18}$$

and that for each natural number k and each (v, Ω)-overtaking optimal program $\{z_t\}_{t=0}^{\infty}$ satisfying (3.16),

$$(\pi^v + g)(x_0^{(k)}) \geq \sup(\pi^v + g) - \delta_k, \tag{3.19}$$

$$\sum_{t=0}^{T_0-1}(v(x_t^{(k)},x_{t+1}^{(k)})-v(\bar{x}_v,\bar{x}_v))-\pi^v(x_0^{(k)})+\pi^v(x_{T_0}^{(k)})\geq-\delta_k, \tag{3.20}$$

$$\max\{\rho(z_t,x_t^{(k)}):\ t=0,\dots,T_0\}>\epsilon. \tag{3.21}$$

By (3.19) and (3.20), for each natural numbers k, the values $\pi^v(x_0^{(k)})$ and $\pi^v(x_{T_0}^{(k)})$ are finite. Extracting a subsequence and re-indexing, if necessary, we may assume without loss of generality that for each integer $t\in[0,T_0]$ there exists

$$x_t=\lim_{k\to\infty}x_t^{(k)}. \tag{3.22}$$

By (3.22) and upper semicontinuity of g and π^v,

$$\pi^v(x_0)\geq\limsup_{k\to\infty}\pi^v(x_0^{(k)}),\ g(x_0)\geq\limsup_{k\to\infty}g(x_0^{(k)}). \tag{3.23}$$

It follows from Proposition 2.26, (3.19), and (3.23) that

$$\sup(\pi^v+g)\geq(\pi^v+g)(x_0)\geq\limsup_{k\to\infty}(\pi^v+g)(x_0^{(k)})\geq\sup(\pi^v+g), \tag{3.24}$$

$$\sup(\pi^v+g)=(\pi^v+g)(x_0)=\lim_{k\to\infty}(\pi^v+g)(x_0^{(k)}). \tag{3.25}$$

Relations (3.23) and (3.25) imply that

$$\pi^v(x_0)=\lim_{k\to\infty}\pi^v(x_0^{(k)}),\ g(x_0)=\lim_{k\to\infty}g(x_0^{(k)}). \tag{3.26}$$

It follows from upper semicontinuity of v and π^v (see Proposition 2.26), (3.18), (3.20), (3.22), and (3.26) that

$$\begin{aligned}&\sum_{t=0}^{T_0-1}(v(x_t,x_{t+1})-v(\bar{x}_v,\bar{x}_v))-\pi^v(x_0)+\pi^v(x_{T_0})\\&\geq\limsup_{k\to\infty}(\sum_{t=0}^{T_0-1}(v(x_t^{(k)},x_{t+1}^{(k)})-v(\bar{x}_v,\bar{x}_v))-\pi^v(x_0^{(k)})+\pi^v(x_{T_0}^{(k)}))\\&\geq\limsup_{k\to\infty}(-\delta_k)=0.\end{aligned} \tag{3.27}$$

Relations (3.25) and (3.27) imply that the values $\pi^v(x_0)$ and $\pi^v(x_{T_0})$ are finite. Combined with Proposition 2.19 and (3.27) this implies that for all integers $t=0,\dots,T_0-1$,

$$v(x_t,x_{t+1})-v(\bar{x}_v,\bar{x}_v)=\pi^v(x_t)-\pi^v(x_{t+1}). \tag{3.28}$$

Since $\pi^v(x_{T_0})$ is finite it follows from Theorem 2.13 that there exists a (v,Ω)-overtaking optimal and (v,Ω)-good program $\{\tilde{x}_t\}_{t=0}^{\infty}$ satisfying

$$\tilde{x}_0=x_{T_0}. \tag{3.29}$$

For all integers $t > T_0$ set

$$x_t = \tilde{x}_{t-T_0}. \tag{3.30}$$

Evidently, $\{x_t\}_{t=0}^{\infty}$ is an (Ω)-program. Since $\{\tilde{x}_t\}_{t=0}^{\infty}$ is (v,Ω)-overtaking optimal and (v,Ω)-good program it follows from Corollary 2.21 and (3.30) that (3.28) holds for all integers $t \geq 0$. Since $\{\tilde{x}_t\}_{t=0}^{\infty}$ is (v,Ω)-good program relations (3.28), (3.29), and Proposition 2.27 imply that $\{x_t\}_{t=0}^{\infty}$ is (v,Ω)-overtaking optimal program satisfying (3.25). By (3.22), for all sufficiently large natural numbers k, $\rho(x_t, x_t^{(k)}) \leq \epsilon/4$, $t = 0, \dots, T_0$. This contradicts (3.16) and (3.21). The contradiction we have reached proves Lemma 3.10. □

3.4 Proof of Theorem 3.4

We may assume that $\bar{r}_v \in (0,1)$. Recall that

$$\{(x,y) \in X \times X : \ \rho(x,\bar{x}_v), \ \rho(y,\bar{x}_v) \leq \bar{r}_v\} \subset \Omega. \tag{3.31}$$

By Lemma 3.10 applied to the function $\bar{v}$ there exists a real number

$$\delta_1 \in (0, \min\{\epsilon, \bar{r}_v/2\})$$

such that the following property holds:

(Pi) for each $(\bar{\Omega})$-program $\{y_t\}_{t=0}^{\tau_0}$ which satisfies

$$(\pi^{\bar{v}} + g)(y_0) \geq \sup(\pi^{\bar{v}} + g) - \delta_1, \tag{3.32}$$

$$\sum_{t=0}^{\tau_0 - 1} (\bar{v}(y_t, y_{t+1}) - \bar{v}(\bar{x}_v, \bar{x}_v)) - \pi^{\bar{v}}(y_0) + \pi^{\bar{v}}(y_{\tau_0}) \geq -\delta_1 \tag{3.33}$$

there exists an $(\bar{v},\bar{\Omega})$-overtaking optimal program $\{z_t\}_{t=0}^{\infty}$ such that

$$(\pi^{\bar{v}} + g)(z_0) = \sup(\pi^{\bar{v}} + g), \tag{3.34}$$

$$\rho(z_t, y_t) \leq \epsilon, \ t = 0, \dots, \tau_0. \tag{3.35}$$

Propositions 2.22 and 2.23 and assumption (A1) imply that there exists a real number $\delta_2 \in (0,\delta_1)$ such that for each point $z \in X$ satisfying $\rho(z,\bar{x}_v) \leq 2\delta_2$,

$$|\pi^{\bar{v}}(z)| = |\pi^{\bar{v}}(z) - \pi^{\bar{v}}(\bar{x}_v)| \leq \delta_1/8 \tag{3.36}$$

and for each $(x,y) \in X \times X$ satisfying $\rho(x,\bar{x}_v) \leq 2\delta_2$, $\rho(y,\bar{x}_v) \leq 2\delta_2$,

$$|v(x,y) - \bar{v}(\bar{x}_v, \bar{x}_v)| \leq \delta_1/8. \tag{3.37}$$

It follows from Theorem 3.2 that there exist a natural number L and a positive number δ_3 such that the following property holds:

(Pii) for each integer $T > 2L$, each $\{u_t\}_{t=0}^{T-1} \subset \mathcal{M}(\Omega)$ satisfying

$$\|u_t - v\| \le \delta_3,\ t = 0\dots, T-1,$$

each $h \in \mathcal{M}(X)$ satisfying

$$\|h - g\| \le 1$$

and each (Ω)-program $\{z_t\}_{t=0}^{T}$ which satisfies

$$z_0 \in \bar{Y}_{L_0},$$

$$h(z_T) + \sum_{t=0}^{T-1} u_t(z_t, z_{t+1}) \ge \sigma(h, \{u_t\}_{t=0}^{T-1}, 0, T, z_0) - M_0,$$

$$\sum_{t=\tau}^{\tau+L-1} u_t(z_t, z_{t+1}) \ge \sigma(\{u_t\}_{t=\tau}^{\tau+L-1}, \tau, \tau+L, z_\tau, z_{\tau+L}) - \delta_3$$

for each integer $\tau \in [0, T-L]$, we have $\rho(z_t, \bar{x}_v) \le \delta_2,\ t = L, \dots, T-L$.

Set

$$\mathcal{D}_0 = \{z \in X :\ (\pi^{\bar{v}} + g)(z) = \sup(\pi^{\bar{v}} + g)\}. \tag{3.38}$$

Proposition 2.22, (2.115), and (2.122) imply that the set $\mathcal{D}_0$ is nonempty and closed.

Let

$$z \in \mathcal{D}_0. \tag{3.39}$$

By (3.38) and (3.39),

$$\pi^{\bar{v}}(z) \ge \sup(\pi^{\bar{v}} + g) - \|g\|. \tag{3.40}$$

It follows from (3.40) and Theorem 2.13 that there exists a $(\bar{v}, \bar{\Omega})$-overtaking optimal and $(\bar{v}, \bar{\Omega})$-good program $\{z_t\}_{t=0}^{\infty}$ such that

$$z_0 = z \in \mathcal{D}_0. \tag{3.41}$$

By Proposition 2.24, for each $(\bar{v}, \bar{\Omega})$-overtaking optimal program $\{\xi_t\}_{t=0}^{\infty}$ such that $\xi_0 \in \mathcal{D}_0$ we have

$$\pi^{\bar{v}}(\xi_0) = \lim_{T\to\infty} \sum_{t=0}^{T-1} (\bar{v}(\xi_t, \xi_{t+1}) - \bar{v}(\bar{x}_v, \bar{x}_v)). \tag{3.42}$$

Lemma 2.8, (3.40), and (3.42) imply that there exists a natural number L_1 such that

$$\mathcal{D}_0 \subset Y_{L_1}. \tag{3.43}$$

By (3.43) and Theorem 2.15 applied to the function $\bar{v}$ there exists a natural number τ_1 such that for each $(\bar{v}, \bar{\Omega})$-overtaking optimal program $\{z_t\}_{t=0}^{\infty}$ satisfying $z_0 \in \mathcal{D}_0$ we have

$$\rho(z_t, \bar{x}_v) \le \delta_2 \text{ for all integers } t \ge \tau_1. \tag{3.44}$$

Choose a number $\delta > 0$ and a natural number T_0 such that

$$\delta < (16(L + \tau_1 + \tau_0 + 8))^{-1} \min\{\delta_1, \delta_2, \delta_3\}, \tag{3.45}$$

$$T_0 > 2L + 2\tau_0 + 2\tau_1 + 4. \tag{3.46}$$

Assume that an integer $T \geq T_0$, $h \in \mathcal{M}(X)$ satisfies

$$\|h - g\| \leq \delta, \tag{3.47}$$

$\{u_t\}_{t=0}^{T-1} \subset \mathcal{M}(\Omega)$ satisfies

$$\|u_t - v\| \leq \delta,\ t = 0 \dots, T - 1 \tag{3.48}$$

and $\{x_t\}_{t=0}^{T}$ is an (Ω)-program which satisfies

$$x_0 \in \bar{Y}_{L_0}, \tag{3.49}$$

$$h(x_T) + \sum_{t=0}^{T-1} u_t(x_t, x_{t+1}) \geq \sigma(h, \{u_t\}_{t=0}^{T-1}, 0, T, x_0) - M_0, \tag{3.50}$$

$$\sum_{t=\tau}^{\tau+T_0-1} u_t(x_t, x_{t+1}) \geq \sigma(\{u_t\}_{t=\tau}^{\tau+T_0-1}, \tau, \tau + T_0, x_\tau, x_{\tau+T_0}) - \delta \tag{3.51}$$

for each integer $\tau \in [0, T - T_0]$,

$$h(x_T) + \sum_{t=T-T_0}^{T-1} u_t(x_t, x_{t+1}) \geq \sigma(h, \{u_t\}_{t=T-T_0}^{T-1}, T - T_0, T, x_{T-T_0}) - \delta. \tag{3.52}$$

It follows from (3.45)–(3.49), (3.51), and the property (Pii) that

$$\rho(x_t, \bar{x}_v) \leq \delta_2,\ t = L, \dots, T - L. \tag{3.53}$$

By (3.46),

$$[T - L - \tau_0 - \tau_1 - 4, T - L - \tau_0 - \tau_1] \subset [L, T - L - \tau_0 - \tau_1]. \tag{3.54}$$

In view of (3.53) and (3.54),

$$\rho(x_t, \bar{x}_v) \leq \delta_2,\ t \in \{T - L - \tau_0 - \tau_1 - 4, T - L - \tau_0 - \tau_1\}. \tag{3.55}$$

There exists (see (3.38)–(3.41)) an $(\bar{v}, \bar{\Omega})$-overtaking optimal program $\{z_t\}_{t=0}^{\infty}$ such that

$$z_0 \in \mathcal{D}_0,\ (\pi^{\bar{v}} + g)(z_0) = \sup(\pi^{\bar{v}} + g). \tag{3.56}$$

Then (3.44) holds.

Define a sequence $\{\tilde{x}_t\}_{t=0}^{T}$ by

$$\tilde{x}_t = x_t,\ t = 0, \dots, T - L - \tau_0 - \tau_1 - 4,$$

$$\tilde{x}_t = z_{T-t},\ t = T - L - \tau_0 - \tau_1 - 3, \dots, T. \tag{3.57}$$

Relations (3.44) and (3.57) imply that

$$\rho(\tilde{x}_{T-L-\tau_0-\tau_1-3}, \bar{x}_v) = \rho(z_{L+\tau_0+\tau_1+3}, \bar{x}_v) \leq \delta_2. \tag{3.58}$$

It follows from (3.31), (3.55), (3.57), and (3.58) that $\{\tilde{x}_t\}_{t=0}^{T}$ is an (Ω)-program. By (3.46)–(3.48), (3.52), and (3.57),

$$\begin{aligned}
\delta &\geq \sigma(h, \{u_t\}_{t=T-T_0}^{T-1}, T-T_0, T, x_{T-T_0}) - h(x_T) - \sum_{t=T-T_0}^{T-1} u_t(x_t, x_{t+1}) \\
&\geq h(\tilde{x}_T) + \sum_{t=T-T_0}^{T-1} u_t(\tilde{x}_t, \tilde{x}_{t+1}) - h(x_T) - \sum_{t=T-T_0}^{T-1} u_t(x_t, x_{t+1}) \\
&= h(\tilde{x}_T) - h(x_T) + \sum_{t=T-L-\tau_0-\tau_1-4}^{T-1} (u_t(\tilde{x}_t, \tilde{x}_{t+1}) - u_t(x_t, x_{t+1})) \\
&\geq g(\tilde{x}_T) - g(x_T) + \sum_{t=T-L-\tau_0-\tau_1-4}^{T-1} v(\tilde{x}_t, \tilde{x}_{t+1}) \\
&\quad - \sum_{t=T-L-\tau_0-\tau_1-4}^{T-1} v(x_t, x_{t+1}) - 2\delta(L+\tau_0+\tau_1+6).
\end{aligned}$$

Combined with (3.55) this implies that

$$\begin{aligned}
&g(x_T) + \sum_{t=T-L-\tau_0-\tau_1-4}^{T-1} v(x_t, x_{t+1}) \\
&\geq g(\tilde{x}_T) + \sum_{t=T-L-\tau_0-\tau_1-4}^{T-1} v(\tilde{x}_t, \tilde{x}_{t+1}) - 2\delta(L+\tau_0+\tau_1+8) \\
&= g(z_0) + v(x_{T-L-\tau_0-\tau_1-4}, z_{L+\tau_0+\tau_1+3}) \\
&\quad + \sum_{t=0}^{L+\tau_0+\tau_1+2} \bar{v}(z_t, z_{t+1}) - 2\delta(L+\tau_0+\tau_1+8).
\end{aligned} \tag{3.59}$$

In view of (3.37), (3.44), and (3.55),

$$|\bar{v}(z_{L+\tau_0+\tau_1+3}, x_{T-L-\tau_0-\tau_1-4}) - \bar{v}(z_{L+\tau_0+\tau_1+3}, z_{L+\tau_0+\tau_1+4})| \leq \delta_1/4.$$

Combined with (3.59) this implies that

$$\begin{aligned}
&g(x_T) + \sum_{t=T-L-\tau_0-\tau_1-4}^{T-1} v(x_t, x_{t+1}) \\
&\geq g(z_0) + \sum_{t=0}^{L+\tau_0+\tau_1+3} \bar{v}(z_t, z_{t+1}) - \delta_1/4 - 2\delta(L+\tau_0+\tau_1+8).
\end{aligned} \tag{3.60}$$

Set

$$y_t = x_{T-t},\ t = 0, \dots, L + \tau_0 + \tau_1 + 4. \tag{3.61}$$

By (3.45), (3.60), and (3.61),

$$\begin{aligned}
&g(y_0) + \sum_{t=0}^{L+\tau_0+\tau_1+3} \bar{v}(y_t, y_{t+1}) \\
&= g(x_T) + \sum_{t=0}^{L+\tau_0+\tau_1+3} v(x_{T-t-1}, x_{T-t}) \\
&= g(x_T) + \sum_{t=T-L-\tau_0-\tau_1-4}^{T-1} v(x_t, x_{t+1}) \\
&\geq g(z_0) + \sum_{t=0}^{L+\tau_0+\tau_1+3} \bar{v}(z_t, z_{t+1}) - 3\delta_1/8.
\end{aligned} \tag{3.62}$$

By (3.56), (3.62), Proposition 2.19, and Corollary 2.21 and $(\bar{v}, \bar{\Omega})$-overtaking optimality of $\{z_t\}_{t=0}^{\infty}$,

$$\begin{aligned}
&(\pi^{\bar{v}} + g)(y_0) - \sup(\pi^{\bar{v}} + g) \\
&+ \sum_{t=0}^{\tau_0-1} (\bar{v}(y_t, y_{t+1}) - \bar{v}(\bar{x}_v, \bar{x}_v)) - \pi^{\bar{v}}(y_0) + \pi^{\bar{v}}(y_{\tau_0}) \\
&\geq (\pi^{\bar{v}} + g)(y_0) - (\pi^{\bar{v}} + g)(z_0) \\
&+ \sum_{t=0}^{L+\tau_0+\tau_1+3} (\bar{v}(y_t, y_{t+1}) - \bar{v}(\bar{x}_v, \bar{x}_v)) - \pi^{\bar{v}}(y_0) + \pi^{\bar{v}}(y_{L+\tau_0+\tau_1+4}) \\
&\geq \pi^{\bar{v}}(y_0) - \pi^{\bar{v}}(z_0) + \sum_{t=0}^{L+\tau_0+\tau_1+3} (\bar{v}(z_t, z_{t+1}) - \bar{v}(\bar{x}_v, \bar{x}_v)) \\
&-\pi^{\bar{v}}(y_0) + \pi^{\bar{v}}(y_{L+\tau_0+\tau_1+4}) - 3\delta_1/8 \\
&= -\pi^{\bar{v}}(z_{L+\tau_0+\tau_1+4}) - \pi^{\bar{v}}(y_{L+\tau_0+\tau_1+4}) - 3\delta_1/8.
\end{aligned} \tag{3.63}$$

In view of (3.36) and (3.44),

$$\pi^{\bar{v}}(z_{L+\tau_0+\tau_1+4}) \leq \delta_1/8. \tag{3.64}$$

By (3.36), (3.55), and (3.61),

$$|\pi^{\bar{v}}(y_{L+\tau_0+\tau_1+4})| = |\pi^{\bar{v}}(x_{T-L-\tau_0-\tau_1-4})| \leq \delta_1/8.$$

Combined with (3.63) and (3.64) this implies that

$$\begin{aligned}
&(\pi^{\bar{v}} + g)(y_0) - \sup(\pi^{\bar{v}} + g) \\
&+ \sum_{t=0}^{\tau_0-1} (\bar{v}(y_t, y_{t+1}) - \bar{v}(\bar{x}_v, \bar{x}_v)) - \pi^{\bar{v}}(y_0) + \pi^{\bar{v}}(y_{\tau_0}) \geq -\delta_1.
\end{aligned}$$

Together with (3.53) and Proposition 2.19 this implies that

$$(\pi^{\bar{v}} + g)(y_0) - \sup(\pi^{\bar{v}} + g) \geq -\delta_1, \tag{3.65}$$

$$\sum_{t=0}^{\tau_0 - 1} (\bar{v}(y_t, y_{t+1}) - \bar{v}(\bar{x}_v, \bar{x}_v)) - \pi^{\bar{v}}(y_0) + \pi^{\bar{v}}(y_{\tau_0}) \geq -\delta_1. \tag{3.66}$$

It follows from (3.65), (3.66), and the property (Pi) that there exists an $(\bar{v}, \bar{\Omega})$-overtaking optimal program $\{\xi_t\}_{t=0}^{\infty}$ such that

$$(\pi^{\bar{v}} + g)(\xi_0) = \sup(\pi^{\bar{v}} + g),$$

$\rho(\xi_t, x_{T-t}) = \rho(\xi_t, y_t) \leq \epsilon$, $t = 0, \dots, \tau_0$. Theorem 3.4 is proved. □

3.5 Proof of Proposition 3.7

Since the mapping $v \to \bar{v}$, $v \in \mathcal{M}(\Omega)$ is an isometry Proposition 3.7 follows from Proposition 2.31 and the following result.

Proposition 3.11. *Suppose that $g \in \mathcal{M}_u(X)$ and $v \in \mathcal{M}(\Omega)$ is an upper semicontinuous function, $\bar{x}_v \in X$, $\bar{r}_v > 0$, $\bar{c}_v > 0$ and that assumptions (A1), (A2) and (A3) hold. Let $\tau_0 \geq 1$ be an integer and $\epsilon > 0$. Then there exist $\delta > 0$ and an integer $T_0 \geq \tau_0$ such that for each $u \in B_d(v, \delta) \cap \mathcal{M}_*(\Omega)$ and each $h \in \mathcal{M}_u(X)$ satisfying $\|h - g\| \leq \delta$ the following property holds:*

for each $\{x_t\}_{t=0}^{\infty} \in \tilde{\mathcal{P}}(h, \bar{u})$ there is $\{y_t\}_{t=0}^{\infty} \in \tilde{\mathcal{P}}(g, \bar{v})$ such that $\rho(x_t, y_t) \leq \epsilon$ for all integers $t = 0, \dots, \tau_0$.

Proof. By (A2), for each $u \in \mathcal{M}_0(\Omega)$ and each integer $T \geq 1$,

$$\sigma(u, T, \bar{x}_u, \bar{x}_u) = Tu(\bar{x}_u, \bar{x}_u), \tag{3.67}$$

$$u(\bar{x}_u, \bar{x}_u) \geq u(z, z) \text{ for all } z \in X \text{ such that } (z, z) \in \Omega. \tag{3.68}$$

We show that together with (3.68) and Assertion 1 of Theorem 2.5 this implies that there exists $\delta_0 \in (0, \epsilon)$ such that for each $u \in B_d(v, \delta_0) \cap \mathcal{M}_0(\Omega)$,

$$\rho(\bar{x}_u, \bar{x}_v) \leq \bar{r}_v / 4. \tag{3.69}$$

By Assertion 1 of Theorem 2.5, there exist a positive number $\tilde{\delta} < 1$ and a natural number $\tilde{L}$ such that the following property holds:

(i) for each integer $T \geq \tilde{L}$, each $\{u_t\}_{t=0}^{T-1} \subset \mathcal{M}(\Omega)$ and each (Ω)-program $\{x_t\}_{t=0}^{T}$ which satisfy

$$\|u_t - v\| \leq \tilde{\delta}, \ t = 0, \dots, T - 1,$$

$$\sum_{t=0}^{T-1} u_t(x_t, x_{t+1}) \geq \sigma(\{u_t\}_{t=0}^{T-1}, 0, T) - 2\bar{c}_v - 4$$

we have

$$\text{Card}(\{t \in \{0,\dots,T\} : \rho(x_t, \bar{x}_v) > 8^{-1}\bar{r}_v\}) < \tilde{L}.$$

Choose a positive number δ_0 such that

$$\delta_0 < \tilde{\delta},\ \delta_0 < \epsilon,\ \delta_0 < (2\tilde{L})^{-1}. \tag{3.70}$$

Assume that

$$u \in B_d(v, \delta_0) \cap \mathcal{M}_0(\Omega). \tag{3.71}$$

In view of (3.71), relation (3.68) holds. Set

$$\xi_t = \bar{x}_u,\ t = 0, 1, \dots. \tag{3.72}$$

We show that (3.69) holds. Assume the contrary. Then

$$\rho(\xi_t, \bar{x}_v) > \bar{r}_v/4,\ t = 0, 1, \dots. \tag{3.73}$$

By (3.70), (3.71), (3.73), and property (i), for each integer $k \geq 0$,

$$\sum_{t=k\tilde{L}}^{(k+1)\tilde{L}-1} u(\xi_t, \xi_{t+1}) < \sigma(u, 0, \tilde{L}) - 2\bar{c}_v - 4. \tag{3.74}$$

It follows from (3.70), (3.71), and (A2) that

$$\sigma(u, 0, \tilde{L}) \leq \sigma(v, 0, \tilde{L}) + \delta_0\tilde{L} \leq \tilde{L}v(\bar{x}_v, \bar{x}_v) + \bar{c}_v + \delta_0\tilde{L}$$

$$\leq \tilde{L}u(\bar{x}_v, \bar{x}_v) + \bar{c}_v + 2\delta_0\tilde{L} \leq \tilde{L}u(\bar{x}_v, \bar{x}_v) + \bar{c}_v + 1. \tag{3.75}$$

Relations (3.74) and (3.75) imply that for each integer $k \geq 0$,

$$\sum_{t=k\tilde{L}}^{(k+1)\tilde{L}-1} u(\xi_t, \xi_{t+1}) \leq \tilde{L}u(\bar{x}_v, \bar{x}_v) - \bar{c}_v - 3.$$

Together with (3.72) this implies that

$$\tilde{L}u(\bar{x}_u, \bar{x}_u) \leq \tilde{L}u(\bar{x}_v, \bar{x}_v) - \bar{c}_v - 3,$$

$$u(\bar{x}_u, \bar{x}_u) < u(\bar{x}_v, \bar{x}_v).$$

This contradicts (3.68). The contradiction we have reached proves (3.69). Thus we have shown that for each function u satisfying (3.71) relation (3.69) is valid.

By Theorem 3.4, there exist $\delta \in (0, \delta_0)$ and a natural number $T_0 \geq \tau_0$ such that the following property holds:

(ii) for each integer $T \geq T_0$, each $h \in \mathcal{M}(X)$ satisfying $\|h - g\| \leq \delta$, each $\{u_t\}_0^{T-1} \subset \mathcal{M}_0(\Omega)$ satisfying

$$\|u_t - v\| \leq \delta,\ t = 0, \dots, T-1$$

and each (Ω)-program $\{z_t\}_{t=0}^T$ which satisfies

$$\rho(z_0, \bar{x}_v) \le \bar{r}_v/2,$$

$$h(z_T) + \sum_{t=0}^{T-1} u_t(z_t, z_{t+1}) \ge \sigma(h, \{u_t\}_{t=0}^{T-1}, 0, T, z_0) - \delta$$

there exists $\{x_t^*\}_{t=0}^\infty \in \tilde{\mathcal{P}}(g, \bar{v})$ such that

$$\rho(z_{T-t}, x_t^*) \le \epsilon, \ t = 0, \dots, \tau_0.$$

Assume that

$$u \in \mathcal{M}_*(\Omega), \ \|u - v\| \le \delta, \ h \in \mathcal{M}_u(X), \ \|h - g\| \le \delta, \tag{3.76}$$

$$\{x_t\}_{t=0}^\infty \in \tilde{\mathcal{P}}(h, \bar{u}). \tag{3.77}$$

In view of (3.76), (3.77) and (A3),

$$\lim_{t \to \infty} x_t = \bar{x}_u. \tag{3.78}$$

It follows from (3.70), (3.76), and the choice of δ_0 that (3.69) holds. By (3.78), there exists an integer $S_0 > T_0$ such that

$$\rho(x_{S_0}, \bar{x}_u) \le \bar{r}_v/4.$$

Together with (3.69) this implies that

$$\rho(x_{S_0}, \bar{x}_v) \le \bar{r}_v/2. \tag{3.79}$$

Set

$$z_t = x_{S_0 - t}, \ t = 0, \dots, S_0. \tag{3.80}$$

By (3.79) and (3.80),

$$\rho(z_0, \bar{x}_v) \le \bar{r}_v/2. \tag{3.81}$$

It is clear that $\{z_t\}_{t=0}^{S_0}$ is an (Ω)-program. We show that

$$h(z_{S_0}) + \sum_{t=0}^{S_0-1} u(z_t, z_{t+1}) = \sigma(h, u, S_0, z_0). \tag{3.82}$$

Let $\{y_t\}_{t=0}^{S_0}$ be an (Ω)-program satisfying

$$y_0 = z_0. \tag{3.83}$$

In order to prove (3.82) it is sufficient to show that

$$h(z_{S_0}) + \sum_{t=0}^{S_0-1} u(z_t, z_{t+1}) \ge h(y_{S_0}) + \sum_{t=0}^{S_0-1} u(y_t, y_{t+1}).$$

It follows from (3.80) that

$$h(z_{S_0}) + \sum_{t=0}^{S_0-1} u(z_t, z_{t+1}) = h(x_0) + \sum_{t=0}^{S_0-1} \bar{u}(x_{S_0-t-1}, x_{S_0-t}). \tag{3.84}$$

Set

$$\bar{y}_t = y_{S_0-t},\ t = 0, \ldots, S_0. \tag{3.85}$$

In view of (3.85),

$$h(y_{S_0}) + \sum_{t=0}^{S_0-1} u(y_t, y_{t+1}) = h(\bar{y}_0) + \sum_{t=0}^{S_0-1} \bar{u}(\bar{y}_t, \bar{y}_{t+1}). \tag{3.86}$$

In view of (3.77), (3.79), (3.80), (3.83)–(3.86), Proposition 2.19, Corollary 2.21, and $(\bar{u}, \bar{\Omega})$-overtaking optimality of $\{x_t\}_{t=0}^{\infty}$ (see (3.77)),

$$h(z_{S_0}) + \sum_{t=0}^{S_0-1} u(z_t, z_{t+1}) - [h(y_{S_0}) + \sum_{t=0}^{S_0-1} u(y_t, y_{t+1})]$$

$$= h(x_0) + \sum_{t=0}^{S_0-1} \bar{u}(x_t, x_{t+1}) - [h(\bar{y}_0) + \sum_{t=0}^{S_0-1} \bar{u}(\bar{y}_t, \bar{y}_{t+1})]$$

$$= h(x_0) + \sum_{t=0}^{S_0-1} [\bar{u}(x_t, x_{t+1}) - u(\bar{x}_u, \bar{x}_u) - \pi^{\bar{u}}(x_t) + \pi^{\bar{u}}(x_{t+1})]$$

$$+\pi^{\bar{u}}(x_0) - \pi^{\bar{u}}(x_{S_0})$$

$$-(h(\bar{y}_0) + \sum_{t=0}^{S_0-1} [\bar{u}(\bar{y}_t, \bar{y}_{t+1}) - u(\bar{x}_u, \bar{x}_u) - \pi^{\bar{u}}(\bar{y}_t) + \pi^{\bar{u}}(\bar{y}_{t+1})]$$

$$+\pi^{\bar{u}}(\bar{y}_0) - \pi^{\bar{u}}(\bar{y}_{S_0}))$$

$$= h(x_0) + \pi^{\bar{u}}(x_0) - h(\bar{y}_0) - \pi^{\bar{u}}(\bar{y}_0)$$

$$- \sum_{t=0}^{S_0-1} [\bar{u}(\bar{y}_t, \bar{y}_{t+1}) - u(\bar{x}_u, \bar{x}_u) - \pi^{\bar{u}}(\bar{y}_t) + \pi^{\bar{u}}(\bar{y}_{t+1})]$$

$$\geq h(x_0) + \pi^{\bar{u}}(x_0) - h(\bar{y}_0) - \pi^{\bar{u}}(\bar{y}_0) \geq 0.$$

Thus (3.82) holds. By (3.76), (3.81), (3.82), the inequality $S_0 > T_0$ and the property (ii) applied to $\{z_t\}_{t=0}^{S_0}$ there exists an $\{x_t^*\}_{t=0}^{\infty} \in \tilde{\mathcal{P}}(g, \bar{v})$ such that for all $t = 0, \ldots, \tau_0$,

$$\epsilon \geq \rho(z_{S_0-t}, x_t^*) = \rho(x_t^*, x_t).$$

Proposition 3.11 is proved. □

3.6 Proof of Theorem 3.6

For any $(x, y) \in X \times X$ and any nonempty set $D \subset X \times X$ put

$$\rho((x,y), D) = \inf\{\rho(x, z_1) + \rho(y, z_2) : \ (z_1, z_2) \in D\}.$$

Since the mapping $v \to \bar{v}$, $v \in \mathcal{M}(\Omega)$ is an isometry Theorem 3.6 follows from Propositions 2.30, 2.31 and the following result.

Proposition 3.12. *Let $\mathfrak{M}$ be either $\bar{\mathcal{M}}_0(\Omega)$ or $\bar{\mathcal{M}}_{c,0}(\Omega)$ and $g \in \mathcal{M}_u(X)$. Then there exists a set $\mathcal{F} \subset \mathfrak{M} \cap \mathcal{M}_*(\Omega)$ which is a countable intersection of open everywhere dense subsets of $\mathfrak{M}$ such that for each $v \in \mathcal{F}$ there exists a unique point $z_v \in X$ such that*

$$(\pi^v + g)(z_v) = \sup(\pi^v + g)$$

and there exist a unique (v, Ω)-overtaking optimal program $\{z_t^v\}_{t=0}^\infty$ satisfying $z_0^v = z_v$.

Let $\mathfrak{M}$ be either $\bar{\mathcal{M}}_0(\Omega)$ or $\bar{\mathcal{M}}_{c,0}(\Omega)$. By Theorem 2.6, there exists a set $\mathcal{F}_0 \subset \mathfrak{M} \cap \mathcal{M}_*(\Omega)$ which is a countable intersection of open everywhere dense subsets of $\mathfrak{M}$.

For each $g \in \mathcal{M}_u(X)$ denote by E_g the set of all $v \in \mathfrak{M} \cap \mathcal{M}_*(\Omega)$ for which the following property holds:

(Pi) there exists a unique point $z_v \in X$ such that

$$(\pi^v + g)(z_v) = \sup(\pi^v + g)$$

and there exists a unique (v, Ω)-overtaking optimal program $\{z_t^v\}_{t=0}^\infty$ satisfying $z_0^v = z_v$.

We proceed the proof of Proposition 3.12 with the following auxiliary result.

Lemma 3.13. *For each $g \in \mathcal{M}_u(X)$, E_g is an everywhere dense subset of $\mathfrak{M}$.*

Proof. Let $v \in \mathfrak{M} \cap \mathcal{M}_*(\Omega)$. It is sufficient to show that for any neighborhood $\mathcal{U}$ of v in $\mathfrak{M}$, $\mathcal{U} \cap E_g \neq \emptyset$. There are two cases:

$$(\pi^v + g)(\bar{x}_v) = \sup(\pi^v + g); \tag{3.87}$$

$$(\pi^v + g)(\bar{x}_v) < \sup(\pi^v + g). \tag{3.88}$$

Assume that (3.87) holds. Let $\gamma \in (0, 1)$. Define

$$v_\gamma(x, y) = v(x, y) - \gamma(\rho(x, \bar{x}_v) + \rho(y, \bar{x}_v)), \ (x, y) \in \Omega. \tag{3.89}$$

It is easy to see that

$$v_\gamma \in \mathfrak{M} \cap \mathcal{M}_0(\Omega) \text{ with } \bar{x}_{v_\gamma} = \bar{x}_v. \tag{3.90}$$

In view of (3.89), every (v_γ, Ω)-good program $\{x_t\}_{t=0}^\infty$ is (v, Ω)-good and

$$\lim_{t\to\infty} x_t = \bar{x}_v.$$

Hence

$$v_\gamma \in \mathcal{M}_*(\Omega). \tag{3.91}$$

Proposition 2.22 and (3.89)–(3.91) imply that

$$v_\gamma(\bar{x}_v, \bar{x}_v) = v(\bar{x}_v, \bar{x}_v),\ \pi^{v_\gamma}(y) \le \pi^v(y),\ y \in X,\ \pi^{v_\gamma}(\bar{x}_v) = \pi^v(\bar{x}_v) = 0,$$

$$(\pi^{v_\gamma} + g)(y) \le (\pi^v + g)(y),\ y \in X,\ (\pi^{v_\gamma} + g)(\bar{x}_v) = (\pi^v + g)(\bar{x}_v). \tag{3.92}$$

Assume that a point $z \in X$ satisfies

$$(\pi^{v_\gamma} + g)(z) = \sup(\pi^{v_\gamma} + g) \tag{3.93}$$

and that $\{z_t\}_{t=0}^\infty$ is a (v_γ, Ω)-overtaking optimal program such that

$$z_0 = z. \tag{3.94}$$

In view of (3.87), (3.89)–(3.94), and Proposition 2.24,

$$\sup(\pi^v + g) - g(z) = \sup(\pi^{v_\gamma} + g) - g(z)$$

$$= \pi^{v_\gamma}(z) = \lim_{T\to\infty} \sum_{t=0}^{T-1} [v_\gamma(z_t, z_{t+1}) - v_\gamma(\bar{x}_v, \bar{x}_v)]$$

$$= \lim_{T\to\infty} [\sum_{t=0}^{T-1} v(z_t, z_{t+1}) - Tv(\bar{x}_v, \bar{x}_v) - \gamma \sum_{t=0}^{T-1} (\rho(z_t, \bar{x}_v) + \rho(z_{t+1}, \bar{x}_v))]$$

$$\le \limsup_{T\to\infty} \sum_{t=0}^{T-1} (v(z_t, z_{t+1}) - v(\bar{x}_v, \bar{x}_v)) - \gamma \sum_{t=0}^{\infty} (\rho(z_t, \bar{x}_v) + \rho(z_{t+1}, \bar{x}_v))$$

$$\le \pi^v(z) - \gamma \sum_{t=0}^{\infty} (\rho(z_t, \bar{x}_v) + \rho(z_{t+1}, \bar{x}_v))$$

$$= -\gamma \sum_{t=0}^{\infty} (\rho(z_t, \bar{x}_v) + \rho(z_{t+1}, \bar{x}_v)) + (\pi^v + g)(z) - g(z).$$

This implies that $z_t = \bar{x}_v$ for all integers $t \ge 0$ and $v_\gamma \in E_g$.

Assume that (3.88) holds. There exist a point $z_* \in X$ which satisfies

$$(\pi^v + g)(z_*) = \sup(\pi^v + g) \tag{3.95}$$

and a (v, Ω)-overtaking optimal program $\{z_t^*\}_{t=0}^\infty$ satisfying

$$z_0^* = z_*.$$

By Proposition 2.24 and (3.95),

$$\pi^v(z_*) = \lim_{T\to\infty} \sum_{t=0}^{T-1} [v(z_t^*, z_{t+1}^*) - v(\bar{x}_v, \bar{x}_v)]. \tag{3.96}$$

Assumption (A3) imply that

$$\lim_{t\to\infty} z_t^* = \bar{x}_v. \tag{3.97}$$

By (3.88), (3.95), (3.97), and Proposition 2.26,

$$(\pi^v + g)(z_t^*) < (\pi^v + g)(z_0^*) \text{ for all large enough integers } t \geq 1. \tag{3.98}$$

In view of (3.98), there exists an integer $\tau_0 \geq 0$ such that

$$(\pi^v + g)(z_{\tau_0}^*) = (\pi^v + g)(z_*)$$

and that

$$(\pi^v + g)(z_t^*) < (\pi^v + g)(z_*)$$

for all integers $t > \tau_0$. We may assume without loss of generality $\tau_0 = 0$. Thus

$$(\pi^v + g)(z_0^*) = \sup(\pi^v + g),\ (\pi^v + g)(z_t^*) < (\pi^v + g)(z_0^*) \text{ for all integers } t \geq 1. \tag{3.99}$$

Let $\gamma \in (0, 1)$. For all $(x, y) \in \Omega$ define

$$v_\gamma(x, y) = v(x, y) - \gamma\rho((x, y), (\{(z_t^*, z_{t+1}^*) : t = 0, 1, \dots\} \cup \{(\bar{x}_v, \bar{x}_v)\})). \tag{3.100}$$

By (3.100),

$$v_\gamma \in \mathfrak{M} \cap \mathcal{M}_0(\Omega),\ v_\gamma(\bar{x}_v, \bar{x}_v) = v(\bar{x}_v, \bar{x}_v),$$

every (v_γ, Ω)-good program is (v, Ω)-good and converges to $\bar{x}_v$. Hence

$$v_\gamma \in \mathcal{M}_*(\Omega),\ \bar{x}_{v_\gamma} = \bar{x}_v. \tag{3.101}$$

It follows from (3.95), (3.96), (3.100), (3.101), and the equality $z_0^* = z_*$ that

$$\pi^{v_\gamma}(y) \leq \pi^v(y),\ y \in X,\ \pi^{v_\gamma}(z_*) = \pi^v(z_*),$$

$$\sup(\pi^v + g) = \sup(\pi^{v_\gamma} + g). \tag{3.102}$$

By (3.101), for all natural numbers T,

$$\sigma(v_\gamma, T, \bar{x}_v, \bar{x}_v) = Tv(\bar{x}_v, \bar{x}_v). \tag{3.103}$$

Proposition 2.22 and (3.101) imply that

$$\pi^{v_\gamma}(\bar{x}_v) = 0. \tag{3.104}$$

Set

$$K = \{(z_t^*, z_{t+1}^*) : t = 0, 1, \dots\} \cup \{(\bar{x}_v, \bar{x}_v)\}. \tag{3.105}$$

Assume that $y \in X$ satisfies

$$(\pi^{v_\gamma} + g)(y) = \sup(\pi^{v_\gamma} + g) \tag{3.106}$$

and that $\{y_t\}_{t=0}^{\infty}$ is a (v_γ, Ω)-overtaking optimal program satisfying

$$y_0 = y. \tag{3.107}$$

By (3.95), (3.100)–(3.102), (3.105)–(3.107), Proposition 2.24 and since the program $\{y_t\}_{t=0}^{\infty}$ is (v_γ, Ω)-overtaking optimal we have

$$\sup(\pi^v + g) = (\pi^v + g)(z_*) = (\pi^{v_\gamma} + g)(z_*) = (\pi^{v_\gamma} + g)(y)$$

$$= \lim_{T\to\infty} \sum_{t=0}^{T-1} [v_\gamma(y_t, y_{t+1}) - v_\gamma(\bar{x}_v, \bar{x}_v)] + g(y)$$

$$= \lim_{T\to\infty} [\sum_{t=0}^{T-1} (v(y_t, y_{t+1}) - v(\bar{x}_v, \bar{x}_v) - \gamma\rho((y_t, y_{t+1}), K)] + g(y)$$

$$\le \pi^v(y) - \gamma \sum_{t=0}^{\infty} \rho((y_t, y_{t+1}), K) + g(y).$$

Together with (3.88) and (3.103) this implies that

$$(\pi^v + g)(y) = (\pi^v + g)(z_*), \tag{3.108}$$

$$(y_t, y_{t+1}) \in K \text{ for all integers } t \ge 0. \tag{3.109}$$

In view of (3.88), (3.95), (3.99), (3.105), (3.107)–(3.109), and the equality $z_0^* = z_*$,

$$y = y_0 = z_*. \tag{3.110}$$

We show by induction that $y_t = z_t^*$ for all integers $t \ge 0$. There are two cases:

$$z_t^* \ne \bar{x}_v \text{ for all integers } t \ge 0; \tag{3.111}$$

$$\bar{x}_v \in \{z_t^* : t = 0, 1, \dots\}. \tag{3.112}$$

Assume that (3.111) holds. By Proposition 2.39, (3.95), (3.96), and (3.111),

$$z_{t_1}^* \ne z_{t_2}^* \text{ for all integers } t_2 > t_1 \ge 0. \tag{3.113}$$

Assume that $T \ge 0$ is an integer and that

$$y_t = z_t^*, \ t = 0, \dots, T. \tag{3.114}$$

(Note that in view of (3.110) and the equality $z_0^* = z_*$ our assumption holds for $T = 0$.) By (3.105), (3.109), (3.111), (3.113), and (3.114),

$$(z_T^*, y_{T+1}) = (y_T, y_{T+1}) \in K = \{(z_t^*, z_{t+1}^*) : t = 0, 1, \dots\} \cup \{(\bar{x}_v, \bar{x}_v)\}$$

and $y_{T+1} = z_{T+1}^*$. Thus $y_t = z_t^*$ for all integers $t \ge 0$.

Assume that (3.112) holds. By (3.88), (3.95), and the equality $z_0^* = z_*$, there is a natural number S such that

$$z_S^* = \bar{x}_v,\ z_t^* \neq \bar{x}_v \text{ for all integers } t \in [0, S). \tag{3.115}$$

Propositions 2.39 and 2.40 imply that

$$z_t^* = \bar{x}_v \ \text{ for all integers } t \geq S, \tag{3.116}$$

$$z_{t_2}^* \neq z_{t_1}^* \text{ for all integers } t_1, t_2 \in [0, S] \text{ such that } t_1 < t_2. \tag{3.117}$$

Assume that $T \geq 0$ is an integer and that

$$y_t = z_t^*,\ t = 0, \dots, T. \tag{3.118}$$

(Note that in view of (3.110), our assumption holds for $T = 0$.) If $T < S$, then by (3.105), (3.109), (3.115), and (3.118),

$$(z_T^*, y_{T+1}) = (y_T, y_{T+1}) \in \{(z_t^*, z_{t+1}^*) : \ t = 0, 1, \dots\} \cup \{(\bar{x}_v, \bar{x}_v)\}$$

and $y_{T+1} = z_{T+1}^*$. If $T \geq S$, then by (3.115), (3.109), (3.105), (3.116) and (3.118),

$$(\bar{x}_v, y_{T+1}) = (y_T, y_{T+1}) \in \{(z_t^*, z_{t+1}^*) : \ t = 0, 1, \dots\} \cup \{(\bar{x}_v, \bar{x}_v)\}$$

and $y_{T+1} = \bar{x}_v = z_{T+1}^*$. Thus $y_t = z_t^*$ for all integers $t \geq 0$ in the both cases (see (3.111), (3.112)). This implies that $v_\gamma \in E_g$. Therefore the inclusion above holds in the both cases (see (3.87), (3.88)). Since $v_\gamma \to v$ as $\gamma \to 0^+$ in $\mathfrak{M}$ we conclude that for any neighborhood $\mathcal{U}$ of v in $\mathfrak{M}$, $\mathcal{U} \cap E_g \neq \emptyset$. Thus E_g is an everywhere dense subset of $\mathfrak{M}$. Lemma 3.13 is proved. □

Completion of the Proof of Proposition 3.12. By definition, for every $v \in E_g$, there exist a unique (v, Ω)-overtaking optimal program $\{z_t^v\}_{t=0}^\infty$ satisfying

$$(\pi^v + g)(z_0^v) = \sup(\pi^v + g).$$

Let $v \in E_g$ and $k \geq 1$ be an integer. By Proposition 3.7, there exist an open neighborhood $\mathcal{U}(v, k)$ of v in $\mathfrak{M}$ and an integer $T(v, k) \geq k$ such that the following property holds:

(Pii) for each $u \in \mathcal{U}(v, k) \cap \mathcal{M}_*(\Omega)$ and each $\{x_t\}_{t=0}^\infty \in \tilde{\mathcal{P}}(g, u)$ we have $\rho(x_t, z_t^v) \leq k^{-1}$, $t = 0, \dots, k$.

Set

$$\mathcal{F}_1 = \cap_{p=1}^\infty \cup \{\mathcal{U}(v, k) : \ v \in E_g,\ k \geq p\},\ \mathcal{F} = \mathcal{F}_1 \cap \mathcal{F}_0. \tag{3.119}$$

Clearly, $\mathcal{F}$ is a countable intersection of open everywhere dense subsets of $\mathfrak{M}$ and $\mathcal{F} \subset \mathcal{F}_0 \subset \mathcal{M}_*(\Omega)$.

Let $u \in \mathcal{F}$, $p \geq 1$ be an integer and $\{x_t^{(i)}\}_{t=0}^{\infty}$, $i = 1, 2$ be (u, Ω)-overtaking optimal programs such that

$$(\pi^v + g)(x_0^{(i)}) = \sup(\pi^v + g), \ i = 1, 2. \tag{3.120}$$

By (3.119), there exist $v_p \in E_g$ and an integer $k_p \geq p$ such that

$$u \in \mathcal{U}(v_p, k_p). \tag{3.121}$$

In view of (3.119), (3.121) and property (Pii), $\rho(x_t^{(i)}, z_t^{v_p}) \leq k_p^{-1} \leq p^{-1}$, $t = 0, \dots, p$, $i = 1, 2$. This implies that $\rho(x_t^{(1)}, x_t^{(2)}) \leq 2p^{-1}$, $t = 0, \dots, p$. Since p is any natural number we conclude that $x_t^{(1)} = x_t^{(2)}$ for all integers $t \geq 0$. Proposition 3.12 is proved. □

3.7 Proof of Proposition 3.9

By Theorem 2.6, there exists a set $\mathcal{G}_0 \subset \mathfrak{M} \cap \mathcal{M}_*(\Omega)$ which is a countable intersection of open everywhere dense subsets of $\mathfrak{M}$.

Denote by E the set of all $(v, g) \in (\mathfrak{M} \cap \mathcal{M}_*(\Omega)) \times \mathfrak{A}$ for which there exists a unique program $\{z_t^{(v,g)}\}_{t=0}^{\infty} \in \tilde{\mathcal{P}}(g, v)$.

By Lemma 3.13, E is an everywhere dense subset of $\mathfrak{M} \times \mathfrak{A}$.

Let $(v, g) \in E$ and $k \geq 1$ be an integer. By Proposition 3.7, there exist an open neighborhood $\mathcal{U}(v, g, k)$ of (v, g) in $\mathfrak{M} \times \mathfrak{A}$ and an integer $T(v, g, k) \geq k$ such that the following property holds:

(i) for each $(u, h) \in \mathcal{U}(v, g, k) \cap (\mathcal{M}_*(\Omega) \times \mathfrak{A})$ and each $\{x_t\}_{t=0}^{\infty} \in \tilde{\mathcal{P}}(h, u)$ we have $\rho(x_t, z_t^{(v,g)}) \leq k^{-1}$, $t = 0, \dots, k$.

Set

$$\mathcal{F} = \cap_{p=1}^{\infty} \cup \{\mathcal{U}(v, g, k) : \ (v, g) \in E, \ k \geq p\} \cap (\mathcal{G}_0 \times \mathfrak{A}). \tag{3.122}$$

Clearly, $\mathcal{F}$ is a countable intersection of open everywhere dense subsets of $\mathfrak{M} \times \mathfrak{A}$.

Let $(u, h) \in \mathcal{F}$, $p \geq 1$ be an integer and

$$\{x_t^{(i)}\}_{t=0}^{\infty} \in \tilde{\mathcal{P}}(h, u), \ i = 1, 2. \tag{3.123}$$

By (3.122), there exist $(v_p, g_p) \in E$ and an integer $k_p \geq p$ such that

$$(u, h) \in \mathcal{U}(v_p, g_p, k_p). \tag{3.124}$$

In view of (3.122), (3.123), and (3.124), $\rho(x_t^{(i)}, z_t^{(g_p, v_p)}) \leq k_p^{-1} \leq p^{-1}$, $t = 0, \dots, p$, $i = 1, 2$. This implies that $\rho(x_t^{(1)}, x_t^{(2)}) \leq 2p^{-1}$, $t = 0, \dots, p$. Since p is any natural number we conclude that $x_t^{(1)} = x_t^{(2)}$ for all integers $t \geq 0$. Proposition 3.9 is proved. □

3.8 The Second Class of Bolza Problems

Let (X, ρ) be a compact metric space and Ω be a nonempty closed subset of $X \times X$. We continue to use the notation, definitions, and assumptions introduced in Chap. 2 and in the previous sections of this chapter. We study the problem

$$g(x_0, x_T) + \sum_{t=0}^{T-1} v(x_t, x_{t+1}) \to \max, \ \{(x_t, x_{t+1})\}_{t=0}^{T-1} \subset \Omega, \qquad (P5)$$

where $T \geq 1$ is an integer number and $v : \Omega \to R^1$ and $g : X \times X \to R^1$ are bounded upper semicontinuous objective functions.

Assume that $v \in \mathcal{M}(\Omega)$ is an upper semicontinuous function. We suppose that there exist a point $\bar{x}_v \in X$ and real positive constants $\bar{c}_v$ and $\bar{r}_v$ such that assumptions (A1), (A2), and (A3) hold.

For every nonempty set Y, every nonempty subset $C \subset Y$, and every function $h : Y \to R^1 \cup \{-\infty\}$ set

$$\sup(h) = \sup\{h(y) : \ y \in Y\}, \ \sup(h; C) = \sup\{h(y) : \ y \in C\}.$$

If (X_i, ρ_i), $i = 1, 2$ are metric spaces, then the product $X_1 \times X_2$ is equipped with the metric

$$\rho_1(x_1, y_1) + \rho_2(x_2, y_2) \text{ for all } (x_1, x_2), \ (y_1, y_2) \in X_1 \times X_2.$$

Denote by $\mathcal{M}(X \times X)$ the collection of all bounded functions $h : X \times X \to R^1$. For every function $h \in \mathcal{M}(X \times X)$ put

$$\|h\| = \sup\{|h(x, y)| : \ x, y \in X\}.$$

It is not difficult to see that $(\mathcal{M}(X \times X), \| \cdot \|)$ is a Banach space.

For every pair of nonnegative integers $T_1 < T_2$, every finite sequence of functions $\{u_t\}_{t=T_1}^{T_2-1} \subset \mathcal{M}(\Omega)$, and every function $h \in \mathcal{M}(X \times X)$ we consider the problem

$$\sum_{t=T_1}^{T_2-1} u_t(x_t, x_{t+1}) + h(x_{T_1}, x_{T_2}) \to \max, \ \{(x_t, x_{t+1})\}_{t=T_1}^{T_2-1} \subset \Omega$$

and define

$$\sigma(h, \{u_t\}_{t=T_1}^{T_2-1}, T_1, T_2) = \sup\{\sum_{t=T_1}^{T_2-1} u_t(x_t, x_{t+1}) + h(x_{T_1}, x_{T_2}) :$$

$$\{x_t\}_{t=T_1}^{T_2} \text{ is an } (\Omega) - \text{program}\}. \qquad (3.125)$$

For every pair of nonnegative integers $T_1 < T_2$, every function $u \in \mathcal{M}(\Omega)$, and every function $h \in \mathcal{M}(X \times X)$ define

$$\sigma(h, u, T_1, T_2) = \sigma(h, \{u_t\}_{t=T_1}^{T_2-1}, T_1, T_2) \text{ where } u_t = u, \ t = T_1, \dots, T_2 - 1.$$

Denote by $\mathcal{M}_u(X \times X)$ the set of all upper semicontinuous functions belonging to the space $\mathcal{M}(X \times X)$ and denote by $\mathcal{M}_c(X \times X)$ the set of all continuous functions belonging to the space $\mathcal{M}(X \times X)$. We consider the complete metric spaces $\mathcal{M}_u(X \times X)$ and $\mathcal{M}_c(X \times X)$ equipped with the metric

$$d_{X \times X}(h_1, h_2) = \|h_1 - h_2\|,\ h_1, h_2 \in \mathcal{M}_u(X \times X).$$

The following result, which is proved in Sect. 3.9, establishes the turnpike property of approximate solutions of problems (P5).

Theorem 3.14. *Suppose that $v \in \mathcal{M}_*(\Omega)$ is an upper semicontinuous function, $\bar{x}_v \in X$, $\bar{r}_v \in (0,1)$, $\bar{c}_v > 0$ and that assumptions (A1), (A2), and (A3) hold. Let $\epsilon \in (0, \bar{r}_v)$ and $M > 0$. Then there exist a natural number L and a number $\delta \in (0, \epsilon)$ such that for every natural number $T > 2L$, every finite sequence of functions $\{u_t\}_{t=0}^{T-1} \subset \mathcal{M}(\Omega)$ which satisfies*

$$\|u_t - v\| \le \delta,\ t = 0 \dots, T-1, \tag{3.126}$$

every function $h \in \mathcal{M}(X \times X)$ satisfying

$$\|h\| \le M \tag{3.127}$$

and every (Ω)-program $\{x_t\}_{t=0}^{T}$ which satisfies

$$h(x_0, x_T) + \sum_{t=0}^{T-1} u_t(x_t, x_{t+1}) \ge \sigma(h, \{u_t\}_{t=0}^{T-1}, 0, T) - M, \tag{3.128}$$

$$\sum_{t=\tau}^{\tau+L-1} u_t(x_t, x_{t+1}) \ge \sigma(\{u_t\}_{t=\tau}^{\tau+L-1}, \tau, \tau+L, x_\tau, x_{\tau+L}) - \delta \tag{3.129}$$

for every integer $\tau \in [0, T-L]$ there exists a pair of integers $\tau_1 \in [0, L]$, $\tau_2 \in [T-L, T]$ such that

$$\rho(x_t, \bar{x}_v) \le \epsilon,\ t = \tau_1, \dots, \tau_2. \tag{3.130}$$

Moreover if $\rho(x_0, \bar{x}_v) \le \delta$, then $\tau_1 = 0$ and if $\rho(x_T, \bar{x}_v) \le \delta$, then $\tau_2 = T$.

Note that in view of (2.115) and Propositions 2.22, 2.26, and 2.31, for every function $v \in \mathcal{M}_*(\Omega)$ and every function $g \in \mathcal{M}_u(X \times X)$, the function

$$(\xi, \eta) \to \pi^v(\xi) + \pi^{\bar{v}}(\eta) + g(\xi, \eta),\ \xi, \eta \in X$$

is upper semicontinuous and bounded from above and has a point of maximum.

Our second result, which is proved in Sect. 3.11, describes the structure of approximate solutions of the problems (P5) in the regions close to the right endpoints.

Theorem 3.15. *Suppose that $g \in \mathcal{M}(X \times X)$ and $v \in \mathcal{M}(\Omega)$ are upper semicontinuous functions, a point $\bar{x}_v \in X$, $\bar{r}_v > 0$, $\bar{c}_v > 0$ and that assumptions (A1), (A2), and (A3) hold. Let τ_0 be a natural number and $\epsilon > 0$. Then there exist a positive number δ and a natural number $T_0 \geq \tau_0$ such that for every natural number $T \geq T_0$, every function $h \in \mathcal{M}(X \times X)$ satisfying $\|h - g\| \leq \delta$, every finite sequence of functions $\{u_t\}_{t=0}^{T-1} \subset \mathcal{M}(\Omega)$ which satisfies*

$$\|u_t - v\| \leq \delta,\ t = 0 \ldots, T-1$$

and every (Ω)-program $\{x_t\}_{t=0}^{T}$ which satisfies

$$h(x_0, x_T) + \sum_{t=0}^{T-1} u_t(x_t, x_{t+1}) \geq \sigma(h, \{u_t\}_{t=0}^{T-1}, 0, T) - \delta$$

there exist a (v, Ω)-overtaking optimal and (v, Ω)-good program $\{x_t^*\}_{t=0}^{\infty}$ and a $(\bar{v}, \bar{\Omega})$-overtaking optimal and $(\bar{v}, \bar{\Omega})$-good program $\{\bar{x}_t^*\}_{t=0}^{\infty}$ such that

$$\pi^v(x_0^*) + \pi^{\bar{v}}(\bar{x}_0^*) + g(x_0^*, \bar{x}_0^*) \geq \pi^v(\xi) + \pi^{\bar{v}}(\eta) + g(\xi, \eta) \text{ for all } \xi, \eta \in X$$

and that for all integers $t = 0, \ldots, \tau_0$, $\rho(x_t, x_t^*) \leq \epsilon$ and $\rho(x_{T-t}, \bar{x}_t^*) \leq \epsilon$.

Definition 3.16. *Let $v \in \mathcal{M}_*(\Omega)$ and $g \in \mathcal{M}_u(X \times X)$. Denote by $\mathcal{L}(g, v)$ the collection of all pairs of sequences $\{x_t\}_{t=0}^{\infty}$, $\{\bar{x}_t\}_{t=0}^{\infty} \subset X$ such that the sequence $\{x_t\}_{t=0}^{\infty}$ is a (v, Ω)-overtaking optimal and (v, Ω)-good program and the sequence $\{\bar{x}_t\}_{t=0}^{\infty}$ is a $(\bar{v}, \bar{\Omega})$-overtaking optimal and $(\bar{v}, \bar{\Omega})$-good program which satisfy*

$$\pi^v(x_0) + \pi^{\bar{v}}(\bar{x}_0) + g(x_0, \bar{x}_0) \geq \pi^v(\xi) + \pi^{\bar{v}}(\eta) + g(\xi, \eta) \text{ for all } \xi, \eta \in X.$$

Since the function

$$(\xi, \eta) \to \pi^v(\xi) + \pi^{\bar{v}}(\eta) + g(\xi, \eta),\ \xi, \eta \in X$$

has a point of maximum, it follows from Propositions 2.22 and 2.31 and Theorem 2.13 that the set $\mathcal{L}(g, v)$ is nonempty.

The following stability result is proved in Sect. 3.12.

Theorem 3.17. *Suppose that $g \in \mathcal{M}_u(X \times X)$, $v \in \mathcal{M}(\Omega)$ is an upper semicontinuous function, a point $\bar{x}_v \in X$, $\bar{r}_v > 0$, $\bar{c}_v > 0$ and that assumptions (A1), (A2), and (A3) hold. Let τ_0 be a natural number and ϵ be a real positive number. Then there exists a positive number δ such that for every function $u \in B_d(v, \delta) \cap \mathcal{M}_*(\Omega)$ and every function $h \in \mathcal{M}_u(X \times X)$ satisfying $\|h - g\| \leq \delta$ the following property holds:*

for every pair of sequences $(\{x_t\}_{t=0}^{\infty}, \{y_t\}_{t=0}^{\infty}) \in \mathcal{L}(h, u)$ there exists a pair of sequences $(\{x_t^*\}_{t=0}^{\infty}, \{y_t^*\}_{t=0}^{\infty}) \in \mathcal{L}(g, v)$ such that for all integers $t = 0, \ldots, \tau_0$,

$$\rho(x_t, x_t^*) \leq \epsilon,\ \rho(y_t, y_t^*) \leq \epsilon.$$

The next theorem, which is proved in Sect. 3.13, establish, for a generic (in the sense of the Baire category) pair (v, g), that the set $\mathcal{L}(g, v)$ is a singleton. In this case approximate solutions of problems (P5) have a simple structure in the regions close to the endpoints.

Theorem 3.18. *Let $\mathfrak{M}$ be either $\bar{\mathcal{M}}_0(\Omega)$ or $\bar{\mathcal{M}}_{c,0}(\Omega)$ and let $\mathfrak{A}$ be either $\mathcal{M}_u(X \times X)$ or $\mathcal{M}_c(X \times X)$. Then there exists a subset $\mathcal{F} \subset (\mathfrak{M} \cap \mathcal{M}_*(\Omega)) \times \mathfrak{A}$ which is a countable intersection of open everywhere dense subsets of $\mathfrak{M} \times \mathfrak{A}$ such that for every pair of functions $(v, g) \in \mathcal{F}$ the set $\mathcal{L}(g, v)$ is a singleton.*

For every point $z = (z_1, z_2) \in X \times X$ and every nonempty set $C \subset X \times X$ define

$$\rho(z, C) = \inf\{\rho(z_1, y_1) + \rho(z_2, y_2) : (y_1, y_2) \in C\}.$$

The results stated in this section were obtained in [91].

3.9 Proof of Theorem 3.14

In view of assertion 1 of Theorem 2.5 there exist a positive number δ_0 and a natural number L_0 such that the following property holds:

(Pi) for every natural number $T \geq L_0$, every finite sequence of functions $\{u_t\}_{t=0}^{T-1} \subset \mathcal{M}(\Omega)$ and every (Ω)-program $\{x_t\}_{t=0}^{T}$ which satisfy

$$\|u_t - v\| \leq \delta_0, \ t = 0, \dots, T-1,$$

$$\sum_{t=0}^{T-1} u_t(x_t, x_{t+1}) \geq \sigma(\{u_t\}_{t=0}^{T-1}, 0, T) - 4M$$

the inequality

$$\mathrm{Card}(\{t \in \{0, \dots, T\} : \rho(x_t, \bar{x}_v) > 4^{-1}\bar{r}_v\}) < L_0$$

holds.

Theorem 2.3 implies there exist a natural number L_1 and a positive number δ_1 such that the following property holds:

(Pii) for every natural number $T > 2L_1$, every finite sequence of functions $\{u_t\}_{t=0}^{T-1} \subset \mathcal{M}(\Omega)$ which satisfies

$$\|u_t - v\| \leq \delta_1, \ t = 0, \dots, T-1$$

and every (Ω)-program $\{x_t\}_{t=0}^{T}$ which satisfies

$$\rho(x_0, \bar{x}_v), \ \rho(x_T, \bar{x}_v) \leq 2^{-1}\bar{r}_v,$$

$$\sum_{t=0}^{T-1} u_t(x_t, x_{t+1}) \geq \sigma(\{u_t\}_{t=0}^{T-1}, 0, T, x_0, x_T) - 4M,$$

$$\sum_{t=\tau}^{\tau+L_1-1} u_t(x_t, x_{t+1}) \geq \sigma(\{u_t\}_{t=\tau}^{\tau+L_1-1}, \tau, \tau+L_1, x_\tau, x_{\tau+L_1}) - \delta_1$$

for every integer $\tau \in [0, T-L_1]$, there exists a pair of integers $\tau_1 \in [0, L_1]$, $\tau_2 \in [T-L_1, T]$ such that $\rho(x_t, \bar{x}_v) \leq \epsilon$, $t = \tau_1, \dots, \tau_2$; moreover if $\rho(x_0, \bar{x}_v) \leq \delta_1$, then $\tau_1 = 0$ and if $\rho(x_T, \bar{x}_v) \leq \delta_1$, then $\tau_2 = T$.

Set

$$\delta = \min\{\delta_0, \delta_1, \epsilon/8, \bar{r}_v/8\}, \tag{3.131}$$

$$L = L_0 + L_1 + 2. \tag{3.132}$$

Assume that a natural number $T > 2L$, a finite sequence of functions $\{u_t\}_{t=0}^{T-1} \subset \mathcal{M}(\Omega)$ satisfies (3.126), a function $h \in \mathcal{M}(X \times X)$ satisfies (3.127) and that an (Ω)-program $\{x_t\}_{t=0}^{T}$ satisfies (4.128) and (3.129) for every integer $\tau \in [0, T-L]$. By (3.127) and (3.128),

$$\sum_{t=0}^{T-1} u_t(x_t, x_{t+1}) \geq \sigma(\{u_t\}_{t=0}^{T-1}, 0, T) - 3M. \tag{3.133}$$

It follows from (3.126), (3.131), (3.132), (3.133), and property (Pi) that there exist integers

$$S_1 \in [0, L_0],\ S_2 \in [T-L_0, T] \tag{3.134}$$

for which

$$\rho(x_{S_i}, \bar{x}_v) \leq 4^{-1}\bar{r}_v,\ i = 1, 2, \tag{3.135}$$

if $\rho(x_0, \bar{x}_v) \leq \delta$, then $S_1 = 0$ and if $\rho(x_T, \bar{x}_v) \leq \delta$, then $S_2 = T$.

In view of (3.126), (3.129), (3.131)–(3.135), and property (Pii) applied to the program $\{x_t\}_{t=S_1}^{S_2}$, there exist integers $\tau_1 \in [S_1, S_1 + L_1]$ and $\tau_2 \in [S_2 - L_1, S_2]$ such that $\rho(x_t, \bar{x}_v) \leq \epsilon$, $t = \tau_1, \dots, \tau_2$, if $\rho(x_{S_1}, \bar{x}_v) \leq \delta$, then $\tau_1 = S_1$ and if $\rho(x_{S_2}, \bar{x}_v) \leq \delta$, then $\tau_2 = S_2$. This completes the proof of Theorem 3.14. □

3.10 Auxiliary Results for Theorem 3.15

Lemma 3.19. *Suppose that $g \in \mathcal{M}(X \times X)$ and $v \in \mathcal{M}(\Omega)$ are upper semicontinuous functions, a point $\bar{x}_v \in X$, $\bar{r}_v$, $\bar{c}_v$ are positive numbers and that assumptions (A1), (A2), and (A3) hold. Let T_0 be a natural number and $\epsilon \in (0,1)$. Then there exists $\delta \in (0, \epsilon)$ such that for every (Ω)-program $\{x_t\}_{t=0}^{T_0}$ and every $(\bar{\Omega})$-program $\{y_t\}_{t=0}^{T_0}$ which satisfy*

$$\pi^v(x_0) + \pi^{\bar{v}}(y_0) + g(x_0, y_0) + \delta \geq \pi^v(\xi) + \pi^{\bar{v}}(\eta) + g(\xi, \eta) \text{ for all } \xi, \eta \in X, \tag{3.136}$$

$$\sum_{t=0}^{T_0-1} (v(x_t, x_{t+1}) - v(\bar{x}_v, \bar{x}_v)) - \pi^v(x_0) + \pi^v(x_{T_0}) \geq -\delta, \tag{3.137}$$

$$\sum_{t=0}^{T_0-1} (\bar{v}(y_t, y_{t+1}) - v(\bar{x}_v, \bar{x}_v)) - \pi^{\bar{v}}(y_0) + \pi^{\bar{v}}(y_{T_0}) \geq -\delta \tag{3.138}$$

there exists a pair $(\{x_t^*\}_{t=0}^{\infty}, \{y_t^*\}_{t=0}^{\infty}) \in \mathcal{L}(g, v)$ *such that for all* $t = 0, \dots, T_0$,

$$\rho(x_t, x_t^*) \leq \epsilon, \ \rho(y_t, y_t^*) \leq \epsilon.$$

Proof. Assume that the lemma does not hold. Then there exist a sequence $\{\delta_k\}_{k=1}^{\infty} \subset (0, 1]$, a sequence of (Ω)-programs $\{x_t^{(k)}\}_{t=0}^{T_0}$, $k = 1, 2, \dots$ and a sequence of $(\bar{\Omega})$-programs $\{y_t^{(k)}\}_{t=0}^{T_0}$, $k = 1, 2, \dots$ such that

$$\lim_{k \to \infty} \delta_k = 0, \tag{3.139}$$

for every integer $k \geq 1$,

$$\pi^v(x_0^{(k)}) + \pi^{\bar{v}}(y_0^{(k)}) + g(x_0^{(k)}, y_0^{(k)}) + \delta_k \geq \pi^v(\xi) + \pi^{\bar{v}}(\eta) + g(\xi, \eta) \text{ for all } \xi, \eta \in X, \tag{3.140}$$

$$\sum_{t=0}^{T_0-1} (v(x_t^{(k)}, x_{t+1}^{(k)}) - v(\bar{x}_v, \bar{x}_v)) - \pi^v(x_0^{(k)}) + \pi^v(x_{T_0}^{(k)}) \geq -\delta_k, \tag{3.141}$$

$$\sum_{t=0}^{T_0-1} (\bar{v}(y_t^{(k)}, y_{t+1}^{(k)}) - v(\bar{x}_v, \bar{x}_v)) - \pi^{\bar{v}}(y_0^{(k)}) + \pi^{\bar{v}}(y_{T_0}^{(k)}) \geq -\delta_k \tag{3.142}$$

and that for every integer $k \geq 1$ and every pair $(\{z_t\}_{t=0}^{\infty}, \{\xi_t\}_{t=0}^{\infty}) \in \mathcal{L}(g, v)$, we have

$$\max\{\rho(z_t, x_t^{(k)}) + \rho(\xi_t, y_t^{(k)}) : \ t = 0, \dots, T_0\} > \epsilon. \tag{3.143}$$

In view of (3.140)–(3.142) and Propositions 2.22 and 2.31, for every integer $k \geq 1$, the values $\pi^v(x_0^{(k)})$, $\pi^{\bar{v}}(y_0^{(k)})$, $\pi^v(x_{T_0}^{(k)})$ and $\pi^{\bar{v}}(y_{T_0}^{(k)})$ are finite. Extracting a subsequence and re-indexing, if necessary, we may assume without loss of generality that for every integer $t \in [0, T_0]$ there exist

$$x_t = \lim_{k \to \infty} x_t^{(k)}, \ y_t = \lim_{k \to \infty} y_t^{(k)}. \tag{3.144}$$

It follows from (3.139), (3.140), (3.144), and upper semicontinuity of the functions g, π^v and $\pi^{\bar{v}}$ that

$$\pi^v(x_0) + \pi^{\bar{v}}(y_0) + g(x_0, y_0) \geq \pi^v(\xi) + \pi^{\bar{v}}(\eta) + g(\xi, \eta) \text{ for all } \xi, \eta \in X, \tag{3.145}$$

$$\pi^v(x_0) \geq \limsup_{k \to \infty} \pi^v(x_0^{(k)}), \ \pi^{\bar{v}}(y_0) \geq \limsup_{k \to \infty} \pi^{\bar{v}}(y_0^{(k)}),$$

$$g(x_0, y_0) \geq \limsup_{k \to \infty} g(x_0^{(k)}, y_0^{(k)}). \tag{3.146}$$

Relations (3.140) and (3.145) imply that

$$\pi^v(x_0) + \pi^{\bar{v}}(y_0) + g(x_0, y_0) = \lim_{k \to \infty} (\pi^v(x_0^{(k)}) + \pi^{\bar{v}}(y_0^{(k)}) + g(x_0^{(k)}, y_0^{(k)})).$$

Combined with (3.146) the equation above implies that

$$\pi^v(x_0) = \lim_{k\to\infty} \pi^v(x_0^{(k)}),\ \pi^{\bar v}(y_0) = \lim_{k\to\infty} \pi^{\bar v}(y_0^{(k)}),$$

$$g(x_0, y_0) = \lim_{k\to\infty} g(x_0^{(k)}, y_0^{(k)}). \tag{3.147}$$

By upper semicontinuity of the functions v, π^v and $\pi^{\bar v}$, (3.139), (3.140), (3.142), (3.144), and (3.147), we have

$$\sum_{t=0}^{T_0-1} (v(x_t, x_{t+1}) - v(\bar x_v, \bar x_v)) - \pi^v(x_0) + \pi^v(x_{T_0})$$

$$\ge \limsup_{k\to\infty}\Big(\sum_{t=0}^{T_0-1} (v(x_t^{(k)}, x_{t+1}^{(k)}) - v(\bar x_v, \bar x_v)) - \pi^v(x_0^{(k)}) + \pi^v(x_{T_0}^{(k)})\Big)$$

$$\ge \limsup_{k\to\infty}(-\delta_k) = 0, \tag{3.148}$$

$$\sum_{t=0}^{T_0-1} (\bar v(y_t, y_{t+1}) - v(\bar x_v, \bar x_v)) - \pi^{\bar v}(y_0) + \pi^{\bar v}(y_{T_0})$$

$$\ge \limsup_{k\to\infty}\Big(\sum_{t=0}^{T_0-1} (v(y_t^{(k)}, y_{t+1}^{(k)}) - v(\bar x_v, \bar x_v)) - \pi^{\bar v}(y_0^{(k)}) + \pi^{\bar v}(y_{T_0}^{(k)})\Big)$$

$$\ge \limsup_{k\to\infty}(-\delta_k) = 0. \tag{3.149}$$

In view of (3.145), (3.148), and (3.149), the values $\pi^v(x_0)$, $\pi^{\bar v}(y_0)$, $\pi^v(x_{T_0})$, and $\pi^{\bar v}(y_{T_0})$ are finite. Together with Proposition 2.19, (3.148), and (3.149) this implies that for all integers $t = 0, \dots, T_0 - 1$, we have

$$v(x_t, x_{t+1}) - v(\bar x_v, \bar x_v) = \pi^v(x_t) - \pi^v(x_{t+1}),$$

$$\bar v(y_t, y_{t+1}) - v(\bar x_v, \bar x_v) = \pi^{\bar v}(y_t) - \pi^{\bar v}(y_{t+1}). \tag{3.150}$$

Since $\pi^v(x_{T_0})$ and $\pi^{\bar v}(y_{T_0})$ are finite it follows from Theorem 2.13 that there exist a (v, Ω)-overtaking optimal and (v, Ω)-good program $\{\tilde x_t\}_{t=0}^\infty$ satisfying

$$\tilde x_0 = x_{T_0} \tag{3.151}$$

and a $(\bar v, \bar\Omega)$-overtaking optimal and $(\bar v, \bar\Omega)$-good program $\{\tilde y_t\}_{t=0}^\infty$ satisfying

$$\tilde y_0 = y_{T_0}. \tag{3.152}$$

For all integers $t > T_0$ put

$$x_t = \tilde x_{t-T_0},\ y_t = \tilde y_{t-T_0}. \tag{3.153}$$

By (3.151)–(3.153), $\{x_t\}_{t=0}^{\infty}$ is an (Ω)-program and $\{y_t\}_{t=0}^{\infty}$ is an $(\bar{\Omega})$-program. Since $\{\tilde{x}_t\}_{t=0}^{\infty}$ is (v, Ω)-overtaking optimal and (v, Ω)-good program and $\{\tilde{y}_t\}_{t=0}^{\infty}$ is $(\bar{v}, \bar{\Omega})$-overtaking optimal and $(\bar{v}, \bar{\Omega})$-good program, it follows from Corollary 2.21 and (3.150)–(3.153) that (3.150) holds for all integers $t \geq 0$. Since $\{\tilde{x}_t\}_{t=0}^{\infty}$ is (v, Ω)-good program and $\{\tilde{y}_t\}_{t=0}^{\infty}$ is $(\bar{v}, \bar{\Omega})$-good program, it follows from (3.145), (3.148), (3.149), (3,150) (holding for all integers $t \geq 0$), (3.152), (3.153), Definition 3.16, and Proposition 2.27 that $\{x_t\}_{t=0}^{\infty}$ is a (v, Ω)-overtaking optimal program and $\{y_t\}_{t=0}^{\infty}$ is a $(\bar{v}, \bar{\Omega})$-overtaking optimal program satisfying

$$(\{x_t\}_{t=0}^{\infty}, \{y_t\}_{t=0}^{\infty}) \in \mathcal{L}(g, v). \tag{3.154}$$

In view of (3.144), for all sufficiently large natural numbers k,

$$\rho(x_t, x_t^{(k)}) \leq \epsilon/4, \ \rho(y_t, y_t^{(k)}) \leq \epsilon/4, \ t = 0, \dots, T_0.$$

Together with (3.154) this contradicts (3.143). The contradiction we have reached proves Lemma 3.19. □

3.11 Proof of Theorem 3.15

We may assume that $\bar{r}_v \in (0, 1)$. Recall that

$$\{(x, y) \in X \times X : \ \rho(x, \bar{x}_v), \ \rho(y, \bar{x}_v) \leq \bar{r}_v\} \subset \Omega. \tag{3.155}$$

Lemma 3.19 implies that there exists a number $\delta_1 \in (0, \min\{\epsilon, \bar{r}_v/2\})$ such that the following property holds:

(Piii) for every (Ω)-program $\{\xi_t\}_{t=0}^{\tau_0}$ and every $(\bar{\Omega})$-program $\{\eta_t\}_{t=0}^{\tau_0}$ which satisfy

$$\pi^v(\xi_0) + \pi^{\bar{v}}(\eta_0) + g(\xi_0, \eta_0) + 2\delta_1 \geq \pi^v(\xi) + \pi^{\bar{v}}(\eta) + g(\xi, \eta) \text{ for all } \xi, \eta \in X, \tag{3.156}$$

$$\sum_{t=0}^{\tau_0 - 1} (v(\xi_t, \xi_{t+1}) - v(\bar{x}_v, \bar{x}_v)) - \pi^v(\xi_0) + \pi^v(\xi_{\tau_0}) \geq -2\delta_1, \tag{3.157}$$

$$\sum_{t=0}^{\tau_0 - 1} (\bar{v}(\eta_t, \eta_{t+1}) - v(\bar{x}_v, \bar{x}_v)) - \pi^{\bar{v}}(\eta_0) + \pi^{\bar{v}}(\eta_{\tau_0}) \geq -2\delta_1 \tag{3.158}$$

there exists a pair $(\{x_t^*\}_{t=0}^{\infty}, \ \{y_t^*\}_{t=0}^{\infty}) \in \mathcal{L}(g, v)$ such that for all $t = 0, \dots, \tau_0$,

$$\rho(\xi_t, x_t^*) \leq \epsilon, \ \rho(\eta_t, y_t^*) \leq \epsilon. \tag{3.159}$$

Propositions 2.22, 2.23, and 2.31 and assumption (A1) imply that there exists a number $\delta_2 \in (0, \delta_1)$ such that for every point $z \in X$ satisfying $\rho(z, \bar{x}_v) \leq 2\delta_2$, we have

$$|\pi^{\bar{v}}(z)| = |\pi^{\bar{v}}(z) - \pi^{\bar{v}}(\bar{x}_v)| \leq \delta_1/8 \tag{3.160}$$

and for every pair $(x, y) \in X \times X$ satisfying $\rho(x, \bar{x}_v) \le 2\delta_2$, $\rho(y, \bar{x}_v) \le 2\delta_2$, we have

$$|v(x, y) - \bar{v}(\bar{x}_v, \bar{x}_v)| \le \delta_1/8. \tag{3.161}$$

Theorem 3.14 implies that there exist a natural number L and a positive number δ_3 such that the following property holds:

(Piv) for every natural number $T > 2L$, every finite sequence of functions $\{u_t\}_{t=0}^{T-1} \subset \mathcal{M}(\Omega)$ which satisfies $\|u_t - v\| \le \delta_3$, $t = 0 \dots, T-1$, every function $h \in \mathcal{M}(X \times X)$ satisfying $\|h - g\| \le 1$ and every (Ω)-program $\{x_t\}_{t=0}^{T}$ which satisfies

$$h(x_0, x_T) + \sum_{t=0}^{T-1} u_t(x_t, x_{t+1}) \ge \sigma(h, \{u_t\}_{t=0}^{T-1}, 0, T) - \delta_3$$

the inequality $\rho(x_t, \bar{x}_v) \le \delta_2$ holds for all $t = L, \dots, T - L$.

Define

$$\mathcal{D}_0 = \{(\xi, \eta) \in X \times X : \pi^v(\xi) + \pi^{\bar{v}}(\eta) + g(\xi, \eta)$$

$$\ge \pi^v(z_1) + \pi^{\bar{v}}(z_2) + g(z_1, z_2) \text{ for all } z_1, z_2 \in X\}. \tag{3.162}$$

We have already noted that the set $\mathcal{D}_0$ is nonempty.

Propositions 2.22 and 2.31, (2.115), and (3.162) imply that for $(\xi, \eta) \in \mathcal{D}_0$, we have

$$\pi^v(\xi) + \pi^{\bar{v}}(\eta) + \|g\| \ge g(\bar{x}_v, \bar{x}_v) + \pi^v(\bar{x}_v) + \pi^{\bar{v}}(\bar{x}_v) \ge -\|g\|,$$

$$\pi^v(\xi),\ \pi^{\bar{v}}(\eta) \ge -2\|g\| - \bar{c}_v. \tag{3.163}$$

Let

$$(\{\xi_t\}_{t=0}^{\infty},\ \{\eta_t\}_{t=0}^{\infty}) \in \mathcal{L}(g, v). \tag{3.164}$$

In view of (3.162)–(3.164), Definition 3.16, and Propositions 2.24 and 2.31, we have

$$\pi^v(\xi_0) = \lim_{T \to \infty} \sum_{t=0}^{T-1} (v(\xi_t, \xi_{t+1}) - v(\bar{x}_v, \bar{x}_v)) \ge -2\|g\| - \bar{c}_v, \tag{3.165}$$

$$\pi^{\bar{v}}(\eta_0) = \lim_{T \to \infty} \sum_{t=0}^{T-1} (\bar{v}(\eta_t, \eta_{t+1}) - \bar{v}(\bar{x}_v, \bar{x}_v)) \ge -2\|g\| - \bar{c}_v. \tag{3.166}$$

By (3.162), (3.164)–(3.166), Definitions 2.2 and 3.16, and Lemma 2.8, there exists an integer $L_1 \ge 1$ for which

$$\mathcal{D}_0 \subset \bar{Y}_{L_1} \times Y_{L_1}. \tag{3.167}$$

It follows from (3.162), (3.167), Definition 3.16, and Theorem 2.15 applied to v and $\bar{v}$ that there exists a natural number τ_1 such that for every $(\{\xi_t\}_{t=0}^{\infty},\ \{\eta_t\}_{t=0}^{\infty}) \in \mathcal{L}(g, v)$,

$$\rho(\xi_t, \bar{x}_v) \le \delta_2,\ \rho(\eta_t, \bar{x}_v) \le \delta_2 \text{ for all integers } t \ge \tau_1. \tag{3.168}$$

Fix a positive number δ and an integer $T_0 \geq 1$ such that

$$\delta < (64(L + \tau_1 + \tau_0 + 8))^{-1} \min\{\delta_1, \delta_2, \delta_3\}, \tag{3.169}$$

$$T_0 > 4L + 4\tau_0 + 4\tau_1 + 16. \tag{3.170}$$

Assume that an integer $T \geq T_0$, a function $h \in \mathcal{M}(X \times X)$ satisfies

$$\|h - g\| \leq \delta, \tag{3.171}$$

a finite sequence of functions $\{u_t\}_{t=0}^{T-1} \subset \mathcal{M}(\Omega)$ satisfies

$$\|u_t - v\| \leq \delta,\ t = 0\ldots, T - 1 \tag{3.172}$$

and that $\{x_t\}_{t=0}^{T}$ is an (Ω)-program which satisfies

$$h(x_0, x_T) + \sum_{t=0}^{T-1} u_t(x_t, x_{t+1}) \geq \sigma(h, \{u_t\}_{t=0}^{T-1}, 0, T) - \delta. \tag{3.173}$$

In view of (3.169)–(3.173) and property (Piv),

$$\rho(x_t, \bar{x}_v) \leq \delta_2,\ t = L, \ldots, T - L. \tag{3.174}$$

Choose

$$(\{\xi_t\}_{t=0}^{\infty},\ \{\eta_t\}_{t=0}^{\infty}) \in \mathcal{L}(g, v). \tag{3.175}$$

By (3.175) and the choice of τ_1, relation (3.168) is true. Define a sequence $\{\tilde{x}_t\}_{t=0}^{T}$ by

$$\tilde{x}_t = \xi_t,\ t = 0, \ldots, L + \tau_0 + \tau_1 + 3,$$

$$\tilde{x}_t = x_t,\ t = L + \tau_0 + \tau_1 + 4, \ldots, T - L - \tau_0 - \tau_1 - 4,$$

$$\tilde{x}_t = \eta_{T-t},\ t = T - L - \tau_0 - \tau_1 - 3, \ldots, T. \tag{3.176}$$

By (3.168) and (3.176),

$$\rho(\tilde{x}_{L+\tau_0+\tau_1+3}, \bar{x}_v) = \rho(\xi_{L+\tau_0+\tau_1+3}, \bar{x}_v) \leq \delta_2,$$

$$\rho(\tilde{x}_{T-L-\tau_0-\tau_1-3}, \bar{x}_v) = \rho(\eta_{L+\tau_0+\tau_1+3}, \bar{x}_v) \leq \delta_2. \tag{3.177}$$

It follows from (3.174), (3.176), (3.177), and the inequality $\delta_2 < \bar{r}_v/2$ that $\{\tilde{x}_t\}_{t=0}^{T}$ is an (Ω)-program. In view (3.172), (3.173), and (3.176), we have

$$\delta \geq \sigma(h, \{u_t\}_{t=0}^{T-1}, 0, T) - h(x_0, x_T) - \sum_{t=0}^{T-1} u_t(x_t, x_{t+1})$$

$$\geq h(\tilde{x}_0, \tilde{x}_T) + \sum_{t=0}^{T-1} u_t(\tilde{x}_t, \tilde{x}_{t+1}) - h(x_0, x_T) - \sum_{t=0}^{T-1} u_t(x_t, x_{t+1})$$

$$= h(\xi_0, \eta_0) - h(x_0, x_T) + \sum_{t=0}^{L+\tau_0+\tau_1+3} (u_t(\tilde{x}_t, \tilde{x}_{t+1}) - u_t(x_t, x_{t+1}))$$

$$+ \sum_{t=T-L-\tau_0-\tau_1-4}^{T-1} (u_t(\tilde{x}_t, \tilde{x}_{t+1}) - u_t(x_t, x_{t+1}))$$

$$\geq g(\xi_0, \eta_0) - g(x_0, x_T)$$

$$+ \sum_{t=0}^{L+\tau_0+\tau_1+3} v(\tilde{x}_t, \tilde{x}_{t+1}) - \sum_{t=0}^{L+\tau_0+\tau_1+3} v(x_t, x_{t+1})$$

$$+ \sum_{t=T-L-\tau_0-\tau_1-4}^{T-1} v(\tilde{x}_t, \tilde{x}_{t+1}) - \sum_{t=T-L-\tau_0-\tau_1-4}^{T-1} v(x_t, x_{t+1}) - 4\delta(L+\tau_0+\tau_1+8). \tag{3.178}$$

It follows from (3.176) and (3.178) that

$$g(x_0, x_T) + \sum_{t=0}^{L+\tau_0+\tau_1+3} v(x_t, x_{t+1}) + \sum_{t=T-L-\tau_0-\tau_1-4}^{T-1} v(x_t, x_{t+1})$$

$$\geq -4\delta(L + \tau_0 + \tau_1 + 8) - \delta + g(\xi_0, \eta_0)$$

$$+ \sum_{t=0}^{L+\tau_0+\tau_1+2} v(\xi_t, \xi_{t+1}) + v(\xi_{L+\tau_0+\tau_1+3}, x_{L+\tau_0+\tau_1+4})$$

$$+ v(x_{T-L-\tau_0-\tau_1-4}, \eta_{L+\tau_0+\tau_1+3}) + \sum_{t=0}^{L+\tau_0+\tau_1+2} \bar{v}(\eta_t, \eta_{t+1}). \tag{3.179}$$

By (3.170), (3.174), (3.177), and the choice of δ_1 (see (3.161)), we have

$$|v(\xi_{L+\tau_0+\tau_1+3}, x_{L+\tau_0+\tau_1+4}) - v(\bar{x}_v, \bar{x}_v)| \leq \delta_1/8, \tag{3.180}$$

$$|v(x_{T-L-\tau_0-\tau_1-4}, \eta_{L+\tau_0+\tau_1+3}) - v(x_v, \bar{x}_v)| \leq \delta_1/8, \tag{3.181}$$

$$|v(x_{L+\tau_0+\tau_1+3}, x_{L+\tau_0+\tau_1+4}) - v(\bar{x}_v, \bar{x}_v)| \leq \delta_1/8, \tag{3.182}$$

$$|v(x_{T-L-\tau_0-\tau_1-4}, x_{T-L-\tau_0-\tau_1-3}) - v(\bar{x}_v, \bar{x}_v)| \leq \delta_1/8. \tag{3.183}$$

In view of (3.169) and (3.179)–(3.183), we have

$$g(x_0, x_T) + \sum_{t=0}^{L+\tau_0+\tau_1+2} v(x_t, x_{t+1}) + \sum_{t=T-L-\tau_0-\tau_1-3}^{T-1} v(x_t, x_{t+1})$$

$$\geq -4\delta(L + \tau_0 + \tau_1 + 8) - 3\delta_1/4 + g(\xi_0, \eta_0)$$

$$+ \sum_{t=0}^{L+\tau_0+\tau_1+2} v(\xi_t, \xi_{t+1}) + \sum_{t=0}^{L+\tau_0+\tau_1+2} \bar{v}(\eta_t, \eta_{t+1}). \tag{3.184}$$

Define

$$y_t = x_{T-t},\ t = 0, \dots, T. \tag{3.185}$$

Relations (3.157) and (3.184) imply that

$$g(x_0, y_0) + \sum_{t=0}^{L+\tau_0+\tau_1+2} v(x_t, x_{t+1}) + \sum_{t=0}^{L+\tau_0+\tau_1+2} \bar{v}(y_t, y_{t+1})$$

$$\geq -4\delta(L + \tau_0 + \tau_1 + 8) - 3\delta_1/4 + g(\xi_0, \eta_0)$$

$$+ \sum_{t=0}^{L+\tau_0+\tau_1+2} v(\xi_t, \xi_{t+1}) + \sum_{t=0}^{L+\tau_0+\tau_1+2} \bar{v}(\eta_t, \eta_{t+1}). \tag{3.186}$$

It follows from (3.168), (3.174), (3.185), (3.186), the inequality $\delta_2 < \bar{r}_v/2$, the choice of δ_2 (see (3.160)), Proposition 2.19, Corollary 2.21, (v, Ω)-overtaking optimality of $\{\xi_t\}_{t=0}^{\infty}$ and $(\bar{v}, \bar{\Omega})$-overtaking optimality of $\{\eta_t\}_{t=0}^{\infty}$ (see (3.175)) that

$$g(x_0, y_0) - g(\xi_0, \eta_0)$$

$$+ \sum_{t=0}^{\tau_0-1} (v(x_t, x_{t+1}) - v(\bar{x}_v, \bar{x}_v)) - \pi^v(x_0) + \pi^v(x_{\tau_0})$$

$$+ \sum_{t=0}^{\tau_0-1} (\bar{v}(y_t, y_{t+1}) - \bar{v}(\bar{x}_v, \bar{x}_v)) - \pi^{\bar{v}}(y_0) + \pi^{\bar{v}}(y_{\tau_0})$$

$$\geq g(x_0, y_0) - g(\xi_0, \eta_0)$$

$$+ \sum_{t=0}^{L+\tau_0+\tau_1+2} (v(x_t, x_{t+1}) - v(\bar{x}_v, \bar{x}_v)) - \pi^v(x_0) + \pi^v(x_{L+\tau_0+\tau_1+3})$$

$$+ \sum_{t=0}^{L+\tau_0+\tau_1+2} (\bar{v}(y_t, y_{t+1}) - \bar{v}(\bar{x}_v, \bar{x}_v)) - \pi^{\bar{v}}(y_0) + \pi^{\bar{v}}(y_{L+\tau_0+\tau_1+3})$$

$$\geq \sum_{t=0}^{L+\tau_0+\tau_1+2} (v(\xi_t, \xi_{t+1}) - v(\bar{x}_v, \bar{x}_v)) - \pi^v(x_0) + \pi^v(x_{L+\tau_0+\tau_1+3})$$

$$+ \sum_{t=0}^{L+\tau_0+\tau_1+2} (\bar{v}(\eta_t, \eta_{t+1}) - \bar{v}(\bar{x}_v, \bar{x}_v)) - \pi^{\bar{v}}(y_0) + \pi^{\bar{v}}(y_{L+\tau_0+\tau_1+3})$$

$$-4\delta(L + \tau_0 + \tau_1 + 8) - 3\delta_1/4$$

$$\geq \sum_{t=0}^{L+\tau_0+\tau_1+2} (v(\xi_t, \xi_{t+1}) - v(\bar{x}_v, \bar{x}_v)) - \pi^v(\xi_0) + \pi^v(\xi_{L+\tau_0+\tau_1+3})$$

$$+\pi^v(\xi_0) - \pi^v(\xi_{L+\tau_0+\tau_1+3}) - \pi^v(x_0) + \pi^v(x_{L+\tau_0+\tau_1+3})$$

$$+\sum_{t=0}^{L+\tau_0+\tau_1+2}(\bar{v}(\eta_t,\eta_{t+1})-\bar{v}(\bar{x}_v,\bar{x}_v))-\pi^{\bar{v}}(\eta_0)+\pi^{\bar{v}}(\eta_{L+\tau_0+\tau_1+3})$$
$$+\pi^{\bar{v}}(\eta_0)-\pi^{\bar{v}}(\eta_{L+\tau_0+\tau_1+3})-\pi^{\bar{v}}(y_0)+\pi^{\bar{v}}(y_{L+\tau_0+\tau_1+3})$$
$$-4\delta(L+\tau_0+\tau_1+8)-3\delta_1/4$$
$$\geq \pi^v(\xi_0)-\pi^v(x_0)+\pi^{\bar{v}}(\eta_0)-\pi^{\bar{v}}(y_0)+\pi^v(x_{L+\tau_0+\tau_1+3})$$
$$-\pi^v(\xi_{L+\tau_0+\tau_1+3})-\pi^{\bar{v}}(\eta_{L+\tau_0+\tau_1+3})+\pi^{\bar{v}}(y_{L+\tau_0+\tau_1+3})$$
$$-4\delta(L+\tau_0+\tau_1+8)-3\delta_1/4$$
$$\geq \pi^v(\xi_0)-\pi^v(x_0)+\pi^{\bar{v}}(\eta_0)-\pi^{\bar{v}}(y_0)-4\delta(L+\tau_0+\tau_1+8)-5\delta_1/4. \quad (3.187)$$

In view of (3.169) and (3.187), we have

$$g(x_0,y_0)+\pi^v(x_0)+\pi^{\bar{v}}(y_0)-(g(\xi_0,\eta_0)+\pi^v(\xi_0)+\pi^{\bar{v}}(\eta_0))$$
$$+\sum_{t=0}^{\tau_0-1}(v(x_t,x_{t+1})-v(\bar{x}_v,\bar{x}_v))-\pi^v(x_0)+\pi^v(x_{\tau_0})$$
$$+\sum_{t=0}^{\tau_0-1}(\bar{v}(y_t,y_{t+1})-\bar{v}(\bar{x}_v,\bar{x}_v))-\pi^{\bar{v}}(y_0)+\pi^{\bar{v}}(y_{\tau_0})\geq -2\delta_1. \quad (3.188)$$

By (3.158), (3.175), Definition 3.16, and Proposition 2.19 applied to the programs $\{x_t\}_{t=0}^{\tau_0}$ and $\{y_t\}_{t=0}^{\tau_0}$, we have

$$g(x_0,y_0)+\pi^v(x_0)+\pi^{\bar{v}}(y_0)-(g(\xi_0,\eta_0)+\pi^v(\xi_0)+\pi^{\bar{v}}(\eta_0))\geq -2\delta_1,$$
$$\sum_{t=0}^{\tau_0-1}(v(x_t,x_{t+1})-v(\bar{x}_v,\bar{x}_v))-\pi^v(x_0)+\pi^v(x_{\tau_0})\geq -2\delta_1,$$
$$\sum_{t=0}^{\tau_0-1}(\bar{v}(y_t,y_{t+1})-\bar{v}(\bar{x}_v,\bar{x}_v))-\pi^{\bar{v}}(y_0)+\pi^{\bar{v}}(y_{\tau_0})\geq -2\delta_1.$$

It follows from the relations above, (3.175), (3.185), Definition 3.16, and property (Piii) that there exists $(\{x_t^*\}_{t=0}^{\infty}, \{y_t^*\}_{t=0}^{\infty})\in\mathcal{L}(g,v)$ such that for all $t=0,\dots,\tau_0$, $\rho(x_t,x_t^*)\leq\epsilon$, $\rho(x_{T-t},y_t^*)=\rho(y_t,y_t^*)\leq\epsilon$. Theorem 3.15 is proved. □

3.12 Proof of Theorem 3.17

In view of (A2), for every function $u \in \mathcal{M}_0(\Omega)$ and every natural number T, we have

$$\sigma(u, T, \bar{x}_u, \bar{x}_u) = Tu(\bar{x}_u, \bar{x}_u). \tag{3.189}$$

We claim that the following property holds:

(Pv) for every $\gamma \in (0, \bar{r}_v/4)$ there exists a number $\delta_\gamma \in (0, \gamma)$ such that for every function $u \in B_d(v, \delta_\gamma) \cap \mathcal{M}_0(\Omega)$ the inequality

$$\rho(\bar{x}_u, \bar{x}_v) \le \gamma \tag{3.190}$$

is valid.

Let $\gamma \in (0, \bar{r}_v/4)$. Assertion 1 of Theorem 2.5 implies that there exists a number $\tilde{\delta} \in (0,1)$ and an integer $\tilde{L} \ge 1$ such that the following property holds:

(Pvi) for every natural number $T \ge \tilde{L}$, every finite sequence of functions $\{u_t\}_{t=0}^{T-1} \subset \mathcal{M}(\Omega)$, and every (Ω)-program $\{x_t\}_{t=0}^{T}$ which satisfy $\|u_t - v\| \le \tilde{\delta}$, $t = 0, \dots, T-1$ and

$$\sum_{t=0}^{T-1} u_t(x_t, x_{t+1}) \ge \sigma(\{u_t\}_{t=0}^{T-1}, 0, T) - 2\bar{c}_v - 4$$

the inequality

$$\mathrm{Card}(\{t \in \{0, \dots, T\} :\ \rho(x_t, \bar{x}_v) > 2^{-1}\gamma\}) < \tilde{L}$$

is true.

Fix a number $\delta_\gamma > 0$ such that

$$\delta_\gamma < \tilde{\delta},\ \delta_\gamma < \gamma,\ \delta_\gamma < (2\tilde{L})^{-1}. \tag{3.191}$$

Assume that

$$u \in B_d(v, \delta_\gamma) \cap \mathcal{M}_0(\Omega). \tag{3.192}$$

By (3.192), relation (3.189) is valid for every natural number T. Set

$$\xi_t = \bar{x}_u,\ t = 0, 1, \dots. \tag{3.193}$$

We show that (3.190) is true. Assume the contrary. Then we have

$$\rho(\xi_t, \bar{x}_v) > \gamma,\ t = 0, 1, \dots. \tag{3.194}$$

By (3.191)–(3.194) and property (Pvi), for every nonnegative integer k, we have

$$\sum_{t=k\tilde{L}}^{(k+1)\tilde{L}-1} u(\xi_t, \xi_{t+1}) < \sigma(u, 0, \tilde{L}) - 2\bar{c}_v - 4. \tag{3.195}$$

In view of (3.191), (3.192), and (A2),

$$\sigma(u,0,\tilde{L}) \le \sigma(v,0,\tilde{L}) + \delta_\gamma \tilde{L} \le \tilde{L}v(\bar{x}_v,\bar{x}_v) + \bar{c}_v + \delta_\gamma \tilde{L}$$

$$\le \tilde{L}u(\bar{x}_v,\bar{x}_v) + \bar{c}_v + 2\delta_\gamma \tilde{L} \le \tilde{L}u(\bar{x}_v,\bar{x}_v) + \bar{c}_v + 1. \quad (3.196)$$

Relations (3.195) and (3.196) imply that for every nonnegative integer k,

$$\sum_{t=k\tilde{L}}^{(k+1)\tilde{L}-1} u(\xi_t,\xi_{t+1}) \le \tilde{L}u(\bar{x}_v,\bar{x}_v) - \bar{c}_v - 3.$$

Combined with (3.193) the inequality above implies that

$$\tilde{L}u(\bar{x}_u,\bar{x}_u) \le \tilde{L}u(\bar{x}_v,\bar{x}_v) - \bar{c}_v - 3, \ u(\bar{x}_u,\bar{x}_u) < u(\bar{x}_v,\bar{x}_v).$$

This contradicts (A2). The contradiction we have reached proves (3.190). Thus we have shown that for every function u satisfying (3.192), inequality (3.190) holds true. Therefore property (Pv) holds as claimed.

It follows from property (Pv) that there exists a number $\delta_0 \in (0,\epsilon)$ such that for every function $u \in B_d(v,\delta_0) \cap \mathcal{M}_0(\Omega)$ the inequality

$$\rho(\bar{x}_u,\bar{x}_v) \le \bar{r}_v/4 \quad (3.197)$$

holds. Theorem 3.15 implies that there exist a number $\delta_1 \in (0,\delta_0)$ and a natural number $T_0 \ge \tau_0$ such that the following property holds:

(Pvii) for every natural number $T \ge T_0$, every function $h \in \mathcal{M}(X \times X)$ which satisfies $\|h-g\| \le \delta_1$, every finite sequence of functions $\{u_t\}_{t=0}^{T-1} \subset \mathcal{M}_0(\Omega)$ which satisfies $\|u_t - v\| \le \delta_1$, $t = 0\dots,T-1$ and every (Ω)-program $\{z_t\}_{t=0}^{T}$ satisfying

$$h(z_0,z_T) + \sum_{t=0}^{T-1} u_t(z_t,z_{t+1}) \ge \sigma(h,\{u_t\}_{t=0}^{T-1},0,T) - 4\delta_1$$

there exist a (v,Ω)-overtaking optimal and (v,Ω)-good program $\{x_t\}_{t=0}^{\infty}$ and a $(\bar{v},\bar{\Omega})$-overtaking optimal and $(\bar{v},\bar{\Omega})$-good program $\{\bar{x}_t\}_{t=0}^{\infty}$ such that

$$\pi^v(x_0) + \pi^{\bar{v}}(\bar{x}_0) + g(x_0,\bar{x}_0) \ge \pi^v(\xi) + \pi^{\bar{v}}(\eta) + g(\xi,\eta) \text{ for all } \xi,\eta \in X$$

and that for all integers $t = 0,\dots,\tau_0$, we have $\rho(x_t,z_t) \le \epsilon$ and $\rho(z_{T-t},\bar{x}_t) \le \epsilon$.

Propositions 2.22, 2.23, and 2.31 and assumption (A1) imply that there exists a number $\delta_2 \in (0,\bar{r}_v/4)$ such that for every point $x \in X$ satisfying $\rho(x,\bar{x}_v) \le 4\delta_2$, we have

$$|\pi^v(x)| \le \delta_1/8, \ |\pi^{\bar{v}}(x)| \le \delta_1/8 \quad (3.198)$$

and that for every point $(x,y) \in X \times X$ which satisfies

$$\rho(x,\bar{x}_v) \le 4\delta_2, \ \rho(y,\bar{x}_v) \le 4\delta_2,$$

we have

$$|v(x,y) - \bar{v}(\bar{x}_v, \bar{x}_v)| \le \delta_1/8. \tag{3.199}$$

In view of property (Pv), there exists a number $\delta_3 \in (0, \delta_1/8)$ such that for every function $u \in B_d(v, \delta_3) \cap \mathcal{M}_0(\Omega)$, we have

$$\rho(\bar{x}_u, \bar{x}_v) \le \delta_2. \tag{3.200}$$

Assume that

$$u \in B_d(v, \delta_3) \cap \mathcal{M}_*(\Omega) \tag{3.201}$$

and that a point $(x, y) \in X \times X$ satisfies

$$\rho(x, \bar{x}_u) \le 2\delta_2,\ \rho(y, \bar{x}_u) \le 2\delta_2. \tag{3.202}$$

By (3.201), relation (3.200) is valid. In view of (3.200) and (3.202),

$$\rho(x, \bar{x}_v) \le 4\delta_2,\ \rho(y, \bar{x}_v) \le 4\delta_2. \tag{3.203}$$

It follows from (3.200), (3.203), and the choice of δ_2 (see (3.198) and (3.199)) that

$$|v(x,y) - \bar{v}(\bar{x}_v, \bar{x}_v)| \le \delta_1/8,\ |v(\bar{x}_u, \bar{x}_u) - \bar{v}(\bar{x}_v, \bar{x}_v)| \le \delta_1/8.$$

In view of the inequalities above,

$$|v(x,y) - v(\bar{x}_u, \bar{x}_u)| \le \delta_1/4.$$

Combined with (3.199) this implies that

$$|u(x,y) - u(\bar{x}_u, \bar{x}_u)| \le \delta_1/2.$$

Thus we have shown that the following property holds:

(Pviii) for every function u satisfying (3.201) and every pair of points $x, y \in X$ which satisfies (3.202),

$$|u(x,y) - u(\bar{x}_u, \bar{x}_u)| \le \delta_1/2.$$

Assume that a function u satisfies (3.201) and points $x, y \in X$ satisfy (3.202). Then relations (3.200) and (3.203) are true. Let $\{y_i\}_{i=0}^{\infty}$ be an (u, Ω)-good program satisfying

$$y_0 = y. \tag{3.204}$$

Define

$$x_0 = x,\ x_i = y_{i-1},\ i = 1, 2, \dots. \tag{3.205}$$

It follows from (3.203)–(3.205) and the inequality $\delta_2 < \bar{r}_v/4$ that $\{x_t\}_{t=0}^{\infty}$ is an (Ω)- program. In view of (2.116), (3.201), (3.202), (3.204), (3.205), and property (Pviii), we have

$$\pi^u(x) \geq \limsup_{k\to\infty} \sum_{i=0}^{k-1} (u(x_i, x_{i+1}) - u(\bar{x}_u, \bar{x}_u))$$

$$= (u(x, y) - u(\bar{x}_u, \bar{x}_u)) + \limsup_{k\to\infty} \sum_{i=0}^{k-1} (u(y_i, y_{i+1}) - u(\bar{x}_u, \bar{x}_u))$$

$$\geq -\delta_1/2 + \limsup_{k\to\infty} \sum_{i=0}^{k-1} (u(y_i, y_{i+1}) - u(\bar{x}_u, \bar{x}_u)).$$

Since $\{y_i\}_{i=0}^{\infty}$ is an arbitrary (u, Ω)-good program satisfying (3.204) we obtain that $\pi^u(x) \geq \pi^u(y) - \delta_1/2$. Thus we have shown that the following property holds:

(Pix) for every function u satisfying (3.201) and every pair of points $x, y \in X$ satisfying (3.202), we have $|\pi^u(x) - \pi^u(y)| \leq \delta_1/2$ and in view of (3.200) and Proposition 2.22, $|\pi^u(x)| \leq \delta_1/2$.

Analogously we can show that the following property holds:

(Px) for every function u satisfying (3.201) and every pair of points $x, y \in X$ satisfying (3.202), we have $|\pi^{\bar{u}}(x) - \pi^{\bar{u}}(y)| \leq \delta_1/2$ and $|\pi^{\bar{u}}(x)| \leq \delta_1/2$.

Theorem 3.14 implies that there exist a positive number δ_4 and a natural number T_1 such that the following property holds:

(Pxi) for every natural number $T > 2T_1$, every function $u \in \mathcal{M}_0(\Omega)$ satisfying $\|u - v\| \leq \delta_4$, every function $h \in \mathcal{M}(X \times X)$ satisfying $\|h - g\| \leq 1$, and every (Ω)-program $\{x_t\}_{t=0}^{T}$ which satisfies $h(x_0, x_T) + \sum_{t=0}^{T-1} u(x_t, x_{t+1}) \geq \sigma(h, u, 0, T) - \delta_4$ the inequality

$$\rho(x_t, \bar{x}_v) \leq \delta_2$$

holds for all $t = T_1, \ldots, T - T_1$.

Choose

$$\delta \in (0, \min\{\delta_0, \delta_1, \delta_2, \delta_3, \delta_4\}). \tag{3.206}$$

Assume that

$$u \in \mathcal{M}_*(\Omega) \cap B_d(v, \delta), \tag{3.207}$$

a function $h \in \mathcal{M}_u(X \times X)$ satisfies

$$\|h - g\| \leq \delta \tag{3.208}$$

and

$$(\{x_t\}_{t=0}^{\infty}, \{y_t\}_{t=0}^{\infty}) \in \mathcal{L}(h, u). \tag{3.209}$$

In view of (3.209), $\{x_t\}_{t=0}^{\infty}$ is a (u, Ω)-overtaking optimal and (u, Ω)-good program, $\{y_t\}_{t=0}^{\infty}$ is a $(\bar{u}, \bar{\Omega})$-overtaking optimal and $(\bar{u}, \bar{\Omega})$-good program such that

$$\pi^u(x_0) + \pi^{\bar{u}}(y_0) + h(x_0, y_0) \geq \pi^u(\xi) + \pi^{\bar{u}}(\eta) + h(\xi, \eta) \text{ for all } \xi, \eta \in X. \tag{3.210}$$

By (3.201), (3.206), (3.207), and the choice of δ_3, relation (3.200) holds. It follows from Proposition 2.31 and (3.207) and (3.209) that

$$\lim_{t\to\infty} x_t = \bar{x}_u,\ \lim_{t\to\infty} y_t = \bar{x}_u. \tag{3.211}$$

In view of (3.211), there exists an integer $S_0 > T_0 + T_1 + 4$ such that

$$\rho(x_t, \bar{x}_u),\ \rho(y_t, \bar{x}_u) \le \delta_2 \text{ for all integers } t \ge S_0. \tag{3.212}$$

By (3.200) and (3.212), for all integers $t \ge S_0$, we have

$$\rho(x_t, \bar{x}_v),\ \rho(y_t, \bar{x}_v) \le 2\delta_2. \tag{3.213}$$

Set

$$z_t = x_t,\ t = 0, \dots, S_0,\ z_t = y_{2S_0+1-t},\ t = S_0 + 1, \dots, 2S_0 + 1. \tag{3.214}$$

It follows from (3.209), (3.213), (3.214), and the inequality $\delta_2 < \bar{r}_v/4$ that $\{z_t\}_{t=0}^{2S_0+1}$ is an (Ω)- program. In view of (3.214),

$$h(z_0, z_{2S_0+1}) + \sum_{t=0}^{2S_0} u(z_t, z_{t+1})$$

$$= h(x_0, y_0) + \sum_{t=0}^{S_0-1} u(x_t, x_{t+1}) + u(x_{S_0}, y_{S_0}) + \sum_{t=S_0+1}^{2S_0} \bar{u}(y_{2S_0-t}, y_{2S_0-t+1})$$

$$= h(x_0, y_0) + \sum_{t=0}^{S_0-1} u(x_t, x_{t+1}) + u(x_{S_0}, y_{S_0}) + \sum_{t=0}^{S_0-1} \bar{u}(y_t, y_{t+1}). \tag{3.215}$$

It follows from Corollary 2.21, Definition 3.16, (3.209), and (3.215) that

$$h(z_0, z_{2S_0+1}) + \sum_{t=0}^{2S_0} u(z_t, z_{t+1})$$

$$= h(x_0, y_0) + S_0 u(\bar{x}_u, \bar{x}_u) + \pi^u(x_0) - \pi^u(x_{S_0})$$
$$+ u(x_{S_0}, y_{S_0}) + S_0 u(\bar{x}_u, \bar{x}_u) + \pi^{\bar{u}}(y_0) - \pi^{\bar{u}}(y_{S_0}). \tag{3.216}$$

By the upper semicontinuity of h, u, there exists an (Ω)- program $\{\xi_t\}_{t=0}^{2S_0+1}$ which satisfies

$$h(\xi_0, \xi_{2S_0+1}) + \sum_{t=0}^{2S_0} u(\xi_t, \xi_{t+1}) = \sigma(h, u, 0, 2S_0 + 1). \tag{3.217}$$

Relations (3.206)–(3.208), (3.217), property (Pxi), and the inequality $S_0 > T_0 + T_1 + 4$ imply that

$$\rho(\xi_t, \bar{x}_v) \le \delta_2,\ t = T_1, \dots, 2S_0 - T_1 + 1. \tag{3.218}$$

In view of (3.218),

$$\rho(\xi_t, \bar{x}_v) \leq \delta_2, \ t = S_0, S_0 + 1. \tag{3.219}$$

By (3.200), (3.207), (3.219), and the choice of δ_3, we have

$$\rho(\xi_t, \bar{x}_u) \leq 2\delta_2, \ t = S_0, S_0 + 1. \tag{3.220}$$

Define

$$\bar{\xi}_t = \xi_{2S_0+1-t}, \ t = 0, \dots, 2S_0 + 1. \tag{3.221}$$

Proposition 2.19, properties (Pviii) and (Pix), (3.207), (3.217), and (3.219)–(3.221) imply that

$$\sigma(h, u, 0, 2S_0 + 1) = \sum_{t=0}^{S_0} u(\xi_t, \xi_{t+1}) + u(\xi_{S_0}, \xi_{S_0+1})$$

$$+ \sum_{t=S_0+1}^{2S_0} \bar{u}(\bar{\xi}_{2S_0-t}, \bar{\xi}_{2S_0-t+1}) + h(\xi_0, \bar{\xi}_0)$$

$$\leq h(\xi_0, \bar{\xi}_0) + S_0 u(\bar{x}_u, \bar{x}_u) + \pi^u(\xi_0) - \pi^u(\xi_{S_0})$$

$$+ u(\xi_{S_0}, \xi_{S_0+1}) + S_0 u(\bar{x}_u, \bar{x}_u) + \pi^{\bar{u}}(\bar{\xi}_0) - \pi^{\bar{u}}(\bar{\xi}_{S_0})$$

$$\leq h(\xi_0, \bar{\xi}_0) + \pi^u(\xi_0) + \pi^{\bar{u}}(\bar{\xi}_0) + (2S_0 + 1)u(\bar{x}_u, \bar{x}_u) + 3\delta_1/2. \tag{3.222}$$

It follows from properties (Pviii) and (Pix), (3.207), (3.212), and (3.216) that

$$h(z_0, z_{2S_0+1}) + \sum_{t=0}^{2S_0} u(z_t, z_{t+1}) \geq h(x_0, y_0)$$

$$+ (2S_0 + 1)u(\bar{x}_u, \bar{x}_u) + \pi^u(x_0) + \pi^{\bar{u}}(y_0) - 3\delta_1/2. \tag{3.223}$$

In view of (3.210), (3.222), and (3.223),

$$h(z_0, z_{2S_0+1}) + \sum_{t=0}^{2S_0} u(z_t, z_{t+1}) - \sigma(h, u, 0, 2S_0 + 1)$$

$$\geq h(x_0, y_0) + \pi^u(x_0) + \pi^{\bar{u}}(y_0) - (h(\xi_0, \bar{\xi}_0) + \pi^u(\xi_0) + \pi^{\bar{u}}(\bar{\xi}_0)) - 3\delta_1 \geq -3\delta_1. \tag{3.224}$$

By (3.206)–(3.208), (3.218), property (Pvii), and the inequality $S_0 > T_0$, there exist a (v, Ω)-overtaking optimal and (v, Ω)-good program $\{x_t^*\}_{t=0}^{\infty}$ and a $(\bar{v}, \bar{\Omega})$-overtaking optimal and $(\bar{v}, \bar{\Omega})$-good program $\{\bar{x}_t^*\}_{t=0}^{\infty}$ such that

$$\pi^v(x_0^*) + \pi^{\bar{v}}(\bar{x}_0^*) + g(x_0^*, \bar{x}_0^*) \geq \pi^v(\xi) + \pi^{\bar{v}}(\eta) + g(\xi, \eta) \text{ for all } \xi, \eta \in X$$

and that for all $t = 0, \dots, \tau_0$, we have $\rho(x_t^*, z_t) \leq \epsilon$ and $\rho(z_{2S_0-t+1}, \bar{x}_t^*) \leq \epsilon$. Combined with (3.214) and the inequality $S_0 > \tau_0 + 4$ this implies that for all integers $t = 0, \dots, \tau_0$, $\rho(x_t, x_t^*) \leq \epsilon$ and $\epsilon \geq \rho(\bar{x}_t^*, z_{2S_0-t+1}) = \rho(y_t, \bar{x}_t^*)$. This completes the proof of Theorem 3.17. □

3.13 Proof of Theorem 3.18

Since the mapping $v \to \bar{v}$, $v \in \mathcal{M}(\Omega)$ is a linear isometry which has the inverse, and a mapping $h \to \bar{h}$, $h \in \mathcal{M}(X \times X)$, where $\bar{h}(x,y) = h(y,x)$, $x, y \in X$, is also a linear isometry which has the inverse, Theorem 3.18 follows from the next result.

Theorem 3.20. *Let $\mathfrak{M}$ be either $\bar{\mathcal{M}}_0(\Omega)$ or $\bar{\mathcal{M}}_{c,0}(\Omega)$ and let $\mathfrak{A}$ be either $\mathcal{M}_u(X \times X)$ or $\mathcal{M}_c(X \times X)$. Then there exists a set $\mathcal{F} \subset (\mathfrak{M} \cap \mathcal{M}_*(\Omega)) \times \mathfrak{A}$ which is a countable intersection of open everywhere dense subsets of $\mathfrak{M} \times \mathfrak{A}$ such that for each $(v,g) \in \mathcal{F}$ there exists a unique $(\bar{v}, \bar{\Omega})$-overtaking optimal and $(\bar{v}, \bar{\Omega})$-good program $\{\bar{x}_t^{(v,g)}\}_{t=0}^\infty$ such that for each $(\{x_t\}_{t=0}^\infty, (\{y_t\}_{t=0}^\infty) \in \mathcal{L}(g,v)$, $y_t = \bar{x}_t^{(v,g)}$ for all integers $t \geq 0$.*

Let $\mathfrak{M}$ be either $\bar{\mathcal{M}}_0(\Omega)$ or $\bar{\mathcal{M}}_{c,0}(\Omega)$ and let $\mathfrak{A}$ be either $\mathcal{M}_u(X \times X)$ or $\mathcal{M}_c(X \times X)$. Theorem 2.6 implies that the set $\mathfrak{M} \cap \mathcal{M}_*(\Omega)$ contains a subset which is a countable intersection of open everywhere dense subsets of $\mathfrak{M}$.

For every pair of functions $(v,g) \in \mathcal{M}_*(\Omega) \times \mathcal{M}_u(X \times X)$ define

$$\phi_{v,g}(x,y) = \pi^v(x) + \pi^{\bar{v}}(y) + g(x,y), \; x, y \in X, \tag{3.225}$$

$$\mathcal{D}(v,g) = \{(x,y) \in X \times X : \; \phi_{v,g}(x,y) = \sup(\phi_{v,g})\}. \tag{3.226}$$

Denote by E_0 the collection of all pairs of functions $(v,g) \in (\mathfrak{M} \cap \mathcal{M}_*(\Omega)) \times \mathfrak{A}$ for which the following property holds:

(Pxii) there exists a unique $(\bar{v}, \bar{\Omega})$-overtaking optimal and $(\bar{v}, \bar{\Omega})$-good program $\{\bar{x}_t^{(v,g)}\}_{t=0}^\infty$ such that for every (v, Ω)-overtaking optimal and (v, Ω)-good program $\{x_t\}_{t=0}^\infty$ and every $(\bar{v}, \bar{\Omega})$-overtaking optimal and $(\bar{v}, \bar{\Omega})$-good program $\{y_t\}_{t=0}^\infty$ which satisfy $\phi_{v,g}(x_0, y_0) = \sup(\phi_{v,g})$ the equality $y_t = \bar{x}_t^{(v,g)}$ holds for all nonnegative integers t.

The proof of Theorem 3.20 is based on the following auxiliary result.

Proposition 3.21. *Let $(v,g) \in (\mathfrak{M} \cap \mathcal{M}_*(\Omega)) \times \mathfrak{A}$ and $\mathcal{U}$ be a neighborhood of $(v,g) \in \mathfrak{M} \times \mathfrak{A}$. Then $\mathcal{U} \cap E_0 \neq \emptyset$.*

Proposition 3.21 implies that E_0 is an everywhere dense subset of $\mathfrak{M} \times \mathfrak{A}$.

Denote by E_1 the collection of all pairs of functions $(v,g) \in (\mathfrak{M} \cap \mathcal{M}_*(\Omega)) \times \mathfrak{A}$ such that $\phi_{v,g}$ has a unique point of maximum. We proceed the proof of Proposition 3.21 with the following lemmas.

Lemma 3.22. *Let $(v,g) \in (\mathfrak{M} \cap \mathcal{M}_*(\Omega)) \times \mathfrak{A}$ and $\mathcal{U}$ be a neighborhood of $(v,g) \in \mathfrak{M} \times \mathfrak{A}$. Then $\mathcal{U} \cap E_1 \neq \emptyset$.*

Proof. Evidently, there exists a pair of points $x_*, y_* \in X$ for which

$$\phi_{v,g}(x_*, y_*) = \sup(\phi_{v,g}). \tag{3.227}$$

Let $\gamma \in (0,1)$. For all points $x, y \in X$ set

$$g_\gamma(x,y) = g(x,y) - \gamma(\rho(x,x_*) + \rho(y,y_*)). \tag{3.228}$$

In view of (3.228), $g_\gamma \in \mathfrak{A}$ and there exists $\gamma_0 \in (0,1)$ such that for all $\gamma \in (0,\gamma_0]$, we have

$$(v, g_\gamma) \in \mathcal{U}.$$

Let $\gamma \in (0,\gamma_0]$. In order to complete the proof it is sufficient to show that $(v, g_\gamma) \in E_1$. Let $x, y \in X$. In view of (3.225), (3.227), and (3.228), we have

$$\phi_{v,g_\gamma}(x,y) = \phi_{v,g}(x,y) - \gamma(\rho(x,x_*) + \rho(y,x_*))$$

$$\leq \phi_{v,g}(x_*,y_*) - \gamma(\rho(x,x_*) + \rho(y,x_*)) = \phi_{v,g_\gamma}(x_*,y_*) - \gamma(\rho(x,x_*) + \rho(y,x_*)).$$

This implies that (x_*, y_*) is a unique point of maximum of ϕ_{v,g_γ} and completes the proof of Lemma 3.22. □

Lemma 3.23. *Let*

$$(v,g) \in E_1 \tag{3.229}$$

and $x_* \in X$ satisfy

$$\phi_{v,g}(x_*, \bar{x}_v) = \sup(\phi_{v,g}). \tag{3.230}$$

Then $(v,g) \in E_1$.

Proof. Let $\{x_t\}_{t=0}^\infty$ be a (v,Ω)-overtaking optimal and (v,Ω)-good program, $\{y_t\}_{t=0}^\infty$ be a $(\bar{v},\bar{\Omega})$-overtaking optimal and $(\bar{v},\bar{\Omega})$-good program for which

$$\phi_{v,g}(x_0, y_0) = \sup(\phi_{v,g}). \tag{3.231}$$

In view of (3.229) and (3.231), we have $x_0 = x_*$, $y_0 = \bar{x}_v$. Combined with Theorem 2.16 this implies that $y_t = \bar{x}_v$ for all nonnegative integers t. Lemma 3.23 is proved. □

Lemma 3.24. *Let*

$$(v,g) \in E_1,\ x_* \in X \setminus \{\bar{x}_v\}, \tag{3.232}$$

$$\phi_{v,g}(\bar{x}_v, x_*) = \sup(\phi_{v,g}) \tag{3.233}$$

and $\mathcal{U}$ be a neighborhood of $(v,g) \in \mathfrak{M} \times \mathfrak{A}$. Then $\mathcal{U} \cap E_0 \neq \emptyset$.

Proof. Define a sequence

$$x_t^* = \bar{x}_v,\ t = 0, 1, \ldots \tag{3.234}$$

which is (v,Ω)-overtaking optimal and (v,Ω)-good program. Theorem 2.13 and (3.233) imply that there exists a $(\bar{v},\bar{\Omega})$-overtaking optimal and $(\bar{v},\bar{\Omega})$-good program $\{\bar{x}_t^*\}_{t=0}^\infty$ satisfying

$$\bar{x}_0^* = x_*. \tag{3.235}$$

Let $\gamma \in (0,1)$. For all points $x, y \in X$ set

$$v_\gamma(x,y) = v(x,y) - \gamma\rho((x,y),(\{(\bar{x}^*_{t+1},\bar{x}^*_t) : t = 0,1,\dots\} \cup \{(\bar{x}_v,\bar{x}_v)\})). \tag{3.236}$$

In view of (3.236),

$$v_\gamma \in \mathfrak{M} \cap \mathcal{M}_0(\Omega),\ \bar{x}_{v_\gamma} = \bar{x}_v \tag{3.237}$$

and in view of Proposition 2.1 and (A3), every (v_γ,Ω)-good program is (v,Ω)-good and converges to $\bar{x}_v$. Thus

$$v_\gamma \in \mathcal{M}_*(\Omega). \tag{3.238}$$

Relations (3.236) and (3.237) imply that

$$v_\gamma(\bar{x}_v,\bar{x}_v) = v(\bar{x}_v,\bar{x}_v),\ \pi^{v_\gamma}(y) \le \pi^v(y),\ y \in X,\ \pi^{\bar{v}_\gamma}(z) \le \pi^{\bar{v}}(z),\ z \in X. \tag{3.239}$$

It follows from Proposition 2.24, (3.235)–(3.237), and $(\bar{v},\bar{\Omega})$-overtaking optimality of the program $\{\bar{x}^*_t\}_{t=0}^\infty$ that

$$\pi^{\bar{v}_\gamma}(x_*) \le \pi^{\bar{v}}(x_*) = \lim_{T\to\infty}\sum_{t=0}^{T-1}(\bar{v}(\bar{x}^*_t,\bar{x}^*_{t+1}) - v(\bar{x}_v,\bar{x}_v))$$

$$= \lim_{T\to\infty}\sum_{t=0}^{T-1}(\bar{v}_\gamma(\bar{x}^*_t,\bar{x}^*_{t+1}) - v_\gamma(\bar{x}_v,\bar{x}_v)) \le \pi^{\bar{v}_\gamma}(x_*). \tag{3.240}$$

By (3.240),

$$\pi^{\bar{v}_\gamma}(x_*) = \pi^{\bar{v}}(x_*). \tag{3.241}$$

Assume that $\{x_t\}_{t=0}^\infty$ is a (v_γ,Ω)-overtaking optimal and (v_γ,Ω)-good program, $\{y_t\}_{t=0}^\infty$ is a $(\bar{v}_\gamma,\bar{\Omega})$-overtaking optimal and $(\bar{v}_\gamma,\bar{\Omega})$-good program such that

$$\phi_{v_\gamma,g}(x_0,y_0) = \sup(\phi_{v_\gamma,g}). \tag{3.242}$$

By (3.233), (3.239), (3.241), and (3.242), we have

$$\sup(\phi_{v_\gamma,g}) = \sup(\phi_{v,g}) = \pi^v(\bar{x}_v) + \pi^{\bar{v}}(x_*) + g(\bar{x}_v,x_*)$$

$$\ge \pi^v(x_0) + \pi^{\bar{v}}(y_0) + g(x_0,y_0) \ge \pi^{v_\gamma}(x_0) + \pi^{\bar{v}_\gamma}(y_0) + g(x_0,y_0) = \sup(\phi_{v_\gamma,g}). \tag{3.243}$$

By (3.232), (3.233), and (3.243),

$$x_0 = \bar{x}_v,\ y_0 = x_*. \tag{3.244}$$

It follows from Proposition 2.24, (3.233), (3.236), (3.241), (3.244), and $(\bar{v}_\gamma,\bar{\Omega})$-overtaking optimality of the program $\{y_t\}_{t=0}^\infty$ that

$$\pi^{\bar{v}_\gamma}(x_*) = \lim_{T\to\infty}\sum_{t=0}^{T-1}(\bar{v}_\gamma(y_t,y_{t+1}) - v(\bar{x}_v,\bar{x}_v))$$

$$= \lim_{T \to \infty} \sum_{t=0}^{T-1} [\bar{v}(y_t, y_{t+1}) - v(\bar{x}_v, \bar{x}_v)$$

$$-\gamma \rho((y_t, y_{t+1}), \{(\bar{x}_t^*, \bar{x}_{t+1}^*) : t = 0, 1, \dots\} \cup \{(\bar{x}_v, \bar{x}_v)\})]$$

$$\leq \pi^{\bar{v}}(x_*) - \gamma \sum_{t=0}^{\infty} \rho((y_t, y_{t+1}), \{(\bar{x}_t^*, \bar{x}_{t+1}^*) : t = 0, 1, \dots\} \cup \{(\bar{x}_v, \bar{x}_v)\}).$$

This implies that

$$(y_j, y_{j+1}) \in \{(\bar{x}_t^*, \bar{x}_{t+1}^*) : t = 0, 1, \dots\} \cup \{(\bar{x}_v, \bar{x}_v)\} \quad (3.245)$$

for all nonnegative integers j. There are two cases:

$$\bar{x}_t^* \neq \bar{x}_v \text{ for all integers } t \geq 0; \quad (3.246)$$

$$\bar{x}_v \in \{\bar{x}_t^* : t = 0, 1, \dots\}. \quad (3.247)$$

Assume that (3.246) is valid. We claim that

$$y_t = \bar{x}_t^* \quad (3.248)$$

for all nonnegative integers t. By (3.244) and (3.245), relation (3.248) is valid for $t = 0$. Assume that T is a nonnegative integer and that (3.248) is true for all integers $t = 0, \dots, T$. In view of (3.245) and (3.248), we have

$$(\bar{x}_T^*, y_{T+1}) = (y_T, y_{T+1}) \in \{(\bar{x}_t^*, \bar{x}_{t+1}^*) : t = 0, 1, \dots\} \cup \{(\bar{x}_v, \bar{x}_v)\}. \quad (3.249)$$

By (3.246) and (3.249) that there exists a nonnegative integer j for which

$$(\bar{x}_T^*, y_{T+1}) = (\bar{x}_j^*, \bar{x}_{j+1}^*). \quad (3.250)$$

It follows from (3.250) that

$$\bar{x}_T^* = \bar{x}_j^*. \quad (3.251)$$

If $T \neq j$, then by Proposition 2.39 and (3.251), we have $\bar{x}_t^* = \bar{x}_v$ for all $t = \min\{j, T\}, \dots, \max\{j, T\}$, $\bar{x}_T^* = \bar{x}_v$ and this contradicts (3.246). Thus $T = j$ and by (3.250), we have $y_{T+1} = \bar{x}_{T+1}^*$. Therefore (3.248) is true for all nonnegative integers t.

Assume that (3.247) is valid. It follows from (3.232), (3.235), and Theorem 2.16 that there exists an integer $S \geq 1$ such that

$$\bar{x}_t^* \neq \bar{x}_v, \; t = 0, \dots, S-1, \; \bar{x}_t^* = \bar{x}_v \text{ for all integers } t \geq S. \quad (3.252)$$

We claim that (3.248) is true for all nonnegative integers t. By (3.235) and (3.244), (3.248) holds for $t = 0$. Assume that T is a nonnegative integer and that (3.248) is valid for all integers $t = 0, \dots, T$. Assume that $T < S$. In view of (3.245), (3.248), and (3.252), we have

$$(\bar{x}_T^*, y_{T+1}) = (y_T, y_{T+1}) \in \{(\bar{x}_t^*, \bar{x}_{t+1}^*) : t = 0, \dots, S-1\} \cup \{(\bar{x}_v, \bar{x}_v)\}. \quad (3.253)$$

It follows from (3.245), (3.252), (3.253), and the inequality $T < S$ that there exists an integer $j \in \{0, \dots, S-1\}$ for which

$$(\bar{x}_T^*, y_{T+1}) = (\bar{x}_j^*, \bar{x}_{j+1}^*). \tag{3.254}$$

If $T \neq j$, then Proposition 2.39 and (3.254) imply that $\bar{x}_t^* = \bar{x}_v$ for all integers $t = \min\{j, T\}, \dots, \max\{j, T\}$ and this contradicts (3.252). Thus $T = j$ and $y_{T+1} = \bar{x}_{T+1}^*$.

Assume that $T \geq S$. By (3.245), (3.248), and (3.252), we have

$$(\bar{x}_v, y_{T+1}) = (y_T, y_{T+1}) \in \{(\bar{x}_t^*, \bar{x}_{t+1}^*) : t = 0, \dots, S-1\} \cup \{(\bar{x}_v, \bar{x}_v)\}.$$

Combined with (3.252) the relation above implies that

$$(\bar{x}_v, y_{T+1}) = (\bar{x}_v, \bar{x}_v), \ y_{T+1} = \bar{x}_v = \bar{x}_{T+1}^*.$$

Therefore in both the cases (3.248) is true for all nonnegative integers t. This implies that $(v_\gamma, g) \in E_0$. Since $v_\gamma \to v$ as $\gamma \to 0^+$ in $\mathfrak{M}$ there exists a number $\gamma_0 \in (0, 1)$ such that $(v_\gamma, g) \in \mathcal{U}$ for all $\gamma \in (0, \gamma_0)$. This completes the proof of Lemma 3.24. □

Lemma 3.25. *Let*

$$(v, g) \in E_1, \tag{3.255}$$

$$x^*, y^* \in X \setminus \{\bar{x}_v\}, \tag{3.256}$$

$$\phi_{v,g}(x^*, y^*) = \sup(\phi_{v,g}) \tag{3.257}$$

and $\mathcal{U}$ *be a neighborhood of* $(v, g) \in \mathfrak{M} \times \mathfrak{A}$. *Then* $\mathcal{U} \cap E_0 \neq \emptyset$.

Proof. For every point $x \in X$ such that $\pi^{\bar{v}}(x) > -\infty$, denote by $\Lambda(x)$ the set of all points $y \in X$ satisfying

$$\bar{v}(x, y) - v(\bar{x}_v, \bar{x}_v) = \pi^{\bar{v}}(x) - \pi^{\bar{v}}(y) \tag{3.258}$$

which is nonempty by Theorem 2.13 and Corollary 2.21. There exists a (v, Ω)-overtaking optimal and (v, Ω)-good program $\{x_t^*\}_{t=0}^\infty$ such that

$$x_0^* = x^*. \tag{3.259}$$

We construct a $(\bar{v}, \bar{\Omega})$-overtaking optimal and $(\bar{v}, \bar{\Omega})$-good program $\{y_t^*\}_{t=0}^\infty$. Define

$$y_0^* = y^*. \tag{3.260}$$

Assume that T is a nonnegative integer, $y_t^* \in X$, $t = 0, \dots, T$ were defined such that: (3.260) holds; if $T \geq 1$, then

$$\pi^{\bar{v}}(y_T^*) > -\infty, \tag{3.261}$$

$$\sum_{t=0}^{T-1} (\bar{v}(y_t^*, y_{t+1}^*) - v(\bar{x}_v, \bar{x}_v)) = \pi^{\bar{v}}(y_0^*) - \pi^{\bar{v}}(y_T^*). \tag{3.262}$$

We define a point y_{T+1} as follows:

(i) if $y_T^* = \bar{x}_v$, then $y_{T+1}^* = \bar{x}_v$;

(ii) if $y_T^* \neq \bar{x}_v$ and $y_T^* \notin \{x_t^* : t = 0, 1, \dots\}$, then choose $y_{T+1}^* \in \Lambda(y_T^*)$;

(iii) assume that $y_T^* \neq \bar{x}_v$ and $y_T^* \in \{x_t^* : t = 0, 1, \dots\}$; then in view of Proposition 2.39, there exists a unique integer $j(T) \geq 0$ such that $y_T^* = x_{j(T)}^*$; if $j(T) \geq 1$ and

$$\bar{v}(x_{j(T)}^*, x_{j(T)-1}^*) - v(\bar{x}_v, \bar{x}_v) = \pi^{\bar{v}}(x_{j(T)}^*) - \pi^{\bar{v}}(x_{j(T)-1}^*) \tag{3.263}$$

then we set $y_{T+1}^* = x_{j(T)-1}^*$; if $j(T) = 0$ or (3.263) does not hold, then choose $y_{T+1}^* \in \Lambda(y_T^*)$.

It is not difficult to see that in all these cases

$$\pi^{\bar{v}}(y_T^*) - \pi^{\bar{v}}(y_{T+1}^*) = \bar{v}(y_T^*, y_{T+1}^*) - v(\bar{x}_v, \bar{x}_v);\ \pi^{\bar{v}}(y_{T+1}^*) > -\infty.$$

Combined with (3.262) the equation above implies that

$$\sum_{t=0}^{T} (\bar{v}(y_t^*, y_{t+1}^*) - v(\bar{x}_v, \bar{x}_v)) = \pi^{\bar{v}}(y_0^*) - \pi^{\bar{v}}(y_{T+1}^*).$$

Thus by induction we constructed the $(\bar{\Omega})$-program $\{y_t^*\}_{t=0}^\infty$ such that (3.260) is true, for every natural number T (3.261) and (3.262) hold and for every nonnegative integer T one of the cases (i)–(iii) holds. By (3.262), $\{y_t^*\}_{t=0}^\infty$ is a $(\bar{v}, \bar{\Omega})$-good program. It follows from (3.261), (3.262), and Propositions 2.19 and 2.27 that $\{y_t^*\}_{t=0}^\infty$ is $(\bar{v}, \bar{\Omega})$-overtaking optimal program.

Let $\gamma \in (0, 1)$ be given. For all points $x, y \in X$ set

$$\begin{aligned} v_\gamma(x, y) = v(x, y) - \gamma\rho((x, y), (\{(x_t^*, x_{t+1}^*) :\\ t = 0, 1, \dots\} \cup \{(y_{t+1}^*, y_t^*) : t = 0, 1, \dots\} \cup \{(\bar{x}_v, \bar{x}_v)\})). \end{aligned} \tag{3.264}$$

In view of (3.264), we have

$$v_\gamma \in \mathfrak{M} \cap \mathcal{M}_0(\Omega),\ \bar{x}_{v_\gamma} = \bar{x}_v \tag{3.265}$$

and Proposition 2.1 and (A3) imply that every (v_γ, Ω)-good program is (v, Ω)-good and converges to the turnpike $\bar{x}_v$. Thus

$$v_\gamma \in \mathcal{M}_*(\Omega). \tag{3.266}$$

By (3.264), we have

$$v_\gamma(\bar{x}_v, \bar{x}_v) = v(\bar{x}_v, \bar{x}_v),\ \pi^{v_\gamma}(y) \leq \pi^v(y),\ y \in X,\ \pi^{\bar{v}_\gamma}(z) \leq \pi^{\bar{v}}(z),\ z \in X. \tag{3.267}$$

Proposition 2.24, (3.259), (3.264), (3.267), and (v, Ω)-overtaking optimality of the program $\{x_t^*\}_{t=0}^\infty$ imply that

$$\pi^{v_\gamma}(x^*) \le \pi^v(x^*) = \lim_{T\to\infty} \sum_{t=0}^{T-1} (v(x_t^*, x_{t+1}^*) - v(\bar{x}_v, \bar{x}_v))$$

$$= \lim_{T\to\infty} \sum_{t=0}^{T-1} (v_\gamma(x_t^*, x_{t+1}^*) - v_\gamma(\bar{x}_v, \bar{x}_v)) \le \pi^{v_\gamma}(x^*). \tag{3.268}$$

By (3.268), we have

$$\pi^{v_\gamma}(x^*) = \pi^v(x^*). \tag{3.269}$$

Since the sequence $\{y_t^*\}_{t=0}^\infty$ is a $(\bar{v}, \bar{\Omega})$-overtaking optimal and $(\bar{v}, \bar{\Omega})$-good program it follows from Proposition 2.24, (3.264), and (3.267) that for all integers $s \ge 0$, we have

$$\pi^{\bar{v}_\gamma}(y_s^*) \le \pi^{\bar{v}}(y_s^*) = \lim_{T\to\infty} \sum_{t=0}^{T-1} (\bar{v}(y_{s+t}^*, y_{s+t+1}^*) - v(\bar{x}_v, \bar{x}_v))$$

$$= \lim_{T\to\infty} \sum_{t=0}^{T-1} (\bar{v}_\gamma(y_{s+t}^*, y_{s+t+1}^*) - v(\bar{x}_v, \bar{x}_v)) \le \pi^{\bar{v}_\gamma}(y_s^*). \tag{3.270}$$

In view of (3.270),

$$\pi^{\bar{v}_\gamma}(y_s^*) = \pi^{\bar{v}}(y_s^*) \text{ for all integers } s \ge 0, \tag{3.271}$$

$$\pi^{\bar{v}_\gamma}(y_0^*) = \pi^{\bar{v}}(y^*). \tag{3.272}$$

Assume that $\{x_t\}_{t=0}^\infty$ is (v_γ, Ω)-overtaking optimal and (v_γ, Ω)-good program and that $\{y_t\}_{t=0}^\infty$ is a $(\bar{v}_\gamma, \bar{\Omega})$-overtaking optimal and $(\bar{v}_\gamma, \bar{\Omega})$-good program such that

$$\phi_{v_\gamma,g}(x_0, y_0) = \sup(\phi_{v_\gamma,g}). \tag{3.273}$$

In view of (3.257), (3.267), (3.269), and (3.272),

$$\sup(\phi_{v_\gamma,g}) = \sup(\phi_{v,g}) = \pi^v(x^*) + \pi^{\bar{v}}(y^*) + g(x^*, y^*). \tag{3.274}$$

By (3.267) and (3.274), we have

$$\phi_{v,g}(x_0, y_0) \ge \phi_{v_\gamma,g}(x_0, y_0) = \sup(\phi_{v,g}). \tag{3.275}$$

By (3.255), (3.257), (3.259), (3.260), and (3.275),

$$x_0 = x^* = x_0^*,\ y_0 = y^* = y_0^*. \tag{3.276}$$

In view of (3.273) and (3.274),

$$\pi^{v_\gamma}(x_0) + \pi^{\bar{v}_\gamma}(y_0) + g(x_0, y_0) = \sup(\phi_{v,g}) \ge \pi^v(x_0) + \pi^{\bar{v}}(y_0) + g(x_0, y_0). \tag{3.277}$$

In view of (3.267) and (3.277),

$$\pi^{\bar{v}_\gamma}(y_0) = \pi^{\bar{v}}(y_0). \tag{3.278}$$

Define

$$K = (\{(x_t^*, x_{t+1}^*) : t = 0, 1, \ldots\} \cup \{(y_{t+1}^*, y_t^*) : t = 0, 1, \ldots\} \cup \{(\bar{x}_v, \bar{x}_v)\}). \tag{3.279}$$

By Proposition 2.24, (3.264), (3.278), (3.279), and $(\bar{v}_\gamma, \bar{\Omega})$-overtaking optimality of the program $\{y_t\}_{t=0}^\infty$, we have

$$\pi^{\bar{v}}(y_0) = \pi^{\bar{v}_\gamma}(y_0) = \lim_{T\to\infty} \sum_{t=0}^{T-1} (\bar{v}_\gamma(y_t, y_{t+1}) - v(\bar{x}_v, \bar{x}_v))$$

$$= \lim_{T\to\infty} \sum_{t=0}^{T-1} (\bar{v}(y_t, y_{t+1}) - v(\bar{x}_v, \bar{x}_v) - \gamma\rho((y_{t+1}, y_t), K))$$

$$\leq \limsup_{T\to\infty} \sum_{t=0}^{T-1} (\bar{v}(y_t, y_{t+1}) - v(\bar{x}_v, \bar{x}_v)) - \gamma \sum_{t=0}^{\infty} \rho((y_{t+1}, y_t), K)$$

$$\leq \pi^{\bar{v}}(y_0) - \gamma \sum_{t=0}^{\infty} \rho((y_{t+1}, y_t), K).$$

This implies that

$$(y_{t+1}, y_t) \in K,\ t = 0, 1, \ldots. \tag{3.280}$$

We claim that

$$y_t = y_t^* \tag{3.281}$$

for all nonnegative integers t. By (3.276), (3.281) is valid for $t = 0$.

Assume that T is a nonnegative integer and that (3.281) is true for all integers $t = 0, \ldots, T$. We claim that it is valid for $t = T + 1$ too. If $y_T^* = \bar{x}_v$, then by the construction of $\{y_t^*\}_{t=0}^\infty$ (case (i)), $y_{T+1}^* = \bar{x}_v$, by Theorem 2.16, $y_{T+1} = \bar{x}_v$ too and (3.281) is true for $t = T + 1$. Assume that

$$y_T^* \neq \bar{x}_v. \tag{3.282}$$

There are two cases: (a)

$$y_T^* \neq x_t^* \text{ for all integers } t \geq 0; \tag{3.283}$$

(b)

$$y_T^* = x_t^* \text{ for some integer } t \geq 0. \tag{3.284}$$

Assume that the case (a) holds. In view of (3.280), we have

$$(y_{T+1}, y_T) \in K. \tag{3.285}$$

By (3.279), (3.282), (3.283), and (3.285), there exists a nonnegative integer j for which

$$(y_{T+1}, y_T) = (y_{j+1}^*, y_j^*). \tag{3.286}$$

In view of (3.286),

$$y_j^* = y_T^*.$$

It follows from Proposition 2.39, (3.282), (3.286), and the equality above that $j = T$, $y_{T+1} = y_{T+1}^*$ and that (3.281) holds for $t = T + 1$.

Assume that the case (b) holds. According to the case (iii) there exists a unique nonnegative integer $j(T)$ for which

$$y_T^* = x_{j(T)}^*. \tag{3.287}$$

If a nonnegative integer i satisfies $y_T = y_i^*$, then in view of (3.281) holding for $t = T$, $y_T^* = y_i^*$, and by Proposition 2.39 and (3.282), $i = T$. Thus

$$\text{if an integer } i \geq 0 \text{ satisfies } y_T = y_i^*, \text{ then } i = T. \tag{3.288}$$

If $j(T) = 0$, then (3.279), (3.280), (3.282), (3.287), and (3.288) imply that $(y_{T+1}, y_T) = (y_{T+1}^*, y_T^*)$ and that (3.281) holds for $t = T + 1$.

Let $j(T)$ be a natural number. It follows from (3.280), (3.282), (3.287), and (3.288) that

$$(y_{T+1}, y_T) \in \{(x_{j(T)-1}^*, x_{j(T)}^*), (y_{T+1}^*, y_T^*)\}. \tag{3.289}$$

If $(y_{T+1}, y_T) = (y_{T+1}^*, y_T^*)$, then (3.281) is valid for $t = T + 1$. Assume that

$$(y_{T+1}, y_T) = (x_{j(T)-1}^*, x_{j(T)}^*). \tag{3.290}$$

In view of (3.290) and (3.281) holding with $t = T$,

$$y_T^* = x_{j(T)}^*, \ y_{T+1} = x_{j(T)-1}^*. \tag{3.291}$$

It follows from Corollary 2.21, (3.264), (3.279), (3.280), (3.290), and (3.291) that

$$\begin{aligned}
&\pi^{\bar{v}_\gamma}(x_{j(T)}^*) - \pi^{\bar{v}_\gamma}(x_{j(T)-1}^*) = \pi^{\bar{v}_\gamma}(y_T) - \pi^{\bar{v}_\gamma}(y_{T+1}) \\
&= \bar{v}_\gamma(y_T, y_{T+1}) - v(\bar{x}_v, \bar{x}_v) = \bar{v}(y_T, y_{T+1}) - v(\bar{x}_v, \bar{x}_v).
\end{aligned} \tag{3.292}$$

Proposition 2.19, (3.271), (3.281) holding with $t = T$, and (3.292) imply that

$$\begin{aligned}
&\pi^{\bar{v}}(y_T) - \pi^{\bar{v}}(y_{T+1}) \geq \bar{v}(y_T, y_{T+1}) - v(\bar{x}_v, \bar{x}_v) = \pi^{\bar{v}_\gamma}(y_T) - \pi^{\bar{v}_\gamma}(y_{T+1}) \\
&= \pi^{\bar{v}_\gamma}(y_T^*) - \pi^{\bar{v}_\gamma}(y_{T+1}) = \pi^{\bar{v}}(y_T^*) - \pi^{\bar{v}_\gamma}(y_{T+1}) = \pi^{\bar{v}}(y_T) - \pi^{\bar{v}_\gamma}(y_{T+1}).
\end{aligned} \tag{3.293}$$

By (3.267) and (3.293),

$$\pi^{\bar{v}_\gamma}(y_{T+1}) = \pi^{\bar{v}}(y_{T+1}). \tag{3.294}$$

In view of (3.287), (3.290), (3.293), and (3.294),

$$\begin{aligned}
&\bar{v}(x_{j(T)}^*, x_{j(T)-1}^*) - v(\bar{x}_v, \bar{x}_v) = \bar{v}(y_T, y_{T+1}) - v(\bar{x}_v, \bar{x}_v) \\
&= \pi^{\bar{v}}(y_T^*) - \pi^{\bar{v}}(y_{T+1}) = \pi^{\bar{v}}(x_{j(T)}^*) - \pi^{\bar{v}}(x_{j(T)-1}^*).
\end{aligned} \tag{3.295}$$

It follows from (3.287), (3.290), (3.295), and the construction of $\{y_t^*\}_{t=0}^{\infty}$ (case (iii)) that $y_{T+1}^* = x_{j(T)-1}^* = y_{T+1}$ and (3.281) is valid for $t = T+1$. Therefore (3.281) is true for all nonnegative integers t and $(v_\gamma, g) \in E_0$. Since $v_\gamma \to v$ as $\gamma \to 0^+$ in $\mathfrak{M}$ there exists $\gamma_0 \in (0,1)$ such that $(v_\gamma, g) \in \mathcal{U}$ for all $g \in (0, \gamma_0)$. This completes the proof of Lemma 3.25. □

Proposition 3.21 follows from Lemmas 3.22, 3.23, 3.24, and 3.25.

Completion of the Proof of Theorem 3.20. For every pair of functions $(v, g) \in E_0$, there exist a unique $(\bar{v}, \bar{\Omega})$-overtaking optimal program $\{x_t^{(v,g)}\}_{t=0}^{\infty}$ for which property (Pxii) holds. Let $(v, g) \in E_0$ be given and k be a natural number. Theorem 3.17 and property (Pxii) imply that there exist an open neighborhood $\mathcal{U}(v, g, k)$ of (v, g) in $\mathfrak{M} \times \mathfrak{A}$ such that the following property holds:

(Pxiii) for every pair of functions $(u, h) \in \mathcal{U}(v, g, k) \cap (\mathcal{M}_*(\Omega) \times \mathfrak{A})$ and every $(\bar{u}, \bar{\Omega})$-overtaking optimal and $(\bar{u}, \bar{\Omega})$-good program $\{x_t\}_{t=0}^{\infty}$ for which there exists a point $y \subset X$ satisfying $\phi_{u,h}(y, x_0) - \sup(\phi_{u,h})$ we have $\rho(x_t, x_t^{(v,g)}) \leq k^{-1}$, $t = 0, \ldots, k$.

Define

$$\mathcal{F} = (\cap_{p=1}^{\infty} \cup \{\mathcal{U}(v, g, k) : (v, g) \in E_0, \text{ an integer } k \geq p\}) \cap (\mathcal{M}_*(\Omega) \times \mathfrak{A}). \tag{3.296}$$

It is clear that $\mathcal{F}$ contains a subset which is a countable intersection of open everywhere dense subsets of $\mathfrak{M} \times \mathfrak{A}$. Let

$$(u, h) \in \mathcal{F} \tag{3.297}$$

be given and let $\{z_t^{(i)}\}_{t=0}^{\infty}$, $i = 1, 2$ be $(\bar{u}, \bar{\Omega})$-overtaking optimal and $(\bar{u}, \bar{\Omega})$-good programs for which there exist $y_i \in X$, $i = 1, 2$ such that $\phi_{u,h}(y_i, z_0^{(i)}) = \sup(\phi_{u,h})$, $i = 1, 2$. Let p be a natural number. In view of (3.296) and (3.297), there exist $(v_p, g_p) \in E_0$ and a natural number $k_p \geq p$ for which $(u, h) \in \mathcal{U}(v_p, g_p, k_p) \cap (\mathcal{M}_*(\Omega) \times \mathfrak{A})$. By the inclusion above, the choice of $\{z_t^{(i)}\}_{t=0}^{\infty}$, $i = 1, 2$ and property (Pxiii), we have that for all integers $t = 0, \ldots, k_p$, $\rho(z_t^{(i)}, x_t^{(v_p, g_p)}) \leq k_p^{-1} \leq p^{-1}$. Therefore for all $t = 0, \ldots, p$, $\rho(z_t^{(1)}, z_t^{(2)}) \leq 2p^{-1}$. Since p is any natural number we conclude that $z_t^{(1)} = z_t^{(2)}$ for all nonnegative integers t. Theorem 3.20 is proved. □

4

Stability Results for Bolza Problems

In this chapter we study the structure of solutions of a discrete-time control system with a compact metric space of states X. This control system is described by a nonempty closed set $\Omega \subset X \times X$ which determines a class of admissible trajectories (programs) and by a a pair of objective functions which determines an optimality criterion. We show that the turnpike phenomenon and the structure of solutions on finite intervals in the regions close to the endpoints are stable under small perturbations of the objective functions and the set Ω.

4.1 Preliminaries and Stability Results for Lagrange Problems

Let (X, ρ) be a compact metric space. For each $x \in X$ and each nonempty set $C \subset X$ set

$$\rho(x, C) = \inf\{\rho(x, y) : \ y \in C\}.$$

For each $x \in X$ and each $r > 0$ set

$$B(x, r) = \{y \in X : \ \rho(x, y) \leq r\}.$$

We equip the space $X \times X$ with the metric ρ_1 defined by

$$\rho_1((x_1, x_2), (y_1, y_2)) = \rho(x_1, y_1) + \rho(x_2, y_2), \ x_1, x_2, y_1, y_2 \in X.$$

For each $(x_1, x_2) \in X \times X$ and each nonempty set $C \subset X \times X$ set

$$\rho_1((x_1, x_2), C) = \inf\{\rho_1((x_1, x_2), (y_1, y_2)) : \ (y_1, y_2) \in C\}.$$

Denote by $\mathcal{M}$ the set of all bounded functions $u : X \times X \to R^1$. For each $w \in \mathcal{M}$ set

A.J. Zaslavski, *Discrete-Time Optimal Control and Games on Large Intervals*, Springer Optimization and Its Applications 119,
DOI 10.1007/978-3-319-52932-5_4

$$\|w\| = \sup\{|w(x,y)| : (x,y) \in X \times X\}.$$

Let Ω be a nonempty closed subset of $X \times X$.

Recall that a sequence $\{x_t\}_{t=0}^{\infty} \subset X$ is called an (Ω)-program if $(x_t, x_{t+1}) \in \Omega$ for all integers $t \geq 0$. A sequence $\{x_t\}_{t=T_1}^{T_2} \subset X$ where integers T_1, T_2 satisfy $0 \leq T_1 < T_2$ is called an (Ω)-program if $(x_t, x_{t+1}) \in \Omega$ for all integers $t \in [T_1, T_2 - 1]$.

Let $v \in \mathcal{M}$ be an upper semicontinuous function. We continue to use the notation, definitions, and assumptions introduced in Chaps. 2 and 3 and discuss Lagrange problems (P1)–(P3) introduced in Chap. 2 and Bolza problems (P4) and (P5) introduced in Chap. 3.

In particular, we suppose that there exist $\bar{x} \in X$ and a constant $\bar{c} > 0$ such that the following assumptions hold.

(A1) $(\bar{x}, \bar{x})$ is an interior point of Ω (there is $\epsilon > 0$ such that $\{(x,y) \in X \times X : \rho(x,\bar{x}),\ \rho(y,\bar{x}) \leq \epsilon\} \subset \Omega$) and v is continuous at $(\bar{x}, \bar{x})$.

(A2) For any integer $T \geq 1$ and any (Ω)-program $\{x_t\}_{t=0}^{T}$,

$$\sum_{t=0}^{T-1} v(x_t, x_{t+1}) \leq T v(\bar{x}, \bar{x}) + \bar{c}.$$

Recall that an (Ω)-program $\{x_t\}_{t=0}^{\infty}$ is called (v, Ω)-good if the sequence

$$\{\sum_{t=0}^{T-1} v(x_t, x_{t+1}) - T v(\bar{x}, \bar{x})\}_{T=1}^{\infty}$$

is bounded.

We also suppose that the following assumption holds.

(A3) (the asymptotic turnpike property) For any (v, Ω)-good program $\{x_t\}_{t=0}^{\infty}$, $\lim_{t\to\infty} \rho(x_t, \bar{x}) = 0$.

For each $x, y \in X$, each integer $T \geq 1$, and each $w \in \mathcal{M}$ set

$$\sigma(w, T, x, y) = \sup\{\sum_{i=0}^{T-1} w(x_i, x_{i+1}) :$$

$$\{x_i\}_{i=0}^{T} \text{ is an } (\Omega) - \text{program and } x_0 = x,\ x_T = y\}.$$

(Here we use the convention that the supremum of an empty set is $-\infty$).

In Chap. 2 of [89] we considered the turnpike properties of approximate solutions of the Lagrange problems (P1) and (P2) and showed that these turnpike properties are stable under small perturbations of the objective function v and the set Ω. These results are stated in this section.

By assumption (A1) there exists $\bar{r} \in (0,1)$ such that

$$B(\bar{x}, \bar{r}) \times B(\bar{x}, \bar{r}) \subset \Omega. \tag{4.1}$$

Fix

$$\bar{\lambda} \in (0, \bar{r}). \tag{4.2}$$

For each $\lambda > 0$ denote by $\mathcal{E}(\lambda)$ the collection of all nonempty sets $\Omega' \subset X \times X$ such that

$$\rho_1(z, \Omega) \leq \lambda \text{ for each } z \in \Omega', \tag{4.3}$$

$$B(\bar{x}, \bar{\lambda}) \times B(\bar{x}, \bar{\lambda}) \subset \Omega'. \tag{4.4}$$

Let integers T_1, T_2 satisfy $0 \leq T_1 < T_2$ and let Ω_t, $t = T_1, \dots, T_2 - 1$ be nonempty subsets of $X \times X$.

A sequence $\{x_t\}_{t=T_1}^{T_2} \subset X$ is called an $(\{\Omega_t\}_{t=T_1}^{T_2-1})$-program if $(x_t, x_{t+1}) \in \Omega_t$ for all integers $t \in [T_1, T_2 - 1]$.

For each $x, y \in X$ and each finite sequence $\{u_t\}_{t=T_1}^{T_2-1} \subset \mathcal{M}$ set

$$\sigma(\{u_t\}_{t=T_1}^{T_2-1}, \{\Omega_t\}_{t=T_1}^{T_2-1}, T_1, T_2, x) = \sup\{\sum_{t=T_1}^{T_2-1} u_t(x_t, x_{t+1}) :$$

$$\{x_t\}_{t=T_1}^{T_2} \text{ is an } (\{\Omega_t\}_{t=T_1}^{T_2-1}) - \text{program and } x_{T_1} = x\},$$

$$\sigma(\{u_t\}_{t=T_1}^{T_2-1}, \{\Omega_t\}_{t=T_1}^{T_2-1}, T_1, T_2, x, y) = \sup\{\sum_{t=T_1}^{T_2-1} u_t(x_t, x_{t+1}) :$$

$$\{x_t\}_{t=T_1}^{T_2} \text{ is an } (\{\Omega_t\}_{t=T_1}^{T_2-1}) - \text{program, } x_{T_1} = x \text{ and } x_{T_2} = y\},$$

$$\sigma(\{u_t\}_{t=T_1}^{T_2-1}, \{\Omega_t\}_{t=T_1}^{T_2-1}, T_1, T_2) = \sup\{\sum_{t=T_1}^{T_2-1} u_t(x_t, x_{t+1}) :$$

$$\{x_t\}_{t=T_1}^{T_2} \text{ is an } (\{\Omega_t\}_{t=T_1}^{T_2-1}) - \text{program}\}. \tag{4.5}$$

(Here we use the convention that the supremum of an empty set is $-\infty$).

Definition 4.1. *Denote by $Y(\{\Omega_t\}_{t=T_1}^{T_2-1}, T_1, T_2)$ the set of all $x \in X$ for which there exists an $(\{\Omega_t\}_{t=T_1}^{T_2-1})$-program $\{x_t\}_{t=T_1}^{T_2}$ such that $x_{T_1} = \bar{x}$ and $x_{T_2} = x$.*

Denote by $\bar{Y}(\{\Omega_t\}_{t=T_1}^{T_2-1}, T_1, T_2)$ the set of all $x \in X$ for which there exists an $(\{\Omega_t\}_{t=T_1}^{T_2-1})$-program $\{x_t\}_{t=T_1}^{T_2}$ such that $x_{T_1} = x$ and $x_{T_2} = \bar{x}$.

For sufficiently small positive numbers δ, we study the structure of approximate solutions of the problems of the type (P2)

$$\sum_{t=0}^{T-1} u_t(x_t, x_{t+1}) \to \max,$$

$$\{x_t\}_{t=0}^{T} \text{ is an } (\{\Omega_t\}_{t=0}^{T-1}) - \text{program and } x_0 = y,$$

and problems of the type (P1)

$$\sum_{t=0}^{T-1} u_t(x_t, x_{t+1}) \to \max,$$

$$\{x_t\}_{t=0}^{T} \text{ is an } (\{\Omega_t\}_{t=0}^{T-1}) - \text{program and } x_0 = y,\ x_T = z,$$

where $T \geq 1$ is an integer, $y, z \in X$ and for all $t = 0, \dots, T-1$, we have

$$\Omega_t \in \mathcal{E}(\delta),\ u_t \in \mathcal{M} \text{ and } \|u_t - v\| \leq \delta.$$

In Chap. 2 of [89] we proved the following four stability results.

Theorem 4.2. *Let ϵ be a positive number and let l_1, l_2 be natural numbers. Then there exist $\delta > 0$ and a natural number $L > l_1 + l_2$ such that each integer $T > 2L$, each*

$$\Omega_t \in \mathcal{E}(\delta),\ t = 0, \dots, T-1,$$

each $u_t \in \mathcal{M}$, $t = 0, \dots, T-1$ satisfying

$$\|u_t - v\| \leq \delta,\ t = 0, \dots, T-1$$

and each $(\{\Omega_t\}_{t=0}^{T-1})$-program $\{x_t\}_{t=0}^{T}$ which satisfies

$$x_0 \in \bar{Y}(\{\Omega_t\}_{t=0}^{l_1-1}, 0, l_1),\ x_T \in Y(\{\Omega_t\}_{t=T-l_2}^{T-1}, T-l_2, T),$$

$$\sigma(\{u_t\}_{t=0}^{T-1}, \{\Omega_t\}_{t=0}^{T-1}, 0, T, x_0, x_T) \leq \sum_{t=0}^{T-1} u_t(x_t, x_{t+1}) + \delta$$

there exist integers $\tau_1 \in [0, L]$, $\tau_2 \in [T-L, T]$ such that

$$\rho(x_t, \bar{x}) \leq \epsilon \text{ for all } t = \tau_1, \dots, \tau_2.$$

Moreover if $\rho(x_0, \bar{x}) \leq \delta$, then $\tau_1 = 0$ and if $\rho(x_T, \bar{x}) \leq \delta$, then $\tau_2 = T$.

Theorem 4.3. *Let ϵ be a positive number and let l_1 be a natural number. Then there exist $\delta > 0$ and a natural number $L > l_1$ such that each integer $T > 2L$, each*

$$\Omega_t \in \mathcal{E}(\delta),\ t = 0, \dots, T-1,$$

each $u_t \in \mathcal{M}$, $t = 0, \dots, T-1$ satisfying

$$\|u_t - v\| \leq \delta,\ t = 0, \dots, T-1$$

and each $(\{\Omega_t\}_{t=0}^{T-1})$-program $\{x_t\}_{t=0}^{T}$ which satisfies

$$x_0 \in \bar{Y}(\{\Omega_t\}_{t=0}^{l_1-1}, 0, l_1),$$

$$\sigma(\{u_t\}_{t=0}^{T-1}, \{\Omega_t\}_{t=0}^{T-1}, 0, T, x_0) \leq \sum_{t=0}^{T-1} u_t(x_t, x_{t+1}) + \delta$$

there exist integers $\tau_1 \in [0, L]$, $\tau_2 \in [T-L, T]$ such that

$$\rho(x_t, \bar{x}) \leq \epsilon \text{ for all } t = \tau_1, \dots, \tau_2.$$

Moreover if $\rho(x_0, \bar{x}) \leq \delta$, then $\tau_1 = 0$ and if $\rho(x_T, \bar{x}) \leq \delta$, then $\tau_2 = T$.

Denote by $\mathrm{Card}(B)$ the cardinality of a set B.

Theorem 4.4. *Let ϵ, M be positive numbers and let l_1, l_2 be natural numbers. Then there exist $\delta > 0$ and a natural number $L > l_1 + l_2$ such that each integer $T > L$, each*

$$\Omega_t \in \mathcal{E}(\delta),\ t = 0, \dots, T-1,$$

each $u_t \in \mathcal{M}$, $t = 0, \dots, T-1$ satisfying

$$\|u_t - v\| \le \delta,\ t = 0, \dots, T-1$$

and each $(\{\Omega_t\}_{t=0}^{T-1})$-program $\{x_t\}_{t=0}^{T}$ which satisfies

$$x_0 \in \bar{Y}(\{\Omega_t\}_{t=0}^{l_1-1}, 0, l_1),\ x_T \in Y(\{\Omega_t\}_{t=T-l_2}^{T-1}, T-l_2, T),$$

$$\sigma(\{u_t\}_{t=0}^{T-1}, \{\Omega_t\}_{t=0}^{T-1}, 0, T, x_0, x_T) \le \sum_{t=0}^{T-1} u_t(x_t, x_{t+1}) + M$$

the inequality

$$Card(\{t \in \{0, \dots, T\} :\ \rho(x_t, \bar{x}) > \epsilon\}) \le L$$

holds.

Theorem 4.5. *Let ϵ, M be positive numbers and let l_1 be a natural number. Then there exist $\delta > 0$ and a natural number $L > l_1$ such that each integer $T > L$, each*

$$\Omega_t \in \mathcal{E}(\delta),\ t = 0, \dots, T-1,$$

each $u_t \in \mathcal{M}$, $t = 0, \dots, T-1$ satisfying

$$\|u_t - v\| \le \delta,\ t = 0, \dots, T-1$$

and each $(\{\Omega_t\}_{t=0}^{T-1})$-program $\{x_t\}_{t=0}^{T}$ which satisfies

$$x_0 \in \bar{Y}(\{\Omega_t\}_{t=0}^{l_1-1}, 0, l_1),$$

$$\sigma(\{u_t\}_{t=0}^{T-1}, \{\Omega_t\}_{t=0}^{T-1}, 0, T, x_0) \le \sum_{t=0}^{T-1} u_t(x_t, x_{t+1}) + M$$

the inequality

$$Card(\{t \in \{0, \dots, T\} :\ \rho(x_t, \bar{x}) > \epsilon\}) \le L$$

holds.

4.2 Extensions

In this section we state the extensions of the turnpike results of the previous section. In these extensions we describe the structure of programs defined on an interval $[0, T]$ with sufficiently large T which are approximate solutions of the corresponding optimal problems on subintervals of the length L, where L is a constant which does not depend on T. These extensions were also obtained in Chap. 2 of [89].

Theorem 4.6. *Let $\epsilon \in (0, \bar{\lambda})$ and M be a positive number. Then there exist $\gamma \in (0, \epsilon)$ and a natural number L_0 such that each integer $L_1 \geq L_0$ there exists a positive number $\delta < \gamma$ such that the following assertion holds.*

Assume that an integer $T > 3L_1$,

$$\Omega_t \in \mathcal{E}(\delta),\ t = 0, \dots, T-1,$$

$u_t \in \mathcal{M}$, $t = 0, \dots, T-1$ satisfy

$$\|u_t - v\| \leq \delta,\ t = 0, \dots, T-1$$

and that an $(\{\Omega_t\}_{t=0}^{T-1})$-program $\{x_t\}_{t=0}^{T}$ and a finite sequence of integers $\{S_i\}_{i=0}^{q}$ satisfy

$$S_0 = 0,\ S_{i+1} - S_i \in [L_0, L_1],\ i = 0, \dots, q-1,\ S_q > T - L_1,$$

$$\sum_{t=S_i}^{S_{i+1}-1} u_t(x_t, x_{t+1}) \geq \sum_{t=S_i}^{S_{i+1}-1} u_t(\bar{x}, \bar{x}) - M$$

for each integer $i \in [0, q-1]$,

$$\sum_{t=S_i}^{S_{i+2}-1} u_t(x_t, x_{t+1}) \geq \sigma(\{u_t\}_{t=S_i}^{S_{i+2}-1}, \{\Omega_t\}_{t=S_i}^{S_{i+2}-1}, S_i, S_{i+2}, x_{S_i}, x_{S_{i+2}}) - \gamma$$

for each integer $i \in [0, q-2]$ and

$$\sum_{t=S_{q-2}}^{T-1} u_t(x_t, x_{t+1}) \geq \sigma(\{u_t\}_{t=S_{q-2}}^{T-1}, \{\Omega_t\}_{t=S_{q-2}}^{T-1}, S_{q-2}, T, x_{S_{q-2}}, x_T) - \gamma.$$

Then there exist integers $\tau_1 \in [0, L_1]$, $\tau_2 \in [T - 2L_1, T]$ such that

$$\rho(x_t, \bar{x}) \leq \epsilon \text{ for all } t = \tau_1, \dots, \tau_2.$$

Moreover if $\rho(x_0, \bar{x}) \leq \gamma$, then $\tau_1 = 0$ and if $\rho(x_T, \bar{x}) \leq \gamma$, then $\tau_2 = T$.

Theorem 4.7. *Let ϵ, M be positive numbers and let l_1, l_2 be natural numbers. Then there exist $\delta > 0$ and a natural number $L > l_1 + l_2$ such that for each integer $T > 2L$, each*

$$\Omega_t \in \mathcal{E}(\delta),\ t = 0, \dots, T-1,$$

each $u_t \in \mathcal{M}$, $t = 0, \dots, T-1$ satisfying

$$\|u_t - v\| \le \delta,\ t = 0, \dots, T-1$$

and each $(\{\Omega_t\}_{t=0}^{T-1})$-program $\{x_t\}_{t=0}^{T}$ which satisfies

$$x_0 \in \bar{Y}(\{\Omega_t\}_{t=0}^{l_1-1}, 0, l_1),\ x_T \in Y(\{\Omega_t\}_{t=T-l_2}^{T-1}, T-l_2, T),$$

$$\sigma(\{u_t\}_{t=0}^{T-1}, \{\Omega_t\}_{t=0}^{T-1}, 0, T, x_0, x_T) \le \sum_{t=0}^{T-1} u_t(x_t, x_{t+1}) + M$$

and

$$\sum_{t=\tau}^{\tau+L-1} u_t(x_t, x_{t+1}) \ge \sigma(\{u_t\}_{t=\tau}^{\tau+L-1}, \{\Omega_t\}_{t=\tau}^{\tau+L-1}, \tau, \tau+L, x_\tau, x_{\tau+L}) - \delta$$

for each integer $\tau \in [0, T-L]$, there exist integers $\tau_1 \in [0, L]$, $\tau_2 \in [T-L, T]$ such that

$$\rho(x_t, \bar{x}) \le \epsilon \text{ for all } t = \tau_1, \dots, \tau_2.$$

Moreover if $\rho(x_0, \bar{x}) \le \delta$, then $\tau_1 = 0$ and if $\rho(x_T, \bar{x}) \le \delta$, then $\tau_2 = T$.

Theorem 4.8. *Let ϵ, M be positive numbers and let l_1 be a natural number. Then there exist $\delta > 0$ and a natural number $L > l_1$ such that for each integer $T > 2L$, each*

$$\Omega_t \in \mathcal{E}(\delta),\ t = 0, \dots, T-1,$$

each $u_t \in \mathcal{M}$, $t = 0, \dots, T-1$ satisfying

$$\|u_t - v\| \le \delta,\ t = 0, \dots, T-1$$

and each $(\{\Omega_t\}_{t=0}^{T-1})$-program $\{x_t\}_{t=0}^{T}$ which satisfies

$$x_0 \in \bar{Y}(\{\Omega_t\}_{t=0}^{l_1-1}, 0, l_1),$$

$$\sigma(\{u_t\}_{t=0}^{T-1}, \{\Omega_t\}_{t=0}^{T-1}, 0, T, x_0) \le \sum_{t=0}^{T-1} u_t(x_t, x_{t+1}) + M$$

and

$$\sum_{t=\tau}^{\tau+L-1} u_t(x_t, x_{t+1}) \ge \sigma(\{u_t\}_{t=\tau}^{\tau+L-1}, \{\Omega_t\}_{t=\tau}^{\tau+L-1}, \tau, \tau+L, x_\tau, x_{\tau+L}) - \delta$$

for each integer $\tau \in [0, T-L]$, there exist integers $\tau_1 \in [0, L]$, $\tau_2 \in [T-L, T]$ such that

$$\rho(x_t, \bar{x}) \le \epsilon \text{ for all } t = \tau_1, \dots, \tau_2.$$

Moreover if $\rho(x_0, \bar{x}) \le \delta$, then $\tau_1 = 0$ and if $\rho(x_T, \bar{x}) \le \delta$, then $\tau_2 = T$.

4.3 Lagrange Problems with Discounting

Denote by $\mathcal{M}_0$ the set of all upper semicontinuous functions $u \in \mathcal{M}$.

It is not difficult to see that the following result holds.

Proposition 4.9. *Let l be a natural number, integers T_1, T_2 satisfy $0 \le T_1 \le T_2 - l$, $\{u_t\}_{t=T_1}^{T_2-1} \subset \mathcal{M}_0$, for any integer $t \in \{T_1, \dots, T_2 - 1\}$, let Ω_t be a closed subset of $X \times X$ such that $(\bar{x}, \bar{x}) \in \Omega_t$ and let*

$$x \in \bar{Y}(\{\Omega_t\}_{t=T_1}^{T_1+l-1}, 0, l).$$

Then there exists an $(\{\Omega_t\}_{t=T_1}^{T_2-1})$-program $\{x_t\}_{t=T_1}^{T_2}$ such that $x_{T_1} = x$ and

$$\sum_{t=T_1}^{T_2-1} u_t(x_t, x_{t+1}) = \sigma(\{u_t\}_{t=T_1}^{T_2-1}, \{\Omega_t\}_{t=T_1}^{T_2-1}, T_1, T_2, x).$$

In Chap. 3 of [89] we proved the following result which shows the stability of the turnpike phenomenon in the case of discounting.

Theorem 4.10. *Let $\epsilon \in (0, \bar{\lambda})$ and let l be a natural number. Then there exist $\delta \in (0, \epsilon)$, a natural number $L > l$ and $\lambda \in (0, 1)$ such that for each integer $T > 2L$, each*

$$\Omega_t \in \mathcal{E}(\delta),\ t = 0, \dots, T - 1,$$

each $u_t \in \mathcal{M}_0$, $t = 0, \dots, T - 1$ satisfying

$$\|u_t - v\| \le \delta,\ t = 0, \dots, T - 1,$$

each sequence $\{\alpha_t\}_{t=0}^{T-1} \subset (0, 1]$ such that

$$\alpha_i \alpha_j^{-1} \ge \lambda \text{ for each } i, j \in \{0, \dots, T - 1\} \text{ satisfying } |i - j| \le L$$

and each $(\{\Omega_t\}_{t=0}^{T-1})$-program $\{x_t\}_{t=0}^{T}$ which satisfies

$$x_0 \in \bar{Y}(\{\Omega_t\}_{t=0}^{l-1}, 0, l),$$

$$\sigma(\{\alpha_t u_t\}_{t=0}^{T-1}, \{\Omega_t\}_{t=0}^{T-1}, 0, T, x_0) = \sum_{t=0}^{T-1} \alpha_t u_t(x_t, x_{t+1})$$

there exist integers $\tau_1 \in [0, L]$, $\tau_2 \in [T - L, T]$ such that

$$\rho(x_t, \bar{x}) \le \epsilon \text{ for all } t = \tau_1, \dots, \tau_2.$$

Moreover if $\rho(x_0, \bar{x}) \le \delta$, then $\tau_1 = 0$ and if $\rho(x_T, \bar{x}) \le \delta$, then $\tau_2 = T$.

Roughly speaking, the turnpike property holds if discount coefficients $\{\alpha_t\}_{t=0}^{T-1} \subset (0,1]$ are changed rather slowly.

Let $\Omega_t \subset X \times X$ be a nonempty set for all nonnegative integers t.

A sequence $\{x_t\}_{t=0}^{\infty} \subset X$ is called an $(\{\Omega_t\}_{t=0}^{\infty})$-program if $(x_t, x_{t+1}) \in \Omega_t$ for all integers $t \geq 0$.

Let $\{u_t\}_{t=0}^{\infty} \subset \mathcal{M}$ be given. An $(\{\Omega_t\}_{t=0}^{\infty})$-program $\{x_t\}_{t=0}^{\infty}$ is called $(\{u_t\}_{t=0}^{\infty}, \{\Omega_t\}_{t=0}^{\infty})$-overtaking optimal if for each $(\{\Omega_t\}_{t=0}^{\infty})$-program $\{y_t\}_{t=0}^{\infty}$ satisfying $x_0 = y_0$, we have

$$\limsup_{T\to\infty}[\sum_{t=0}^{T-1} u_t(y_t, y_{t+1}) - \sum_{t=0}^{T-1} u_t(x_t, x_{t+1})] \leq 0.$$

The following result establishes the stability of turnpike phenomenon for overtaking optimal programs. It was obtained in Chap. 3 of [89].

Theorem 4.11. *Let $\epsilon \in (0, \bar{\lambda})$ and let l be a natural number. Then there exist $\delta \in (0,\epsilon)$, a natural number $L > l$ and $\lambda \in (0,1)$ such that for each*

$$\Omega_t \in \mathcal{E}(\delta),\ t = 0, 1, \dots,$$

each $u_t \in \mathcal{M}_0$, $t = 0, 1, \dots$, satisfying

$$\|u_t - v\| \leq \delta,\ t = 0, 1, \dots,$$

each sequence $\{\alpha_t\}_{t=0}^{\infty} \subset (0,1]$ such that

$$\alpha_i \alpha_j^{-1} \geq \lambda \text{ for each pair of nonnegative integers } i, j \text{ satisfying } |i-j| \leq L$$

and each $(\{\alpha_t u_t\}_{t=0}^{\infty}, \{\Omega_t\}_{t=0}^{\infty})$-overtaking optimal program $\{x_t\}_{t=0}^{\infty}$ which satisfies

$$x_0 \in \bar{Y}(\{\Omega_t\}_{t=0}^{l-1}, 0, l)$$

the following inequality holds:

$$\rho(x_t, \bar{x}) \leq \epsilon \text{ for all integers } t \geq L.$$

Moreover, if $\rho(x_0, \bar{x}) \leq \delta$, then

$$\rho(x_t, \bar{x}) \leq \epsilon \text{ for all integers } t \geq 0.$$

The following existence result was obtained in Chap. 3 of [89].

Theorem 4.12. *Let $l \geq 1$ be an integer, $\epsilon = \bar{\lambda}/4$ and let $\delta \in (0,\epsilon)$, an integer $L > l$ and $\lambda \in (0,1)$ be as guaranteed by Theorem 4.10.*

Let

$$u_t \in \mathcal{M}_0 \text{ and } \|u_t - v\| \leq \delta,\ t = 0, 1, \dots,$$

$$\Omega_t \in \mathcal{E}(\delta),\ t = 0, 1, \dots$$

be closed subsets of $X \times X$ *for all nonnegative integers* t, *and let* $\{\alpha_t\}_{t=0}^{\infty} \subset (0,1]$ *satisfy the relations*

$$\lim_{t\to\infty} \alpha_t = 0,$$

$$\alpha_i \alpha_j^{-1} \ge \lambda \text{ for all nonnegative integers } i,j \text{ satisfying } |i-j| \le L.$$

Then for each $z \in \bar{Y}(\{\Omega_t\}_{t=0}^{l-1}, 0, l)$ *there exists an* $(\{\Omega_t\}_{t=0}^{\infty})$*-program* $\{x_t^{(z)}\}_{t=0}^{\infty}$ *such that* $x_0^{(z)} = z$ *and that the following property holds:*

For each real number $\gamma > 0$ *there exists an integer* $n_0 \ge 1$ *such that for each integer* $T \ge n_0$ *and each point* $z \in \bar{Y}(\{\Omega_t\}_{t=0}^{l-1}, 0, l)$ *the inequality*

$$|\sigma(\{\alpha_t u_t\}_{t=0}^{T-1}, \{\Omega_t\}_{t=0}^{T-1}, 0, T, z) - \sum_{t=0}^{T-1} \alpha_t u_t(x_t^{(z)}, x_{t+1}^{(z)})| \le \gamma$$

holds.

It is clear that Theorem 4.12 establishes the existence of $(\{\alpha_t u_t\}_{t=0}^{\infty})$-overtaking optimal program when the sequence of the discount coefficients $\{\alpha_t\}_{t=0}^{\infty}$ tends to zero slowly.

Note that the existence of an $(\{\alpha_t u_t\}_{t=0}^{\infty})$-overtaking optimal program when the discount coefficients $\{\alpha_t\}_{t=0}^{\infty}$ tends to zero rapidly is a well-known fact.

4.4 Turnpike Results for Perturbed Problems (P4)

For each nonempty set Y and each function $h : Y \to R^1 \cup \{-\infty\}$ set

$$\sup(h) = \sup\{h(y) : y \in Y\}.$$

Denote by $\mathfrak{M}$ the set of all bounded functions $h : X \to R^1$. For each $h \in \mathfrak{M}$ set

$$\|h\| = \sup\{|h(x)| : x \in X\}.$$

Clearly, $(\mathfrak{M}, \|\cdot\|)$ is a Banach space. For each $h_1, h_2 \in \mathfrak{M}$ set

$$d_X(h_1, h_2) = \|h_1 - h_2\|.$$

Let integers T_1, T_2 satisfy $0 \le T_1 < T_2$ and let Ω_t, $t = T_1, \dots, T_2 - 1$ be nonempty subsets of $X \times X$.

For each $x \in X$, each finite sequence $\{u_t\}_{t=T_1}^{T_2-1} \subset \mathcal{M}$, and each $h \in \mathfrak{M}$ we consider the problem

$$\sum_{t=T_1}^{T_2-1} u_t(x_t, x_{t+1}) + h(x_{T_2}) \to \max,$$

$$\{x_t\}_{t=T_1}^{T_2} \text{ is an } (\{\Omega_t\}_{t=T_1}^{T_2-1}) - \text{program and } x_{T_1} = x$$

and set

$$\sigma(\{u_t\}_{t=T_1}^{T_2-1}, h, \{\Omega_t\}_{t=T_1}^{T_2-1}, T_1, T_2, x) = \sup\{\sum_{t=T_1}^{T_2-1} u_t(x_t, x_{t+1}) + h(x_{T_2}) :$$

$$\{x_t\}_{t=T_1}^{T_2} \text{ is an } (\{\Omega_t\}_{t=T_1}^{T_2-1}) - \text{program and } x_{T_1} = x\}. \tag{4.6}$$

(Recall that we use the convention that the supremum of an empty set is $-\infty$).

For each $x \in X$, each $u \in \mathcal{M}$, and each $h \in \mathfrak{M}$ set

$$\sigma(u, h, \{\Omega_t\}_{t=T_1}^{T_2-1}, T_1, T_2, x) = \sigma(\{u_t\}_{t=T_1}^{T_2-1}, h, \{\Omega_t\}_{t=T_1}^{T_2-1}, T_1, T_2, x),$$

where $u_t = u$, $t = T_1, \dots, T_2 - 1$.

In the next section we will prove the following result.

Theorem 4.13. *Let ϵ, M be positive numbers and let l_1 be a natural number. Then there exist $\delta > 0$ and a natural number $L > l_1$ such that each integer $T > L$, each*

$$\Omega_t \in \mathcal{E}(\delta),\ t = 0, \dots, T-1, \tag{4.7}$$

each $u_t \in \mathcal{M}$, $t = 0, \dots, T-1$ satisfying

$$\|u_t - v\| \le \delta,\ t = 0, \dots, T-1, \tag{4.8}$$

each $h \in \mathfrak{M}$ satisfying

$$\|h\| \le M \tag{4.9}$$

and each $(\{\Omega_t\}_{t=0}^{T-1})$-program $\{x_t\}_{t=0}^{T}$ which satisfies

$$x_0 \in \bar{Y}(\{\Omega_t\}_{t=0}^{l_1-1}, 0, l_1),$$

$$\sigma(\{u_t\}_{t=0}^{T-1}, h, \{\Omega_t\}_{t=0}^{T-1}, 0, T, x_0) \le \sum_{t=0}^{T-1} u_t(x_t, x_{t+1}) + h(x(T_2)) + M$$

the inequality

$$Card(\{t \in \{0, \dots, T\} :\ \rho(x_t, \bar{x}) > \epsilon\}) \le L$$

holds.

The next theorem is proved in Sect. 4.6.

Theorem 4.14. *Let ϵ be a positive number and let l_1 be a natural number, $M_0 > 0$. Then there exist $\delta > 0$ and a natural number $L > l_1$ such that each integer $T > 2L$, each*

$$\Omega_t \in \mathcal{E}(\delta),\ t = 0, \dots, T-1,$$

each $u_t \in \mathcal{M},\ t = 0,\dots,T-1$ *satisfying*

$$\|u_t - v\| \le \delta,\ t = 0,\dots,T-1,$$

each $h \in \mathfrak{M}$ *satisfying*

$$\|h\| \le M_0$$

and each $(\{\Omega_t\}_{t=0}^{T-1})$*-program* $\{x_t\}_{t=0}^{T}$ *which satisfies*

$$x_0 \in \bar{Y}(\{\Omega_t\}_{t=0}^{l_1-1}, 0, l_1),$$

$$\sigma(\{u_t\}_{t=0}^{T-1}, h, \{\Omega_t\}_{t=0}^{T-1}, 0, T, x_0) \le \sum_{t=0}^{T-1} u_t(x_t, x_{t+1}) + h(x(T_2)) + M_0$$

and

$$\sigma(\{u_t\}_{t=\tau}^{\tau+L-1}, \{\Omega_t\}_{t=\tau}^{\tau+L-1}, \tau, \tau+L, x_\tau, x_{\tau+L}) \le \sum_{t=\tau}^{\tau+L-1} u_t(x_t, x_{t+1}) + \delta$$

for each integer $\tau \in [0, T-L]$*, there exist integers* $\tau_1 \in [0, L],\ \tau_2 \in [T-L, T]$ *such that*

$$\rho(x_t, \bar{x}) \le \epsilon \text{ for all } t = \tau_1, \dots, \tau_2.$$

Moreover if $\rho(x_0, \bar{x}) \le \delta$*, then* $\tau_1 = 0$ *and if* $\rho(x_T, \bar{x}) \le \delta$*, then* $\tau_2 = T$.

4.5 Proof of Theorem 4.13

By Theorem 4.5, there exist $\delta > 0$ and a natural number $L > l_1$ such that the following property holds:

(P1) for each integer $T > L$, each

$$\Omega_t \in \mathcal{E}(\delta),\ t = 0,\dots,T-1,$$

each $u_t \in \mathcal{M},\ t = 0,\dots,T-1$ satisfying

$$\|u_t - v\| \le \delta,\ t = 0,\dots,T-1$$

and each $(\{\Omega_t\}_{t=0}^{T-1})$-program $\{x_t\}_{t=0}^{T}$ which satisfies

$$x_0 \in \bar{Y}(\{\Omega_t\}_{t=0}^{l_1-1}, 0, l_1),$$

$$\sigma(\{u_t\}_{t=0}^{T-1}, \{\Omega_t\}_{t=0}^{T-1}, 0, T, x_0) \le \sum_{t=0}^{T-1} u_t(x_t, x_{t+1}) + 3M$$

the inequality

$$\mathrm{Card}(\{t \in \{0,\dots,T\} :\ \rho(x_t, \bar{x}) > \epsilon\}) \le L \tag{4.10}$$

holds.

Assume that an integer $T > L$, (4.7) holds, $u_t \in \mathcal{M}$, $t = 0, \dots, T-1$ satisfy (4.8), $h \in \mathfrak{M}$ satisfies (4.9) and that an $(\{\Omega_t\}_{t=0}^{T-1})$-program $\{x_t\}_{t=0}^{T}$ satisfies

$$x_0 \in \bar{Y}(\{\Omega_t\}_{t=0}^{l_1-1}, 0, l_1), \tag{4.11}$$

$$\sigma(\{u_t\}_{t=0}^{T-1}, h, \{\Omega_t\}_{t=0}^{T-1}, 0, T, x_0) \le \sum_{t=0}^{T-1} u_t(x_t, x_{t+1}) + h(x(T_2)) + M.$$

By the inequality above, (4.5) and (4.9),

$$\sigma(\{u_t\}_{t=0}^{T-1}, \{\Omega_t\}_{t=0}^{T-1}, 0, T, x_0)$$
$$\le \sigma(\{u_t\}_{t=0}^{T-1}, h, \{\Omega_t\}_{t=0}^{T-1}, 0, T, x_0) + \|h\|$$
$$\le \sum_{t=0}^{T-1} u_t(x_t, x_{t+1}) + 2\|h\| + M \le \sum_{t=0}^{T-1} u_t(x_t, x_{t+1}) + 3M.$$

It follows from the relation above, (4.7), (4.11), and (P1) that (4.10) is true. Theorem 4.13 is proved. □

4.6 Proof of Theorem 4.14

By Theorem 4.8, there exist $\delta > 0$ and a natural number $L > l_1$ such that the following problem holds:

(P2) for each integer $T > 2L$, each

$$\Omega_t \in \mathcal{E}(\delta),\ t = 0, \dots, T-1,$$

each $u_t \in \mathcal{M}$, $t = 0, \dots, T-1$ satisfying

$$\|u_t - v\| \le \delta,\ t = 0, \dots, T-1$$

and each $(\{\Omega_t\}_{t=0}^{T-1})$-program $\{x_t\}_{t=0}^{T}$ which satisfies

$$x_0 \in \bar{Y}(\{\Omega_t\}_{t=0}^{l_1-1}, 0, l_1),$$

$$\sigma(\{u_t\}_{t=0}^{T-1}, \{\Omega_t\}_{t=0}^{T-1}, 0, T, x_0) \le \sum_{t=0}^{T-1} u_t(x_t, x_{t+1}) + 3M_0$$

and

$$\sum_{t=\tau}^{\tau+L-1} u_t(x_t, x_{t+1}) \ge \sigma(\{u_t\}_{t=\tau}^{\tau+L-1}, \{\Omega_t\}_{t=\tau}^{\tau+L-1}, \tau, \tau+L, x_\tau, x_{\tau+L}) - \delta$$

for each integer $\tau \in [0, T-L]$, there exist integers $\tau_1 \in [0, L]$, $\tau_2 \in [T-L, T]$ such that

$$\rho(x_t, \bar{x}) \le \epsilon \text{ for all } t = \tau_1, \dots, \tau_2;$$

if $\rho(x_0, \bar{x}) \le \delta$, then $\tau_1 = 0$ and if $\rho(x_T, \bar{x}) \le \delta$, then $\tau_2 = T$.

Assume that an integer $T > 2L$,

$$\Omega_t \in \mathcal{E}(\delta),\ t = 0, \dots, T-1,$$

$u_t \in \mathcal{M},\ t = 0, \dots, T-1$ satisfy

$$\|u_t - v\| \le \delta,\ t = 0, \dots, T-1,$$

$h \in \mathfrak{M}$ satisfies

$$\|h\| \le M_0 \tag{4.12}$$

and an $(\{\Omega_t\}_{t=0}^{T-1})$-program $\{x_t\}_{t=0}^{T}$ satisfies

$$x_0 \in \bar{Y}(\{\Omega_t\}_{t=0}^{l_1-1}, 0, l_1),$$

$$\sigma(\{u_t\}_{t=0}^{T-1}, h, \{\Omega_t\}_{t=0}^{T-1}, 0, T, x_0) \le \sum_{t=0}^{T-1} u_t(x_t, x_{t+1}) + h(x(T_2)) + M_0$$

and

$$\sigma(\{u_t\}_{t=\tau}^{\tau+L-1}, \{\Omega_t\}_{t=\tau}^{\tau+L-1}, \tau, \tau+L, x_\tau, x_{\tau+L}) \le \sum_{t=\tau}^{\tau+L-1} u_t(x_t, x_{t+1}) + \delta$$

for each integer $\tau \in [0, T-L]$.

By (4.12), (4.5), and the relation above,

$$\begin{aligned} &\sigma(\{u_t\}_{t=0}^{T-1}, \{\Omega_t\}_{t=0}^{T-1}, 0, T, x_0) \\ &\le \sigma(\{u_t\}_{t=0}^{T-1}, h, \{\Omega_t\}_{t=0}^{T-1}, 0, T, x_0) + \|h\| \\ &\le \sum_{t=0}^{T-1} u_t(x_t, x_{t+1}) + 2\|h\| + M_0 \le \sum_{t=0}^{T-1} u_t(x_t, x_{t+1}) + 2\|h\| + 3M_0. \end{aligned}$$

The validity of Theorem 4.14 follows from property (P2) and the relation above. □

4.7 Turnpike Results for Perturbed Problems (P3)

In Sect. 4.9 we prove the following result.

Theorem 4.15. *Let ϵ, M be positive numbers. Then there exist $\delta > 0$ and a natural number L such that for each integer $T > L$, each*

$$\Omega_t \in \mathcal{E}(\delta),\ t = 0, \dots, T-1,$$

each $u_t \in \mathcal{M},\ t = 0, \dots, T-1$ satisfying

$$\|u_t - v\| \le \delta,\ t = 0, \dots, T-1$$

and each $(\{\Omega_t\}_{t=0}^{T-1})$*-program* $\{x_t\}_{t=0}^{T}$ *which satisfies*

$$\sigma(\{u_t\}_{t=0}^{T-1}, \{\Omega_t\}_{t=0}^{T-1}, 0, T) \le \sum_{t=0}^{T-1} u_t(x_t, x_{t+1}) + M$$

the inequality

$$Card(\{t \in \{0, \dots, T\} :\ \rho(x_t, \bar{x}) > \epsilon\}) \le L$$

holds.

The next result is proved in Sect. 4.10.

Theorem 4.16. *Let* ϵ, M *be positive numbers. Then there exist* $\delta > 0$ *and a natural number* L *such that for each integer* $T > 2L$*, each*

$$\Omega_t \in \mathcal{E}(\delta),\ t = 0, \dots, T-1,$$

each $u_t \in \mathcal{M},\ t = 0, \dots, T-1$ *satisfying*

$$\|u_t - v\| \le \delta,\ t = 0, \dots, T-1$$

and each $(\{\Omega_t\}_{t=0}^{T-1})$*-program* $\{x_t\}_{t=0}^{T}$ *which satisfies*

$$\sigma(\{u_t\}_{t=0}^{T-1}, \{\Omega_t\}_{t=0}^{T-1}, 0, T) \le \sum_{t=0}^{T-1} u_t(x_t, x_{t+1}) + M_0$$

and

$$\sigma(\{u_t\}_{t=\tau}^{\tau+L-1}, \{\Omega_t\}_{t=\tau}^{\tau+L-1}, \tau, \tau+L, x_\tau, x_{\tau+L}) \le \sum_{t=\tau}^{\tau+L-1} u_t(x_t, x_{t+1}) + \delta \quad (4.13)$$

for each integer $\tau \in [0, T-L]$*, there exist integers* $\tau_1 \in [0, L]$*,* $\tau_2 \in [T-L, T]$ *such that*

$$\rho(x_t, \bar{x}) \le \epsilon \text{ for all } t = \tau_1, \dots, \tau_2.$$

Moreover if $\rho(x_0, \bar{x}) \le \delta$*, then* $\tau_1 = 0$ *and if* $\rho(x_T, \bar{x}) \le \delta$*, then* $\tau_2 = T$*.*

4.8 Auxiliary Results

Lemma 4.17. *There exist* $\delta \in (0,1)$*, a natural number* L *and* $M > 0$ *such that each integer* $T \ge L$*, each sequence of upper semicontinuous functions* $\{u_t\}_{t=0}^{T-1} \subset \mathcal{M}$ *satisfying*

$$\|u_t - v\| \le \delta,\ t = 0, \dots, T-1,$$

each pair of points $\xi_0, \xi_1 \in X$ *satisfying*

$$\rho(\xi_i, \bar{x}) \le \bar{r},\ i = 0, 1$$

and each (Ω)*-program* $\{y_t\}_{t=0}^{T}$ *the inequality*

$$\sum_{t=0}^{T-1} u(y_t, y_{t+1}) \le \sigma(\{u_t\}_{t=0}^{T-1}, 0, T, \xi_0, \xi_1) + M \tag{4.14}$$

holds.

Proof. By Theorem 2.3, there exist an integer $L_0 \ge 1$ and $\delta_0 \in (0,1)$ such that the following property holds:

(i) for each integer $T > 2L_0$, each $\{u_t\}_{t=0}^{T-1} \subset \mathcal{M}$ satisfying

$$\|u_t - v\| \le \delta_0,\ t = 0 \dots T-1,$$

and each (Ω)-program $\{x_t\}_{t=0}^{T}$ which satisfies

$$\rho(x_0, \bar{x}) \le \bar{r},\ \rho(x_T, \bar{x}) \le \bar{r},$$

$$\sum_{t=0}^{T-1} u_t(x_t, x_{t+1}) \ge \sigma(\{u_t\}_{t=0}^{T-1}, 0, T, x_0, x_T) - \delta_0$$

we have

$$\rho(x_t, \bar{x}) \le \bar{r}/4,\ t = L_0, \dots, T - L_0.$$

By Theorem 2.5, there exist a natural number L_1 and $\delta_1 \in (0,1)$ such that the following property holds:

(ii) for each integer $T \ge 2L_1$, each $\{u_t\}_{t=0}^{T-1} \subset \mathcal{M}$ satisfying

$$\|u_t - v\| \le \delta_1,\ t = 0 \dots T-1,$$

and each (Ω)-program $\{x_t\}_{t=0}^{T}$ which satisfies

$$\sum_{t=0}^{T-1} u_t(x_t, x_{t+1}) = \sigma(\{u_t\}_{t=0}^{T-1}, 0, T)$$

we have

$$\rho(x_t, \bar{x}) \le \bar{r}/4,\ t = L_1, \dots, T - L_1.$$

Set

$$L = 2(L_0 + L_1) + 4,\ d = \min\{\delta_0, \delta_1\},$$
$$M = 4(L_0 + L_1 + 1)(\|v\| + 1). \tag{4.15}$$

Let $\xi_0, \xi_1 \in X$ satisfy

$$\rho(\xi_i, \bar{x}) \le \bar{r},\ i = 0, 1, \tag{4.16}$$

$T \geq L$ be an integer and let a sequence of upper semicontinuous functions $\{u_t\}_{t=0}^{T-1} \subset \mathcal{M}$ satisfy

$$\|u_t - v\| \leq \delta, \ t = 0, \dots, T-1. \tag{4.17}$$

Clearly, there exists an (Ω)-program $\{x_t\}_{t=0}^{T}$ such that

$$x_0 = \xi_0, \ x_T = \xi_1, \tag{4.18}$$

$$\sum_{t=0}^{T-1} u_t(x_t, x_{t+1}) = \sigma(\{u_t\}_{t=0}^{T-1}, 0, T, \xi_0, \xi_1). \tag{4.19}$$

Let $\{y_t\}_{t=0}^{T}$ be an (Ω)-program. We show that (4.14) holds. We may assume without loss of generality that

$$\sum_{t=0}^{T-1} u_t(y_t, y_{t+1}) = \sigma(\{u_t\}_{t=0}^{T-1}, 0, T). \tag{4.20}$$

By property (i) and (4.15)–(4.19),

$$\rho(x_t, \bar{x}) \leq \bar{r}/4, \ t = L_0, \dots, T - L_0. \tag{4.21}$$

In view of property (ii), (4.15), (4.17), and (4.20),

$$\rho(y_t, \bar{x}) \leq \bar{r}/4, \ t = L_1, \dots, T - L_1. \tag{4.22}$$

It follows from (4.1), (4.15), (4.21), and (4.22) that there exists an (Ω)-program $\{z_t\}_{t=0}^{T}$ such that

$$z_t = x_t, \ t = 0, \dots, L_0 + L_1,$$

$$z_t = x_t, \ t = T - L_0 - L_1, \dots, T,$$

$$z_t = y_t, \ t = L_0 + L_1 + 1, \dots, T - L_0 - L_1 - 1. \tag{4.23}$$

By (4.15), (4.17), and (4.23),

$$\left|\sum_{t=0}^{T-1} u_t(z_t, z_{t+1}) - \sum_{t=0}^{T-1} u_t(y_t, y_{t+1})\right| \leq 4(L_0 + L_1 + 1)(\|v\| + 1). \tag{4.24}$$

In view of (4.15), (4.18), (4.19), and (4.23),

$$\sum_{t=0}^{T-1} u_t(x_t, x_{t+1}) \geq \sum_{t=0}^{T-1} u_t(z_t, z_{t+1})$$

$$\geq -4(L_0 + L_1 + 1)(\|v\| + 1) + \sum_{t=0}^{T-1} u_t(y_t, y_{t+1})$$

$$= \sum_{t=0}^{T-1} u_t(y_t, y_{t+1}) - M.$$

Lemma 4.17 is proved. □

Lemma 4.17 implies the following result.

Lemma 4.18. *There exist $\delta \in (0,1)$ and $M > 0$ such that for each integer $T \geq 1$, each sequence of upper semicontinuous functions $\{u_t\}_{t=0}^{T-1} \subset \mathcal{M}$ satisfying*

$$\|u_t - v\| \leq \delta, \ t = 0, \dots, T-1,$$

each pair of points $\xi_0, \xi_1 \in X$ satisfying

$$\rho(\xi_i, \bar{x}) \leq \bar{r}, \ i = 0, 1$$

and each (Ω)-program $\{y_t\}_{t=0}^{T}$ the inequality

$$\sum_{t=0}^{T-1} u(y_t, y_{t+1}) \leq \sigma(\{u_t\}_{t=0}^{T-1}, 0, T, \xi_0, \xi_1) + M$$

holds.

Lemma 4.19 (Lemma 2.12 of [89]). *Let ϵ be a positive number and let L be a natural number. Then there exists $\delta > 0$ such that for each*

$$\Omega_t \in \mathcal{E}(\delta), \ t = 0, \dots, L-1,$$

each $u_t \in \mathcal{M}$, $t = 0, \dots, L-1$ satisfying

$$\|u_t - v\| \leq \delta, \ t = 0, \dots, L-1$$

and each $(\{\Omega_t\}_{t=0}^{L-1})$-program $\{x_t\}_{t=0}^{L}$ there exists an (Ω)-program $\{y_t\}_{t=0}^{L}$ such that

$$\rho(x_t, y_t) \leq \epsilon \text{ for all } t = 0, \dots, L$$

and

$$\sum_{t=0}^{L-1} v(y_t, y_{t+1}) \geq \sum_{t=0}^{L-1} u_t(x_t, x_{t+1}) - \epsilon.$$

Lemma 4.20. *Let ϵ, M be positive numbers. Then there exist $\delta > 0$ and a natural number L such that for each integer $T \geq L$, each*

$$\Omega_t \in \mathcal{E}(\delta), \ t = 0, \dots, T-1, \tag{4.25}$$

each $u_t \in \mathcal{M}$, $t = 0, \dots, T-1$ satisfying

$$\|u_t - v\| \leq \delta, \ t = 0, \dots, T-1 \tag{4.26}$$

and each $(\{\Omega_t\}_{t=0}^{T-1})$-program $\{x_t\}_{t=0}^{T}$ which satisfies

$$\sigma(\{u_t\}_{t=0}^{T-1}, \{\Omega_t\}_{t=0}^{T-1}, 0, T) \leq \sum_{t=0}^{T-1} u_t(x_t, x_{t+1}) + M \tag{4.27}$$

the following inequality holds:

$$\min\{\rho(x_t, \bar{x}) : \ t = 0, \dots, L-1\} \leq \epsilon.$$

Proof. We may assume without loss of generality that

$$\epsilon < 2^{-1}\bar{\lambda}. \tag{4.28}$$

By Lemma 2.7, there exists a natural number L_0 such that the following property holds:

(i) for each (Ω)-program $\{z_t\}_{t=0}^{L_0}$ which satisfies

$$\sum_{t=0}^{L_0-1} v(z_t, z_{t+1}) \geq L_0 v(\bar{x}, \bar{x}) - 4$$

we have

$$\min\{\rho(z_i, \bar{x}) : i = 1, \dots, L_0\} \leq \epsilon/4.$$

By Lemma 4.19, there exists $\delta_0 > 0$ such that the following property holds:

(ii) for each

$$\Omega_t \in \mathcal{E}(\delta_0),\ t = 0, \dots, L_0 - 1,$$

each $u_t \in \mathcal{M},\ t = 0, \dots, L_0 - 1$ satisfying

$$\|u_t - v\| \leq \delta_0,\ t = 0, \dots, L_0 - 1$$

and each $(\{\Omega_t\}_{t=0}^{L_0-1})$-program $\{z_t\}_{t=0}^{L_0}$ there exists an (Ω)-program $\{\tilde{z}_t\}_{t=0}^{L_0}$ such that

$$\rho(z_t, \tilde{z}_t) \leq \epsilon/4 \text{ for all } t = 0, \dots, L_0$$

and

$$\sum_{t=0}^{L_0-1} v(\tilde{z}_t, \tilde{z}_{t+1}) \geq \sum_{t=0}^{L_0-1} u_t(z_t, z_{t+1}) - 1.$$

Choose a positive number δ such that

$$\delta < \delta_0,\ L_0\delta < 1, \tag{4.29}$$

a natural number

$$q_0 > M + 2(\|v\| + 1)(L_0 + 1) \tag{4.30}$$

and set

$$L = q_0 L_0 + 1. \tag{4.31}$$

Assume that the lemma does not hold with our choice of δ, L. Then there exist an integer

$$T \geq L, \tag{4.32}$$

$$\Omega_t \in \mathcal{E}(\delta),\ t = 0, \dots, T - 1, \tag{4.33}$$

$u_t \in \mathcal{M},\ t = 0, \dots, T-1$ satisfying (4.26) and an $(\{\Omega_t\}_{t=0}^{T-1})$-program $\{x_t\}_{t=0}^{T}$ which satisfies (4.27) and such that

$$\rho(x_t, \bar{x}) > \epsilon,\ t = 0, \dots, L - 1. \tag{4.34}$$

Clearly, there exists an integer

$$S \in [L-1, T] \tag{4.35}$$

such that

$$\rho(x_t, \bar{x}) > \epsilon,\ t = 0, \dots, S \tag{4.36}$$

and at least one of the following relations holds:

$$S = T; \tag{4.37}$$

$$S < T \text{ and } \rho(x_{S+1}, \bar{x}) \le \epsilon. \tag{4.38}$$

Define

$$y_t = \bar{x},\ t = 0, \dots, S \tag{4.39}$$

and if (4.38) holds, then

$$y_t = x_t \text{ for all integers } t = S+1, \dots, T. \tag{4.40}$$

It follows from (4.4), (4.28), and (4.37)–(4.40) that $\{y_t\}_{t=0}^{T}$ is an $(\{\Omega_t\}_{t=0}^{T-1})$-program. In view of (4.27),

$$M \ge \sum_{t=0}^{T-1} u_t(y_t, y_{t+1}) - \sum_{t=0}^{T-1} u_t(x_t, x_{t+1}). \tag{4.41}$$

Relations (4.31) and (4.35) imply that

$$S \ge L - 1 \ge q_0 L_0 \tag{4.42}$$

and there exists an integer $q \ge q_0$ such that

$$qL_0 \le S < (q+1)L_0. \tag{4.43}$$

By (4.26), (4.29), (4.37)–(4.41), and (4.43),

$$\begin{aligned}
&\sum_{t=0}^{qL_0-1} u_t(y_t, y_{t+1}) - \sum_{t=0}^{qL_0-1} u_t(x_t, x_{t+1}) \\
&\le \sum_{t=0}^{S-1} u_t(y_t, y_{t+1}) - \sum_{t=0}^{S-1} u_t(x_t, x_{t+1}) + 2(\|v\| + \delta)L_0 \\
&\le \sum_{t=0}^{T-1} u_t(y_t, y_{t+1}) - \sum_{t=0}^{T-1} u_t(x_t, x_{t+1}) + 2(\|v\| + \delta) + 2(\|v\| + \delta)L_0 \\
&\le M + 2(\|v\| + 1)(L_0 + 1).
\end{aligned} \tag{4.44}$$

Let $k \in \{0, \dots, q-1\}$. We estimate

$$\sum_{t=kL_0}^{(k+1)L_0-1} u_t(y_t, y_{t+1}) - \sum_{t=kL_0}^{(k+1)L_0-1} u_t(x_t, x_{t+1}).$$

Property (ii), (4.26), (4.29), and (4.33) imply that there exists an (Ω)-program $\{\tilde{x}_t\}_{t=kL_0}^{(k+1)L_0}$ such that

$$\rho(x_t, \tilde{x}_t) \le \epsilon/4 \text{ for all } t = L_0 k, \dots, L_0(k+1) \tag{4.45}$$

and

$$\sum_{t=kL_0}^{(k+1)L_0-1} v(\tilde{x}_t, \tilde{x}_{t+1}) \ge \sum_{t=kL_0}^{(k+1)L_0-1} u_t(x_t, x_{t+1}) - 1. \tag{4.46}$$

By (4.36), (4.43), and (4.45), for all $t = kL_0, \dots, (k+1)L_0$,

$$\rho(\tilde{x}_t, \bar{x}) \ge \rho(x_t, \bar{x}) - \rho(\tilde{x}_t, x_t) > \epsilon - \epsilon/4. \tag{4.47}$$

Property (i) and (4.47) imply that

$$\sum_{t=kL_0}^{(k+1)L_0-1} v(\tilde{x}_t, \tilde{x}_{t+1}) < L_0 v(\bar{x}, \bar{x}) - 4. \tag{4.48}$$

In view of (4.46) and (4.48),

$$\sum_{t=kL_0}^{(k+1)L_0-1} u_t(x_t, x_{t+1}) < L_0 v(\bar{x}, \bar{x}) - 3. \tag{4.49}$$

It follows from (4.26), (4.39), and (4.33) that

$$\sum_{t=kL_0}^{(k+1)L_0-1} u_t(y_t, y_{t+1}) \ge (v(\bar{x}, \bar{x}) - \delta)L_0. \tag{4.50}$$

By (4.29), (4.49), and (4.50),

$$\sum_{t=kL_0}^{(k+1)L_0-1} u_t(y_t, y_{t+1}) - \sum_{t=kL_0}^{(k+1)L_0-1} u_t(x_t, x_{t+1})$$

$$\ge L_0 v(\bar{x}, \bar{x}) - L_0\delta - L_0 v(\bar{x}, \bar{x}) + 3 = 3 - L_0\delta \ge 2. \tag{4.51}$$

In view of (4.44) and (4.51),

$$M + 2(\|v\| + 1)(L_0 + 1) \ge \sum_{t=0}^{qL_0-1} u_t(y_t, y_{t+1}) - \sum_{t=0}^{qL_0-1} u_t(x_t, x_{t+1}) \ge 2q \ge 2q_0$$

and

$$q_0 \le M + 2(\|v\| + 1)(L_0 + 1).$$

This contradicts (4.30). The contradiction we have reached completes the proof of Lemma 4.20. □

4.9 Proof of Theorem 4.15

By Lemma 4.20, there exist $\delta_0 > 0$ and a natural number L_0 such that the following property holds:

(i) for each integer $T \geq L_0$, each

$$\Omega_t \in \mathcal{E}(\delta_0),\ t = 0, \dots, T-1,$$

each $u_t \in \mathcal{M},\ t = 0, \dots, T-1$ satisfying

$$\|u_t - v\| \leq \delta_0,\ t = 0, \dots, T-1$$

and each $(\{\Omega_t\}_{t=0}^{T-1})$-program $\{x_t\}_{t=0}^{T}$ which satisfies

$$\sigma(\{u_t\}_{t=0}^{T-1}, \{\Omega_t\}_{t=0}^{T-1}, 0, T) \leq \sum_{t=0}^{T-1} u_t(x_t, x_{t+1}) + M \tag{4.52}$$

we have

$$\min\{\rho(x_t, \bar{x}) :\ t = 0, \dots, L_0 - 1\} \leq \bar{\lambda}/4. \tag{4.53}$$

By Theorem 4.5, there exist $\delta \in (0, \delta_0)$ and a natural number $L > L_0$ such that the following property holds:

(ii) for each integer $T > L$, each

$$\Omega_t \in \mathcal{E}(\delta),\ t = 0, \dots, T-1,$$

each $u_t \in \mathcal{M},\ t = 0, \dots, T-1$ satisfying

$$\|u_t - v\| \leq \delta,\ t = 0, \dots, T-1$$

and each $(\{\Omega_t\}_{t=0}^{T-1})$-program $\{x_t\}_{t=0}^{T}$ which satisfies

$$x_0 \in \bar{Y}(\{\Omega_t\}_{t=0}^{L_0-1}, 0, L_0), \tag{4.54}$$

$$\sigma(\{u_t\}_{t=0}^{T-1}, \{\Omega_t\}_{t=0}^{T-1}, 0, T, x_0) \leq \sum_{t=0}^{T-1} u_t(x_t, x_{t+1}) + M \tag{4.55}$$

we have

$$\mathrm{Card}(\{t \in \{0, \dots, T\} :\ \rho(x_t, \bar{x}) > \epsilon\}) \leq L. \tag{4.56}$$

Assume that an integer $T > L$,

$$\Omega_t \in \mathcal{E}(\delta),\ t = 0, \dots, T-1, \tag{4.57}$$

$u_t \in \mathcal{M},\ t = 0, \dots, T-1$ satisfy

$$\|u_t - v\| \leq \delta,\ t = 0, \dots, T-1 \tag{4.58}$$

and $(\{\Omega_t\}_{t=0}^{T-1})$-program $\{x_t\}_{t=0}^{T}$ satisfies (4.52). Property (i), (4.52), (4.57), and (4.58) imply that (4.53) holds. It follows from (4.4), (4.53), and (4.57) that (4.54) is valid. Relation (4.52) implies (4.55). Relations (4.54), (4.55), (4.57), and (4.58) imply (4.56). Theorem 4.15 is proved. □

4.10 Proof of Theorem 4.16

By Lemma 4.20, there exist $\delta_0 > 0$ and a natural number L_0 such that the following property holds:

(i) for each integer $T \geq L_0$, each

$$\Omega_t \in \mathcal{E}(\delta_0),\ t = 0, \dots, T-1,$$

each $u_t \in \mathcal{M}$, $t = 0, \dots, T-1$ satisfying

$$\|u_t - v\| \leq \delta_0,\ t = 0, \dots, T-1$$

and each $(\{\Omega_t\}_{t=0}^{T-1})$-program $\{x_t\}_{t=0}^{T}$ which satisfies

$$\sigma(\{u_t\}_{t=0}^{T-1}, \{\Omega_t\}_{t=0}^{T-1}, 0, T) \leq \sum_{t=0}^{T-1} u_t(x_t, x_{t+1}) + M \tag{4.59}$$

we have

$$\min\{\rho(x_t, \bar{x}) :\ t = 0, \dots, L_0 - 1\} \leq \bar{\lambda}/4.$$

By Theorem 4.8, there exist a positive number

$$\delta < \min\{\delta_0, \bar{\lambda}/4\}$$

and a natural number L_1 such that the following property holds:

(ii) each integer $T > 2L_1$, each

$$\Omega_t \in \mathcal{E}(\delta),\ t = 0, \dots, T-1,$$

each $u_t \in \mathcal{M}$, $t = 0, \dots, T-1$ satisfying

$$\|u_t - v\| \leq \delta,\ t = 0, \dots, T-1$$

and each $(\{\Omega_t\}_{t=0}^{T-1})$-program $\{x_t\}_{t=0}^{T}$ which satisfies

$$\rho(x_0, \bar{x}) \leq \bar{\lambda},$$

$$\sigma(\{u_t\}_{t=0}^{T-1}, \{\Omega_t\}_{t=0}^{T-1}, 0, T, x_0) \leq \sum_{t=0}^{T-1} u_t(x_t, x_{t+1}) + M \tag{4.60}$$

and

$$\sum_{t=\tau}^{\tau+L_1-1} u_t(x_t, x_{t+1}) \geq \sigma(\{u_t\}_{t=\tau}^{\tau+L_1-1}, \{\Omega_t\}_{t=\tau}^{\tau+L_1-1}, \tau, \tau + L_1, x_\tau, x_{\tau+L_1}) - \delta \tag{4.61}$$

for each integer $\tau \in [0, T-L_1]$, there exist integers $\tau_1 \in [0, L_1]$, $\tau_2 \in [T-L_1, T]$ such that

$$\rho(x_t, \bar{x}) \leq \epsilon \text{ for all } t = \tau_1, \dots, \tau_2;$$

moreover if $\rho(x_0, \bar{x}) \le \delta$, then $\tau_1 = 0$ and if $\rho(x_T, \bar{x}) \le \delta$, then $\tau_2 = T$.

Set

$$L = 2L_0 + 2L_1. \tag{4.62}$$

Assume that an integer $T > 2L$,

$$\Omega_t \in \mathcal{E}(\delta),\ t = 0, \dots, T-1, \tag{4.63}$$

$u_t \in \mathcal{M}$, $t = 0, \dots, T-1$ satisfy

$$\|u_t - v\| \le \delta,\ t = 0, \dots, T-1 \tag{4.64}$$

and an $(\{\Omega_t\}_{t=0}^{T-1})$-program $\{x_t\}_{t=0}^{T}$ satisfies (4.59) and (4.13) for all integers $\tau \in [0, T-L]$. Property (i), (4.59), and (4.62)–(4.64) imply that there exists an integer $S_0 \in [0, L_0 - 1]$ such that

$$\rho(x_{S_0}, \bar{x}) \le \bar{\lambda}/4. \tag{4.65}$$

If $\rho(x_0, \bar{x}) \le \delta$, then we set $S_0 = 0$. It follows from (4.13) that (4.61) holds for all $\tau \in \{0, \dots, T - L_1\}$. By (4.59), (4.62)–(4.65), and property (ii) applied to the $(\{\Omega_{S_0+t}\}_{t=0}^{T-1-S_0})$-program $\{x_{S_0+t}\}_{t=0}^{T-S_0}$, there exist integers

$$\tau_1 \in [0, L_1],\ \tau_2 \in [T - S_0 - L_1, T - S_0] \tag{4.66}$$

such that

$$\rho(x_{S_0+t}, \bar{x}) \le \epsilon \text{ for all } t = \tau_1, \dots, \tau_2. \tag{4.67}$$

Moreover if $\rho(x_0, \bar{x}) \le \delta$, then $\tau_1 = 0$ and if $\rho(x_T, \bar{x}) \le \delta$, then $\tau_2 = T - S_0$. In view of (4.67),

$$\rho(x_t, \bar{x}) \le \epsilon \text{ for all } t = \tau_1 + S_0, \dots, \tau_2 + S_0.$$

By (4.62) and (4.66),

$$\tau_1 + S_0 \le L_1 + L_0 \le L,\ T - (\tau_2 + S_0) \le L_1 \le L.$$

Clearly, if if $\rho(x_0, \bar{x}) \le \delta$, then $S_0 = 0$, $\tau_1 = 0$ and if $\rho(x_T, \bar{x}) \le \delta$, then $\tau_2 + S_0 = T$. This completes the proof of Theorem 4.16. □

4.11 Turnpike Results for Perturbed Problems (P5)

Denote by $\mathfrak{A}$ the set of all bounded functions $h : X \times X \to R^1$. For each $h \in \mathfrak{A}$ set

$$\|h\| = \sup\{|h(x_1, x_2)| :\ x_1, x_2 \in X\}.$$

Clearly, $(\mathfrak{A}, \|\cdot\|)$ is a Banach space. For each $h_1, h_2 \in \mathfrak{A}$ set

$$d_{X \times X}(h_1, h_2) = \|h_1 - h_2\|.$$

Let integers T_1, T_2 satisfy $0 \leq T_1 < T_2$ and let Ω_t, $t = T_1, \dots, T_2 - 1$ be nonempty subsets of $X \times X$.

For each finite sequence $\{u_t\}_{t=T_1}^{T_2-1} \subset \mathcal{M}$ and each $h \in \mathfrak{A}$ we consider the problem

$$\sum_{t=T_1}^{T_2-1} u_t(x_t, x_{t+1}) + h(x_{T_1}, x_{T_2}) \to \max,$$

$$\{x_t\}_{t=T_1}^{T_2} \text{ is an } (\{\Omega_t\}_{t=T_1}^{T_2-1}) - \text{program}$$

and set

$$\sigma(\{u_t\}_{t=T_1}^{T_2-1}, h, \{\Omega_t\}_{t=T_1}^{T_2-1}, T_1, T_2) = \sup\{\sum_{t=T_1}^{T_2-1} u_t(x_t, x_{t+1}) + h(x_{T_1}, x_{T_2}) :$$

$$\{x_t\}_{t=T_1}^{T_2} \text{ is an } (\{\Omega_t\}_{t=T_1}^{T_2-1}) - \text{program}\}. \tag{4.68}$$

(Recall that we use the convention that the supremum of an empty set is $-\infty$).

For each $u \in \mathcal{M}$ and each $h \in \mathfrak{A}$ set

$$\sigma(u, h, \{\Omega_t\}_{t=T_1}^{T_2-1}, T_1, T_2) = \sigma(\{u_t\}_{t=T_1}^{T_2-1}, h, \{\Omega_t\}_{t=T_1}^{T_2-1}, T_1, T_2),$$

where $u_t = u$, $t = T_1, \dots, T_2 - 1$.

In the next section we will prove the following result.

Theorem 4.21. *Let ϵ, M be positive numbers. Then there exist $\delta > 0$ and a natural number L such that for each integer $T > L$, each*

$$\Omega_t \in \mathcal{E}(\delta),\ t = 0, \dots, T - 1,$$

each $u_t \in \mathcal{M}$, $t = 0, \dots, T - 1$ satisfying

$$\|u_t - v\| \leq \delta,\ t = 0, \dots, T - 1,$$

each $h \in \mathfrak{A}$ satisfying

$$\|h\| \leq M$$

and each $(\{\Omega_t\}_{t=0}^{T-1})$-program $\{x_t\}_{t=0}^{T}$ which satisfies

$$\sigma(\{u_t\}_{t=0}^{T-1}, h, \{\Omega_t\}_{t=0}^{T-1}, 0, T) \leq \sum_{t=0}^{T-1} u_t(x_t, x_{t+1}) + h(x_0, x_T) + M$$

the inequality

$$Card(\{t \in \{0, \dots, T\} : \rho(x_t, \bar{x}) > \epsilon\}) \leq L$$

holds.

The next theorem is proved in Sect. 4.13.

Theorem 4.22. *Let ϵ, M be positive numbers. Then there exist $\delta > 0$ and a natural number L such that for each integer $T > 2L$, each*

$$\Omega_t \in \mathcal{E}(\delta),\ t = 0, \dots, T-1,$$

each $u_t \in \mathcal{M}$, $t = 0, \dots, T-1$ satisfying

$$\|u_t - v\| \le \delta,\ t = 0, \dots, T-1,$$

each $h \in \mathfrak{A}$ satisfying

$$\|h\| \le M$$

and each $(\{\Omega_t\}_{t=0}^{T-1})$-program $\{x_t\}_{t=0}^{T}$ which satisfies

$$\sigma(\{u_t\}_{t=0}^{T-1}, h, \{\Omega_t\}_{t=0}^{T-1}, 0, T) \le \sum_{t=0}^{T-1} u_t(x_t, x_{t+1}) + h(x_0, x_T) + M$$

and

$$\sigma(\{u_t\}_{t=\tau}^{\tau+L-1}, \{\Omega_t\}_{t=\tau}^{\tau+L-1}, \tau, \tau+L, x_\tau, x_{\tau+L}) \le \sum_{t=\tau}^{\tau+L-1} u_t(x_t, x_{t+1}) + \delta$$

for each integer $\tau \in [0, T-L]$, there exist integers $\tau_1 \in [0, L]$, $\tau_2 \in [T-L, T]$ such that

$$\rho(x_t, \bar{x}) \le \epsilon \text{ for all } t = \tau_1, \dots, \tau_2.$$

Moreover if $\rho(x_0, \bar{x}) \le \delta$, then $\tau_1 = 0$ and if $\rho(x_T, \bar{x}) \le \delta$, then $\tau_2 = T$.

4.12 Proof of Theorem 4.21

By Theorem 4.15, there exist $\delta > 0$ and a natural number L such that the following property holds:

(i) for each integer $T > L$, each

$$\Omega_t \in \mathcal{E}(\delta),\ t = 0, \dots, T-1,$$

each $u_t \in \mathcal{M}$, $t = 0, \dots, T-1$ satisfying

$$\|u_t - v\| \le \delta,\ t = 0, \dots, T-1$$

and each $(\{\Omega_t\}_{t=0}^{T-1})$-program $\{x_t\}_{t=0}^{T}$ which satisfies

$$\sigma(\{u_t\}_{t=0}^{T-1}, \{\Omega_t\}_{t=0}^{T-1}, 0, T) \le \sum_{t=0}^{T-1} u_t(x_t, x_{t+1}) + 3M \tag{4.69}$$

the inequality

$$\mathrm{Card}(\{t \in \{0, \dots, T\} :\ \rho(x_t, \bar{x}) > \epsilon\}) \le L \tag{4.70}$$

holds.

Assume that an integer $T > L$,

$$\Omega_t \in \mathcal{E}(\delta),\ t = 0, \dots, T-1, \tag{4.71}$$

$u_t \in \mathcal{M},\ t = 0, \dots, T-1$ satisfy

$$\|u_t - v\| \le \delta,\ t = 0, \dots, T-1, \tag{4.72}$$

$h \in \mathfrak{A}$ satisfies

$$\|h\| \le M \tag{4.73}$$

and that an $(\{\Omega_t\}_{t=0}^{T-1})$-program $\{x_t\}_{t=0}^{T}$ satisfies

$$\sigma(\{u_t\}_{t=0}^{T-1}, h, \{\Omega_t\}_{t=0}^{T-1}, 0, T) \le \sum_{t=0}^{T-1} u_t(x_t, x_{t+1}) + h(x_0, x_T) + M. \tag{4.74}$$

By (4.73) and (4.74),

$$\begin{aligned}
&\sigma(\{u_t\}_{t=0}^{T-1}, \{\Omega_t\}_{t=0}^{T-1}, 0, T) \\
&\le \sigma(\{u_t\}_{t=0}^{T-1}, h, \{\Omega_t\}_{t=0}^{T-1}, 0, T) + \|h\| \\
&\le \sum_{t=0}^{T-1} u_t(x_t, x_{t+1}) + 2\|h\| + M \le \sum_{t=0}^{T-1} u_t(x_t, x_{t+1}) + 3M
\end{aligned}$$

and (4.69) holds. Property (i), (4.69), (4.71), and (4.72) imply (4.70). This completes the proof of Theorem 4.21. □

4.13 Proof of Theorem 4.22

By Theorem 4.16, there exist $\delta > 0$ and a natural number L such that the following property holds:

(i) for each integer $T > 2L$, each

$$\Omega_t \in \mathcal{E}(\delta),\ t = 0, \dots, T-1,$$

each $u_t \in \mathcal{M},\ t = 0, \dots, T-1$ satisfying

$$\|u_t - v\| \le \delta,\ t = 0, \dots, T-1$$

and each $(\{\Omega_t\}_{t=0}^{T-1})$-program $\{x_t\}_{t=0}^{T}$ which satisfies

$$\sigma(\{u_t\}_{t=0}^{T-1}, \{\Omega_t\}_{t=0}^{T-1}, 0, T) \le \sum_{t=0}^{T-1} u_t(x_t, x_{t+1}) + 3M$$

and

$$\sigma(\{u_t\}_{t=\tau}^{\tau+L-1}, \{\Omega_t\}_{t=\tau}^{\tau+L-1}, \tau, \tau+L, x_\tau, x_{\tau+L}) \le \sum_{t=\tau}^{\tau+L-1} u_t(x_t, x_{t+1}) + \delta \tag{4.75}$$

for each integer $\tau \in [0, T-L]$, there exist integers $\tau_1 \in [0, L]$, $\tau_2 \in [T-L, T]$ such that

$$\rho(x_t, \bar{x}) \le \epsilon \text{ for all } t = \tau_1, \dots, \tau_2;$$

if $\rho(x_0, \bar{x}) \le \delta$, then $\tau_1 = 0$ and if $\rho(x_T, \bar{x}) \le \delta$, then $\tau_2 = T$.

Assume that an integer $T > 2L$,

$$\Omega_t \in \mathcal{E}(\delta),\ t = 0, \dots, T-1, \tag{4.76}$$

$u_t \in \mathcal{M}$, $t = 0, \dots, T-1$ satisfy

$$\|u_t - v\| \le \delta,\ t = 0, \dots, T-1, \tag{4.77}$$

$h \in \mathfrak{A}$ satisfies

$$\|h\| \le M \tag{4.78}$$

and that an $(\{\Omega_t\}_{t=0}^{T-1})$-program $\{x_t\}_{t=0}^{T}$ satisfies

$$\sigma(\{u_t\}_{t=0}^{T-1}, h, \{\Omega_t\}_{t=0}^{T-1}, 0, T) \le \sum_{t=0}^{T-1} u_t(x_t, x_{t+1}) + h(x_0, x_T) + M \tag{4.79}$$

and (4.75) for each integer $\tau \in [0, T-L]$.

By (4.78) and (4.79),

$$\sigma(\{u_t\}_{t=0}^{T-1}, \{\Omega_t\}_{t=0}^{T-1}, 0, T)$$
$$\le \sigma(\{u_t\}_{t=0}^{T-1}, h, \{\Omega_t\}_{t=0}^{T-1}, 0, T) + \|h\|$$
$$\le \sum_{t=0}^{T-1} u_t(x_t, x_{t+1}) + 2\|h\| + M \le \sum_{t=0}^{T-1} u_t(x_t, x_{t+1}) + 3M.$$

The validity of Theorem 4.22 follows from property (i), the relation above and (4.75)–(4.76). □

4.14 Perturbed Problems (P4)

Define

$$\bar{v}(x, y) = v(y, x),\ (x, y) \in X \times X,$$
$$\bar{\Omega} = \{(x, y) \in X \times X :\ (y, x) \in \Omega\}. \tag{4.80}$$

Consider the function $\pi^{\bar{v}} : X \to R^1 \cup \{-\infty\}$ see ((2.113)–(2.116)). Let $g \in \mathfrak{M}$ be an upper semicontinuous function. It is known (see Sect. 3.2) that $\pi^v + g$, $\pi^{\bar{v}} + g : X \to R^1 \cup \{-\infty\}$ are upper semicontinuous and bounded from above functions satisfying, in view of Proposition 2.22,

$$(\pi^v + g)(\bar{x}) = g(\bar{x}),\ (\pi^{\bar{v}} + g)(\bar{x}) = g(\bar{x})$$

which have maximizers. By Theorem 2.13, there exists an $(\bar{v}, \bar{\Omega})$-overtaking optimal program $\{x_t^*\}_{t=0}^{\infty}$ such that

$$(\pi^{\bar{v}} + g)(x_0^*) = \sup(\pi^{\bar{v}} + g), \ \lim_{t\to\infty} x_t^* = \bar{x}. \tag{4.81}$$

Let $\lambda > 0$ and $T \geq L$ be natural numbers. Denote by $\mathcal{E}(\lambda, L, T)$ the collection of all sequences

$$\{\Omega_t\}_{t=0}^{T-1} \subset \mathcal{E}(\lambda) \tag{4.82}$$

for which there exists an $(\{\Omega_t\}_{t=T-L}^{T-1})$-program $\{y_t\}_{t=T-L}^{T}$ such that

$$\rho(x_{T-t}^*, y_t) \leq \lambda, \ t = T-L, \dots, T, \tag{4.83}$$

$$|v(y_t, y_{t+1}) - v(x_{T-t}^*, x_{T-t-1}^*)| \leq \lambda, \ t = T-L, \dots, T-1 \tag{4.84}$$

and

$$|g(y_T) - g(x_0^*)| \leq \lambda. \tag{4.85}$$

Example 4.23. Let $\lambda > 0$ and $T \geq L$ be natural numbers and let

$$\{\Omega_t\}_{t=0}^{T-1} \subset \mathcal{E}(\lambda).$$

Assume that for all integers $t = T-L, \dots, T-1$,

$$(x_{T-t}^*, x_{T-t-1}^*) \in \Omega_t.$$

Then

$$\{\Omega_t\}_{t=0}^{T-1} \in \mathcal{E}(\delta, L, T).$$

Example 4.24. Let $\lambda > 0$ and $T \geq L$ be natural numbers, the function g be continuous at x_0^*, and the functions $\bar{v}$ be continuous at (x_t^*, x_{t+1}^*) for all integers $t = 0, \dots, L-1$. Then there exists $\tilde{\lambda} \in (0, \lambda)$ such that for a sequence

$$\{\Omega_t\}_{t=0}^{T-1} \subset \mathcal{E}(\tilde{\lambda})$$

for which there exists an $(\{\Omega_t\}_{t=T-L}^{T-1})$-program $\{y_t\}_{t=T-L}^{T}$ such that

$$\rho(x_{T-t}^*, y_t) \leq \tilde{\lambda}, \ t = T-L, \dots, T$$

the inclusion

$$\{\Omega_t\}_{t=0}^{T-1} \in \mathcal{E}(\lambda, L, T)$$

is true.

Example 4.25. For each $x \in X$ set

$$\Omega(x) = \{y \in X : \ (x, y) \in \Omega\}.$$

Let $T \geq L$ be natural numbers, the function g be continuous at x_0^*, and the functions $\bar{v}$ be continuous at (x_t^*, x_{t+1}^*) for all integers $t = 0, \dots, L-1$. Suppose that there exists $\bar{r}_0 \in (0, \bar{r})$ such that for each integer t satisfying $1 \leq t < L$,

$$\Omega(x) \neq \emptyset \text{ for each } x \in B(x_t^*, \bar{r}_0)$$

and that the following property holds:

(C1) for each $\epsilon > 0$ there exists $\delta(\epsilon) \in (0, \bar{r}_0)$ such that for each $x \in B(x_t^*, \delta(\epsilon))$,

$$\Omega(x) \cap B(x_{t-1}^*, \epsilon) \neq \emptyset.$$

Proposition 4.26. *Let $L > 2$ be an integer and $\lambda > 0$. Then there exists a positive number*

$$\lambda_0 < \min\{\lambda, \bar{r}_0\}$$

such that the following assertion holds.

Assume that a finite sequence

$$\{\Omega_t\}_{t=0}^{T-1} \subset \mathcal{E}(\lambda_0)$$

satisfies

$$\Omega_{T-L}(x_L^*) \cap B(x_{L-1}^*, \lambda_0) \neq \emptyset \tag{4.86}$$

and possesses the property

(C2) for each integer t satisfying $0 < t \leq L - 1$, each $x \in B(x_t^*, \bar{r}_0)$, and each $y \in \Omega(x)$, the inclusion

$$\rho(y, \Omega_{T-t}(x)) \leq \lambda_0$$

is true.

Then

$$\{\Omega_t\}_{t=0}^{T-1} \in \mathcal{E}(\lambda, L, T).$$

Proof. In view of Example 4.24, there exists a positive number

$$\tilde{\lambda} < \min\{\lambda, \bar{r}_0\}$$

such that the following property holds:

(C3) for each $\{\Omega_t\}_{t=0}^{T-1} \subset \mathcal{E}(\tilde{\lambda})$ for which there exists an $(\{\Omega_t\}_{t=T-L}^{T-1})$-program $\{y_t\}_{t=T-L}^{T}$ such that

$$\rho(x_{T-t}^*, y_t) \leq \tilde{\lambda},\ t = T - L, \ldots, T \tag{4.87}$$

we have

$$\{\Omega_t\}_{t=0}^{T-1} \in \mathcal{E}(\lambda, L, T).$$

Set

$$\delta_0 = \tilde{\lambda}. \tag{4.88}$$

In view of (C1), there exists

$$\delta_1 \in (0, \delta_0/4) \tag{4.89}$$

such that

$$\Omega(x) \cap B(x_0^*, \delta_0/4) \neq \emptyset \text{ for each } x \in B(x_1^*, \delta_1). \tag{4.90}$$

Assume that a natural number $k < L-1$ and that we have defined a finite sequence of positive numbers $\{\delta_i\}_{i=0}^k$ such that (4.89) holds and that for each $i = 1, \dots, k$,

$$\delta_i < \delta_{i-1}/4, \tag{4.91}$$

$$\Omega(x) \cap B(x_{i-1}^*, \delta_{i-1}/4) \neq \emptyset \text{ for each } x \in B(x_i^*, \delta_i). \tag{4.92}$$

(In view of (4.89) and (4.90), our assumption holds for $k = 1$.) By (C1), there exists

$$\delta_{k+1} \in (0, \delta_k/4)$$

such that

$$\Omega(x) \cap B(x_k^*, \delta_k/4) \neq \emptyset \text{ for each } x \in B(x_{k+1}^*, \delta_{k+1}).$$

Thus by induction we showed that our assumption holds for $k = L-1$. Therefore there exists a finite sequence of positive numbers $\{\delta_i\}_{i=0}^{L-1}$ such that (4.89) holds and that for every $i = 1, \dots, L-1$, (4.91) and (4.92) hold.

Set

$$\lambda_0 = \delta_{L-1}/4. \tag{4.93}$$

Assume that

$$\{\Omega_t\}_{t=0}^{T-1} \subset \mathcal{E}(\lambda_0)$$

satisfies (4.86) and (C2). In view of property (C3), in order to complete the proof it is sufficient to construct an $(\{\Omega_t\}_{t=T-L}^{T-1})$-program $\{y_t\}_{t=T-L}^{T}$ satisfying (4.87). We set

$$y_{T-L} = x_L^*. \tag{4.94}$$

By (4.86) and (4.94), there exists

$$y_{T-L+1} \in \Omega_{T-L}(y_{T-L}) \cap B(x_{L-1}^*, \lambda_0). \tag{4.95}$$

Assume that an integer S satisfies $T - L + 1 \leq S < T$ and that we defined a sequence $\{y_i\}_{i=T-L}^{S} \subset X$ such that (4.94) and (4.95) hold and that for each integer t satisfying $T - L + 1 \leq t \leq S$,

$$y_t \in \Omega_{t-1}(y_{t-1}) \cap B(x_{T-t}^*, \delta_{T-t}). \tag{4.96}$$

(Note that in view of (4.93) and (4.95) our assumption holds for $S = T-L+1$.) By (4.92) (with $i = T - S$), there exists

$$\tilde{y}_{S+1} \in \Omega(y_S) \cap B(x_{T-S-1}^*, \delta_{T-S-1}/4). \tag{4.97}$$

Property (C2) with $t = T - S$, $x = y_S$, $y = \tilde{y}_{S+1}$, (4.93), and (4.97) imply that there exists a point y_{S+1} such that

$$y_{S+1} \in \Omega_S(y_S),\ \rho(y_{S+1}, \tilde{y}_{S+1}) \leq 2\lambda_0 \leq \delta_{T-S-1}/2. \tag{4.98}$$

It follows from (4.97) and (4.98) such that

$$y_{S+1} \in \Omega_S(y_S) \cap B(x^*_{T-S-1}, \delta_{T-S-1}).$$

Thus our assumption holds for $S+1$ too. By induction we constructed a finite sequence $\{y_t\}_{t=T-L}^{T} \subset X$ satisfying (4.94) and (4.95), and such that for all $t = T-L+1, \dots, T$ (4.96) holds which together with (4.88) and (4.91) imply

$$y_t \in \Omega_{t-1}(y_{t-1}) \cap B(x^*_{T-t}, \tilde{\lambda}).$$

This completes the proof of Proposition 4.26. □

4.15 An Auxiliary Result for Problem (P4)

Lemma 4.27. *Let $h \in \mathfrak{M}$ be an upper supercontinuous function, ϵ be a positive number, and let L be a natural number. Then there exists $\delta > 0$ such that for each*

$$\Omega_t \in \mathcal{E}(\delta),\ t = 0, \dots, L-1,$$

each $u_t \in \mathcal{M}$, $t = 0, \dots, L-1$ satisfying

$$\|u_t - v\| \le \delta,\ t = 0, \dots, L-1,$$

each $\xi \in \mathfrak{M}$ satisfying

$$\|h - \xi\| \le \delta$$

and each $(\{\Omega_t\}_{t=0}^{L-1})$-program $\{x_t\}_{t=0}^{L}$ there exists an (Ω)-program $\{y_t\}_{t=0}^{L}$ such that

$$\rho(x_t, y_t) \le \epsilon \text{ for all } t = 0, \dots, L$$

and

$$h(y_L) + \sum_{t=0}^{L-1} v(y_t, y_{t+1}) \ge \xi(x_L) + \sum_{t=0}^{L-1} u_t(x_t, x_{t+1}) - \epsilon.$$

Proof. Assume that the lemma does not hold. Then for each natural number k there exist sets

$$\Omega_t^{(k)} \in \mathcal{E}(k^{-1}),\ t = 0, \dots, L-1, \tag{4.99}$$

functions $u_t^{(k)} \in \mathcal{M}$, $t = 0, \dots, L-1$ satisfying

$$\|u_t^{(k)} - v\| \le k^{-1},\ t = 0, \dots, L-1, \tag{4.100}$$

$h_k \in \mathfrak{M}$ satisfying

$$\|h_k - h\| \le k^{-1} \tag{4.101}$$

and an $(\{\Omega_t^{(k)}\}_{t=0}^{L-1})$-program $\{x_t^{(k)}\}_{t=0}^{L}$ such that the following property holds:

(i) if an (Ω)-program $\{y_t\}_{t=0}^{L}$ satisfies

$$\rho(x_t^{(k)}, y_t) \le \epsilon \text{ for all } t = 0, \dots, L,$$

then

$$h(y_L) + \sum_{t=0}^{L-1} v(y_t, y_{t+1}) < h_k(x_L^{(k)}) + \sum_{t=0}^{L-1} u_t^{(k)}(x_t^{(k)}, x_{t+1}^{(k)}) - \epsilon.$$

Extracting subsequences and re-indexing, if necessary, we may assume without loss of generality that for all integers $t = 0, \dots, L$ there exists

$$x_t = \lim_{k \to \infty} x_t^{(k)}. \tag{4.102}$$

By (4.3), (4.99), and (4.102), for each integer $t \in \{0, \dots, L-1\}$ and each natural number k,

$$\rho_1((x_t, x_{t+1}), \Omega) \le \rho_1((x_t, x_{t+1}), (x_t^{(k)}, x_{t+1}^{(k)})) + \rho_1((x_t^{(k)}, x_{t+1}^{(k)}), \Omega)$$

$$\le \rho_1((x_t, x_{t+1}), (x_t^{(k)}, x_{t+1}^{(k)})) + k^{-1} \to 0 \text{ as } k \to \infty$$

and

$$(x_t, x_{t+1}) \in \Omega.$$

Therefore $\{x_t\}_{t=0}^L$ is an (Ω)-program.

By (4.102) and upper semicontinuity of the function v, h, there exists a natural number k_0 such that for each integer $k \ge k_0$,

$$\rho(x_t, x_t^{(k)}) \le \epsilon,\ t = 0, \dots, L \tag{4.103}$$

and

$$h(x_L) + \sum_{t=0}^{L-1} v(x_t, x_{t+1}) \ge h(x_L^{(k)}) + \sum_{t=0}^{L-1} v(x_t^{(k)}, x_{t+1}^{(k)}) - \epsilon/4. \tag{4.104}$$

Choose a natural number $k_1 > k_0$ such that

$$16 k_1^{-1} L < \epsilon. \tag{4.105}$$

Assume that an integer $k \ge k_1$. Then (4.103) and (4.104) hold. In view of (4.100), (4.101), (4.104), and (4.105),

$$h(x_L) + \sum_{t=0}^{L-1} v(x_t, x_{t+1}) \ge h(x_L^{(k)}) + \sum_{t=0}^{L-1} v(x_t^{(k)}, x_{t+1}^{(k)}) - \epsilon/4$$

$$\ge h_k(x_L^{(k)}) + \sum_{t=0}^{L-1} u_t^{(k)}(x_t^{(k)}, x_{t+1}^{(k)}) - \epsilon/4$$

$$-\|h - h_k\| - L \max\{\|u_t^{(k)} - v\| : t = 0, \dots, L-1\}$$

$$\geq h_k(x_L^{(k)}) + \sum_{t=0}^{L-1} u_t^{(k)}(x_t^{(k)}, x_{t+1}^{(k)}) - \epsilon/4 - k_1^{-1}(L+1)$$

$$\geq h_k(x_L^{(k)}) + \sum_{t=0}^{L-1} u_t^{(k)}(x_t^{(k)}, x_{t+1}^{(k)}) - \epsilon/2. \tag{4.106}$$

When combined with (4.103) relation (4.106) contradicts property (i). The contradiction we have reached proves Lemma 4.27. □

4.16 Structure of Solutions of Perturbed Problems (P4)

We use the notation, definitions, and assumptions introduced in the previous section of this chapter and in Chaps. 2 and 3. Note that the function $v \in \mathcal{M}$ was introduced in Sect. 4.1 while the function $g \in \mathfrak{M}$ was introduced in Sect. 4.14.

We prove the following result which describes the structure of approximate solutions of the perturbed problems of the type (P4) in the regions close to the right endpoints.

Theorem 4.28. *Let $\epsilon, M > 0$ and L_0, l_0 be natural numbers. Then there exist $\delta > 0$ and an integer $L \geq L_0$ such that for each integer $T \geq L$, each*

$$\{\Omega_t\}_{t=0}^{T-1} \in \mathcal{E}(\delta, L, T), \tag{4.107}$$

each $h \in \mathfrak{M}$ satisfying

$$\|h - g\| \leq \delta,$$

each $\{u_t\}_{t=0}^{T-1} \subset \mathcal{M}$ satisfying

$$\|u_t - v\| \leq \delta,\ t = 0 \ldots, T-1 \tag{4.108}$$

and each $(\{\Omega_t\}_{t=0}^{T-1})$-program $\{x_t\}_{t=0}^{T}$ which satisfies

$$x_0 \in \bar{Y}(\{\Omega_t\}_{t=0}^{l_0-1}, 0, l_0),$$

$$\sigma(\{u_t\}_{t=0}^{T-1}, h, \{\Omega_t\}_{t=0}^{T-1}, 0, T, x_0) - M \leq \sum_{t=0}^{T-1} u_t(x_t, x_{t+1}) + h(x_T),$$

$$\sigma(\{u_t\}_{t=\tau}^{\tau+L-1}, \{\Omega_t\}_{t=\tau}^{\tau+L-1}, \tau, \tau+L, x_\tau, x_{\tau+L}) \leq \sum_{t=\tau}^{\tau+L-1} u_t(x_t, x_{t+1}) + \delta$$

for each $\tau \in \{0, \ldots, T-L\}$ and

$$\sigma(\{u_t\}_{t=T-L}^{T-1}, h, \{\Omega_t\}_{t=T-L}^{T-1}, T-L, T, x_{T-L}) \leq \sum_{t=T-L}^{T-1} u_t(x_t, x_{t+1}) + h(x_T) + \delta$$

there exists an $(\bar{v}, \bar{\Omega})$*-overtaking optimal program* $\{z_t^*\}_{t=0}^{\infty}$ *such that*

$$(\pi^{\bar{v}} + g)(z_0^*) = \sup(\pi^{\bar{v}} + g),$$

$$\rho(x_{T-t}, z_t^*) \leq \epsilon,\ t = 0, \dots, L_0.$$

Proof. By Lemma 3.10 applied to the pair $(\bar{v}, \bar{\Omega})$ there exists a real number

$$\delta_0 \in (0, \min\{\epsilon, \bar{\lambda}/4\}) \tag{4.109}$$

such that the following property holds:

(i) for each $(\bar{\Omega})$-program $\{y_t\}_{t=0}^{L_0}$ which satisfies

$$(\pi^{\bar{v}} + g)(y_0) \geq \sup(\pi^{\bar{v}} + g) - \delta_0,$$

$$\sum_{t=0}^{L_0-1} (\bar{v}(y_t, y_{t+1}) - \bar{v}(\bar{x}, \bar{x})) - \pi^{\bar{v}}(y_0) + \pi^{\bar{v}}(y_{L_0}) \geq -\delta_0$$

there exists an $(\bar{v}, \bar{\Omega})$-overtaking optimal program $\{z_t\}_{t=0}^{\infty}$ such that

$$(\pi^{\bar{v}} + g)(z_0) = \sup(\pi^{\bar{v}} + g),$$

$$\rho(z_t, y_t) \leq \epsilon/8,\ t = 0, \dots, L_0.$$

Propositions 2.22 and 2.23 and assumption (A1) imply that there exists a real number $\delta_1 \in (0, \delta_0)$ such that for each point $z \in X$ satisfying $\rho(z, \bar{x}) \leq 2\delta_1$,

$$|\pi^{\bar{v}}(z)| = |\pi^{\bar{v}}(z) - \pi^{\bar{v}}(\bar{x})| \leq \delta_0/8 \tag{4.110}$$

and that for each $(x, y) \in X \times X$ satisfying $\rho(x, \bar{x}) \leq 2\delta_1$, $\rho(y, \bar{x}) \leq 2\delta_1$,

$$|v(x, y) - v(\bar{x}, \bar{x})| \leq \delta_0/8. \tag{4.111}$$

By Theorem 4.14, there exist $\delta_2 \in (0, \delta_0)$ and a natural number $L_1 > l_0$ such that the following property holds:

(ii) for each integer $T > 2L_1$, each

$$\Omega_t \in \mathcal{E}(\delta_2),\ t = 0, \dots, T-1,$$

each $u_t \in \mathcal{M}$, $t = 0, \dots, T-1$ satisfying

$$\|u_t - v\| \leq \delta_2,\ t = 0, \dots, T-1,$$

each $h \in \mathfrak{M}$ satisfying

$$\|h - g\| \leq \delta_2$$

and each $(\{\Omega_t\}_{t=0}^{T-1})$-program $\{x_t\}_{t=0}^{T}$ which satisfies

$$x_0 \in \bar{Y}(\{\Omega_t\}_{t=0}^{l_0-1}, 0, l_0),$$

$$\sigma(\{u_t\}_{t=0}^{T-1}, h, \{\Omega_t\}_{t=0}^{T-1}, 0, T, x_0) \le \sum_{t=0}^{T-1} u_t(x_t, x_{t+1}) + h(x_T) + M$$

and

$$\sigma(\{u_t\}_{t=\tau}^{\tau+L_1-1}, \{\Omega_t\}_{t=\tau}^{\tau+L_1-1}, \tau, \tau+L_1, x_\tau, x_{\tau+L_1}) \le \sum_{t=\tau}^{\tau+L_1-1} u_t(x_t, x_{t+}) + \delta_2$$

for each $\tau \in \{0, \dots, T-L_1\}$, we have

$$\rho(x_t, \bar{x}) \le \delta_1 \text{ for all } t = L_1, \dots, t - L_1.$$

Since the program $\{x_t^*\}_{t=0}^{\infty}$ is $(\bar{v}, \bar{\Omega})$-overtaking optimal it follows from Theorem 2.14 that

$$\lim_{t\to\infty} \rho(x_t^*, \bar{x}) = 0. \tag{4.112}$$

In view of (4.112), there exists a natural number L_2 such that

$$\rho(x_t^*, \bar{x}) \le \delta_1/2 \text{ for all integers } t \ge L_2. \tag{4.113}$$

Set

$$\delta_3 = 24^{-1}\delta_0(2L_0 + 2L_1 + 2L_2 + 8)^{-1}. \tag{4.114}$$

By Lemma 4.27, there exists $\delta_4 > 0$ such that the following property holds:

(iii) for each

$$\Omega_t \in \mathcal{E}(\delta_4),\ t = 0, \dots, L_0 + L_1 + L_2 + 3$$

and each $(\{\Omega_t\}_{t=0}^{L_0+L_1+L_2+3})$-program $\{x_t\}_{t=0}^{L_0+L_1+L_2+4}$ there exists an (Ω)-program $\{y_t\}_{t=0}^{L_0+L_1+L_2+4}$ such that

$$\rho(x_t, y_t) \le \delta_1/4,\ t = 0, \dots, L_0 + L_1 + L_2 + 4$$

and

$$g(y_{L_0+L_1+L_2+4}) + \sum_{t=0}^{L_0+L_1+L_2+3} v(y_t, y_{t+1}) \ge g(x_{L_0+L_1+L_2+4}) + \sum_{t=0}^{L_0+L_1+L_2+3} v(x_t, x_{t+1}) - \delta_0/8.$$

Choose a number $\delta > 0$ and a natural number L such that

$$L \ge 2(L_0 + L_1 + L_2 + 8), \tag{4.115}$$

$$\delta \le 32^{-1}L^{-1}\min\{\delta_0, \delta_1, \delta_2, \delta_3, \delta_4\}. \tag{4.116}$$

Assume that an integer $T \ge L$,

$$\{\Omega_t\}_{t=0}^{T-1} \in \mathcal{E}(\delta, L, T), \tag{4.117}$$

$\{u_t\}_{t=0}^{T-1} \subset \mathcal{M}$ satisfies

$$\|u_t - v\| \le \delta,\ t = 0\dots,T-1, \tag{4.118}$$

$h \in \mathfrak{M}$ satisfies

$$\|h - g\| \le \delta \tag{4.119}$$

and that an $(\{\Omega_t\}_{t=0}^{T-1})$-program $\{x_t\}_{t=0}^{T}$ satisfies

$$x_0 \in \bar{Y}(\{\Omega_t\}_{t=0}^{l_0-1}, 0, l_0), \tag{4.120}$$

$$\sigma(\{u_t\}_{t=0}^{T-1}, h, \{\Omega_t\}_{t=0}^{T-1}, 0, T, x_0) \le \sum_{t=0}^{T-1} u_t(x_t, x_{t+1}) + h(x_T) + M, \tag{4.121}$$

$$\sigma(\{u_t\}_{t=\tau}^{\tau+L-1}, \{\Omega_t\}_{t=\tau}^{\tau+L-1}, \tau, \tau+L, x_\tau, x_{\tau+L}) \le \sum_{t=\tau}^{\tau+L-1} u_t(x_t, x_{t+}) + \delta \tag{4.122}$$

for each $\tau \in \{0, \dots, T-L\}$ and

$$\sigma(\{u_t\}_{t=T-L}^{T-1}, h, \{\Omega_t\}_{t=T-L}^{T-1}, T-L, T, x_{T-L}) \le \sum_{t=T-L}^{T-1} u_t(x_t, x_{t+1}) + h(x_T) + \delta. \tag{4.123}$$

It follows from (4.115) to (4.122) and the property (ii) that

$$\rho(x_t, \bar{x}) \le \delta_1,\ t = L_1, \dots, T-L_1. \tag{4.124}$$

By (4.115),

$$[T - L_0 - L_1 - L_2 - 4, T - L_0 - L_1 - L_2] \subset [L_1, T - L_0 - L_1 - L_2]. \tag{4.125}$$

In view of (4.124) and (4.125),

$$\rho(x_t, \bar{x}) \le \delta_1,\ t \in \{T - L_0 - L_1 - L_2 - 4, T - L_0 - L_1 - L_2\}. \tag{4.126}$$

By (4.117) (see (4.82)–(4.85)), there exists an $(\{\Omega_t\}_{t=T-L}^{T-1})$-program $\{y_t\}_{t=T-L}^{T}$ such that

$$\rho(y_t, x^*_{T-t}) \le \delta,\ t = T-L, \dots, T, \tag{4.127}$$

$$|v(y_t, y_{t+1}) - v(x^*_{T-t}, x^*_{T-t-1})| \le \delta,\ t = T-L, \dots, T-1 \tag{4.128}$$

and

$$|g(y_T) - g(x^*_0)| \le \delta. \tag{4.129}$$

Define a sequence $\{\tilde{x}_t\}_{t=0}^T$ by

$$\tilde{x}_t = x_t,\ t = 0, \dots, T - L_0 - L_1 - L_2 - 4,$$
$$\tilde{x}_t = y_t,\ t = T - L_0 - L_1 - L_2 - 3, \dots, T. \tag{4.130}$$

Relations (4.109), (4.113), (4.115), (4.116), (4.127), and (4.130) imply that

$$\rho(\tilde{x}_{T-L_0-L_1-L_2-3}, \bar{x}) = \rho(y_{T-L_0-L_1-L_2-3}, \bar{x})$$
$$\le \rho(y_{T-L_0-L_1-L_2-3}, x^*_{L_0+L_1+L_2+3}) + \rho(x^*_{L_0+L_1+L_2+3}, \bar{x})$$
$$\le \delta + \delta_1/2 \le \delta_1 < \delta_0 < \bar{\lambda}/4. \tag{4.131}$$

By (4.4), (4.109), (4.126), (4.130), and (4.131), $\{\tilde{x}_t\}_{t=0}^T$ is an $(\{\Omega_t\}_{t=0}^{T-1})$-program. It follows from (4.115), (4.118), (4.119), (4.123), and (4.130) that

$$\delta \ge \sigma(\{u_t\}_{t=T-L}^{T-1}, h, \{\Omega_t\}_{t=T-L}^{T-1}, T-L, T, x_{T-L})$$
$$-h(x_T) - \sum_{t=T-L}^{T-1} u_t(x_t, x_{t+1})$$
$$\ge h(\tilde{x}_T) + \sum_{t=T-L}^{T-1} u_t(\tilde{x}_t, \tilde{x}_{t+1}) - h(x_T) - \sum_{t=T-L}^{T-1} u_t(x_t, x_{t+1})$$
$$= h(\tilde{x}_T) - h(x_T)$$
$$+ \sum_{t=T-L_0-L_1-L_2-4}^{T-1} u_t(\tilde{x}_t, \tilde{x}_{t+1}) - \sum_{t=T-L_0-L_1-L_2-4}^{T-1} u_t(x_t, x_{t+1})$$
$$\ge g(\tilde{x}_T) - g(x_T) + \sum_{t=T-L_0-L_1-L_2-4}^{T-1} v(\tilde{x}_t, \tilde{x}_{t+1})$$
$$- \sum_{t=T-L_0-L_1-L_2-4}^{T-1} v(x_t, x_{t+1}) - 2\delta(L_0 + L_1 + L_2 + 6).$$

Combined with (4.130) this implies that

$$g(x_T) + \sum_{t=T-L_0-L_1-L_2-4}^{T-1} v(x_t, x_{t+1})$$
$$\ge g(\tilde{x}_T) + \sum_{t=T-L_0-L_1-L_2-4}^{T-1} v(\tilde{x}_t, \tilde{x}_{t+1}) - 2\delta(L_0 + L_1 + L_2 + 8)$$
$$= g(y_T) + v(x_{T-L_0-L_1-L_2-4}, y_{T-L_0-L_1-L_2-3})$$
$$+ \sum_{t=T-L_0-L_1-L_2-3}^{T-1} v(y_t, y_{t+1}) - 2\delta(L_0 + L_1 + L_2 + 8). \tag{4.132}$$

In view of (4.131),

$$\rho(y_{T-L_0-L_1-L_2-3}, \bar{x}) \le \delta + \delta_1/2 \le \delta_1. \tag{4.133}$$

By (4.111), (4.126), and (4.133),

$$|v(x_{T-L_0-L_1-L_2-4}, y_{T-L_0-L_1-L_2-3}) - v(\bar{x}, \bar{x})| \le \delta_0/8. \tag{4.134}$$

It follows from (4.115), (4.129), (4.132), and (4.134) that

$$\begin{aligned}
&g(x_T) + \sum_{t=T-L_0-L_1-L_2-4}^{T-1} v(x_t, x_{t+1}) \\
&\ge g(y_T) + \sum_{t=T-L_0-L_1-L_2-3}^{T-1} v(y_t, y_{t+1}) \\
&-2\delta(L_0 + L_1 + L_2 + 8) - \delta_0/8 + v(\bar{x}, \bar{x}) \\
&\ge g(x_0^*) + \sum_{t=T-L_0-L_1-L_2-3}^{T-1} v(x_{T-t}^*, x_{T-t-1}^*) \\
&-3\delta(L_0 + L_1 + L_2 + 8) - \delta_0/8 + v(\bar{x}, \bar{x}) \\
&= g(x_0^*) + \sum_{t=0}^{L_0+L_1+L_2+2} \bar{v}(x_t^*, x_{t+1}^*) \\
&- 3\delta(L_0 + L_1 + L_2 + 8) - \delta_0/8 + v(\bar{x}, \bar{x}).
\end{aligned} \tag{4.135}$$

Property (iii), (4.116), and (4.117) imply that there there exists an (Ω)-program $\{\widehat{x}_t\}_{t=T-L_0-L_1-L_2-4}^{T}$ such that

$$\rho(x_t, \widehat{x}_t) \le \delta_1/4, \ t = T - L_0 - L_1 - L_2 - 4, \dots, T \tag{4.136}$$

and

$$\begin{aligned}
&g(\widehat{x}_T) + \sum_{t=T-L_0-L_1-L_2-4}^{T-1} v(\widehat{x}_t, \widehat{x}_{t+1}) \\
&\ge g(x_T) + \sum_{t=T-L_0-L_1-L_2-4}^{T-1} v(x_t, x_{t+1}) - \delta_0/8.
\end{aligned} \tag{4.137}$$

By (4.135) and (4.137),

$$g(\widehat{x}_T) + \sum_{t=T-L_0-L_1-L_2-4}^{T-1} v(\widehat{x}_t, \widehat{x}_{t+1})$$

$$\geq g(x_0^*) + \sum_{t=0}^{L_0+L_1+L_2+2} \bar{v}(x_t^*, x_{t+1}^*)$$

$$- 3\delta(L_0 + L_1 + L_2 + 8) - \delta_0/4 + v(\bar{x}, \bar{x}). \tag{4.138}$$

It follows from (4.111), (4.113), and (4.116) that

$$g(\widehat{x}_T) + \sum_{t=T-L_0-L_1-L_2-4}^{T-1} v(\widehat{x}_t, \widehat{x}_{t+1})$$

$$\geq g(x_0^*) + \sum_{t=0}^{L_0+L_1+L_2+3} \bar{v}(x_t^*, x_{t+1}^*) - 3\delta(L_0 + L_1 + L_2 + 8) - 3\delta_0/8$$

$$\geq g(x_0^*) + \sum_{t=0}^{L_0+L_1+L_2+3} \bar{v}(x_t^*, x_{t+1}^*) - \delta_0/2. \tag{4.139}$$

Set

$$\widehat{y}_t = \widehat{x}_{T-t},\ t = 0, \dots, L_0 + L_1 + L_2 + 4. \tag{4.140}$$

By (4.139) and (4.140),

$$g(\widehat{y}_0) + \sum_{t=0}^{L_0+L_1+L_2+3} \bar{v}(\widehat{y}_t, \widehat{y}_{t+1})$$

$$= g(\widehat{x}_T) + \sum_{t=0}^{L_0+L_1+L_2+3} v(\widehat{x}_{T-t-1}, \widehat{x}_{T-t})$$

$$= g(\widehat{x}_T) + \sum_{t=T-L_0-L_1-L_2-4}^{T-1} v(\widehat{x}_t, \widehat{x}_{t+1})$$

$$\geq g(x_0^*) + \sum_{t=0}^{L_0+L_1+L_2+3} \bar{v}(x_t^*, x_{t+1}^*) - \delta_0/2. \tag{4.141}$$

By (4.81), (4.115), (4.126), (4.136), (4.140), (4.141), Proposition 2.19, and Corollary 2.21 and $(\bar{v}, \bar{\Omega})$-overtaking optimality of $\{x_t^*\}_{t=0}^{\infty}$,

$$(\pi^{\bar{v}} + g)(\widehat{y}_0) - \sup(\pi^{\bar{v}} + g)$$

$$+ \sum_{t=0}^{L_0-1} (\bar{v}(\widehat{y}_t, \widehat{y}_{t+1}) - \bar{v}(\bar{x}, \bar{x})) - \pi^{\bar{v}}(\widehat{y}_0) + \pi^{\bar{v}}(\widehat{y}_{L_0})$$

$$\geq (\pi^{\bar{v}} + g)(\widehat{y}_0) - (\pi^{\bar{v}} + g)(x_0^*)$$

$$+ \sum_{t=0}^{L_0+L_1+L_2+3} (\bar{v}(\widehat{y}_t, \widehat{y}_{t+1}) - \bar{v}(\bar{x}, \bar{x})) - \pi^{\bar{v}}(\widehat{y}_0) + \pi^{\bar{v}}(\widehat{y}_{L_0+L_1+L_2+4})$$

$$\geq \pi^{\bar{v}}(\widehat{y}_0) - \pi^{\bar{v}}(x_0^*) + \sum_{t=0}^{L_0+L_1+L_2+3} (\bar{v}(x_t^*, x_{t+1}^*) - \bar{v}(\bar{x}, \bar{x}))$$
$$-\pi^{\bar{v}}(\widehat{y}_0) + \pi^{\bar{v}}(\widehat{y}_{L+\tau_0+\tau_1+4}) - \delta_0/2$$
$$= -\pi^{\bar{v}}(x^*_{L_0+L_1+L_2+4}) - \pi^{\bar{v}}(\widehat{y}_{L_0+L_1+L_2+4}) - \delta_0/2. \tag{4.142}$$

In view of (4.110) and (4.113),

$$|\pi^{v}(x^*_{L_0+L_1+L_2+4})| \leq \delta_0/8. \tag{4.143}$$

By (4.140),

$$|\pi^{\bar{v}}(\widehat{y}_{L_0+L_1+L_2+4})| = |\pi^{\bar{v}}(\widehat{x}_{T-L_0-L_1-L_2-4})|. \tag{4.144}$$

In view of (4.126) and (4.136),

$$\rho(\widehat{x}_{T-L_0-L_1-L_2-4}, \bar{x})$$
$$\leq \rho(\widehat{x}_{T-L_0-L_1-L_2-4}, x_{T-L_0-L_1-L_2-4}) + \rho(x_{T-L_0-L_1-L_2-4}, \bar{x}) \leq \delta_1/4 + \delta_1. \tag{4.145}$$

Relations (4.110), (4.144), and (4.145) imply that

$$|\pi^{\bar{v}}(\widehat{y}_{L_0+L_1+L_2+4})| \leq \delta_0/8. \tag{4.146}$$

It follows from (4.142), (4.144), and (4.146) that

$$(\pi^{\bar{v}} + g)(\widehat{y}_0) - \sup(\pi^{\bar{v}} + g)$$
$$+ \sum_{t=0}^{L_0-1} (\bar{v}(\widehat{y}_t, \widehat{y}_{t+1}) - \bar{v}(\bar{x}, \bar{x}))$$
$$-\pi^{\bar{v}}(\widehat{y}_0) + \pi^{\bar{v}}(\widehat{y}_{L_0}) > -\delta_0.$$

Together with Proposition 2.19 this implies that

$$(\pi^{\bar{v}} + g)(\widehat{y}_0) - \sup(\pi^{\bar{v}} + g) \geq -\delta_0, \tag{4.147}$$

$$\sum_{t=0}^{L_0-1} (\bar{v}(\widehat{y}_t, \widehat{y}_{t+1}) - \bar{v}(\bar{x}, \bar{x})) - \pi^{\bar{v}}(\widehat{y}_0) + \pi^{\bar{v}}(\widehat{y}_{L_0}) \geq -\delta_0. \tag{4.148}$$

It follows from (4.140), (4.147), (4.148), and the property (i) that there exists an $(\bar{v}, \bar{\Omega})$-overtaking optimal program $\{\xi_t\}_{t=0}^{\infty}$ such that

$$(\pi^{\bar{v}} + g)(\xi_0) = \sup(\pi^{\bar{v}} + g), \tag{4.149}$$

$$\rho(\xi_t, \widehat{y}_t) \leq \epsilon/8,\ t = 0, \dots, L_0. \tag{4.150}$$

By (4.136), (4.140), and (4.150), for all $t = 0, \dots, L_0$,

$$\rho(x_{T-t}, \xi_t) \leq \rho(x_{T-t}, \widehat{x}_{T-t}) + \rho(\widehat{x}_{T-t}, \xi_t)$$
$$\leq \delta_1/4 + \rho(\widehat{y}_t, \xi_t) \leq \delta_1/4 + \epsilon/8 < \epsilon.$$

Theorem 4.28 is proved. □

4.17 Perturbed Problems (P5)

We use the notation, definitions, and assumptions introduced in the previous sections of this chapter and in Chaps. 2 and 3.

Let $g \in \mathfrak{A}$ be an upper semicontinuous function. For each $\xi, \eta \in X$ set

$$\psi(\xi, \eta) = \pi^v(\xi) + \pi^{\bar{v}}(\eta) + g(\xi, \eta). \tag{4.151}$$

It is known (see Sect. 3.8) that ψ is an upper semicontinuous and bounded from above function which has maximizers. By Theorem 2.13, there exist a (v, Ω)-overtaking optimal program $\{x_t^*\}_{t=0}^{\infty}$ and a $(\bar{v}, \bar{\Omega})$-overtaking optimal program $\{\bar{x}_t^*\}_{t=0}^{\infty}$ such that

$$\psi(x_0^*, \bar{x}_0^*) = \sup(\psi). \tag{4.152}$$

By Theorem 2.14,

$$\lim_{t\to\infty} x_t^* = \bar{x}, \ \lim_{t\to\infty} \bar{x}_t^* = \bar{x}. \tag{4.153}$$

Let $\lambda > 0$ and L be a natural number and $T \geq 2L$ be an integer. Denote by $\mathcal{F}(\lambda, L, T)$ the collection of all finite sequences

$$\{\Omega_t\}_{t=0}^{T-1} \subset \mathcal{E}(\lambda)$$

for which there exist an $(\{\Omega_t\}_{t=0}^{L-1})$-program $\{y_t\}_{t=0}^{L-1}$ and an $(\{\Omega_t\}_{t=T-L}^{T-1})$-program $\{\tilde{y}_t\}_{t=T-L}^{T}$ such that

$$\rho(y_t, x_t^*) \leq \lambda, \ t = 0, \ldots, L, \tag{4.154}$$

$$\rho(\tilde{y}_t, \bar{x}_{T-t}^*) \leq \lambda, \ t = T - L, \ldots, T, \tag{4.155}$$

$$|v(y_t, y_{t+1}) - v(x_t^*, x_{t+1}^*)| \leq \lambda, \ t = 0, \ldots, L - 1, \tag{4.156}$$

$$|v(\tilde{y}_t, \tilde{y}_{t+1}) - v(\bar{x}_{T-t}^*, \bar{x}_{T-t-1}^*)| \leq \lambda, \ t = T - L, \ldots, T - 1 \tag{4.157}$$

and

$$|g(y_0, \tilde{y}_T) - g(x_0^*, \bar{x}_0^*)| \leq \lambda. \tag{4.158}$$

Example 4.29. Let $\lambda > 0$ and L be a natural number, $T \geq 2L$ be an integer and let

$$\{\Omega_t\}_{t=0}^{T-1} \subset \mathcal{E}(\lambda).$$

Assume that for all integers $t = 0, \ldots, L - 1$,

$$(x_t^*, x_{t+1}^*) \in \Omega_t$$

and that for all integers $t = T - L, \ldots, T - 1$,

$$(\bar{x}_{T-t}^*, \bar{x}_{T-t-1}^*) \in \Omega_t.$$

Then

$$\{\Omega_t\}_{t=0}^{T-1} \in \mathcal{F}(\lambda, L, T).$$

Example 4.30. Let $\lambda > 0$ and L be a natural number, $T \geq 2L$ be an integer, the function g be continuous at $(x_0^*, \bar{x}_0^*)$, the function v is continuous at (x_t^*, x_{t+1}^*) for all integers $t = 0, \dots, L-1$ and the function $\bar{v}$ be continuous at $(\bar{x}_t^*, \bar{x}_{t+1}^*)$ for all integers $t = 0, \dots, L-1$. Then there exists $\tilde{\lambda} \in (0, \lambda)$ such that for every sequence

$$\{\Omega_t\}_{t=0}^{T-1} \subset \mathcal{E}(\tilde{\lambda})$$

for which there exist an $(\{\Omega_t\}_{t=0}^{L-1})$-program $\{y_t\}_{t=0}^{L}$ and an $(\{\Omega_t\}_{t=T-L}^{T-1})$-program $\{\tilde{y}_t\}_{t=T-L}^{T}$ such that

$$\rho(x_t^*, y_t) \leq \tilde{\lambda},\ t = 0, \dots, L,$$

$$\rho(\bar{x}_{T-t}^*, \tilde{y}_t) \leq \tilde{\lambda},\ t = T-L, \dots, T$$

the inclusion

$$\{\Omega_t\}_{t=0}^{T-1} \in \mathcal{F}(\lambda, L, T)$$

is true.

Example 4.31. For each $x \in X$ set

$$\Omega(x) = \{y \in X :\ (x, y) \in \Omega\}.$$

Let L be a natural number, $T \geq 2L$ be an integer, the function g be continuous at $(x_0^*, \bar{x}_0^*)$, the function v be continuous at (x_t^*, x_{t+1}^*) for all integers $t = 0, \dots, L-1$, and the function $\bar{v}$ be continuous at $(\bar{x}_t^*, \bar{x}_{t+1}^*)$ for all integers $t = 0, \dots, L-1$. Suppose that there exists $\bar{r}_0 \in (0, \bar{r})$ such that for each integer $t \in \{1, \dots, L\} \setminus \{L\}$,

$$\Omega(x) \neq \emptyset \text{ for each } x \in B(\bar{x}_t^*, \bar{r}_0), \tag{4.159}$$

for each integer $t \in \{0, \dots, L-1\}$,

$$\Omega(x) \neq \emptyset \text{ for each } x \in B(x_t^*, \bar{r}_0) \tag{4.160}$$

and the following property holds:

(D1) for each $\epsilon > 0$ there exists $\delta(\epsilon) \in (0, \bar{r}_0)$ such that for each $t \in \{0, \dots, L-1\}$ and each $x \in B(x_t^*, \delta(\epsilon))$,

$$\Omega(x) \cap B(x_{t+1}^*, \epsilon) \neq \emptyset$$

and for each $t \in \{1, \dots, L\} \setminus \{L\}$ and each $x \in B(\bar{x}_t^*, \delta(\epsilon))$,

$$\Omega(x) \cap B(\bar{x}_{t-1}^*, \epsilon) \neq \emptyset.$$

Proposition 4.32. *Let $L > 2$ be an integer and $\lambda > 0$. Then there exists a positive number*

$$\lambda_0 < \min\{\lambda, \bar{r}_0\}$$

such that the following assertion holds.

Assume that a finite sequence

$$\{\Omega_t\}_{t=0}^{T-1} \subset \mathcal{E}(\lambda_0)$$

satisfies

$$\Omega_{T-L}(\bar{x}_L^*) \cap B(\bar{x}_{L-1}^*, \lambda_0) \neq \emptyset, \tag{4.161}$$

$$\Omega_0(x_0^*) \cap B(x_1^*, \lambda_0) \neq \emptyset \tag{4.162}$$

and possesses the properties

(D2) for each integer t satisfying $0 < t \leq L-1$, each $x \in B(\bar{x}_t^*, \bar{r}_0)$, and each $y \in \Omega(x)$, the inclusion

$$\rho(y, \Omega_{T-t}(x)) \leq \lambda_0 \tag{4.163}$$

is true;

(D3) for each integer $t \in [0, L-1]$, each $x \in B(x_t^*, \bar{r}_0)$, and each $y \in \Omega(x)$, the inclusion

$$\rho(y, \Omega_t(x)) \leq \lambda_0 \tag{4.164}$$

is true.

Then

$$\{\Omega_t\}_{t=0}^{T-1} \in \mathcal{F}(\lambda, L, T).$$

Proof. In view of Example 4.30, there exists a positive number

$$\tilde{\lambda} < \min\{\lambda, \bar{r}_0\}$$

such that the following property holds:

(D4) for each $\{\Omega_t\}_{t=0}^{T-1} \subset \mathcal{E}(\tilde{\lambda})$ for which there exist an $(\{\Omega_t\}_{t=0}^{L-1})$-program $\{y_t\}_{t=0}^{L}$ and an $(\{\Omega_t\}_{t=T-L}^{T-1})$-program $\{\tilde{y}_t\}_{t=T-L}^{T}$ such that

$$\rho(x_t^*, y_t) \leq \tilde{\lambda},\ t = 0, \dots, L,$$

$$\rho(\bar{x}_{T-t}^*, \tilde{y}_t) \leq \tilde{\lambda},\ t = T-L, \dots, T$$

we have

$$\{\Omega_t\}_{t=0}^{T-1} \in \mathcal{F}(\lambda, L, T).$$

Denote by $\mathcal{F}_1$ the set of all $\{\Omega_t\}_{t=0}^{T-1} \subset \mathcal{E}(\tilde{\lambda})$ for which there exists an $(\{\Omega_t\}_{t=0}^{L-1})$-program $\{y_t\}_{t=0}^{L}$ such that

$$\rho(x_t^*, y_t) \leq \tilde{\lambda},\ t = 0, \dots, L, \tag{4.165}$$

and denote by $\mathcal{F}_2$ the set of all $\{\Omega_t\}_{t=0}^{T-1} \subset \mathcal{E}(\tilde{\lambda})$ for which there exists an $(\{\Omega_t\}_{t=T-L}^{T-1})$-program $\{\tilde{y}_t\}_{t=T-L}^{T}$ such that

$$\rho(\bar{x}_{T-t}^*, \tilde{y}_t) \leq \tilde{\lambda},\ t = T-L, \dots, T. \tag{4.166}$$

Property (D4) implies that

$$\mathcal{F}_1 \cap \mathcal{F}_2 \subset \mathcal{F}(\lambda, L, T). \tag{4.167}$$

By (4.167), in order to prove the proposition it is sufficient to show the existence of a positive number

$$\lambda_0 < \min\{\lambda, \bar{r}_0\}$$

such that for each

$$\{\Omega_t\}_{t=0}^{T-1} \subset \mathcal{E}(\lambda_0)$$

satisfying (4.161), (4.162), (D2), and (D3) we have

$$\{\Omega_t\}_{t=0}^{T-1} \in \mathcal{F}_1 \cap \mathcal{F}_2.$$

Proposition 4.26 implies the existence of a positive number

$$\lambda_0 < \min\{\lambda, \bar{r}_0\}$$

such that for each

$$\{\Omega_t\}_{t=0}^{T-1} \subset \mathcal{E}(\lambda_0)$$

satisfying (4.161) and (D2) we have

$$\{\Omega_t\}_{t=0}^{T-1} \in \mathcal{F}_2.$$

Therefore it is sufficient to show the existence of a positive number

$$\lambda_0 < \min\{\lambda, \bar{r}_0\}$$

such that for each

$$\{\Omega_t\}_{t=0}^{T-1} \subset \mathcal{E}(\lambda_0)$$

satisfying (4.162) and (D3) we have

$$\{\Omega_t\}_{t=0}^{T-1} \in \mathcal{F}_1.$$

Set

$$\delta_0 = \tilde{\lambda}. \tag{4.168}$$

In view of (D1), there exists

$$\delta_1 \in (0, \delta_0/4)$$

such that

$$\Omega(x) \cap B(x_L^*, \delta_0/4) \neq \emptyset \text{ for each } x \in B(x_{L-1}^*, \delta_1). \tag{4.169}$$

Assume that a natural number $k < L-1$ and that we have defined a finite sequence of positive numbers $\{\delta_i\}_{i=0}^{k}$ such that (4.168) holds and that for each $i = 1, \dots, k$,

$$\delta_i < \delta_{i-1}/4, \tag{4.170}$$

$$\Omega(x) \cap B(x_{L-i+1}^*, \delta_{i-1}/4) \neq \emptyset \text{ for each } x \in B(x_{L-i}^*, \delta_i). \tag{4.171}$$

(In view of (4.169), our assumption holds for $k = 1$.) By (D1), there exists

$$\delta_{k+1} \in (0, \delta_k/4)$$

such that

$$\Omega(x) \cap B(x^*_{L-k}, \delta_k/4) \neq \emptyset \text{ for each } x \in B(x^*_{L-k-1}, \delta_{k+1}).$$

Thus by induction we showed that our assumption holds for $k = L-1$. Therefore there exists a finite sequence of positive numbers $\{\delta_i\}_{i=0}^{L-1}$ such that (4.170) and (4.171) hold for every $i = 1, \dots, L-1$.

Set

$$\lambda_0 = \delta_{L-1}/4. \tag{4.172}$$

Assume that

$$\{\Omega_t\}_{t=0}^{T-1} \subset \mathcal{E}(\lambda_0)$$

satisfies (4.162) and (D3). In order to complete the proof it is sufficient to show that

$$\{\Omega_t\}_{t=0}^{T-1} \in \mathcal{F}_1.$$

In order to meet this goal it is sufficient to construct an $(\{\Omega_t\}_{t=0}^{L-1})$-program $\{y_t\}_{t=0}^{L}$ such that

$$\rho(x^*_t, y_t) \leq \tilde{\lambda},\ t = 0, \dots, L.$$

We set

$$y_0 = x^*_0. \tag{4.173}$$

By (4.162) and (4.173), there exists

$$y_1 \in \Omega_0(y_0) \cap B(x^*_1, \lambda_0). \tag{4.174}$$

Assume that an integer S satisfies $1 \leq S < L$ and that we defined a sequence $\{y_i\}_{i=0}^{S} \subset X$ such that (4.173) and (4.174) hold and that for each integer $t \in \{1, \dots, S\}$,

$$y_t \in \Omega_{t-1}(y_{t-1}) \cap B(x^*_t, \delta_{L-t}). \tag{4.175}$$

(Note that in view of (4.172) and (4.174) our assumption holds for $S = 1$.) By (4.171) (with $i = L - S$) and (4.175), there exists

$$\tilde{y}_{S+1} \in \Omega(y_S) \cap B(x^*_{S+1}, \delta_{L-S-1}/4). \tag{4.176}$$

Property (D3), (4.172), (4.175), and (4.176) imply that there exists a point y_{S+1} such that

$$y_{S+1} \in \Omega_S(y_S),\ \rho(y_{S+1}, \tilde{y}_{S+1}) \leq 2\lambda_0 \leq \delta_{L-S-1}/2. \tag{4.177}$$

It follows from (4.176) and (4.177) such that

$$y_{S+1} \in \Omega_S(y_S) \cap B(x^*_{S+1}, \delta_{L-S-1}).$$

Thus our assumption holds for $S+1$ too. By induction we constructed a finite sequence $\{y_t\}_{t=0}^{L} \subset X$ satisfying (4.173) and (4.174), and such that for all $t = 1, \dots, L$

$$y_t \in \Omega_{t-1}(y_{t-1})$$

and

$$\rho(y_t, x_t^*) \le \tilde{\lambda},\ t = 0, \dots, L.$$

This completes the proof of Proposition 4.32. □

4.18 An Auxiliary Result for Problem (P5)

Lemma 4.33. *Let $f \in \mathfrak{A}$ be an upper supercontinuous function, ϵ be a positive number and let L_1, L_2 be natural numbers. Then there exists $\delta > 0$ such that for each*

$$\Omega_t \in \mathcal{E}(\delta),\ t = 0, \dots, L_1 - 1,$$

each

$$\tilde{\Omega}_t \in \mathcal{E}(\delta),\ t = 0, \dots, L_2 - 1,$$

each $u_t \in \mathcal{M}$, $t = 0, \dots, L_1 - 1$ satisfying

$$\|u_t - v\| \le \delta,\ t = 0, \dots, L_1 - 1,$$

each $\tilde{u}_t \in \mathcal{M}$, $t = 0, \dots, L_2 - 1$ satisfying

$$\|\tilde{u}_t - v\| \le \delta,\ t = 0, \dots, L_2 - 1,$$

each $h \in \mathfrak{A}$ satisfying

$$\|h - f\| \le \delta,$$

each $(\{\Omega_t\}_{t=0}^{L_1-1})$-program $\{x_t\}_{t=0}^{L_1}$ and each $(\{\tilde{\Omega}_t\}_{t=0}^{L_2-1})$-program $\{\tilde{x}_t\}_{t=0}^{L_2}$ there exist an (Ω)-program $\{y_t\}_{t=0}^{L_1}$ and an (Ω)-program $\{\tilde{y}_t\}_{t=0}^{L_2}$ such that

$$\rho(x_t, y_t) \le \epsilon \text{ for all } t = 0, \dots, L_1,$$

$$\rho(\tilde{x}_t, \tilde{y}_t) \le \epsilon \text{ for all } t = 0, \dots, L_2$$

and

$$f(y_0, \tilde{y}_{L_2}) + \sum_{t=0}^{L_1-1} v(y_t, y_{t+1}) + \sum_{t=0}^{L_2-1} v(\tilde{y}_t, \tilde{y}_{t+1})$$

$$\ge h(x_0, \tilde{x}_{L_2}) + \sum_{t=0}^{L_1-1} u_t(x_t, x_{t+1}) + \sum_{t=0}^{L_2-1} u_t(\tilde{x}_t, \tilde{x}_{t+1}) - \epsilon.$$

Proof. Assume that the lemma does not hold. Then for each natural number k there exist sets

$$\Omega_t^{(k)} \in \mathcal{E}(k^{-1}),\ t = 0, \dots, L_1 - 1, \tag{4.178}$$

$$\tilde{\Omega}_t^{(k)} \in \mathcal{E}(k^{-1}),\ t = 0, \dots, L_2 - 1, \tag{4.179}$$

functions $u_t^{(k)} \in \mathcal{M}$, $t = 0, \dots, L_1 - 1$ satisfying

$$\|u_t^{(k)} - v\| \le k^{-1},\ t = 0, \dots, L_1 - 1, \tag{4.180}$$

functions $\tilde{u}_t^{(k)} \in \mathcal{M}$, $t = 0, \dots, L_2 - 1$ satisfying

$$\|\tilde{u}_t^{(k)} - v\| \le k^{-1},\ t = 0, \dots, L_2 - 1, \tag{4.181}$$

$h_k \in \mathfrak{A}$ satisfying

$$\|h_k - f\| \le k^{-1}, \tag{4.182}$$

an $(\{\Omega_t^{(k)}\}_{t=0}^{L_1-1})$-program $\{x_t^{(k)}\}_{t=0}^{L_1}$ and an $(\{\tilde{\Omega}_t^{(k)}\}_{t=0}^{L_2-1})$-program $\{\tilde{x}_t^{(k)}\}_{t=0}^{L_2}$ such that the following property holds:

(i) if an (Ω)-program $\{y_t\}_{t=0}^{L_1}$ and an (Ω)-program $\{\tilde{y}_t\}_{t=0}^{L_2}$ satisfy

$$\rho(x_t^{(k)}, y_t) \le \epsilon \text{ for all } t = 0, \dots, L_1,$$

$$\rho(\tilde{x}_t^{(k)}, \tilde{y}_t) \le \epsilon \text{ for all } t = 0, \dots, L_2$$

then

$$f(y_0, \tilde{y}_{L_2}) + \sum_{t=0}^{L_1-1} v(y_t, y_{t+1}) + \sum_{t=0}^{L_2-1} v(\tilde{y}_t, \tilde{y}_{t+1})$$

$$< h_k(x_0^{(k)}, \tilde{x}_{L_2}^{(k)}) + \sum_{t=0}^{L_1-1} u_t^{(k)}(x_t^{(k)}, x_{t+1}^{(k)}) + \sum_{t=0}^{L_2-1} \tilde{u}_t^{(k)}(\tilde{x}_t^{(k)}, \tilde{x}_{t+1}^{(k)}) - \epsilon.$$

Extracting subsequences and re-indexing, if necessary, we may assume without loss of generality that for all integers $t = 0, \dots, L_1$ there exists

$$x_t = \lim_{k \to \infty} x_t^{(k)} \tag{4.183}$$

and that for all integers $t = 0, \dots, L_2$ there exists

$$\tilde{x}_t = \lim_{k \to \infty} \tilde{x}_t^{(k)}. \tag{4.184}$$

By (4.178) and (4.183), for each integer $t \in \{0, \dots, L_1 - 1\}$ and each natural number k,

$$\rho_1((x_t, x_{t+1}), \Omega) \le \rho_1((x_t, x_{t+1}), (x_t^{(k)}, x_{t+1}^{(k)})) + \rho_1((x_t^{(k)}, x_{t+1}^{(k)}), \Omega)$$

$$\le \rho_1((x_t, x_{t+1}), (x_t^{(k)}, x_{t+1}^{(k)})) + k^{-1} \to 0 \text{ as } k \to \infty$$

and

$$(x_t, x_{t+1}) \in \Omega, \tag{4.185}$$

for each integer $t \in \{0, \ldots, L_2 - 1\}$ and each natural number k,

$$\rho_1((\tilde{x}_t, \tilde{x}_{t+1}), \Omega) \le \rho_1((\tilde{x}_t, \tilde{x}_{t+1}), (\tilde{x}_t^{(k)}, \tilde{x}_{t+1}^{(k)})) + \rho_1((\tilde{x}_t^{(k)}, \tilde{x}_{t+1}^{(k)}), \Omega)$$

$$\le \rho_1((\tilde{x}_t, \tilde{x}_{t+1}), (\tilde{x}_t^{(k)}, \tilde{x}_{t+1}^{(k)})) + k^{-1} \to 0 \text{ as } k \to \infty$$

and

$$(\tilde{x}_t, \tilde{x}_{t+1}) \in \Omega. \tag{4.186}$$

By (4.183), (4.184), and upper semicontinuity of the functions v, g, there exists a natural number k_0 such that for each integer $k \ge k_0$,

$$\rho(x_t, x_t^{(k)}) \le \epsilon, \ t = 0, \ldots, L_1,$$

$$\rho(\tilde{x}_t, \tilde{x}_t^{(k)}) \le \epsilon, \ t = 0, \ldots, L_2, \tag{4.187}$$

and

$$f(x_0, \tilde{x}_{L_2}) + \sum_{t=0}^{L_1-1} v(x_t, x_{t+1}) + \sum_{t=0}^{L_2-1} v(\tilde{x}_t, \tilde{x}_{t+1})$$

$$\ge f(x_0^{(k)}, \tilde{x}_{L_2}^{(k)}) + \sum_{t=0}^{L_1-1} v(x_t^{(k)}, x_{t+1}^{(k)}) + \sum_{t=0}^{L_2-1} v(\tilde{x}_t^{(k)}, \tilde{x}_{t+1}^{(k)}) - \epsilon/4. \tag{4.188}$$

Choose a natural number $k_1 > k_0$ such that

$$16k_1^{-1}(L_1 + L_2 + 1) < \epsilon. \tag{4.189}$$

Assume that an integer $k \ge k_1$. Then (4.187) and (4.188) hold. In view of (4.180)–(4.182), (4.188), and (4.189),

$$f(x_0, \tilde{x}_{L_2}) + \sum_{t=0}^{L_1-1} v(x_t, x_{t+1}) + \sum_{t=0}^{L_2-1} v(\tilde{x}_t, \tilde{x}_{t+1})$$

$$\ge h_k(x_0^{(k)}, \tilde{x}_{L_2}^{(k)}) + \sum_{t=0}^{L_1-1} u_t^{(k)}(x_t^{(k)}, x_{t+1}^{(k)})$$

$$+ \sum_{t=0}^{L_2-1} \tilde{u}_t^{(k)}(\tilde{x}_t^{(k)}, \tilde{x}_{t+1}^{(k)}) - \epsilon/4 - k_1^{-1}(L_1 + L_2 + 1)$$

$$\ge h_k(x_0^{(k)}, \tilde{x}_{L_2}^{(k)}) + \sum_{t=0}^{L_1-1} u_t^{(k)}(x_t^{(k)}, x_{t+1}^{(k)})$$

$$+ \sum_{t=0}^{L_2-1} \tilde{u}_t^{(k)}(\tilde{x}_t^{(k)}, \tilde{x}_{t+1}^{(k)}) - \epsilon/2.$$

When combined with (4.185)–(4.187) this contradicts property (i). The contradiction we have reached proves Lemma 4.33. □

4.19 Structure of Solutions of Perturbed Problems (P5)

Recall that the function $v \in \mathcal{M}$ was introduced in Sect. 4.1 while the function $g \in \mathfrak{A}$ was introduced in Sect. 4.17.

We prove the following result which describes the structure of approximate solutions of the perturbed problems of the type (P5) in the regions close to the right endpoints.

Theorem 4.34. *Let $\epsilon > 0$ and L_0 be a natural number. Then there exist $\delta > 0$ and an integer $L \geq L_0$ such that for each integer $T \geq 2L$, each*

$$\{\Omega_t\}_{t=0}^{T-1} \in \mathcal{F}(\delta, L, T),$$

each $h \in \mathfrak{A}$ satisfying

$$\|h - g\| \leq \delta,$$

each $\{u_t\}_{t=0}^{T-1} \subset \mathcal{M}$ satisfying

$$\|u_t - v\| \leq \delta, \; t = 0 \ldots, T - 1$$

and each $(\{\Omega_t\}_{t=0}^{T-1})$-program $\{x_t\}_{t=0}^{T}$ which satisfies

$$h(x_0, x_T) + \sum_{t=0}^{T-1} u_t(x_t, x_{t+1}) \geq \sigma(\{u_t\}_{t=0}^{T-1}, h, \{\Omega_t\}_{t=0}^{T-1}, 0, T) - \delta$$

there exists a (v, Ω)-overtaking optimal and (v, Ω)-good program $\{\xi_t\}_{t=0}^{\infty}$ and a $(\bar{v}, \bar{\Omega})$-overtaking optimal and $(\bar{v}, \bar{\Omega})$-good program $\{\bar{\xi}_t\}_{t=0}^{\infty}$ such that

$$\psi(\xi_0, \bar{\xi}_0) = \sup(\psi)$$

and for all $t = 0, \ldots, L_0$,

$$\rho(x_t, \xi_t) \leq \epsilon, \; \rho(x_{T-t}, \bar{\xi}_t) \leq \epsilon.$$

Proof. By Lemma 3.19, there exists a real number

$$\delta_0 \in (0, \min\{\epsilon, \bar{\lambda}/4\})$$

such that the following property holds:

(Pi) for each (Ω)-program $\{y_t\}_{t=0}^{L_0}$ and each $(\bar{\Omega})$-program $\{\bar{y}_t\}_{t=0}^{L_0}$ which satisfy

$$\psi(y_0, \bar{y}_0) \geq \sup(\psi) - 2\delta_0,$$

$$\sum_{t=0}^{L_0-1} (v(y_t, y_{t+1}) - v(\bar{x}, \bar{x})) - \pi^v(y_0) + \pi^v(y_{L_0}) \geq -2\delta_0$$

$$\sum_{t=0}^{L_0-1} (\bar{v}(\bar{y}_t, \bar{y}_{t+1}) - \bar{v}(\bar{x}, \bar{x})) - \pi^{\bar{v}}(\bar{y}_0) + \pi^{\bar{v}}(\bar{y}_{L_0}) \geq -2\delta_0$$

there exist a (v, Ω)-overtaking optimal and (v, Ω)-good program $\{\xi_t\}_{t=0}^{\infty}$ and a $(\bar{v}, \bar{\Omega})$-overtaking optimal and $(\bar{v}, \bar{\Omega})$-good program $\{\bar{\xi}_t\}_{t=0}^{\infty}$ such that

$$\psi(\xi_0, \bar{\xi}_0) = \sup(\psi)$$

and for all $t = 0, \dots, L_0$,

$$\rho(\xi_t, y_t) \le \epsilon/4, \ \rho(\bar{\xi}_t, \bar{y}_t) \le \epsilon/4.$$

Propositions 2.22 and 2.23 and assumption (A1) imply that there exists a real number $\delta_1 \in (0, \delta_0)$ such that for each point $z \in X$ satisfying $\rho(z, \bar{x}) \le 2\delta_1$,

$$|\pi^{\bar{v}}(z)| = |\pi^{\bar{v}}(z) - \pi^{\bar{v}}(\bar{x})| \le \delta_0/8, \tag{4.190}$$

$$|\pi^{v}(z)| = |\pi^{v}(z) - \pi^{v}(\bar{x})| \le \delta_0/8 \tag{4.191}$$

and that for each $(x, y) \in X \times X$ satisfying

$$\rho(x, \bar{x}) \le 2\delta_1, \ \rho(y, \bar{x}) \le 2\delta_1,$$

we have

$$|v(x, y) - \bar{v}(\bar{x}, \bar{x})| \le \delta_0/8. \tag{4.192}$$

By Theorem 4.22, there exist $\delta_2 \in (0, \delta_0)$ and a natural number L_1 such that the following property holds:

(Pii) for each integer $T > 2L_1$, each

$$\Omega_t \in \mathcal{E}(\delta_2), \ t = 0, \dots, T-1,$$

each $u_t \in \mathcal{M}$, $t = 0, \dots, T-1$ satisfying

$$\|u_t - v\| \le \delta_2, \ t = 0, \dots, T-1,$$

each $h \in \mathfrak{A}$ satisfying

$$\|h - g\| \le 1$$

and each $(\{\Omega_t\}_{t=0}^{T-1})$-program $\{x_t\}_{t=0}^{T}$ which satisfies

$$\sum_{t=0}^{T-1} u_t(x_t, x_{t+1}) + h(x_0, x_T)$$

$$\ge \sigma(\{u_t\}_{t=0}^{T-1}, h, \{\Omega_t\}_{t=0}^{T-1}, 0, T) - \delta_2,$$

we have

$$\rho(x_t, \bar{x}) \le \delta_1 \text{ for all } t = L_1, \dots, t - L_1.$$

By (A3), there exists a natural number L_2 such that for all integers $t \ge L_2$,

$$\rho(x_t^*, \bar{x}) \le \delta_1/2, \ \rho(\bar{x}_t^*, \bar{x}) \le \delta_1/2. \tag{4.193}$$

Set

$$\delta_3 = 48^{-1}\delta_0(2L_0 + 2L_1 + 2L_2 + 9)^{-1}. \tag{4.194}$$

By Lemma 4.33, there exists $\delta_4 > 0$ such that the following property holds:

(Piii) for each

$$\Omega_t \in \mathcal{E}(\delta_4),\ t = 0, \dots, L_0 + L_1 + L_2 + 2,$$

each

$$\tilde{\Omega}_t \in \mathcal{E}(\delta_4),\ t = 0, \dots, L_0 + L_1 + L_2 + 2,$$

each $\{\Omega_t\}_{t=0}^{L_0+L_1+L_2+2}$-program $\{x_t\}_{t=0}^{L_0+L_1+L_2+3}$ and each $\{\tilde{\Omega}_t\}_{t=0}^{L_0+L_1+L_2+2}$-program $\{\tilde{x}_t\}_{t=0}^{L_0+L_1+L_2+3}$ there exists an (Ω)-program $\{y_t\}_{t=0}^{L_0+L_1+L_2+3}$ and an (Ω)-program $\{\tilde{y}_t\}_{t=0}^{L_0+L_1+L_2+3}$ such that for all integers $t = 0, \dots, L_0 + L_1 + L_2 + 3$,

$$\rho(x_t, y_t) \le \delta_1/4,\ \rho(\tilde{x}_t, \tilde{y}_t) \le \delta_1/4$$

and

$$g(y_0, \tilde{y}_{L_0+L_1+L_2+3}) + \sum_{t=0}^{L_0+L_1+L_2+2} v(y_t, y_{t+1}) + \sum_{t=0}^{L_0+L_1+L_2+2} v(\tilde{y}_t, \tilde{y}_{t+1})$$

$$\ge g(x_0, \tilde{x}_{L_0+L_1+L_2+3}) + \sum_{t=0}^{L_0+L_1+L_2+2} v(x_t, x_{t+1}) + \sum_{t=0}^{L_0+L_1+L_2+2} v(\tilde{x}_t, \tilde{x}_{t+1}) - \delta_0/8.$$

Choose a number $\delta > 0$ and a natural number L such that

$$L \ge 2(L_0 + L_1 + L_2 + 4), \tag{4.195}$$

$$\delta \le 4^{-1} \min\{\delta_0, \delta_1, \delta_2, \delta_3, \delta_4\}. \tag{4.196}$$

Assume that an integer $T \ge 2L$,

$$\{\Omega_t\}_{t=0}^{T-1} \in \mathcal{F}(\delta, L, T), \tag{4.197}$$

$h \in \mathfrak{A}$ satisfies

$$\|h - g\| \le \delta \tag{4.198}$$

$\{u_t\}_{t=0}^{T-1} \subset \mathcal{M}$ satisfies

$$\|u_t - v\| \le \delta,\ t = 0 \dots, T - 1, \tag{4.199}$$

and that an $(\{\Omega_t\}_{t=0}^{T-1})$-program $\{x_t\}_{t=0}^{T}$ satisfies

$$h(x_0, x_T) + \sum_{t=0}^{T-1} u_t(x_t, x_{t+1})$$

$$\ge \sigma(\{u_t\}_{t=0}^{T-1}, h, \{\Omega_t\}_{t=0}^{T-1}, 0, T) - \delta. \tag{4.200}$$

It follows from (4.195) to (4.200) and the property (Pii) that

$$\rho(x_t, \bar{x}) \le \delta_1,\ t = L_1, \dots, T - L_1. \tag{4.201}$$

By the definition of $\mathcal{F}(\cdot,\cdot,\cdot)$ (see (4.155)–(4.158)), there exist an $(\{\Omega_t\}_{t=0}^{L-1})$-program $\{y_t^{(1)}\}_{t=0}^{L}$ and an $(\{\Omega_t\}_{t=T-L}^{T-1})$-program $\{y_t^{(2)}\}_{t=T-L}^{T}$ such that

$$\rho(y_t^{(1)}, x_t^*) \le \delta,\ t = 0, \dots, L, \tag{4.202}$$

$$\rho(y_t^{(2)}, \bar{x}_{T-t}^*) \le \delta,\ t = T-L, \dots, T, \tag{4.203}$$

$$|g(y_0^{(1)}, y_T^{(2)}) - g(x_0^*, \bar{x}_0^*)| \le \delta, \tag{4.204}$$

$$|v(y_t^{(1)}, y_{t+1}^{(1)}) - v(x_t^*, x_{t+1}^*)| \le \delta,\ t = 0, \dots, L-1, \tag{4.205}$$

$$|v(y_t^{(2)}, y_{t+1}^{(2)}) - v(\bar{x}_{T-t}^*, \bar{x}_{T-t-1}^*)| \le \delta,\ t = T-L, \dots, T-1. \tag{4.206}$$

Define a sequence $\{\tilde{x}_t\}_{t=0}^{T}$ by

$$\tilde{x}_t = y_t^{(1)},\ t = 0, \dots, L_0 + L_1 + L_2 + 3,$$

$$\tilde{x}_t = x_t,\ t = L_0 + L_1 + L_2 + 4, \dots, T - L_0 - L_1 - L_2 - 4,$$

$$\tilde{x}_t = y_t^{(2)},\ t = T - L_0 - L_1 - L_2 - 3, \dots, T. \tag{4.207}$$

Relations (4.193), (4.195), (4.196), (4.202), (4.203), and (4.207) imply that

$$\begin{aligned} \rho(\tilde{x}_{L_0+L_1+L_2+3}, \bar{x}) &= \rho(y_{L_0+L_1+L_2+3}^{(1)}, \bar{x}) \\ &\le \rho(y_{L_0+L_1+L_2+3}^{(1)}, x_{L_0+L_1+L_2+3}^*) + \rho(x_{L_0+L_1+L_2+3}^*, \bar{x}) \\ &\le \delta + \delta_1/2 \le \delta_1 \end{aligned} \tag{4.208}$$

and

$$\begin{aligned} \rho(\tilde{x}_{T-L_0-L_1-L_2-3}, \bar{x}) &= \rho(y_{T-L_0-L_1-L_2-3}^{(2)}, \bar{x}) \\ &\le \rho(y_{T-L_0-L_1-L_2-3}^{(2)}, \bar{x}_{L_0+L_1+L_2+3}^*) + \rho(\bar{x}_{L_0+L_1+L_2+3}^*, \bar{x}) \\ &\le \delta + \delta_1/2 \le \delta_1. \end{aligned} \tag{4.209}$$

By (4.4), (4.181), (4.207)–(4.209), $\{\tilde{x}_t\}_{t=0}^{T}$ is an $(\{\Omega_t\}_{t=0}^{T-1})$-program. It follows from (4.198), (4.199), (4.200), and (4.207) that

$$\begin{aligned} \delta &\ge \sigma(\{u_t\}_{t=0}^{T-1}, h, \{\Omega_t\}_{t=0}^{T-1}, 0, T) - h(x_0, x_T) - \sum_{t=0}^{T-1} u_t(x_t, x_{t+1}) \\ &\ge h(\tilde{x}_0, \tilde{x}_T) + \sum_{t=0}^{T-1} u_t(\tilde{x}_t, \tilde{x}_{t+1}) - h(x_0, x_T) - \sum_{t=0}^{T-1} u_t(x_t, x_{t+1}) \\ &= h(y_0^{(1)}, y_T^{(2)}) - h(x_0, x_T) \\ &+ \sum_{t=0}^{L_0+L_1+L_2+3} (u_t(\tilde{x}_t, \tilde{x}_{t+1}) - u_t(x_t, x_{t+1})) \end{aligned}$$

$$+ \sum_{t=T-L_0-L_1-L_2-4}^{T-1} (u_t(\tilde{x}_t, \tilde{x}_{t+1}) - u_t(x_t, x_{t+1}))$$
$$\geq g(y_0^{(1)}, y_T^{(2)}) - g(x_0, x_T)$$
$$+ \sum_{t=0}^{L_0+L_1+L_2+3} (v(\tilde{x}_t, \tilde{x}_{t+1}) - v(x_t, x_{t+1}))$$
$$+ \sum_{t=T-L_0-L_1-L_2-4}^{T-1} (v(\tilde{x}_t, \tilde{x}_{t+1}) - v(x_t, x_{t+1}))$$
$$-4\delta(L_0 + L_1 + L_2 + 6)$$
$$= g(y_0^{(1)}, y_T^{(2)}) - g(x_0, x_T)$$
$$+ \sum_{t=0}^{L_0+L_1+L_2+2} (v(y_t^{(1)}, y_{t+1}^{(1)}) - v(x_t, x_{t+1}))$$
$$+v(y^{(1)}_{L_0+L_1+L_2+3}, x_{L_0+L_1+L_2+4}) - v(x_{L_0+L_1+L_2+3}, x_{L_0+L_1+L_2+4})$$
$$+ \sum_{t=T-L_0-L_1-L_2-3}^{T-1} (v(y_t^{(2)}, y_{t+1}^{(2)}) - v(x_t, x_{t+1}))$$
$$+v(x_{T-L_0-L_1-L_2-4}, y^{(2)}_{T-L_0-L_1-L_2-3}) - v(x_{T-L_0-L_1-L_2-4}, x_{T-L_0-L_1-L_2-3})$$
$$- 4\delta(L_0 + L_1 + L_2 + 6). \tag{4.210}$$

By (4.193), (4.195), (4.196), (4.201), (4.203), and the choice of δ_1 (see (4.192)),

$$|v(y^{(1)}_{L_0+L_1+L_2+3}, x_{L_0+L_1+L_2+4}) - v(\bar{x}, \bar{x})| \leq \delta_0/8, \tag{4.211}$$
$$|v(x_{L_0+L_1+L_2+3}, x_{L_0+L_1+L_2+4}) - v(\bar{x}, \bar{x})| \leq \delta_0/8, \tag{4.212}$$
$$|v(x_{T-L_0-L_1-L_2-4}, y^{(2)}_{T-L_0-L_1-L_2-3}) - v(\bar{x}, \bar{x})| \leq \delta_0/8. \tag{4.213}$$
$$|v(x_{T-L_0-L_1-L_2-4}, x_{T-L_0-L_1-L_2-3}) - v(\bar{x}, \bar{x})| \leq \delta_0/8. \tag{4.214}$$

It follows from (4.210) to (4.214) that

$$\delta \geq g(y_0^{(1)}, y_T^{(2)}) - g(x_0, x_T)$$
$$+ \sum_{t=0}^{L_0+L_1+L_2+2} (v(y_t^{(1)}, y_{t+1}^{(1)}) - v(x_t, x_{t+1}))$$
$$+ \sum_{t=T-L_0-L_1-L_2-3}^{T-1} (v(y_t^{(2)}, y_{t+1}^{(2)}) - v(x_t, x_{t+1}))$$

$$- \delta_0/2 - 4\delta(L_0 + L_1 + L_2 + 6). \tag{4.215}$$

In view of (4.195), (4.204)–(4.206), and (4.215),

$$\begin{aligned}
&g(x_0, x_T) + \sum_{t=0}^{L_0+L_1+L_2+2} v(x_t, x_{t+1}) + \sum_{t=T-L_0-L_1-L_2-3}^{T-1} v(x_t, x_{t+1}) \\
&\geq g(y_0^{(1)}, y_T^{(2)}) + \sum_{t=0}^{L_0+L_1+L_2+2} v(y_t^{(1)}, y_{t+1}^{(1)}) + \sum_{t=T-L_0-L_1-L_2-3}^{T-1} v(y_t^{(2)}, y_{t+1}^{(2)}) \\
&\quad -\delta_0/2 - 4\delta(L_0 + L_1 + L_2 + 6) - \delta \\
&\geq g(x_0^*, \bar{x}_0^*) + \sum_{t=0}^{L_0+L_1+L_2+2} v(x_t^*, x_{t+1}^*) + \sum_{t=T-L_0-L_1-L_2-3}^{T-1} v(\bar{x}_{T-t}^*, \bar{x}_{T-t-1}^*) \\
&\quad -\delta(1 + 2(L_0 + L_1 + L_2 + 3)) - \delta_0/2 - 4\delta(L_0 + L_1 + L_2 + 7) \\
&\geq g(x_0^*, \bar{x}_0^*) + \sum_{t=0}^{L_0+L_1+L_2+2} v(x_t^*, x_{t+1}^*) + \sum_{t=0}^{L_0+L_1+L_2+2} \bar{v}(\bar{x}_t^*, \bar{x}_{t+1}^*) \\
&\quad - \delta_0/2 - 6\delta(L_0 + L_1 + L_2 + 8).
\end{aligned} \tag{4.216}$$

Property (Piii), (4.196), and (4.197) imply that there exist an (Ω)-program $\{x_t^{(1)}\}_{t=0}^{L_0+L_1+L_2+3}$ and an (Ω)-program $\{x_t^{(2)}\}_{t=T-L_0-L_1-L_2-3}^{T}$ such that

$$\rho(x_t, x_t^{(1)}) \leq \delta_1/4, \ t = 0, \dots, L_0 + L_1 + L_2 + 3, \tag{4.217}$$

$$\rho(x_t, x_t^{(2)}) \leq \delta_1/4, \ t = T - L_0 - L_1 - L_2 - 3, \dots, T \tag{4.218}$$

and

$$\begin{aligned}
&g(x_0^{(1)}, x_T^{(2)}) + \sum_{t=0}^{L_0+L_1+L_2+2} v(x_t^{(1)}, x_{t+1}^{(1)}) + \sum_{t=T-L_0-L_1-L_2-3}^{T-1} v(x_t^{(2)}, x_{t+1}^{(2)}) \\
&\geq g(x_0, x_T) + \sum_{t=0}^{L_0+L_1+L_2+2} v(x_t, x_{t+1}) + \sum_{t=T-L_0-L_1-L_2-3}^{T-1} v(x_t, x_{t+1}) - \delta_0/8.
\end{aligned} \tag{4.219}$$

By (4.216) and (4.219),

$$\begin{aligned}
&g(x_0^{(1)}, x_T^{(2)}) + \sum_{t=0}^{L_0+L_1+L_2+2} v(x_t^{(1)}, x_{t+1}^{(1)}) + \sum_{t=T-L_0-L_1-L_2-3}^{T-1} v(x_t^{(2)}, x_{t+1}^{(2)}) \\
&\geq g(x_0^*, \bar{x}_0^*) + \sum_{t=0}^{L_0+L_1+L_2+2} v(x_t^*, x_{t+1}^*) + \sum_{t=0}^{L_0+L_1+L_2+2} \bar{v}(\bar{x}_t^*, \bar{x}_{t+1}^*)
\end{aligned}$$

$$-5\delta_0/8 - 6\delta(L_0 + L_1 + L_2 + 8). \tag{4.220}$$

Set

$$\bar{x}_t^{(2)} = x_{T-t}^{(2)},\ t = 0, \dots, L_0 + L_1 + L_2 + 3. \tag{4.221}$$

In view of the relation above,

$$\sum_{t=T-L_0-L_1-L_2-3}^{T-1} v(x_t^{(2)}, x_{t+1}^{(2)}) = \sum_{t=T-L_0-L_1-L_2-3}^{T-1} v(\bar{x}_{T-t}^{(2)}, \bar{x}_{T-t-1}^{(2)})$$

$$= \sum_{t=T-L_0-L_1-L_2-3}^{T-1} \bar{v}(\bar{x}_{T-t-1}^{(2)}, \bar{x}_{T-t}^{(2)}) = \sum_{t=0}^{L_0+L_1+L_2+2} \bar{v}(\bar{x}_t^{(2)}, \bar{x}_{t+1}^{(2)}).$$

It follows from (4.194), (4.196), (4.220), (4.221), and the relation above that

$$g(x_0^{(1)}, \bar{x}_0^{(2)}) + \sum_{t=0}^{L_0+L_1+L_2+2} v(x_t^{(1)}, x_{t+1}^{(1)}) + \sum_{t=0}^{L_0+L_1+L_2+2} \bar{v}(\bar{x}_t^{(2)}, \bar{x}_{t+1}^{(2)})$$

$$\geq g(x_0^*, \bar{x}_0^*) + \sum_{t=0}^{L_0+L_1+L_2+2} v(x_t^*, x_{t+1}^*) + \sum_{t=0}^{L_0+L_1+L_2+2} \bar{v}(\bar{x}_t^*, \bar{x}_{t+1}^*) - (3/4)\delta_0. \tag{4.222}$$

By (4.195), (4.201), (4.217), (4.218), (4.220)–(4.222), Proposition 2.19, Corollary 2.21, and the choice of $x_t^*, \bar{x}_t^*$, $t = 0, 1, \dots,$

$$g(x_0^{(1)}, \bar{x}_0^{(2)}) - g(x_0^*, \bar{x}_0^*)$$

$$+ \sum_{t=0}^{L_0-1} (v(x_t^{(1)}, x_{t+1}^{(1)}) - v(\bar{x}, \bar{x})) - \pi^v(x_0^{(1)}) + \pi^v(x_{L_0}^{(1)})$$

$$+ \sum_{t=0}^{L_0-1} (\bar{v}(\bar{x}_t^{(2)}, \bar{x}_{t+1}^{(2)}) - v(\bar{x}, \bar{x})) - \pi^{\bar{v}}(\bar{x}_0^{(2)}) + \pi^{\bar{v}}(\bar{x}_{L_0}^{(2)})$$

$$\geq g(x_0^{(1)}, \bar{x}_0^{(2)}) - g(x_0^*, \bar{x}_0^*)$$

$$+ \sum_{t=0}^{L_0+L_1+L_2+2} (v(x_t^{(1)}, x_{t+1}^{(1)}) - v(\bar{x}, \bar{x}))$$

$$-\pi^v(x_0^{(1)}) + \pi^v(x_{L_0+L_1+L_2+3}^{(1)})$$

$$+ \sum_{t=0}^{L_0+L_1+L_2+2} (\bar{v}(\bar{x}_t^{(2)}, \bar{x}_{t+1}^{(2)}) - v(\bar{x}, \bar{x}))$$

$$-\pi^{\bar{v}}(\bar{x}_0^{(2)}) + \pi^{\bar{v}}(\bar{x}_{L_0+L_1+L_2+3}^{(2)})$$

$$\geq \sum_{t=0}^{L_0+L+1+L_2+2} (v(x_t^*, x_{t+1}^*) - v(\bar{x}, \bar{x}))$$

$$+ \sum_{t=0}^{L_0+L_1+L_2+2} (\bar{v}(\bar{x}_t^*, \bar{x}_{t+1}^*) - v(\bar{x}, \bar{x}))$$

$$-\pi^v(x_0^{(1)}) + \pi^v(x_{L_0+L_1+L_2+3}^{(1)}) - \pi^{\bar{v}}(\bar{x}_0^{(2)}) + \pi^{\bar{v}}(\bar{x}_{L_0+L_1+L_2+3}^{(2)}) - (3/4)\delta_0$$

$$= \pi^v(x_0^*) + \pi^v(x_{L_0+L_1+L_2+3}^*) + \pi^{\bar{v}}(\bar{x}_0^*) - \pi^{\bar{v}}(\bar{x}_{L_0+L_1+L_2+3}^*)$$

$$-\pi^v(x_0^{(1)}) + \pi^v(x_{L_0+L_1+L_2+3}^{(1)}) - \pi^{\bar{v}}(\bar{x}_0^{(2)}) + \pi^{\bar{v}}(\bar{x}_{L_0+L_1+L_2+3}^{(2)}) - (3/4)\delta_0. \quad (4.223)$$

In view of (4.190), (4.191), and (4.193),

$$|\pi^v(x_{L_0+L_1+L_2+3}^*)| \leq \delta_1/8, \ |\pi^{\bar{v}}(\bar{x}_{L_0+L_1+L_2+3}^*)| \leq \delta_1/8. \quad (4.224)$$

By (4.201), (4.217), (4.218), and (4.221),

$$\rho(x_{L_0+L_1+L_2+3}^{(1)}, \bar{x})$$

$$\leq \rho(x_{L_0+L_1+L_2+3}^{(1)}, x_{L_0+L_1+L_2+3}) + \rho(x_{L_0+L_1+L_2+3}, \bar{x})$$

$$\leq \delta_1/4 + \delta_1, \quad (4.225)$$

$$\rho(\bar{x}_{L_0+L_1+L_2+3}^{(2)}, \bar{x}) = \rho(x_{T-L_0-L_1-L_2-3}^{(2)}, \bar{x})$$

$$\leq \rho(x_{T-L_0-L_1-L_2-3}^{(2)}, x_{T-L_0-L_1-L_2-3}) + \rho(x_{T-L_0-L_1-L_2-3}, \bar{x})$$

$$\leq \delta_1/4 + \delta_1. \quad (4.226)$$

It follows from (4.190), (4.191), (4.225), and (4.226) that

$$|\pi^v(x_{L_0+L_1+L_2+3}^{(1)})| \leq \delta_0/8, \ |\pi^{\bar{v}}(\bar{x}_{L_0+L_1+L_2+3}^{(2)})| \leq \delta_0/8. \quad (4.227)$$

By (4.223), (4.224), and (4.227),

$$g(x_0^{(1)}, \bar{x}_0^{(2)}) - g(x_0^*, \bar{x}_0^*)$$

$$+ \sum_{t=0}^{L_0-1} (v(x_t^{(1)}, x_{t+1}^{(1)}) - v(\bar{x}, \bar{x})) - \pi^v(x_0^{(1)}) + \pi^v(x_{L_0}^{(1)})$$

$$+ \sum_{t=0}^{L_0-1} (\bar{v}(\bar{x}_t^{(2)}, \bar{x}_{t+1}^{(2)}) - v(\bar{x}, \bar{x})) - \pi^{\bar{v}}(\bar{x}_0^{(2)}) + \pi^{\bar{v}}(\bar{x}_{L_0}^{(2)})$$

$$\geq \pi^v(x_0^*) + \pi^{\bar{v}}(\bar{x}_0^*) - \pi^v(x_0^{(1)}) - \pi^{\bar{v}}(\bar{x}_0^{(2)}) - (5/4)\delta_0. \quad (4.228)$$

It follows from (4.151), (4.152), and (4.228) that

$$\psi(x_0^{(1)}, \bar{x}_0^{(2)}) - \sup(\psi)$$

$$+\sum_{t=0}^{L_0-1}(v(x_t^{(1)},x_{t+1}^{(1)})-v(\bar{x},\bar{x}))-\pi^v(x_0^{(1)})+\pi^v(x_{L_0}^{(1)})$$

$$+\sum_{t=0}^{L_0-1}(\bar{v}(\bar{x}_t^{(2)},\bar{x}_{t+1}^{(2)})-v(\bar{x},\bar{x}))-\pi^{\bar{v}}(\bar{x}_0^{(2)})+\pi^{\bar{v}}(\bar{x}_{L_0}^{(2)})\geq-(5/4)\delta_0. \quad (4.229)$$

Proposition 2.19, (4.225), (4.226), and (4.229) imply that

$$\psi(x_0^{(1)},\bar{x}_0^{(2)})\geq\sup(\psi)-(5/4)\delta_0, \quad (4.230)$$

$$\sum_{t=0}^{L_0-1}(v(x_t^{(1)},x_{t+1}^{(1)})-v(\bar{x},\bar{x}))-\pi^v(x_0^{(1)})+\pi^v(x_{L_0}^{(1)})\geq-(5/4)\delta_0, \quad (4.231)$$

$$\sum_{t=0}^{L_0-1}(\bar{v}(\bar{x}_t^{(2)},\bar{x}_{t+1}^{(2)})-v(\bar{x},\bar{x}))-\pi^{\bar{v}}(\bar{x}_0^{(2)})+\pi^{\bar{v}}(\bar{x}_{L_0}^{(2)})\geq-(5/4)\delta_0. \quad (4.232)$$

By (4.221), (4.230)–(4.232), and property (Pi), there exist a (v,Ω)-overtaking optimal and (v,Ω)-good program $\{\xi_t\}_{t=0}^{\infty}$ and a $(\bar{v},\bar{\Omega})$-overtaking optimal and $(\bar{v},\bar{\Omega})$-good program $\{\bar{\xi}_t\}_{t=0}^{\infty}$ such that

$$\psi(\xi_0,\bar{\xi}_0)=\sup(\psi)$$

and for all $t=0,\dots,L_0$,

$$\rho(\xi_t,x_t^{(1)})\leq\epsilon/4,\ \rho(\bar{\xi}_t,\bar{x}_t^{(2)})\leq\epsilon/4. \quad (4.233)$$

In view of (4.217), (4.218), (4.221), and (4.233), for all $t=0,\dots,L_0$,

$$\rho(\xi_t,x_t)\leq\rho(\xi_t,x_t^{(1)})+\rho(x_t^{(1)},x_t)\leq\epsilon/4+\delta_1/4\leq\epsilon,$$

$$\rho(\bar{\xi}_t,x_{T-t})\leq\rho(\bar{\xi}_t,x_{T-t}^{(2)})+\rho(x_{T-t}^{(2)},x_{T-t})\leq\epsilon/4+\delta_1/4\leq\epsilon.$$

Theorem 4.34 is proved. □

4.20 Perturbed Problems (P5) with Discounting

In this section we assume that the function $g\in\mathfrak{M}$, introduced in Sect. 4.14, is identically zero. Denote by $\mathcal{M}_0$ the set of all upper semicontinuous functions $u\in\mathcal{M}$.

Theorem 4.35. *Let $\epsilon\in(0,\bar{\lambda})$ and l_0, L_0 be natural numbers. Then there exist $\delta>0$, an integer $L>L_0+l_0$ and $\lambda\in(0,1)$ such that for each integer $T\geq L$, each*

$$\{\Omega_t\}_{t=0}^{T-1}\in\mathcal{E}(\delta,L,T),$$

each $u_t \in \mathcal{M}_0$, $t = 0, \dots, T-1$ *satisfying*

$$\|u_t - v\| \le \delta,\ t = 0, \dots, T-1,$$

each sequence $\{\alpha_t\}_{t=0}^{T-1} \subset (0, 1]$ *such that*

$$\alpha_i \alpha_j^{-1} \ge \lambda \text{ for each } i, j \in \{0, \dots, T-1\} \text{ satisfying } |i - j| \le L$$

and each $(\{\Omega_t\}_{t=0}^{T-1})$*-program* $\{x_t\}_{t=0}^{T}$ *which satisfies*

$$x_0 \in \bar{Y}(\{\Omega_t\}_{t=0}^{l_0-1}, 0, l_0),$$

$$\sigma(\{\alpha_t u_t\}_{t=0}^{T-1}, \{\Omega_t\}_{t=0}^{T-1}, 0, T, x_0) = \sum_{t=0}^{T-1} \alpha_t u_t(x_t, x_{t+1})$$

there exists a $(\bar{v}, \bar{\Omega})$*-overtaking optimal program* $\{z_t^*\}_{t=0}^{\infty}$ *such that*

$$\pi^{\bar{v}}(z_0^*) = \sup(\pi^{\bar{v}}),$$

and for all $t = 0, \dots, L_0$,

$$\rho(x_{T-t}, z_t^*) \le \epsilon.$$

Proof. By Theorem 4.28, there exist $\delta_0 > 0$ and an integer $L_1 \ge L_0$ such that the following property holds:

(Pi) for each integer $T \ge L_1$, each

$$\{\Omega_t\}_{t=0}^{T-1} \in \mathcal{E}(\delta_0, L_1, T),$$

each $\{u_t\}_{t=0}^{T-1} \subset \mathcal{M}$ satisfying

$$\|u_t - v\| \le \delta_0,\ t = 0 \dots, T-1$$

and each $(\{\Omega_t\}_{t=0}^{T-1})$-program $\{x_t\}_{t=0}^{T}$ which satisfies

$$x_0 \in \bar{Y}(\{\Omega_t\}_{t=0}^{l_0-1}, 0, l_0),$$

$$\sigma(\{u_t\}_{t=0}^{T-1}, \{\Omega_t\}_{t=0}^{T-1}, 0, T, x_0) \le \sum_{t=0}^{T-1} u_t(x_t, x_{t+1}) + \delta_0$$

there exists a $(\bar{v}, \bar{\Omega})$-overtaking optimal program $\{z_t^*\}_{t=0}^{\infty}$ such that

$$\pi^{\bar{v}}(z_0^*) = \sup(\pi^{\bar{v}}),$$

$$\rho(x_{T-t}, z_t^*) \le \epsilon,\ t = 0, \dots, L_0.$$

By Theorem 4.10, there exist $\delta_1 > 0$ a natural number $L_2 > l_0$ and $\lambda_1 \in (0, 1)$ such that the following property holds:

(Pii) for each integer $T > 2L_2$, each

$$\Omega_t \in \mathcal{E}(\delta_1),\ t = 0, \dots, T-1,$$

each $u_t \in \mathcal{M}_0,\ t = 0, \dots, T-1$ satisfying

$$\|u_t - v\| \le \delta_1,\ t = 0, \dots, T-1,$$

each sequence $\{\alpha_t\}_{t=0}^{T-1} \subset (0,1]$ such that

$$\alpha_i \alpha_j^{-1} \ge \lambda_1 \text{ for each } i,j \in \{0,\dots,T-1\} \text{ satisfying } |i-j| \le L_2$$

and each $(\{\Omega_t\}_{t=0}^{T-1})$-program $\{x_t\}_{t=0}^{T}$ which satisfies

$$x_0 \in \bar{Y}(\{\Omega_t\}_{t=0}^{l_0-1}, 0, l_0),$$

$$\sigma(\{\alpha_t u_t\}_{t=0}^{T-1}, \{\Omega_t\}_{t=0}^{T-1}, 0, T, x_0) = \sum_{t=0}^{T-1} \alpha_t u_t(x_t, x_{t+1})$$

we have

$$\rho(x_t, \bar{x}) \le \bar{\lambda}/4,\ t = L_2, \dots, T-L_2.$$

Choose an integer

$$L > 2(L_0 + L_1 + L_2 + l_0) + 4, \tag{4.234}$$

a positive number

$$\delta < 4^{-1} \min\{1, \delta_0, \delta_1\} \tag{4.235}$$

and a positive number $\lambda \in (\lambda_1, 1)$ such that

$$(1-\lambda)\lambda^{-1}(\|v\| + 1) < \delta. \tag{4.236}$$

Assume that an integer

$$T \ge L, \tag{4.237}$$

$$\{\Omega_t\}_{t=0}^{T-1} \in \mathcal{E}(\delta, L, T), \tag{4.238}$$

$u_t \in \mathcal{M}_0,\ t = 0, \dots, T-1$ satisfy

$$\|u_t - v\| \le \delta,\ t = 0, \dots, T-1, \tag{4.239}$$

a sequence $\{\alpha_t\}_{t=0}^{T-1} \subset (0,1]$ satisfies

$$\alpha_i \alpha_j^{-1} \ge \lambda \text{ for each } i,j \in \{0,\dots,T-1\} \text{ satisfying } |i-j| \le L \tag{4.240}$$

and that $(\{\Omega_t\}_{t=0}^{T-1})$-program $\{x_t\}_{t=0}^{T}$ satisfies

$$x_0 \in \bar{Y}(\{\Omega_t\}_{t=0}^{l_0-1}, 0, l_0), \tag{4.241}$$

$$\sigma(\{\alpha_t u_t\}_{t=0}^{T-1}, \{\Omega_t\}_{t=0}^{T-1}, 0, T, x_0) = \sum_{t=0}^{T-1} \alpha_t u_t(x_t, x_{t+1}). \tag{4.242}$$

Property (Pii), (4.234), (4.235), (4.237), (4.238), and (4.240)–(4.242) imply that

$$\rho(x_t, \bar{x}) \le \bar{\lambda}/4,\ t = L_2, \dots, T - L_2$$

and in particular,

$$\rho(x_{T-L+L_2}, \bar{x}) \le \bar{\lambda}/4. \tag{4.243}$$

In view of (4.242),

$$\sigma(\{\alpha_t u_t\}_{t=T-L+L_2}^{T-1}, \{\Omega_t\}_{t=T-L+L_2}^{T-1}, T - L + L_2, T, x_{T-L+L_2})$$

$$= \sum_{t=T-L+L_2}^{T-1} \alpha_t u_t(x_t, x_{t+1}). \tag{4.244}$$

By (4.244),

$$\sigma(\{\alpha_{T-1}^{-1}\alpha_t u_t\}_{t=T-L+L_2}^{T-1}, \{\Omega_t\}_{t=T-L+L_2}^{T-1}, T - L + L_2, T, x_{T-L+L_2})$$

$$= \sum_{t=T-L+L_2}^{T-1} \alpha_{T-1}^{-1}\alpha_t u_t(x_t, x_{t+1}). \tag{4.245}$$

It follows from (4.234)–(4.236), (4.239), and (4.240) that for all integers $t = T - L + L_2, \dots, T - 1$,

$$\|\alpha_{T-1}\alpha_t u_t - v\| \le \|\alpha_{T-1}\alpha_t u_t - u_t\| + \|u_t - v\|$$

$$\le |\alpha_{T-1}\alpha_t - 1|\|u_t\| + \delta$$

$$\le (\|v\| + 1)|\lambda - 1|\lambda^{-1} + \delta < 2\delta < \delta_0. \tag{4.246}$$

Property (Pi), (4.234), (4.235), (4.237)–(4.239), (4.242), (4.243), (4.245), and (4.246) imply that there exists a $(\bar{v}, \bar{\Omega})$-overtaking optimal program $\{z_t^*\}_{t=0}^{\infty}$ such that

$$\pi^{\bar{v}}(z_0^*) = \sup(\pi^{\bar{v}}),$$

and for all $t = 0, \dots, L_0$,

$$\rho(x_{T-t}, z_t^*) \le \epsilon.$$

Theorem 4.35 is proved. □

5

Unconstrained Games Without Convexity–Concavity Assumptions

In this chapter we study the structure of approximate solutions for a class of unconstrained dynamic discrete-time two-player zero-sum games without using standard convexity–concavity assumptions and establish two turnpike results. These results describe the structure of approximate solutions which is independent of the length of domains.

5.1 Preliminaries and Main Results

Let (X, ρ_X) and (Y, ρ_Y) be compact metric spaces equipped with the metrics ρ_X and ρ_Y respectively. We consider the set $X \times X \times Y \times Y$ equipped with the product topology induced by the metric

$$\rho((x_1, x_2, y_1, y_2), (x_1', x_2', y_1', y_2'))$$

$$= \rho_X(x_1, x_1') + \rho_X(x_2, x_2') + \rho_Y(y_1, y_1') + \rho_Y(y_2, y_2'),$$

$$x_1, x_2, x_1', x_2' \in X, \; y_1, y_2, y_1', y_2' \in Y.$$

Denote by $C(X \times X \times Y \times Y)$ the set of all continuous functions $f : X \times X \times Y \times Y \to R^1$. For every function $f \in C(X \times X \times Y \times Y)$ set

$$\|f\| = \sup\{|f(x_1, x_2, y_1, y_2)| : \; x_1, x_2 \in X, \; y_1, y_2 \in Y\}.$$

It is clear that $(C(X \times X \times Y \times Y), \|\cdot\|)$ is a Banach space.

Let a function $f \in C(X \times X \times Y \times Y)$ be given. We associate with the function f a dynamic discrete-time two-player zero-sum game and study its optimal solutions. Namely, given an integer $n \geq 1$ we consider a discrete-time two-player zero-sum game over the interval $[0, n]$. For this game the collection $\{\{x_i\}_{i=0}^n : x_i \in X, \; i = 0, \dots n\}$ is the set of strategies for the first player, the

A.J. Zaslavski, *Discrete-Time Optimal Control and Games on Large Intervals*, Springer Optimization and Its Applications 119,
DOI 10.1007/978-3-319-52932-5_5

collection $\{\{y_i\}_{i=0}^n : y_i \in Y, i = 0, \dots n\}$ is the set of strategies for the second player, and the cost for the first player associated with the pair of strategies $\{x_i\}_{i=0}^n$, $\{y_i\}_{i=0}^n$ is defined by

$$\sum_{i=0}^{n-1} f_i(x_i, x_{i+1}, y_i, y_{i+1}).$$

Denote by $\mathcal{A}_0$ the set of all functions $f \in C(X \times X \times Y \times Y)$ such that there exists a pair of points $x_f \in X$ and $y_f \in Y$ for which

$$f(x_f, x_f, y, y) \le f(x_f, x_f, y_f, y_f) \le f(x, x, y_f, y_f)$$
$$\text{for all } x \in X \text{ and all } y \in Y. \tag{5.1}$$

The following result is proved in Sect. 5.2.

Proposition 5.1. *$\mathcal{A}_0$ is a closed subset of the Banach space $(C(X \times X \times Y \times Y), \|\cdot\|)$.*

We study a dynamic discrete-time two-player zero-sum game associated with $f \in \mathcal{A}_0$. If a function $f \in \mathcal{A}_0$ and a pair of points $x_f \in X$ and $y_f \in Y$ satisfy (5.1), then the pair (x_f, y_f) is a saddle point for the function

$$\bar{f}(x, y) := f(x, x, y, y),\ x \in X,\ y \in Y.$$

Let us now define approximate solutions (saddle points) of our dynamic games.

Let $f \in C(X \times X \times Y \times Y)$, $n_2 > n_1$ be integers and let M be a nonnegative number. A pair of finite sequences $\{\bar{x}_i\}_{i=n_1}^{n_2} \subset X$, $\{\bar{y}_i\}_{i=n_1}^{n_2} \subset Y$ is called (f, M)-good if the following properties hold:

(i) for every finite sequence $\{x_i\}_{i=n_1}^{n_2} \subset X$ which satisfies

$$x_{n_1} = \bar{x}_{n_1},\ x_{n_2} = \bar{x}_{n_2}$$

the inequality

$$M + \sum_{i=n_1}^{n_2-1} f(x_i, x_{i+1}, \bar{y}_i, \bar{y}_{i+1}) \ge \sum_{i=n_1}^{n_2-1} f(\bar{x}_i, \bar{x}_{i+1}, \bar{y}_i, \bar{y}_{i+1}) \tag{5.2}$$

is valid;

(ii) for every finite sequence $\{y_i\}_{i=n_1}^{n_2} \subset Y$ which satisfies

$$y_{n_1} = \bar{y}_{n_1},\ y_{n_2} = \bar{y}_{n_2}$$

the inequality

$$M + \sum_{i=n_1}^{n_2-1} f(\bar{x}_i, \bar{x}_{i+1}, \bar{y}_i, \bar{y}_{i+1}) \ge \sum_{i=n_1}^{n_2-1} f(\bar{x}_i, \bar{x}_{i+1}, y_i, y_{i+1}) \tag{5.3}$$

is valid.

If a pair of finite sequences $\{x_i\}_{i=n_1}^{n_2} \subset X$, $\{y_i\}_{i=n_1}^{n_2} \subset Y$ is $(f,0)$-good then it is called (f)-optimal.

A pair of sequences $\{\tilde{x}_i\}_{i=0}^{\infty} \subset X$, $\{\tilde{y}_i\}_{i=0}^{\infty} \subset Y$ is called (f,M)-good if for every integer $n \geq 1$ the pair of sequences $\{\tilde{x}_i\}_{i=0}^{n}$, $\{\tilde{y}_i\}_{i=0}^{n}$ is (f,M)-good.

A pair of sequences $\{\tilde{x}_i\}_{i=0}^{\infty} \subset X$, $\{\tilde{y}_i\}_{i=0}^{\infty} \subset Y$ is called (f)-good if it is (f,M)-good with some positive number M.

A pair of sequences $\{\tilde{x}_i\}_{i=0}^{\infty} \subset X$, $\{\tilde{y}_i\}_{i=0}^{\infty} \subset Y$ is called (f)-optimal if for every integer $n \geq 1$ the pair of sequences $\{\tilde{x}_i\}_{i=0}^{n}$, $\{\tilde{y}_i\}_{i=0}^{n}$ is (f)-optimal.

In Sect. 5.3 we prove the following result.

Proposition 5.2. *Let $f \in \mathcal{A}_0$, $x_f \in X$ and $y_f \in Y$ satisfy (5.1) and let $\bar{x}_i^f = x_f$, $\bar{y}_i^f = y_f$ for all nonnegative integers i. Then the pair $\{\bar{x}_i^f\}_{i=0}^{\infty}$, $\{\bar{y}_i^f\}_{i=0}^{\infty}$ is (f)-optimal if and only if it is (f)-good.*

In Sect. 5.4 we prove the next result.

Proposition 5.3. *Let $f \in \mathcal{A}_0$, $x_f \in X$ and $y_f \in Y$ satisfy (5.1) and let $\bar{x}_i^f = x_f$, $\bar{y}_i^f = y_f$ for all nonnegative integers i. Then the following properties are equivalent:*

(i) the pair $\{\bar{x}_i^f\}_{i=0}^{\infty}$, $\{\bar{y}_i^f\}_{i=0}^{\infty}$ is (f)-optimal;

(ii) for every integer $n \geq 1$ and every pair of finite sequences $\{x_i\}_{i=0}^{n} \subset X$, $\{y_i\}_{i=0}^{n} \subset Y$ the inequality

$$-4\|f\| + \sum_{i=0}^{n-1} f(x_f, x_f, y_i, y_{i+1})$$

$$\leq n f(x_f, x_f, y_f, y_f) \leq \sum_{i=0}^{n-1} f(x_i, x_{i+1}, y_f, y_f) + 4\|f\|$$

is valid;

(iii) there exists a positive number c such that for every integer $n \geq 1$ and every pair of finite sequences $\{x_i\}_{i=0}^{n} \subset X$, $\{y_i\}_{i=0}^{n} \subset Y$ the inequality

$$-c + \sum_{i=0}^{n-1} f(x_f, x_f, y_i, y_{i+1})$$

$$\leq n f(x_f, x_f, y_f, y_f) \leq \sum_{i=0}^{n-1} f(x_i, x_{i+1}, y_f, y_f) + c$$

is valid.

It should be mentioned that analogs of properties (i)–(iii) are used in the infinite horizon optimal control and they are usually posed, when one obtains a turnpike result where the turnpike is a singleton. See, for example, [84]. It was shown in [67] that these properties hold if X and Y are convex sets in finite-dimensional Euclidean spaces and f satisfies the convexity–concavity assumptions.

The next result is easily deduced from Proposition 5.3.

Proposition 5.4. *Let $f \in \mathcal{A}_0$, $x_f \in X$ and $y_f \in Y$ satisfy (5.1) and let property (iii) of Proposition 5.3 hold. Then the following assertions hold.*

1. For every sequence $\{x_i\}_{i=0}^{\infty} \subset X$ either the sequence

$$\{|\sum_{i=0}^{n-1}[f(x_i, x_{i+1}, y_f, y_f) - f(x_f, x_f, y_f, y_f)]|\}_{n=1}^{\infty}$$

is bounded or

$$\lim_{n\to\infty} \sum_{i=0}^{n-1}[f(x_i, x_{i+1}, y_f, y_f) - f(x_f, x_f, y_f, y_f)] = \infty.$$

2. For every sequence $\{y_i\}_{i=0}^{\infty} \subset Y$ either the sequence

$$\{|\sum_{i=0}^{n-1}[f(x_f, x_f, y_i, y_{i+1}) - f(x_f, x_f, y_f, y_f)]|\}_{n=1}^{\infty}$$

is bounded or

$$\lim_{n\to\infty} \sum_{i=0}^{n-1}[f(x_f, x_f, y_i, y_{i+1}) - f(x_f, x_f, y_f, y_f)] = -\infty.$$

Denote by $\mathcal{A}$ the set of all function $f \in \mathcal{A}_0$ for which there exist a pair of points $x_f \in X$ and $y_f \in Y$ satisfying (5.1) and a positive number c_f such that for every integer $n \geq 1$ and every pair of finite sequences $\{x_i\}_{i=0}^{n} \subset X$, $\{y_i\}_{i=0}^{n} \subset Y$ the following inequality holds:

$$-c_f + \sum_{i=0}^{n-1} f(x_f, x_f, y_i, y_{i+1})$$

$$\leq nf(x_f, x_f, y_f, y_f) \leq \sum_{i=0}^{n-1} f(x_i, x_{i+1}, y_f, y_f) + c_f. \tag{5.4}$$

Remark 5.5. In view of Proposition 5.3 we may assume without loss of generality that $c_f = 4\|f\|$. In the sequel we associate with each $f \in \mathcal{A}$ points $x_f \in X$ and $y_f \in Y$ satisfying (5.1) and such that (5.4) holds with $c_f = 4\|f\|$ for each natural number n and each pair of sequences $\{x_i\}_{i=0}^{n} \subset X$, $\{y_i\}_{i=0}^{n} \subset Y$.

In Sect. 5.6 we prove the following proposition.

Proposition 5.6. *$\mathcal{A}$ is a closed subset of the Banach space $(C(X \times X \times Y \times Y), \|\cdot\|)$.*

The following proposition is proved in Sect. 5.7.

Proposition 5.7. *Let $f \in \mathcal{A}$, $z_1, z_2 \in X$, $\xi_1, \xi_2 \in Y$, n be a natural number and let*

$$x_0 = z_1,\ x_n = z_2,\ x_i = x_f \text{ for all integers } i \text{ satisfying } 0 < i < n, \tag{5.5}$$

$$y_0 = \xi_1,\ y_n = \xi_2,\ y_i = y_f \text{ for all integers } i \text{ satisfying } 0 < i < n. \tag{5.6}$$

Then the pair of sequences $\{x_i\}_{i=0}^n, \{y_i\}_{i=0}^n$ is $(f, 12\|f\|)$-good.

In this chapter we establish a turnpike property of (f)-good pairs of sequences which means that they spend most of the time in a small neighborhood of the pair (x_f, y_f). It is known in the optimal control theory that turnpike properties of approximately optimal solutions are deduced from an asymptotic turnpike property of solutions of corresponding infinite horizon optimal control problems [70, 84].

Let $f \in \mathcal{A}$. We say that the function f has the asymptotic turnpike property (or, briefly, (ATP)) if for every pair of points $\tilde{x} \in X$, $\tilde{y} \in Y$ which satisfies

$$f(\tilde{x}, \tilde{x}, y, y) \le f(\tilde{x}, \tilde{x}, \tilde{y}, \tilde{y}) \le f(x, x, \tilde{y}, \tilde{y}) \text{ for all } x \in X \text{ and all } y \in Y$$

we have $\tilde{x} = x_f$ and $\tilde{y} = y_f$ and if for every pair of sequences $\{x_i\}_{i=0}^\infty \subset X$, $\{y_i\}_{i=0}^\infty \subset Y$ which satisfies

$$\sup\{\sum_{i=0}^{n-1} f(x_i, x_{i+1}, y_f, y_f) - nf(x_f, x_f, y_f, y_f) :\ n \text{ is a natural number}\} < \infty$$

and

$$\inf\{\sum_{i=0}^{n-1} f(x_f, x_f, y_i, y_{i+1}) - nf(x_f, x_f, y_f, y_f) :\ n \text{ is a natural number}\} > -\infty$$

we have

$$\lim_{i \to \infty} \rho_X(x_i, x_f) = 0,\ \lim_{i \to \infty} \rho_Y(y_i, y_f) = 0.$$

Let $f \in \mathcal{A}$. Denote by $S(f)$ the set of all pairs of points $(x, y) \in X \times Y$ such that

$$f(x, x, \xi, \xi) \le f(x, x, y, y) \le f(z, z, y, y) \text{ for all } z \in X \text{ and all } \xi \in Y. \tag{5.7}$$

It is clear that $S(f) \neq \emptyset$ and for all $(x_1, y_1),\ (x_2, y_2) \in S(f)$, we have

$$f(x_1, x_1, y_1, y_1) = f(x_2, x_2, y_2, y_2). \tag{5.8}$$

We consider the topological subspace $\mathcal{A} \subset C(X \times X \times Y \times Y)$ with the relative topology induced by the metric $d(f, g) = \|f - g\|$, $f, g \in \mathcal{A}$.

The next two theorems are the main results of this chapter.

Theorem 5.8. *Let f possess (ATP) and M, ϵ be positive numbers. Then there exist natural numbers l and Q and a positive number δ such that for every function $g \in \mathcal{A}$ which satisfies $\|f - g\| \le \delta$, every natural number $T > Ql$ and every (g, M)-good pair of finite sequences $\{x_i\}_{i=0}^T \subset X$, $\{y_i\}_{i=0}^T \subset Y$ there exist a natural number $q \le Q$ and finite sequences of integers $\{a_i\}_{i=1}^q$, $\{b_i\}_{i=1}^q \subset [0, T]$ such that*

$$0 \le b_i - a_i \le l, \ i = 1, \dots, q,$$

$$\rho_X(x_i, x_f) \le \epsilon, \ \rho_Y(y_i, y_f) \le \epsilon$$

for all integers $i \in [0, T] \setminus \cup_{j=1}^q [a_j, b_j]$.

It should be mentioned that Theorem 5.8 establishes the stability of the turnpike phenomenon under small perturbations of the objective function f.

Theorem 5.9. *There exists a set $\mathcal{F} \subset \mathcal{A}$ which is a countable intersection of open everywhere dense subsets of $\mathcal{A}$ such that each $f \in \mathcal{F}$ possesses (ATP).*

Theorem 5.9 shows that a generic (typical) function $f \in \mathcal{A}$ possesses (ATP). Results of this kind for classes of single-player control systems have been established in [70, 83].

Theorem 5.8 is proved in Sect. 5.9 while the proof of Theorem 5.9 is given in Sect. 5.10. Section 5.8 contains auxiliary results.

5.2 Proof of Proposition 5.1

Assume that $\{f_n\}_{n=1}^\infty \subset \mathcal{A}_0$, $f \in C(X \times X \times Y \times Y)$ and

$$\lim_{n \to \infty} \|f_n - f\| = 0. \tag{5.9}$$

For every natural number n there exists a pair of points $x_n \in X$ and $y_n \in Y$ such that for every point $x \in X$ and every point $y \in Y$, we have

$$f_n(x_n, x_n, y, y) \le f_n(x_n, x_n, y_n, y_n) \le f_n(x, x, y_n, y_n). \tag{5.10}$$

Extracting a subsequence and re-indexing, if necessary, we may assume without loss of generality that there exist

$$x_* = \lim_{n \to \infty} x_n, \ y_* = \lim_{n \to \infty} y_n. \tag{5.11}$$

In view of (5.9), (5.10), and (5.11), for every point $x \in X$ and every point $y \in Y$, we have

$$f(x_*, x_*, y, y) = \lim_{n \to \infty} f(x_n, x_n, y, y) = \lim_{n \to \infty} f_n(x_n, x_n, y, y)$$

$$\le \lim_{n \to \infty} f_n(x_n, x_n, y_n, y_n) = \lim_{n \to \infty} f(x_n, x_n, y_n, y_n) = f(x_*, x_*, y_*, y_*),$$

$$f(x_*, x_*, y_*, y_*) = \lim_{n \to \infty} f_n(x_n, x_n, y_n, y_n) \le \lim_{n \to \infty} f_n(x, x, y_n, y_n)$$

$$= \lim_{n \to \infty} f(x, x, y_n, y_n) = f(x, x, y_*, y_*)$$

and

$$f(x_*, x_*, y, y) \le f(x_*, x_*, y_*, y_*) \le f(x, x, y_*, y_*).$$

Therefore $f \in \mathcal{A}_0$. This completes the proof of Proposition 5.1. □

5.3 Proof of Proposition 5.2

It is clear that if the pair of sequences $\{\bar{x}_i^f\}_{i=0}^\infty$, $\{\bar{y}_i^f\}_{i=0}^\infty$ is (f)-optimal it is also (f)-good.

Assume that the pair of sequences $\{\bar{x}_i^f\}_{i=0}^\infty$, $\{\bar{y}_i^f\}_{i=0}^\infty$ is (f)-good. We claim that it is (f)-optimal. There exists a positive number M such that the pair of sequences $\{\bar{x}_i^f\}_{i=0}^\infty$, $\{\bar{y}_i^f\}_{i=0}^\infty$ is (f, M)-good.

Let $n \ge 1$ be an integer. In order to complete the proof of the proposition it is sufficient to show that the pair of sequences $\{\bar{x}_i^f\}_{i=0}^n$, $\{\bar{y}_i^f\}_{i=0}^n$ is (f)-optimal.

Assume that

$$\{z_i\}_{i=0}^n \subset X,\ \{\xi_i\}_{i=0}^n \subset Y,\ z_0 = z_n = x_f,\ \xi_0 = \xi_n = y_f. \tag{5.12}$$

There exists a pair of sequences $\{z_i\}_{i=n+1}^\infty \subset X$, $\{\xi_i\}_{i=n+1}^\infty \subset Y$ such that

$$z_{i+n} = z_i,\ \xi_{i+n} = \xi_i \text{ for all nonnegative integers } i. \tag{5.13}$$

Since the pair of sequences $\{\bar{x}_i^f\}_{i=0}^\infty$, $\{\bar{y}_i^f\}_{i=0}^\infty$ is (f, M)-good relations (5.2), (5.3), (5.12) and (5.13) imply that for every integer $k \ge 1$, we have

$$M \ge \sum_{i=0}^{kn-1} [f(x_f, x_f, y_f, y_f) - f(z_i, z_i, y_f, y_f)]$$

$$= k \sum_{i=0}^{n-1} [f(x_f, x_f, y_f, y_f) - f(z_i, z_i, y_f, y_f)],$$

$$\sum_{i=0}^{n-1} f(z_i, z_{i+1}, y_f, y_f) \ge n f(x_f, x_f, y_f, y_f)$$

and

$$M \ge \sum_{i=0}^{kn-1} [f(x_f, x_f, \xi_i, \xi_{i+1}) - f(x_f, x_f, y_f, y_f)]$$

$$= k \sum_{i=0}^{n-1} [f(x_f, x_f, \xi_i, \xi_{i+1}) - f(x_f, x_f, y_f, y_f)],$$

$$\sum_{i=0}^{n-1} f(x_f, x_f, \xi_i, \xi_{i+1}) \le n f(x_f, x_f, y_f, y_f).$$

Thus the pair of sequences $\{\bar{x}_i^f\}_{i=0}^n$, $\{\bar{y}_i^f\}_{i=0}^n$ is (f)-optimal. This completes the proof of Proposition 5.2. □

5.4 Proof of Proposition 5.3

We claim that property (i) implies property (ii). Assume that property (i) holds, $n \geq 1$ is an integer and $\{x_i\}_{i=0}^n \subset X$, $\{y_i\}_{i=0}^n \subset Y$. Define

$$x'_0 = x_f,\ x'_i = x_{i-1},\ i = 1, \dots, n+1,\ x'_{n+2} = x_f, \tag{5.14}$$

$$y'_0 = y_f,\ y'_i = y_{i-1},\ i = 1, \dots, n+1,\ y'_{n+2} = y_f.$$

Property (i) and (5.14) imply that

$$(n+2)f(x_f, x_f, y_f, y_f) = \sum_{i=0}^{n+1} f(\bar{x}_i^f, \bar{x}_{i+1}^f, \bar{y}_i^f, \bar{y}_{i+1}^f) \leq \sum_{i=0}^{n+1} f(x'_i, x'_{i+1}, y_f, y_f)$$

$$\leq \sum_{i=0}^{n+1} f(x_i, x_{i+1}, y_f, y_f) + 2\|f\|,$$

$$\sum_{i=0}^{n-1} f(x_i, x_{i+1}, y_f, y_f) \geq nf(x_f, x_f, y_f, y_f) - 4\|f\|,$$

$$(n+2)f(x_f, x_f, y_f, y_f) = \sum_{i=0}^{n+1} f(\bar{x}_i^f, \bar{x}_{i+1}^f, \bar{y}_i^f, \bar{y}_{i+1}^f) \geq \sum_{i=0}^{n+1} f(x_f, x_f, y'_i, y'_{i+1})$$

$$\geq \sum_{i=0}^{n-1} f(x_f, x_f, y_i, y_{i+1}) - 2\|f\|,$$

$$\sum_{i=0}^{n-1} f(x_f, x_f, y_i, y_{i+1}) \leq nf(x_f, x_f, y_f, y_f) + 4\|f\|.$$

Hence property (ii) holds.

It is clear that property (ii) implies property (iii).

Assume that property (iii) holds. In order to complete the proof of the proposition it is sufficient to show that property (i) holds. Let $n \geq 1$ be an integer. We show that the pair of finite sequences $\{\bar{x}_i^f\}_{i=0}^n$, $\{\bar{y}_i^f\}_{i=0}^n$ is (f)-optimal.

Assume that

$$\{x_i\}_{i=0}^n \subset X,\ \{y_i\}_{i=0}^n \subset Y,\ x_0 = x_n = x_f,\ y_0 = y_n = y_f. \tag{5.15}$$

There exists a pair of sequences $\{x_i\}_{i=n+1}^\infty \subset X$, $\{y_i\}_{i=n+1}^\infty \subset Y$ such that for every nonnegative integer i, we have

$$x_{i+n} = x_i,\ y_{i+n} = y_i. \tag{5.16}$$

Property (iii), (5.15), and (5.16) imply that for every integer $k \geq 1$, we have

$$
\begin{aligned}
&-c + k\sum_{i=0}^{n-1} f(x_f, x_f, y_i, y_{i+1}) \\
&= -c + \sum_{i=0}^{nk-1} f(x_f, x_f, y_i, y_{i+1}) \leq nkf(x_f, x_f, y_f, y_f) \\
&\leq \sum_{i=0}^{nk-1} f(x_i, x_{i+1}, y_f, y_f) + c \\
&= k\sum_{i=0}^{n-1} f(x_i, x_{i+1}, y_f, y_f) + c,
\end{aligned}
$$

$$
\begin{aligned}
-c/k + \sum_{i=0}^{n-1} f(x_f, x_f, y_i, y_{i+1}) &\leq nf(x_f, x_f, y_f, y_f) \\
&\leq \sum_{i=0}^{n-1} f(x_i, x_{i+1}, y_f, y_f) + c/k.
\end{aligned}
$$

Since k is an arbitrary natural number we obtain that

$$
\sum_{i=0}^{n-1} f(x_f, x_f, y_i, y_{i+1}) \leq nf(x_f, x_f, y_f, y_f) \leq \sum_{i=0}^{n-1} f(x_i, x_{i+1}, y_f, y_f)
$$

and that $\{\bar{x}_i^f\}_{i=0}^n$, $\{\bar{y}_i^f\}_{i=0}^n$ is an (f)-optimal pair of sequences. This completes the proof of Proposition 5.3. □

5.5 Good Pairs of Sequences

Proposition 5.10. *Let $f \in \mathcal{A}_0$, $x_f \in X$ and $y_f \in Y$ satisfy*

$$
f(x_f, x_f, y, y) \leq f(x_f, x_f, y_f, y_f) \leq f(x, x, y_f, y_f)
$$

$$
\text{for all } x \in X \text{ and all } y \in Y \tag{5.17}
$$

and let a positive number c be such that for every integer $k \geq 1$ and every pair of finite sequences $\{x_i\}_{i=0}^k \subset X$, $\{y_i\}_{i=0}^k \subset Y$ the following inequality holds:

$$
\begin{aligned}
-c + \sum_{i=0}^{k-1} f(x_f, x_f, y_i, y_{i+1}) &\leq kf(x_f, x_f, y_f, y_f) \\
&\leq \sum_{i=0}^{k-1} f(x_i, x_{i+1}, y_f, y_f) + c. \qquad (5.18)
\end{aligned}
$$

Let M be a positive number, $n \geq 1$ be an integer and let $\{u_i\}_{i=0}^n \subset X$, $\{v_i\}_{i=0}^n \subset Y$ be an (f, M)-good pair of sequences. Then

$$|\sum_{i=0}^{n-1} f(u_i, u_{i+1}, y_f, y_f) - nf(x_f, x_f, y_f, y_f)| \leq c + 2M + 8\|f\|,$$

$$|\sum_{i=0}^{n-1} f(x_f, x_f, v_i, v_{i+1}) - nf(x_f, x_f, y_f, y_f)| \leq c + 2M + 8\|f\|.$$

Proof. In view of (5.18),

$$-c + \sum_{i=0}^{n-1} f(x_f, x_f, v_i, v_{i+1}) \leq nf(x_f, x_f, y_f, y_f)$$
$$\leq \sum_{i=0}^{n-1} f(u_i, u_{i+1}, y_f, y_f) + c. \quad (5.19)$$

Define

$$u'_0 = u_0,\ u'_n = u_n,\ u'_i = x_f \text{ for all integers } i \text{ satisfying } 0 < i < n,$$
$$v'_0 = v_0,\ v'_n = v_n,\ v'_i = y_f \text{ for all integers } i \text{ satisfying } 0 < i < n. \quad (5.20)$$

Since $\{u_i\}_{i=0}^n, \{v_i\}_{i=0}^n$ is an (f, M)-good pair of sequences relations (5.19) and (5.20) imply that

$$nf(x_f, x_f, y_f, y_f) - c - M - 4\|f\| \leq \sum_{i=0}^{n-1} f(u_i, u_{i+1}, y_f, y_f) - M - 4\|f\|$$
$$\leq -M + \sum_{i=0}^{n-1} f(u_i, u_{i+1}, v'_i, v'_{i+1}) \leq \sum_{i=0}^{n-1} f(u_i, u_{i+1}, v_i, v_{i+1})$$
$$\leq \sum_{i=0}^{n-1} f(u'_i, u'_{i+1}, v_i, v_{i+1}) + M$$
$$\leq \sum_{i=0}^{n-1} f(x_f, x_f, v_i, v_{i+1}) + M + 4\|f\| \leq nf(x_f, x_f, y_f, y_f) + c + M + 4\|f\|.$$

This implies that

$$|\sum_{i=0}^{n-1} f(u_i, u_{i+1}, y_f, y_f) - nf(x_f, x_f, y_f, y_f)| \leq c + 2M + 8\|f\|,$$

$$|\sum_{i=0}^{n-1} f(x_f, x_f, v_i, v_{i+1}) - nf(x_f, x_f, y_f, y_f)| \leq c + 2M + 8\|f\|.$$

This completes the proof of Proposition 5.10. □

5.6 Proof of Proposition 5.6

Assume that $\{f_n\}_{n=1}^{\infty} \subset \mathcal{A}$, $f \in C(X \times X \times Y \times Y)$ and that

$$\lim_{n\to\infty} \|f_n - f\| = 0. \tag{5.21}$$

Proposition 5.1 implies that $f \in \mathcal{A}_0$. It follows from definition of $\mathcal{A}$ (see (5.4)), Proposition 5.3, and Remark 5.5 that for every integer number $n \geq 1$ there exists a pair of points $x_{f_n} \in X$ and $y_{f_n} \in Y$ such that

$$f_n(x_{f_n}, x_{f_n}, y, y) \leq f_n(x_{f_n}, x_{f_n}, y_{f_n}, y_{f_n})$$

$$\leq f_n(x, x, y_{f_n}, y_{f_n}) \text{ for all } x \in X \text{ and all } y \in Y \tag{5.22}$$

and that for every integer $k \geq 1$ and every pair of finite sequences $\{x_i\}_{i=0}^{k} \subset X$, $\{y_i\}_{i=0}^{k} \subset Y$, we have

$$-4\|f_n\| + \sum_{i=0}^{k-1} f_n(x_{f_n}, x_{f_n}, y_i, y_{i+1}) \leq k f_n(x_{f_n}, x_{f_n}, y_{f_n}, y_{f_n})$$

$$\leq \sum_{i=0}^{k-1} f_n(x_i, x_{i+1}, y_{f_n}, y_{f_n}) + 4\|f_n\|. \tag{5.23}$$

Extracting a subsequence and re-indexing, if necessary, we may assume without loss of generality that there exist

$$x^* = \lim_{n\to\infty} x_{f_n}, \ y^* = \lim_{n\to\infty} y_{f_n}. \tag{5.24}$$

In view of (5.21), (5.22), and (5.24), for every pair of points $x \in X$ and $y \in Y$,

$$f(x^*, x^*, y, y) = \lim_{n\to\infty} f(x_{f_n}, x_{f_n}, y, y) = \lim_{n\to\infty} f_n(x_{f_n}, x_{f_n}, y, y)$$

$$\leq \lim_{n\to\infty} f_n(x_{f_n}, x_{f_n}, y_{f_n}, y_{f_n}) = \lim_{n\to\infty} f(x_{f_n}, x_{f_n}, y_{f_n}, y_{f_n}) = f(x^*, x^*, y^*, y^*)$$

$$= \lim_{n\to\infty} f_n(x_{f_n}, x_{f_n}, y_{f_n}, y_{f_n})$$

$$\leq \lim_{n\to\infty} f_n(x, x, y_{f_n}, y_{f_n}) = \lim_{n\to\infty} f(x, x, y_{f_n}, y_{f_n}) = f(x, x, y^*, y^*)$$

and

$$f(x^*, x^*, y, y) \leq f(x^*, x^*, y^*, y^*) \leq f(x, x, y^*, y^*)$$

$$\text{for all } x \in X \text{ and all } y \in Y.$$

Let $k \geq 1$ be an integer and let $\{x_i\}_{i=0}^{k} \subset X$, $\{y_i\}_{i=0}^{k} \subset Y$. It follows from (6.21), (5.23), and (5.24) that

$$-4\|f\| + \sum_{i=0}^{k-1} f(x^*, x^*, y_i, y_{i+1})$$

$$= \lim_{n\to\infty} [-4\|f_n\| + \sum_{i=0}^{k-1} f(x_{f_n}, x_{f_n}, y_i, y_{i+1})]$$

$$= \lim_{n\to\infty}[-4\|f_n\| + \sum_{i=0}^{k-1} f_n(x_{f_n}, x_{f_n}, y_i, y_{i+1})]$$

$$\le \lim_{n\to\infty} kf_n(x_{f_n}, x_{f_n}, y_{f_n}, y_{f_n})$$

$$= \lim_{n\to\infty} kf(x_{f_n}, x_{f_n}, y_{f_n}, y_{f_n}) = kf(x^*, x^*, y^*, y^*)$$

$$= \lim_{n\to\infty} kf_n(x_{f_n}, x_{f_n}, y_{f_n}, y_{f_n})$$

$$\le \lim_{n\to\infty}[4\|f_n\| + \sum_{i=0}^{k-1} f_n(x_i, x_{i+1}, y_{f_n}, y_{f_n})]$$

$$= 4\|f\| + \lim_{n\to\infty}\sum_{i=0}^{k-1} f(x_i, x_{i+1}, y_{f_n}, y_{f_n}) = 4\|f\| + \sum_{i=0}^{k-1} f(x_i, x_{i+1}, y^*, y^*).$$

Therefore $f \in \mathcal{A}$ with $x_f = x^*$ and $y_f = y^*$. This completes the proof of Proposition 5.6. □

5.7 Proof of Proposition 5.7

In view of (5.4), (5.5), and (5.6), for every pair of finite sequences $\{u_i\}_{i=0}^n \subset X$, $\{v_i\}_{i=0}^n \subset Y$, the following inequalities hold:

$$\sum_{i=0}^{n-1} f(u_i, u_{i+1}, y_i, y_{i+1}) - \sum_{i=0}^{n-1} f(x_i, x_{i+1}, y_i, y_{i+1})$$

$$\ge \sum_{i=0}^{n-1} f(u_i, u_{i+1}, y_f, y_f) - 4\|f\| - nf(x_f, x_f, y_f, y_f) - 4\|f\| \ge -12\|f\|,$$

$$\sum_{i=0}^{n-1} f(x_i, x_{i+1}, v_i, v_{i+1}) - \sum_{i=0}^{n-1} f(x_i, x_{i+1}, y_i, y_{i+1})$$

$$\le \sum_{i=0}^{n-1} f(x_f, x_f, v_i, v_{i+1}) + 4\|f\| - nf(x_f, x_f, y_f, y_f) + 4\|f\| \le 12\|f\|.$$

This completes the proof of Proposition 5.7. □

5.8 Auxiliary Results

Assume that $f \in \mathcal{A}$ has (ATP). We suppose that the sum over empty set is zero.

Lemma 5.11. *Let ϵ be a positive number. Then there exists a positive number δ such that for every function $g \in \mathcal{A}_0$ satisfying $\|g - f\| \le \delta$, every point $x \in X$ and every point $y \in Y$ satisfying*

$$g(x, x, \xi, \xi) \le g(x, x, y, y) \le g(z, z, y, y) \text{ for all } z \in X \text{ and all } \xi \in Y$$

the following inequalities hold:

$$\rho_X(x, x_f) \le \epsilon, \ \rho_Y(y, y_f) \le \epsilon.$$

Proof. Assume the contrary. Then there exist sequences

$$\{g_n\}_{n=1}^{\infty} \subset \mathcal{A}_0, \ \{x_n\}_{n=1}^{\infty} \subset X, \ \{y_n\}_{n=1}^{\infty} \subset Y$$

such that

$$\lim_{n \to \infty} \|f - g_n\| = 0 \tag{5.25}$$

and for all natural numbers n, we have

$$g_n(x_n, x_n, \xi, \xi) \le g_n(x_n, x_n, y_n, y_n) \le g_n(z, z, y_n, y_n)$$

$$\text{for all } z \in X \text{ and all } \xi \in Y, \tag{5.26}$$

$$\rho_X(x_n, x_f) + \rho_Y(y_n, y_f) \ge \epsilon. \tag{5.27}$$

Extracting a subsequence and re-indexing, if necessary, we may assume without loss of generality that there exist $\lim_{n\to\infty} x_n$ and $\lim_{n\to\infty} y_n$. Arguing as in the proof of Proposition 5.6 and using (5.25) and (5.26) we conclude that for all points $z \in X$ and all points $\xi \in Y$, the following inequality holds:

$$f(\lim_{n\to\infty} x_n, \lim_{n\to\infty} x_n, \xi, \xi) \le f(\lim_{n\to\infty} x_n, \lim_{n\to\infty} x_n, \lim_{n\to\infty} y_n, \lim_{n\to\infty} y_n)$$

$$\le f(z, z, \lim_{n\to\infty} y_n, \lim_{n\to\infty} y_n).$$

Since the function f has (ATP) we conclude that $\lim_{n\to\infty} x_n = x_f$ and $\lim_{n\to\infty} y_n = y_f$. This contradicts (5.27). The contradiction we have reached proves Lemma 5.11. □

Lemma 5.12. *Let M, ϵ be positive numbers and let $\tau_0 \ge 1$ be an integer. Then there exists a natural number $\tau > \tau_0$ such that for every natural number $T \ge \tau$, every finite sequence $\{x_i\}_{i=0}^{T} \subset X$ satisfying*

$$\sum_{i=0}^{T-1} f(x_i, x_{i+1}, y_f, y_f) \le T f(x_f, x_f, y_f, y_f) + M$$

and every integer $s \in [0, T - \tau]$ there exists an integer s_0 such that

$$[s_0, s_0 + \tau_0] \subset [s, s + \tau]$$

and

$$\rho_X(x_i, x_f) \le \epsilon, \ i = s_0, \dots, s_0 + \tau_0.$$

Proof. Let us assume the contrary. Then for every natural number $k > \tau_0$ there exist a natural number $T_k \geq k$, a finite sequence $\{x_i^{(k)}\}_{i=0}^{T_k} \subset X$ which satisfies

$$\sum_{i=0}^{T_k-1} f(x_i^{(k)}, x_{i+1}^{(k)}, y_f, y_f) \leq T_k f(x_f, x_f, y_f, y_f) + M, \tag{5.28}$$

and an integer $s_k \in [0, T_k - k]$ such that the following property holds:

(P1) for every integer p which satisfies

$$[p, p + \tau_0] \subset [s_k, s_k + k]$$

we have

$$\max\{\rho_X(x_i^{(k)}, x_f) : i = p, \dots, p + \tau_0\} > \epsilon. \tag{5.29}$$

For every natural number $k > \tau_0$ define

$$u_i^{(k)} = x_{i+s_k}^{(k)}, \ i = 0, \dots, k. \tag{5.30}$$

In view of (5.30) and property (P1), for every natural number $k > \tau_0$ the following property holds:

(P2) for every integer p which satisfies

$$[p, p + \tau_0] \subset [0, k]$$

the inequality

$$\max\{\rho_X(u_i^{(k)}, x_f) : i = p, \dots, p + \tau_0\} > \epsilon \tag{5.31}$$

is valid.

Let $k > \tau_0$ be a natural number. It follows from (5.28), (5.30), and the definition of $\mathcal{A}$ (see (5.4)) that

$$\sum_{i=0}^{k-1} f(u_i^{(k)}, u_{i+1}^{(k)}, y_f, y_f)$$

$$= \sum_{i=s_k}^{s_k+k-1} f(x_i^{(k)}, x_{i+1}^{(k)}, y_f, y_f)$$

$$= \sum_{i=0}^{T_k-1} f(x_i^{(k)}, x_{i+1}^{(k)}, y_f, y_f)$$

$$-\sum\{f(x_i^{(k)}, x_{i+1}^{(k)}, y_f, y_f) : \ i \text{ is an integer such that } 0 \leq i < s_k\}$$

$$-\sum\{f(x_i^{(k)}, x_{i+1}^{(k)}, y_f, y_f) : \ i \text{ is an integer such that } s_k + k \leq i < T\}$$

$$\leq T_k f(x_f, x_f, y_f, y_f) + M - s_k f(x_f, x_f, y_f, y_f) + c_f$$

$$-(T_k - s_k - k)f(x_f, x_f, y_f, y_f) + c_f$$
$$= kf(x_f, x_f, y_f, y_f) + M + 2c_f. \tag{5.32}$$

Let $q > \tau_0$ be a natural number. In view of (5.32) and the definition of $\mathcal{A}$ (see (5.4)), for every natural number $k > q$, we have

$$\sum_{i=0}^{q-1} f(u_i^{(k)}, u_{i+1}^{(k)}, y_f, y_f) = \sum_{i=0}^{k-1} f(u_i^{(k)}, u_{i+1}^{(k)}, y_f, y_f) - \sum_{i=q}^{k-1} f(u_i^{(k)}, u_{i+1}^{(k)}, y_f, y_f)$$

$$\leq kf(x_f, x_f, y_f, y_f) + M + 2c_f - (k-q)f(x_f, x_f, y_f, y_f)$$
$$+ c_f = qf(x_f, x_f, y_f, y_f) + M + 3c_f. \tag{5.33}$$

It is clear that there exists a strictly increasing sequence of natural numbers $\{k_j\}_{j=1}^{\infty}$ with $k_1 > \tau_0$ such that for every nonnegative integer i, the sequence $\{u_i^{(k_j)}\}_{j=1}^{\infty}$ converges. For every nonnegative integer i define

$$u_i = \lim_{j \to \infty} u_i^{(k_j)}. \tag{5.34}$$

It follows from (5.33) and (5.34) that for every natural number $q > \tau_0$, we have

$$\sum_{i=0}^{q-1} f(u_i, u_{i+1}, y_f, y_f) \leq qf(x_f, x_f, y_f, y_f) + M + 3c_f$$

and together with (ATP) the relation above implies that

$$\lim_{i \to \infty} u_i = x_f.$$

Evidently, there exists a natural number p such that

$$\rho_X(u_i, x_f) \leq \epsilon/4 \text{ for all integers } i \geq p. \tag{5.35}$$

In view of (5.34), there exists a natural number j such that

$$k_j > p + \tau_0 + 4,$$

$$\rho_X(u_i, u_i^{(k_j)}) \leq \epsilon/4, \ i = 0, \dots, p + \tau_0 + 4. \tag{5.36}$$

It follows from (5.35) and (5.36) that for all integers $i = p, \dots, p + \tau_0$, we have

$$\rho_X(u_i^{(k_j)}, x_f) \leq \rho_X(u_i^{(k_j)}, u_i) + \rho_X(u_i, x_f) \leq \epsilon/2.$$

This contradicts property (P2). The contradiction we have reached proves Lemma 5.12. □

Analogously to Lemma 5.12 we can prove the following lemma.

Lemma 5.13. *Let M, ϵ be positive numbers and let $\tau_0 \geq 1$ be an integer. Then there exists a natural number $\tau > \tau_0$ such that for every natural number $T \geq \tau$, every sequence $\{y_i\}_{i=0}^T \subset Y$ which satisfies*

$$\sum_{i=0}^{T-1} f(x_f, x_f, y_i, y_{i+1}) \geq T f(x_f, x_f, y_f, y_f) - M$$

and every integer $s \in [0, T - \tau]$ there exists an integer s_0 such that

$$[s_0, s_0 + \tau_0] \subset [s, s + \tau]$$

and

$$\rho_Y(y_i, y_f) \leq \epsilon,\ i = s_0, \dots, s_0 + \tau_0.$$

Lemma 5.14. *Let M, ϵ be positive numbers. Then there exists a natural number $\tau_1 > 1$ such that for every natural number $T \geq \tau_1$, every pair of finite sequences $\{x_i\}_{i=0}^T \subset X$ and $\{y_i\}_{i=0}^T \subset Y$ which satisfies*

$$\sum_{i=0}^{T-1} f(x_i, x_{i+1}, y_f, y_f) \leq T f(x_f, x_f, y_f, y_f) + M, \tag{5.37}$$

$$\sum_{i=0}^{T-1} f(x_f, x_f, y_i, y_{i+1}) \geq T f(x_f, x_f, y_f, y_f) - M \tag{5.38}$$

and every integer $s \in [0, T - \tau_1]$ there exists an integer s_0 for which

$$\{s_0, s_0 + 1\} \subset [s, s + \tau_1],$$

$$\rho_X(x_i, x_f) \leq \epsilon \text{ and } \rho_Y(y_i, y_f) \leq \epsilon,\ i = s_0, s_0 + 1.$$

Proof. Lemma 5.13 (with $\tau_0 = 1$) implies that there exists a natural number $\tau_0 > 1$ such that the following property holds:

(P3) for every natural number $T \geq \tau_0$, every finite sequence $\{y_i\}_{i=0}^T \subset Y$ which satisfies

$$\sum_{i=0}^{T-1} f(x_f, x_f, y_i, y_{i+1}) \geq T f(x_f, x_f, y_f, y_f) - M - 2c_f$$

and every integer $s \in [0, T - \tau_0]$ there exists an integer s_0 for which

$$\{s_0, s_0 + 1\} \subset [s, s + \tau_0]$$

and

$$\rho_Y(y_i, y_f) \leq \epsilon,\ i = s_0, s_0 + 1.$$

Lemma 5.12 implies that there exists a natural number $\tau_1 > \tau_0$ such that the following property holds:

(P4) for every natural number $T \geq \tau_1$, every finite sequence $\{x_i\}_{i=0}^{T} \subset X$ which satisfies

$$\sum_{i=0}^{T-1} f(x_i, x_{i+1}, y_f, y_f) \leq Tf(x_f, x_f, y_f, y_f) + M$$

and every integer $s \in [0, T - \tau_1]$ there exists an integer s_0 for which

$$[s_0, s_0 + \tau_0] \subset [s, s + \tau_1]$$

and

$$\rho_X(x_i, x_f) \leq \epsilon, \ i = s_0, \dots, s_0 + \tau_0.$$

Assume that $T \geq \tau_1$ is a natural number, $\{x_i\}_{i=0}^{T} \subset X$, $\{y_i\}_{i=0}^{T} \subset Y$, (5.37) and (5.38) are valid and that an integer $s \in [0, T - \tau_1]$. Property (P4) and (5.37) imply that there exists an integer q_0 such that

$$[q_0, q_0 + \tau_0] \subset [s, s + \tau_1], \tag{5.39}$$

$$\rho_X(x_i, x_f) \leq \epsilon, \ i = q_0, \dots, q_0 + \tau_0.$$

In view of (5.38) and the definition of $\mathcal{A}$ (see (5.4)),

$$\sum_{i=q_0}^{q_0+\tau_0-1} f(x_f, x_f, y_i, y_{i+1}) = \sum_{i=0}^{T-1} f(x_f, x_f, y_i, y_{i+1})$$

$$-\sum \{f(x_f, x_f, y_i, y_{i+1}) : \ i \text{ is an integer such that } 0 \leq i < q_0\}$$

$$-\sum \{f(x_f, x_f, y_i, y_{i+1}) : \ i \text{ is an integer such that } q_0 + \tau_0 \leq i < T\}$$

$$\geq Tf(x_f, x_f, y_f, y_f) - M - q_0 f(x_f, x_f, y_f, y_f) - c_f$$

$$-(T - q_0 - \tau_0) f(x_f, x_f, y_f, y_f) - c_f = \tau_0 f(x_f, x_f, y_f, y_f) - M - 2c_f. \tag{5.40}$$

It follows from (5.40) and property (P3) (with $T = \tau_0$) that there exists an integer s_0 for which

$$\{s_0, s_0 + 1\} \subset [q_0, q_0 + \tau_0]$$

and

$$\rho_Y(y_i, y_f) \leq \epsilon, \ i = s_0, s_0 + 1.$$

Lemma 5.14 is proved. □

The next auxiliary result easily follows from Lemma 5.14 and Proposition 5.10.

Lemma 5.15. *Let M, ϵ be positive numbers. Then there exists a natural number $\tau_1 > 1$ such that for every natural number $T \geq \tau_1$, every (f, M)-good pair of sequences $\{x_i\}_{i=0}^T \subset X$ and $\{y_i\}_{i=0}^T \subset Y$ and every integer $s \in [0, T - \tau_1]$ there exists an integer s_0 such that*

$$\{s_0, s_0 + 1\} \subset [s, s + \tau_1],$$

$$\rho_X(x_i, x_f) \leq \epsilon \text{ and } \rho_Y(y_i, y_f) \leq \epsilon,\ i = s_0, s_0 + 1.$$

Lemma 5.16. *Let M, ϵ be positive numbers. Then there exist a natural number $\tau > 1$ and a positive number δ such that for every function $g \in C(X \times X \times Y \times Y)$ which satisfies $\|g - f\| \leq \delta$ and every (g, M)-good pair of sequences $\{x_i\}_{i=0}^\tau \subset X$ and $\{y_i\}_{i=0}^\tau \subset Y$ there exists an integer s_0 such that*

$$\{s_0, s_0 + 1\} \subset [0, \tau], \tag{5.41}$$

$$\rho_X(x_i, x_f) \leq \epsilon \text{ and } \rho_Y(y_i, y_f) \leq \epsilon,\ i = s_0, s_0 + 1.$$

Proof. Lemma 5.15 implies that there exists a natural number $\tau > 1$ such that the following property holds:

(P5) for every $(f, M+1)$-good pair of sequences $\{x_i\}_{i=0}^\tau \subset X$ and $\{y_i\}_{i=0}^\tau \subset Y$ there exists an integer s_0 such that (5.41 holds.

Put

$$\delta = (8\tau)^{-1}. \tag{5.42}$$

Let $g \in C(X \times X \times Y \times Y)$ satisfy

$$\|g - f\| \leq \delta \tag{5.43}$$

and let $\{x_i\}_{i=0}^\tau \subset X$ and $\{y_i\}_{i=0}^\tau \subset Y$ be a (g, M)-good pair of sequences. In view of (5.42) and (5.43), $\{x_i\}_{i=0}^\tau$ and $\{y_i\}_{i=0}^\tau$ is an $(f, M + 1)$-good pair of sequences. Combined with property (P5) this implies that there exists an integer s_0 such that (5.41) holds. This completes the proof of Lemma 5.16.

The next auxiliary result easily follows from Lemma 5.16.

Lemma 5.17. *Let M, ϵ be positive numbers. Then there exist a natural number $\tau > 1$ and a positive number δ such that for every function $g \in C(X \times X \times Y \times Y)$ which satisfies $\|g - f\| \leq \delta$, every natural number $T \geq \tau$, every (g, M)-good pair of sequences $\{x_i\}_{i=0}^T \subset X$ and $\{y_i\}_{i=0}^T \subset Y$ and every integer $s \in [0, T - \tau]$ there exists an integer s_0 such that*

$$\{s_0, s_0 + 1\} \subset [s, s + \tau],$$

$$\rho_X(x_i, x_f) \leq \epsilon \text{ and } \rho_Y(y_i, y_f) \leq \epsilon,\ i = s_0, s_0 + 1.$$

Lemma 5.18. *Let M, ϵ be positive numbers. Then there exist a natural number $\tau > 1$ and a positive number δ such that for every function $g \in \mathcal{A}$ which satisfies $\|g - f\| \le \delta$, every pair of points $(z, \xi) \in S(g)$ and every finite sequence $\{x_i\}_{i=0}^{\tau} \subset X$ which satisfies*

$$\sum_{i=0}^{\tau-1} g(x_i, x_{i+1}, \xi, \xi) \le \tau g(z, z, \xi, \xi) + M \tag{5.44}$$

there exists an integer $s_0 \in [0, \tau - 1]$ such that

$$\rho_X(x_i, x_f) \le \epsilon, \; i = s_0, s_0 + 1. \tag{5.45}$$

Proof. Lemma 5.12 (with $\tau_0 = 1$) implies that there exists a natural number $\tau > 1$ such that the following property holds:

(P6) for every finite sequence $\{x_i\}_{i=0}^{\tau} \subset X$ which satisfies

$$\sum_{i=0}^{\tau-1} f(x_i, x_{i+1}, y_f, y_f) \le \tau f(x_f, x_f, y_f, y_f) + M + 4$$

there exists an integer $s_0 \in [0, \tau - 1]$ for which (5.45) is valid.

Since the function f is uniformly continuous on the space $X \times X \times Y \times Y$ there exists a positive number δ_1 such that

$$|f(u_1, u_2, v_1, v_2) - f(u_1', u_2', v_1', v_2')| \le (8\tau)^{-1} \tag{5.46}$$

for all $u_1, u_2, u_1', u_2' \in X$ and all $v_1, v_2, v_1', v_2' \in Y$ which satisfy

$$\rho_X(u_i, u_i') \le \delta_1, \; \rho_Y(v_i, v_i') \le \delta_1, \; i = 1, 2.$$

Lemma 5.11 implies that there exists a number

$$\delta \in (0, (8\tau)^{-1}) \tag{5.47}$$

such that for every function $g \in \mathcal{A}$ which satisfies $\|g - f\| \le \delta$ and every pair of points $(z, \xi) \in S(g)$, we have

$$\rho_X(z, x_f) \le \delta_1, \; \rho_Y(\xi, y_f) \le \delta_1. \tag{5.48}$$

Assume that a functions $g \in \mathcal{A}$ satisfies

$$\|g - f\| \le \delta, \tag{5.49}$$

$$(z, \xi) \in S(g) \tag{5.50}$$

and that a finite sequence $\{x_i\}_{i=0}^{\tau} \subset X$ satisfies (5.44). In view of (5.44), (5.47), and (5.49),

$$\sum_{i=0}^{\tau-1} f(x_i, x_{i+1}, \xi, \xi) \le \tau f(z, z, \xi, \xi) + M + 2\delta\tau$$

$$\leq \tau f(z, z, \xi, \xi) + M + 1. \tag{5.51}$$

It follows from (5.48), (5.49), (5.50), and the choice of δ (see (5,47)) that

$$\rho_X(z, x_f) \leq \delta_1,\ \rho_Y(\xi, y_f) \leq \delta_1. \tag{5.52}$$

It follows from (5.51), (5.52), and the choice of δ_1 (see (5.46)) that

$$\sum_{i=0}^{\tau-1} f(x_i, x_{i+1}, y_f, y_f) \leq \sum_{i=0}^{\tau-1} f(x_i, x_{i+1}, \xi, \xi) + 8^{-1}$$

$$\leq \tau f(z, z, \xi, \xi) + M + 8^{-1} + 1 \leq \tau f(x_f, x_f, y_f, y_f) + M + 2. \tag{5.53}$$

Property (P6) and (5.53) imply that there exists an integer $s_0 \in [0, \tau - 1]$ such that (5.45) holds. This completes the proof of Lemma 5.18. □

Analogously to Lemma 5.18 we can prove the next auxiliary result.

Lemma 5.19. *Let M, ϵ be positive numbers. Then there exist a natural number $\tau > 1$ and a positive number δ such that for every function $g \in \mathcal{A}$ which satisfies $\|g - f\| \leq \delta$, every pair of points $(z, \xi) \in S(g)$ and every finite sequence $\{y_i\}_{i=0}^{\tau} \subset Y$ which satisfies*

$$\sum_{i=0}^{\tau-1} g(z, z, y_i, y_{i+1}) \geq \tau g(z, z, \xi, \xi) - M$$

there exists an integer $s_0 \in [0, \tau - 1]$ such that

$$\rho_Y(y_i, y_f) \leq \epsilon,\ i = s_0, s_0 + 1.$$

Let $g \in \mathcal{A}$, integers $T_2 > T_1 \geq 0$, $z_1, z_2 \in X$ and $\xi_1, \xi_2 \in Y$. Define

$$\sigma_X(g, T_1, T_2, z_1, z_2, \xi_1, \xi_2) = \inf\{ \sum_{i=T_1}^{T_2-1} g(x_i, x_{i+1}, \xi, \xi) :$$

$$\{x_i\}_{i=T_1}^{T_2} \subset X,\ x_{T_1} = z_1,\ x_{T_2} = z_2\}, \tag{5.54}$$

$$\sigma_Y(g, T_1, T_2, z_1, z_2, \xi_1, \xi_2) = \sup\{ \sum_{i=T_1}^{T_2-1} g(z_1, z_2, y_i, y_{i+1}) :$$

$$\{y_i\}_{i=T_1}^{T_2} \subset Y,\ y_{T_1} = \xi_1,\ y_{T_2} = \xi_2\}. \tag{5.55}$$

It is not difficult to see that the next auxiliary result holds.

Lemma 5.20. *Let $g \in \mathcal{A}$, $T_2 > T_1 \geq 0$ be integers and let*

$$z_1, z_2, \tilde{z}_1, \tilde{z}_2 \in X,$$

$$\xi_1, \xi_2, \tilde{\xi}_1, \tilde{\xi}_2 \in Y.$$

Then

$$|\sigma_X(g, T_1, T_2, z_1, z_2, \xi_1, \xi_2) - \sigma_X(g, T_1, T_2, \tilde{z}_1, \tilde{z}_2, \xi_1, \xi_2)| \leq 4\|g\|,$$

$$|\sigma_Y(g, T_1, T_2, z_1, z_2, \xi_1, \xi_2) - \sigma_Y(g, T_1, T_2, z_1, z_2, \tilde{\xi}_1, \tilde{\xi}_2)| \leq 4\|g\|,$$

$$|\sigma_X(g, T_1, T_2, z_1, z_2, y_g, y_g) - (T_2 - T_1)g(x_g, x_g, y_g, y_g)| \leq 4\|g\|,$$

$$|\sigma_Y(g, T_1, T_2, x_g, x_g, \xi_1, \xi_2) - (T_2 - T_1)g(x_g, x_g, y_g, y_g)| \leq 4\|g\|.$$

The next lemma easily follows from Lemmas 5.18 and 5.20.

Lemma 5.21. *Let M, ϵ be positive numbers. Then there exist a number $\delta \in (0, 1)$ and a natural number $\tau > 1$ such that for every function $g \in \mathcal{A}$ which satisfies $\|g - f\| \leq \delta$, every point $(x_g, y_g) \in S(g)$, every natural number $T \geq \tau$, every finite $\{x_i\}_{i=0}^T \subset X$ which satisfies*

$$\sum_{i=0}^{T-1} g(x_i, x_{i+1}, y_g, y_g) \leq \sigma_X(g, 0, T, x_0, x_T, y_g, y_g) + M$$

and every integer $p \in [0, T - \tau]$ there exists an integer j such that

$$\{j, j+1\} \subset [p, p+\tau]$$

and

$$\rho_X(x_i, x_f) \leq \epsilon, \ i = j, j+1.$$

The next auxiliary result is easily deduced from Lemmas 5.19 and 5.21.

Lemma 5.22. *Let M, ϵ be positive numbers. Then there exist a number $\delta \in (0, 1)$ and a natural number $\tau > 1$ such that for every function $g \in \mathcal{A}$ which satisfies $\|g - f\| \leq \delta$, every point $(x_g, y_g) \in S(g)$, every natural number $T \geq \tau$, every finite sequence $\{y_i\}_{i=0}^T \subset Y$ which satisfies*

$$\sum_{i=0}^{T-1} g(x_g, x_g, y_i, y_{i+1}) \geq \sigma_Y(g, 0, T, x_g, x_g, y_0, y_T) - M$$

and every integer $p \in [0, T - \tau]$ there exists an integer j such that

$$\{j, j+1\} \subset [p, p+\tau]$$

and

$$\rho_Y(y_i, y_f) \leq \epsilon, \ i = j, j+1.$$

It is not difficult to see that the following auxiliary result holds.

Lemma 5.23. *Let ϵ be a positive number. Then there exists a positive number δ such that for every pair of nonnegative integers $T_2 > T_1$, each $z_1, z_2, \tilde{z}_1, \tilde{z}_2 \in X$ and each $\xi_1, \xi_2, \tilde{\xi}_1, \tilde{\xi}_2 \in Y$ which satisfy*

$$\rho_X(z_i, \tilde{z}_i) \leq \delta, \ i = 1, 2, \ \rho_Y(\xi_i, \tilde{\xi}_i) \leq \delta, \ i = 1, 2$$

the following inequalities hold:

$$|\sigma_X(f, T_1, T_2, z_1, z_2, \xi_1, \xi_2) - \sigma_X(f, T_1, T_2, \tilde{z}_1, \tilde{z}_2, \xi_1, \xi_2)| \leq \epsilon,$$
$$|\sigma_Y(f, T_1, T_2, z_1, z_2, \xi_1, \xi_2) - \sigma_Y(f, T_1, T_2, z_1, z_2, \tilde{\xi}_1, \tilde{\xi}_2)| \leq \epsilon.$$

Lemma 5.24. *Let ϵ be a positive number. Then there exists a positive number δ such that for every natural number T and every finite sequence $\{x_i\}_{i=0}^{T} \subset X$ which satisfies*

$$x_0 = x_f, \ x_T = x_f,$$

$$\sum_{i=0}^{T-1} f(x_i, x_{i+1}, y_f, y_f) \leq \sigma_X(f, 0, T, x_f, x_f, y_f, y_f) + \delta$$

the inequality

$$\rho_X(x_i, x_f) \leq \epsilon$$

holds for all integers $i = 0, \dots, T$.

Proof. Assume the contrary. Then for every integer $k \geq 1$ there exists a finite sequence $\{x_i^{(k)}\}_{i=0}^{T_k} \subset X$ where $T_k \geq 1$ is an integer such that

$$x_0^{(k)} = x_f, \ x_{T_k}^{(k)} = x_f, \tag{5.56}$$

$$\sum_{i=0}^{T_k-1} f(x_i^{(k)}, x_{i+1}^{(k)}, y_f, y_f) \leq \sigma_X(f, 0, T_k, x_f, x_f, y_f, y_f) + 2^{-k}, \tag{5.57}$$

$$\max\{\rho_X(x_i^{(k)}, x_f) : \ i = 0, \dots, T_k\} > \epsilon. \tag{5.58}$$

There exists a sequence $\{x_i\}_{i=0}^{\infty} \subset X$ satisfying

$$x_i = x_i^{(1)}, \ i = 0, \dots, T_1 \tag{5.59}$$

and such that for every natural number k, we have

$$x_{\sum_{i=1}^{k} T_i + j} = x_j^{(k+1)}, \ j = 1, \dots, T_{k+1}. \tag{5.60}$$

Proposition 5.3 and (5.4) imply that for every natural number $k \geq 1$,

$$\sigma_X(f, 0, T_k, x_f, x_f, y_f, y_f) = T_k f(x_f, x_f, y_f, y_f). \tag{5.61}$$

In view of (5.57), (5.59), (5.60), and (5.61),

$$\sum\{f(x_i, x_{i+1}, y_f, y_f) : i = 0, \dots, \sum_{j=1}^{k} T_j - 1\}$$
$$\leq \sum_{j=1}^{k} (T_j f(x_f, x_f, y_f, y_f) + 2^{-j})$$
$$\leq (\sum_{j=1}^{k} T_j) f(x_f, x_f, y_f, y_f) + 1. \tag{5.62}$$

By (5.62) and (ATP), we have

$$\lim_{i \to \infty} \rho_X(x_i, x_f) = 0.$$

Then there exists an integer $i_0 \geq 1$ such that

$$\rho_X(x_i, x_f) \leq \epsilon/2$$

for all integers $i \geq i_0$. Combined with (5.59) and (5.60) this implies that for all natural number $k \geq i_0$, we have

$$\rho_X(x_i^{(k)}, x_f) \leq \epsilon/2, \ i = 0, \dots, T_k.$$

This contradicts (5.60). The contradiction we have reached completes the proof of Lemma 5.24. □

Analogously to Lemma 5.24 we can prove the next auxiliary result.

Lemma 5.25. *Let ϵ be a positive number. Then there exists a positive number δ such that for every natural number T and every finite sequence $\{y_i\}_{i=0}^{T} \subset Y$ which satisfies*

$$y_0 = y_f, \ y_T = y_f$$

and

$$\sum_{i=0}^{T-1} f(x_f, x_f, y_i, y_{i+1}) \geq \sigma_Y(f, 0, T, x_f, x_f, y_f, y_f) - \delta$$

the following inequality holds:

$$\rho_Y(y_i, y_f) \leq \epsilon, \ i = 0, \dots, T.$$

It is not difficult to see that the next lemma holds.

Lemma 5.26. *Let ϵ be a positive number and $\tau \geq 1$ be an integer. Then there exists a positive number δ such that for each $z_1, z_2, \tilde{z}_1, \tilde{z}_2 \in X$ and each $\xi_1, \xi_2, \tilde{\xi}_1, \tilde{\xi}_2 \in Y$ which satisfy*

$$\rho_X(z_i, \tilde{z}_i) \leq \delta, \ i = 1, 2, \ \rho_Y(\xi_i, \tilde{\xi}_i) \leq \delta, \ i = 1, 2$$

the inequalities

$$|\sigma_X(f,0,\tau,z_1,z_2,\xi_1,\xi_2) - \sigma_X(f,0,\tau,\tilde{z}_1,\tilde{z}_2,\tilde{\xi}_1,\tilde{\xi}_2)| \le \epsilon,$$

$$|\sigma_Y(f,0,\tau,z_1,z_2,\xi_1,\xi_2) - \sigma_Y(f,0,\tau,\tilde{z}_1,\tilde{z}_2,\tilde{\xi}_1,\tilde{\xi}_2)| \le \epsilon$$

are valid.

The next lemma follows from Lemmas 5.23 and 5.24.

Lemma 5.27. *Let ϵ be a positive number. Then there exists a positive number δ such that for every natural number T and every finite sequence $\{x_i\}_{i=0}^T \subset X$ which satisfies*

$$\rho_X(x_0,x_f) \le \delta,\ \rho_X(x_T,x_f) \le \delta,$$

$$\sum_{i=0}^{T-1} f(x_i,x_{i+1},y_f,y_f) \le \sigma_X(f,0,T,x_0,x_T,y_f,y_f) + \delta$$

the inequality

$$\rho_X(x_i,x_f) \le \epsilon$$

holds for all integers $i = 0,\dots,T$.

The following auxiliary result is easily deduced from Lemmas 5.23 and 5.25.

Lemma 5.28. *Let ϵ be a positive number. Then there exists a positive number δ such that for every natural number T and every finite sequence $\{y_i\}_{i=0}^T \subset Y$ which satisfies*

$$\rho_Y(y_0,y_f) \le \delta,\ \rho_Y(y_T,y_f) \le \delta,$$

$$\sum_{i=0}^{T-1} f(x_f,x_f,y_i,y_{i+1}) \ge \sigma_Y(f,0,T,x_f,x_f,y_0,y_T) - \delta$$

the inequality

$$\rho_Y(y_i,y_f) \le \epsilon$$

holds for all integers $i = 0,\dots,T$.

Lemma 5.29. *Let ϵ be a positive number. Then there exists a number $\delta \in (0,1)$ such that for every function $g \in \mathcal{A}$ which satisfies $\|f-g\| \le \delta$, every point $(x_g,y_g) \in S(g)$, every natural number T and every finite sequence $\{x_i\}_{i=0}^T \subset X$ which satisfies*

$$\rho_X(x_0,x_f) \le \delta,\ \rho_X(x_T,x_f) \le \delta, \tag{5.63}$$

$$\sum_{i=0}^{T-1} g(x_i,x_{i+1},y_g,y_g) \le \sigma_X(g,0,T,x_0,x_T,y_g,y_g) + \delta \tag{5.64}$$

the inequality:

$$\rho_X(x_i,x_f) \le \epsilon$$

holds for all integers $i = 0,\dots,T$.

Proof. Lemma 5.27 implies that there exists a number $\delta_1 \in (0,1)$ such that the following property holds:

(P7) for every natural number T and every finite sequence $\{x_i\}_{i=0}^T \subset X$ which satisfies

$$\rho_X(x_0, x_f) \le \delta_1, \ \rho_X(x_T, x_f) \le \delta_1,$$

$$\sum_{i=0}^{T-1} f(x_i, x_{i+1}, y_f, y_f) \le \sigma_X(f, 0, T, x_0, x_T, y_f, y_f) + \delta_1$$

the inequality

$$\rho_X(x_i, x_f) \le \epsilon$$

holds for all integers $i = 0, \dots, T$.

Lemma 5.21 implies that there exist a number

$$\delta_2 \in (0, \delta_1) \tag{5.65}$$

and a natural number τ_0 such that the following property holds:

(P8) for every function $g \in \mathcal{A}$ which satisfies $\|g - f\| \le \delta_2$, every point $(x_g, y_g) \in S(g)$, every natural number $T \ge \tau_0$, every finite sequence $\{x_i\}_{i=0}^T \subset X$ which satisfies

$$\sum_{i=0}^{T-1} g(x_i, x_{i+1}, y_g, y_g) \le \sigma_X(g, 0, T, x_0, x_T, y_g, y_g) + 4$$

and every integer $p \in [0, T - \tau_0]$ there exists an integer j such that

$$\{j, j+1\} \subset [p, p + \tau_0]$$

and

$$\rho_X(x_i, x_f) \le \delta_1, \ i = j, j+1.$$

Since the function f is continuous there exists a number

$$\delta_3 \in (0, (2\tau_0 + 1)^{-1} 8^{-1} \delta_2) \tag{5.66}$$

such that the following property holds:

(P9) for each $z_1, z_2, \tilde{z}_1, \tilde{z}_2 \in X$ and each $\xi_1, \xi_2, \tilde{\xi}_1, \tilde{\xi}_2 \in Y$ which satisfy

$$\rho_X(z_i, \tilde{z}_i) \le \delta_3, \ i = 1, 2, \ \rho_Y(\xi_i, \tilde{\xi}_i) \le \delta_3 \ i = 1, 2$$

the inequality

$$|f(z_1, z_2, \xi_1, \xi_2) - f(\tilde{z}_1, \tilde{z}_2, \tilde{\xi}_1, \tilde{\xi}_2)| \le 8^{-1}(2\tau_0 + 1)^{-1}\delta_2$$

is true.

Lemma 5.11 implies that there exists a number

$$\delta \in (0, \delta_3) \tag{5.67}$$

such that the following property holds:

(P10) for every function $g \in \mathcal{A}$ which satisfies $\|g - f\| \le \delta$ and every point $(x, y) \in S(g)$ the inequalities

$$\rho_X(x, x_f) \le \delta_3, \ \rho_Y(y, y_f) \le \delta_3$$

are true.

Assume that a function $g \in \mathcal{A}$,

$$\|f - g\| \le \delta, \ (x_g, y_g) \in S(g), \tag{5.68}$$

T is a natural number and a that a finite sequence $\{x_i\}_{i=0}^T \subset X$ satisfies (5.63) and (5.64). In view of (5.64), (5.66), (5.67), (5.68), and property (P8), there exist a strictly increasing sequence of nonnegative integers $\{t_i\}_{i=0}^q$, where $q \ge 1$ is an integer, such that

$$t_0 = 0, \ t_q = T, \ t_{i+1} - t_i \le \tau_0 \text{ for all } i = 0, \dots, q-1, \tag{5.69}$$

$$\rho_X(x_{t_i}, x_f) \le \delta_1, \ i = 0, \dots, q. \tag{5.70}$$

Let $j \in \{0, \dots, T\}$ be given. We claim that

$$\rho_X(x_j, x_f) \le \epsilon.$$

In view of (5.69), there exists an integer $k \in [0, q-1]$ for which

$$j \in [t_k, t_{k+1}]. \tag{5.71}$$

It follows from (5.64) and (5.70) that

$$\rho_X(x_{t_k}, x_f) \le \delta_1, \ \rho_X(x_{t_k+1}, x_f) \le \delta_1, \tag{5.72}$$

$$\sum_{i=t_k}^{t_{k+1}-1} g(x_i, x_{i+1}, y_g, y_g) \le \sigma_X(g, 0, t_{k+1} - t_k, x_{t_k}, x_{t_{k+1}}, y_g, y_g) + \delta. \tag{5.73}$$

In view of (5.68), (5.69), and (5.73), we have

$$\sum_{i=t_k}^{t_{k+1}-1} f(x_i, x_{i+1}, y_g, y_g) \le \sum_{i=t_k}^{t_{k+1}-1} g(x_i, x_{i+1}, y_g, y_g) + \delta\tau_0$$

$$\le \sigma_X(g, 0, t_{k+1} - t_k, x_{t_k}, x_{t_{k+1}}, y_g, y_g) + \delta(\tau_0 + 1)$$

$$\le \sigma_X(f, 0, t_{k+1} - t_k, x_{t_k}, x_{t_{k+1}}, y_g, y_g) + \delta(2\tau_0 + 1). \tag{5.74}$$

Property (P10) and (5.68) imply that

$$\rho_X(x_g, x_f) \le \delta_3, \ \rho_Y(y_g, y_f) \le \delta_3. \tag{5.75}$$

By (5.69), (5.75), and property (P9), we have

$$|\sum_{i=t_k}^{t_{k+1}-1} f(x_i, x_{i+1}, y_g, y_g) - \sum_{i=t_k}^{t_{k+1}-1} f(x_i, x_{i+1}, y_f, y_f)| \le \tau_0(2\tau_0+1)^{-1}8^{-1}\delta_2, \tag{5.76}$$

$$|\sigma_X(f, 0, t_{k+1}-t_k, x_{t_k}, x_{t_{k+1}}, y_g, y_g) - \sigma_X(f, 0, t_{k+1}-t_k, x_{t_k}, x_{t_{k+1}}, y_f, y_f) \\ \le \tau_0(2\tau_0+1)^{-1}8^{-1}\delta_2. \tag{5.77}$$

It follows from (5.76), (5.74), (5.77), (5.66), (5.67), and (5.65) that

$$\sum_{i=t_k}^{t_{k+1}-1} f(x_i, x_{i+1}, y_f, y_f) \le \sum_{i=t_k}^{t_{k+1}-1} f(x_i, x_{i+1}, y_g, y_g) + 16^{-1}\delta_2$$

$$\le \sigma_X(f, 0, t_{k+1}-t_k, x_{t_k}, x_{t_{k+1}}, y_g, y_g) + \delta(2\tau_0+1) + 16^{-1}\delta_2$$

$$\le \sigma_X(f, 0, t_{k+1}-t_k, x_{t_k}, x_{t_{k+1}}, y_f, y_f) + 16^{-1}\delta_2 + \delta(2\tau_0+1) + 16^{-1}\delta_2$$

$$\le \sigma_X(f, 0, t_{k+1}-t_k, x_{t_k}, x_{t_k+1}, y_f, y_f) + \delta_1. \tag{5.78}$$

Property (P7), (5.71), (5.72), (5.78) imply that $\rho_X(x_g, x_f) \le \epsilon$. This completes the proof of Lemma 5.29. □

Analogously to Lemma 5.29 we can prove the following result.

Lemma 5.30. *Let ϵ be a positive number. Then there exists a number $\delta \in (0,1)$ such that for every function $g \in \mathcal{A}$ which satisfies $\|f-g\| \le \delta$, every point $(x_g, y_g) \in S(g)$, every natural number T and every finite sequence $\{y_i\}_{i=0}^T \subset Y$ which satisfies*

$$\rho_Y(y_0, y_f) \le \delta, \ \rho_Y(y_T, y_f) \le \delta,$$

$$\sum_{i=0}^{T-1} g(x_g, x_g, y_i, y_{i+1}) \ge \sigma_Y(g, 0, T, x_g, x_g, y_0, y_T) - \delta$$

the inequality

$$\rho_Y(y_i, y_f) \le \epsilon$$

holds for all integers $i = 0, \dots, T$.

Lemma 5.31. *Let M, ϵ be positive numbers. Then there exist integers $l \ge 1$ and $Q \ge 1$ and a number $\delta > 0$ such that for every function $g \in \mathcal{A}$ which satisfies $\|f-g\| \le \delta$, every point $(x_g, y_g) \in S(g)$, every natural number $T > Ql$, and every finite sequence $\{x_i\}_{i=0}^T \subset X$ which satisfies*

$$\sum_{i=0}^{T-1} g(x_i, x_{i+1}, y_g, y_g) \le Tg(x_g, x_g, y_g, y_g) + M \tag{5.79}$$

there exist a natural number $q \le Q$ and finite sequences of integers $\{a_i\}_{i=1}^q$, $\{b_i\}_{i=1}^q \subset [0, T]$ such that

$$0 \le b_i - a_i \le l, \ i = 1, \dots, q,$$

$$\rho_X(x_i, x_f) \le \epsilon$$

for all integers $i \in [0, T] \setminus \cup_{j=1}^{q} [a_j, b_j]$.

Proof. Lemma 5.29 implies that there exists a number $\delta_0 \in (0, 1)$ such that the following property holds:

(P11) for every $g \in \mathcal{A}$ which satisfies $\|f - g\| \le \delta_0$, every point $(x_g, y_g) \in S(g)$, every natural number T, and every finite sequence $\{x_i\}_{i=0}^{T} \subset X$ which satisfies

$$\rho_X(x_0, x_f) \le \delta_0, \ \rho_X(x_T, x_f) \le \delta_0,$$

$$\sum_{i=0}^{T-1} g(x_i, x_{i+1}, y_g, y_g) \le \sigma_X(g, 0, T, x_0, x_T, y_g, y_g) + \delta_0$$

the inequality

$$\rho_X(x_i, x_f) \le \epsilon$$

holds for all integers $i = 0, \dots, T$.

Lemma 5.21 there exist a positive number

$$\delta < \min\{\delta_0, M/8\}$$

and a natural number τ_0 such that the following property holds:

(P12) for every function $g \in \mathcal{A}$ which satisfies $\|g - f\| \le \delta$, every $(x_g, y_g) \in S(g)$, every natural $T \ge \tau_0$, and every finite sequence $\{x_i\}_{i=0}^{T} \subset X$ which satisfies

$$\sum_{i=0}^{T-1} g(x_i, x_{i+1}, y_g, y_g) \le \sigma_X(g, 0, T, x_0, x_T, y_g, y_g) + M$$

and every integer $p \in [0, T - \tau_0]$ there exists an integer j such that

$$\{j, j+1\} \subset [p, p + \tau_0]$$

and

$$\rho_X(x_i, x_f) \le \delta_0, \ i = j, j+1.$$

Choose natural numbers

$$Q > 6 + 3\delta_0^{-1}(4(\|f\| + 1) + M), \ l > 4(\tau_0 + 1). \tag{5.80}$$

Assume that a function $g \in \mathcal{A}$,

$$\|f - g\| \le \delta, \ (x_g, y_g) \in S(g), \tag{5.81}$$

$T > Ql$ is a natural number and that a finite sequence $\{x_i\}_{i=0}^{T} \subset X$ satisfies (5.79).

Set $t_0 = 0$. Assume that $k \ge 0$ is an integer and that we have defined a strictly increasing sequence of integers t_p, $p = 0, \dots, k$ such that $t_k < T$ and that for every nonnegative integer p satisfying $p < k$, we have

$$t_{p+1} - t_p \ge 2, \tag{5.82}$$

$$\sum_{j=t_p}^{t_{p+1}-1} g(x_j, x_{j+1}, y_g, y_g) > \sigma_X(g, 0, t_{p+1} - t_p, x_{t_p}, x_{t_{p+1}}, y_g, y_g) + \delta_0, \tag{5.83}$$

$$\sum_{j=t_p}^{t_{p+1}-2} g(x_j, x_{j+1}, y_g, y_g) \le \sigma_X(g, 0, t_{p+1} - t_p - 1, x_{t_p}, x_{t_{p+1}-1}, y_g, y_g) + \delta_0. \tag{5.84}$$

(It is clear that for $k = 0$ this assumption holds.) If

$$\sum_{j=t_k}^{T-1} g(x_j, x_{j+1}, y_g, y_g) \le \sigma_X(g, 0, T - t_k, x_{t_k}, x_T, y_g, y_g) + \delta_0,$$

then set $q = k + 1$, $t_q = T$ and the construction is completed.

Assume that

$$\sum_{j=t_k}^{T-1} g(x_j, x_{j+1}, y_g, y_g) > \sigma_X(g, 0, T - t_k, x_{t_k}, x_T, y_g, y_g) + \delta_0.$$

Then there exists an integer t_{k+1} such that $t_k + 1 < t_{k+1} \le T$ (it is possible that $t_{k+1} = T$) such that

$$\sum_{j=t_k}^{t_{k+1}-1} g(x_j, x_{j+1}, y_g, y_g) > \sigma_X(g, 0, t_{k+1} - t_k, x_{t_k}, x_{t_{k+1}}, y_g, y_g) + \delta_0,$$

$$\sum_{j=t_k}^{t_{k+1}-2} g(x_j, x_{j+1}, y_g, y_g) \le \sigma_X(g, 0, t_{k+1} - t_k - 1, x_{t_k}, x_{t_{k+1}-1}, y_g, y_g) + \delta_0.$$

It is clear that the assumption made for k also holds for $k + 1$ (if $k + 1 < q$) and the construction is completed after a finite number of steps and the last element of the sequence satisfies $t_q = T$.

It follows from the construction of the sequence $\{t_i\}_{i=0}^q$ that for every integer p which satisfies $0 \le p < q - 1$, relations (5.82)–(5.84) are valid and if $t_q - t_{q-1} > 1$, then (5.84) is true with $p = q - 1$.

In view of (5.79), (5.54), (5.55), Proposition 5.3, (5.4), Lemma 5.20, and (5.81), we have

$$\sum_{i=0}^{T-1} g(x_i, x_{i+1}, y_g, y_g) \le M + Tg(x_g, x_g, y_g, y_g)$$

$$= M + \sigma_X(g, 0, T, x_g, x_g, y_g, y_g) \le M + \sigma_X(g, 0, T, x_0, x_T, y_g, y_g) + 4\|g\|$$
$$\le M + 4(\|f\| + 1) + \sigma_X(g, 0, T, x_0, x_T, y_g, y_g). \quad (5.85)$$

It follows from (5.83) and (5.85) that

$$4(\|f\| + 1) + M \ge \sum_{i=0}^{T-1} g(x_i, x_{i+1}, y_g, y_g) - \sigma_X(g, 0, T, x_0, x_T, y_g, y_g)$$
$$\ge \sum \{ \sum_{i=t_p}^{t_{p+1}-1} g(x_i, x_{i+1}, y_g, y_g) - \sigma_X(g, t_p, t_{p+1}, x_{t_p}, x_{t_{p+1}}, y_g, y_g) :$$
$$p \in \{0, \dots, q-1\} \text{ and } p < q-1\} \ge \delta_0(q-2),$$
$$q \le 2 + \delta_0^{-1}(4(\|f\| + 1) + M). \quad (5.86)$$

Define

$$E = \{p \in \{0, \dots, q-1\} : t_{p+1} - t_p > 4(\tau_0 + 1)\}. \quad (5.87)$$

Let $p \in E$ be given. Then (5.84) is true. In view of (5.81), (5.84), (5.87), and property (P12), there exist integers

$$j_{1,p} \in [t_p, t_p + \tau_0], \ j_{2,p} \in [t_{p+1} - 1 - \tau_0, \ t_{p+1} - 1] \quad (5.88)$$

such that

$$\rho_X(x_{j_{s,p}}, x_f) \le \delta_0, \ s = 1, 2. \quad (5.89)$$

It follows from (5.88), the inclusion $p \in E$, (5.87), (5.89), (5.84), (5.81), and property (P11) that

$$\rho_X(x_i, x_f) \le \epsilon, \ i \in \{j_{1,p}, \dots, j_{2,p}\},$$
$$\rho_X(x_i, x_f) \le \epsilon, \ i \in \{t_p + \tau_0, \dots, t_{p+1} - 1 - \tau_0\} \quad (5.90)$$

for all integers $p \in E$. In view of (5.90), we have

$$\{i \in \{0, \dots, T\} : \rho_X(x_i, x_f) > \epsilon\} \subset \cup\{\{t_i, \dots, t_{i+1}\} : \ i \in \{0, \dots, q-1\} \setminus E\}$$
$$\cup (\cup\{(\{t_p, \dots, t_p + \tau_0 - 1\} \cup \{t_{p+1} - \tau_0, \dots, t_{p+1}\}) : \ p \in E\}). \quad (5.91)$$

Consider the collection of closed intervals

$$[t_i, t_{i+1}], \ i \in \{0, \dots, q-1\} \setminus E, \ [t_p, t_p + \tau_0 - 1], \ p \in E, \ [t_{p+1} - \tau_0, t_{p+1}], \ p \in E.$$

It is clear that the number of intervals in this collection does not exceed

$$3q \le 6 + 3\delta_0^{-1}(4(\|f\| + 1) + M) < Q$$

(see (5.80) and (5.86)) and in view of (5.80) and (5.87) the lengths of all these intervals do not exceed $4(\tau_0 + 1) \le l$. This completes the proof of Lemma 5.31.
□

Analogously to Lemma 5.31 we can prove the following result.

Lemma 5.32. *Let M, ϵ be positive numbers. Then there exist integers $l \geq 1$ and $Q \geq 1$ and a positive number δ such that for every function $g \in \mathcal{A}$ which satisfies $\|f - g\| \leq \delta$, every point $(x_g, y_g) \in S(g)$, every integer $T > Ql$, and every finite sequence $\{y_i\}_{i=0}^T \subset Y$ which satisfies*

$$\sum_{i=0}^{T-1} g(x_g, x_g, y_i, y_{i+1}) \geq Tg(x_g, x_g, y_g, y_g) - M$$

there exist a natural number $q \leq Q$ and finite sequences of integers $\{a_i\}_{i=1}^q$, $\{b_i\}_{i=1}^q \subset [0, T]$ such that

$$0 \leq b_i - a_i \leq l,\ i = 1, \ldots, q,$$

$$\rho_Y(y_i, y_f) \leq \epsilon$$

for all integers $i \in [0, T] \setminus \cup_{j=1}^q [a_j, b_j]$.

5.9 Proof of Theorem 5.8

Lemma 5.31 implies that there exist integers $l_1 \geq 1$ and $Q_1 \geq 1$ and a positive number $\delta_1 < 1$ such that the following property holds:

(P13) for every function $g \in \mathcal{A}$ which satisfies $\|f - g\| \leq \delta_1$, every point $(x_g, y_g) \in S(g)$, every integer $T > Q_1 l_1$ and every finite sequence $\{x_i\}_{i=0}^T \subset X$ which satisfies

$$\sum_{i=0}^{T-1} g(x_i, x_{i+1}, y_g, y_g) \leq Tg(x_g, x_g, y_g, y_g) + 2M + 12(\|f\| + 1)$$

there exist a natural number $q \leq Q_1$ and finite sequences of integers $\{a_i\}_{i=1}^q$, $\{b_i\}_{i=1}^q \subset [0, T]$ such that

$$0 \leq b_i - a_i \leq l_1,\ i = 1, \ldots, q,$$

$$\rho_X(x_i, x_f) \leq \epsilon$$

for all integers $i \in [0, T] \setminus \cup_{j=1}^q [a_j, b_j]$.

Lemma 5.32 implies that there exist integers $l_2 \geq 1$ and $Q_2 \geq 1$ and a positive number $\delta_2 < 1$ such that the following property holds:

(P14) for every function $g \in \mathcal{A}$ which satisfies $\|f - g\| \leq \delta_2$, every point $(x_g, y_g) \in S(g)$, every integer $T > Q_2 l_2$ and every finite sequence $\{y_i\}_{i=0}^T \subset Y$ which satisfies

$$\sum_{i=0}^{T-1} g(x_g, x_g, y_i, y_{i+1}) \geq Tg(x_g, x_g, y_g, y_g) - 2M - 12(\|f\| + 1)$$

there exist a natural number $q \le Q_2$ and finite sequences of integers $\{a_i\}_{i=1}^q$, $\{b_i\}_{i=1}^q \subset [0, T]$ such that

$$0 \le b_i - a_i \le l_2,\ i = 1, \dots, q,$$

$$\rho_Y(y_i, y_f) \le \epsilon$$

for all integers $i \in [0, T] \setminus \cup_{j=1}^q [a_j, b_j]$.

Set

$$\delta = \min\{\delta_1, \delta_2\},\ l = \max\{l_1, l_2\},\ Q = Q_1 + Q_2. \tag{5.92}$$

Assume that

$$g \in \mathcal{A},\ \|f - g\| \le \delta, \tag{5.93}$$

an integer $T > Ql$ and that $\{x_i\}_{i=0}^T \subset X$, $\{y_i\}_{i=0}^T \subset Y$ is a (g, M)-good pair of sequences. Let a point

$$(x_g, y_g) \in S(g) \tag{5.94}$$

be an element of the set $S(g)$ associated with the function g (see the definition of $\mathcal{A}$, relation (5.4) and Remark 5.5). Then we have

$$g(x_g, x_g, \xi, \xi) \le g(x_g, x_g, y_g, y_g) \le g(z, z, y_g, y_g)$$
$$\text{for all } z \in X \text{ and all } \xi \in Y \tag{5.95}$$

and for every integer $n \ge 1$ and every pair of finite sequences

$$\{z_i\}_{i=0}^n \subset X,\ \{\xi_i\}_{i=0}^n \subset Y$$

the inequality

$$-4\|g\| + \sum_{i=0}^{n-1} g(x_g, x_g, \xi_i, \xi_{i+1}) \le n g(x_g, x_g, y_g, y_g)$$
$$\le \sum_{i=0}^{n-1} g(z_i, z_{i+1}, y_g, y_g) + 4\|g\| \tag{5.96}$$

holds. In view of (5.93), (5.94), (5.95), (5.96), and Proposition 5.10, we have

$$|\sum_{i=0}^{T-1} g(x_i, x_{i+1}, y_g, y_g) - T g(x_g, x_g, y_g, y_g)| \le 2M + 12\|g\| \le 2M + 12(\|f\| + 1), \tag{5.97}$$

$$|\sum_{i=0}^{T-1} g(x_g, x_g, y_i, y_{i+1}) - T g(x_g, x_g, y_g, y_g)| \le 2M + 12\|g\| \le 2M + 12(\|f\| + 1). \tag{5.98}$$

Properties (P13) and (P14), (5.92), (5.93), (5.94), (5.97), and (5.98) imply that there exist natural numbers $q_1 \le Q_1$, $q_2 \le Q_2$ and finite sequences of integers

$$\{a_{1,i}\}_{i=1}^{q_1},\ \{b_{1,i}\}_{i=1}^{q_1}, \{a_{2,i}\}_{i=1}^{q_2},\ \{b_{2,i}\}_{i=1}^{q_2} \subset [0,T]$$

such that

$$0 \le b_{1,i} - a_{1,i} \le l_1,\ i = 1, \dots, q_1,$$

$$0 \le b_{2,i} - a_{2,i} \le l_2,\ i = 1, \dots, q_2,$$

$$\rho_X(x_i, x_f) \le \epsilon,\ \rho_Y(y_i, y_f) \le \epsilon$$

for all integers

$$i \in [0,T] \setminus [\cup(\cup_{j=1}^{q_1}[a_{1,j}, b_{1,j}]) \cup (\cup_{j=1}^{q_2}[a_{2,j}, b_{2,j}])].$$

This completes the proof of Theorem 5.8. □

5.10 Proof of Theorem 5.9

Denote by $\mathcal{E}$ the set of all function $f \in \mathcal{A}$ which have (ATP).

Lemma 5.33. *$\mathcal{E}$ is an everywhere dense subset of $\mathcal{A}$.*

Proof. Let $f \in \mathcal{A}$ be given, $r \in (0,1]$ and let $(x_f, y_f) \in S(f)$ be an element of the set $S(f)$ associated with the function f such that (5.1) is valid and (5.4) is true with $c_f = 4\|f\|$. Define

$$f_r(x_1, x_2, y_1, y_2) = f(x_1, x_2, y_1, y_2) + r\rho_X(x_1, x_f)$$

$$+ r\rho_X(x_2, x_f) - r\rho_Y(y_1, y_f) - r\rho_Y(y_2, y_f). \tag{5.99}$$

It is clear that

$$f_r \in \mathcal{A}_0,\ (x_f, y_f) \in S(f_r) \text{ and } f_r \in \mathcal{A}. \tag{5.100}$$

Let

$$(x, y) \in S(f_r). \tag{5.101}$$

It follows from (5.100), (5.101), and the definition of $S(\cdot)$ (see (5.7)) that

$$f_r(x, x, y_f, y_f) \le f_r(x, x, y, y) \le f_r(x_f, x_f, y, y)$$

$$\le f_r(x_f, x_f, y_f, y_f) \le f_r(x, x, y_f, y_f). \tag{5.102}$$

In view of (5.99), (5.102), and the inclusion $(x_f, y_f) \in S(f)$, we have

$$f(x, x, y_f, y_f) + 2r\rho_X(x, x_f) = f(x_f, x_f, y_f, y_f) \le f(x, x, y_f, y_f),$$

$$x = x_f$$

and

$$f(x_f, x_f, y, y) \le f(x_f, x_f, y_f, y_f) = f(x_f, x_f, y, y) - 2r\rho_Y(y, y_f),$$

$$y = y_f.$$

Hence $S(f_r) = \{(x_f, y_f)\}$. We claim that the function f_r has (ATP).

Let a sequence $\{x_i\}_{i=0}^{\infty} \subset X$ satisfy

$$\sup\{\sum_{i=0}^{n-1} f_r(x_i, x_{i+1}, y_f, y_f) - nf_r(x_f, x_f, y_f, y_f) : \ n = 1, 2, \dots\} < \infty.$$

It follows from the inequality above, (5.99) and Proposition 5.4 that

$$\sum_{i=0}^{\infty}(\rho_X(x_i, x_f) + \rho_X(x_{i+1}, x_f)) < \infty.$$

This implies that $\rho_X(x_i, x_f) \to 0$ as $i \to \infty$.

Analogously we can show that for every sequence $\{y_i\}_{i=0}^{\infty} \subset Y$ which satisfies

$$\inf\{\sum_{i=0}^{n-1} f_r(x_f, x_f, y_i, y_{i+1}) - nf_r(x_f, x_f, y_f, y_f) : \ n = 1, 2, \dots\} > -\infty$$

we have $\rho_Y(y_i, y_f) \to 0$ as $i \to \infty$. Hence f_r has (ATP). Since $\|f - f_r\| \to 0$ as $r \to 0^+$ this completes the proof of Lemma 5.33.

Let a function $f \in \mathcal{E}$ be given, $(x_f, y_f) \in S(f)$ and $n \geq 1$ be an integer. It follows from Lemmas 5.11, 5.31, and 5.32 that there exists an open neighborhood $V(f, n)$ of the function f in the space $\mathcal{A}$ such that the following properties hold:

(i) for every function $g \in V(f, n)$ and every point $(z, \xi) \in S(g)$ we have

$$\rho_X(z, x_f) \leq (4n)^{-1}, \ \rho_Y(\xi, y_f) \leq (4n)^{-1};$$

(ii) for every function $g \in V(f, n)$, every point $(x_g, y_g) \in S(g)$, and every sequence $\{x_i\}_{i=0}^{\infty} \subset X$ which satisfies

$$\sup\{\sum_{i=0}^{T-1} g(x_i, x_{i+1}, y_g, y_g) - Tg(x_g, x_g, y_g, y_g) : \ T = 1, 2, \dots\} \leq n,$$

for all sufficiently large natural numbers i, we have

$$\rho_X(x_i, x_f) \leq (4n)^{-1};$$

(iii) for every function $g \in V(f, n)$, every point $(x_g, y_g) \in S(g)$, and every sequence $\{y_i\}_{i=0}^{\infty} \subset Y$ which satisfies

$$\inf\{\sum_{i=0}^{T-1} g(x_g, x_g, y_i, y_{i+1}) - Tg(x_g, x_g, y_g, y_g) : \ T = 1, 2, \dots\} \geq -n,$$

for all sufficiently large natural numbers i, we have

$$\rho_Y(y_i, y_f) \le (4n)^{-1}.$$

Define

$$\mathcal{F} = \cap_{n=1}^{\infty} \cup \{V(f,n) : n = 1, 2, \dots\}. \tag{5.103}$$

It is clear that $\mathcal{F}$ is a countable intersection of open everywhere dense subsets of the space $\mathcal{A}$.

Let $g \in \mathcal{F}$ be given. In view of (5.103), for every integer $n \ge 1$ there exists a function $f_n \in \mathcal{E}$ such that

$$g \in V(f_n, n). \tag{5.104}$$

Let

$$(z_1, \xi_1),\ (z_2, \xi_2) \in S(g) \tag{5.105}$$

and let $n \ge 1$ be an integer. It follows from (5.104), (5.105), and property (i) that

$$\rho_X(z_i, x_{f_n}),\ \rho_Y(\xi_i, y_{f_n}) \le (4n)^{-1},\ i = 1, 2,$$
$$\rho_X(z_1, z_2),\ \rho_Y(\xi_1, \xi_2) \le (2n)^{-1}$$

and since n is an arbitrary natural number we conclude that $z_1 = z_2$, $\xi_1 = \xi_2$ and $S(g)$ is a singleton: $S(g) = \{(x_g, y_g)\}$ and

$$\rho_X(x_g, x_{f_n}),\ \rho_Y(y_g, y_{f_n}) \le (4n)^{-1} \text{ for all integers } n \ge 1. \tag{5.106}$$

Assume that a sequence $\{x_i\}_{i=0}^{\infty} \subset X$ satisfies

$$\sup\{\sum_{i=0}^{T-1} g(x_i, x_{i+1}, y_g, y_g) - Tg(x_g, x_g, y_g, y_g) : T = 1, 2, \dots\} < \infty.$$

Fix a natural number n such that

$$2n^{-1} < \epsilon, \tag{5.107}$$

$$\sup\{\sum_{i=0}^{T-1} g(x_i, x_{i+1}, y_g, y_g) - Tg(x_g, x_g, y_g, y_g) : T = 1, 2, \dots\} \le n.$$

It follows from (5.104), (5.106), (5.107) and property (ii) that for all sufficiently large natural numbers i, we have

$$\rho_X(x_i, x_{f_n}) \le (4n)^{-1},$$
$$\rho_X(x_i, x_g) \le \rho_X(x_i, x_{f_n}) + \rho_X(x_{f_n}, x_g) \le (2n)^{-1} < \epsilon.$$

Since ϵ is an arbitrary positive number we conclude that $\lim_{i\to\infty} \rho_X(x_i, x_g) = 0$.

Analogously we can show that if $\{y_i\}_{i=0}^{\infty} \subset Y$ satisfies

$$\inf\{\sum_{i=0}^{T-1} g(x_g, x_g, y_i, y_{i+1}) - Tg(x_g, x_g, y_g, y_g) : T = 1, 2, \dots\} > -\infty,$$

then $\lim_{i\to\infty} \rho_Y(y_i, y_g) = 0$. Therefore the function g has (ATP). This completes the proof of Theorem 5.9. □

6

Constrained Games Without Convexity–Concavity Assumptions

In this chapter we study existence and turnpike properties of approximate solutions for a class of dynamic constrained discrete-time two-player zero-sum games without using convexity–concavity assumptions. We describe the structure of approximate solutions which is independent of the length of the interval, for all sufficiently large intervals and show that approximate solutions are determined mainly by the objective function, and are essentially independent of the choice of interval and endpoint conditions.

6.1 Preliminaries

Let X, Y, U, V be nonempty sets equipped with metrics ρ_X, ρ_Y, ρ_U, and ρ_V, respectively. Suppose that (X, ρ_X), (Y, ρ_Y), (U, ρ_U), and (V, ρ_V) are compact metric spaces.

For each $x \in X$, let $U(x)$ be a subset of U which can be empty and for each $y \in Y$, let $V(y)$ be a subset of V which can also be empty. We suppose that the sets

$$\{(x, u) \in X \times U : \ u \in U(x)\} \text{ and } \{(y, v) \in Y \times V : \ v \in V(y)\}$$

are closed in the product topologies.

Let

$$F_X : \ \{(x, u) \in X \times U : \ u \in U(x)\} \to 2^X \setminus \{\emptyset\}$$

and

$$F_Y : \ \{(y, v) \in Y \times V : \ v \in V(y)\} \to 2^Y \setminus \{\emptyset\}$$

be such that the sets

$$\{(x, u, x') \in X \times U \times X : \ u \in U(x), \ x' \in F_X(x, u)\},$$

A.J. Zaslavski, *Discrete-Time Optimal Control and Games on Large Intervals*, Springer Optimization and Its Applications 119,
DOI 10.1007/978-3-319-52932-5_6

$$\{(y, v, y') \in Y \times V \times Y : v \in V(y), \ y' \in F_Y(y, v)\}$$

are closed in the product topologies.

Suppose that

$$f : \{(x, u, y, v) \in X \times U \times Y \times V : u \in U(x), \ v \in V(y)\} \to R^1$$

is a bounded function and set

$$\|f\| = \sup\{|f(x, u, y, v)| : (x, y) \in X \times Y, \ u \in U(x), \ v \in V(y)\}. \quad (6.1)$$

Let $T_1, T_2 > T_1$ be integers. A pair of finite sequences $(\{x_t\}_{t=T_1}^{T_2}, \{u_t\}_{t=T_1}^{T_2-1})$ is called a strategy of the first player if

$$\{x_t\}_{t=T_1}^{T_2} \subset X, \ \{u_t\}_{t=T_1}^{T_2-1} \subset U \quad (6.2)$$

and for all integers $t = T_1, \dots, T_2 - 1$,

$$u_t \in U(x_t), \ x_{t+1} \in F_X(x_t, u_t). \quad (6.3)$$

We denote by $\mathrm{Str}_1(T_1, T_2)$ the set of all the strategies of the first player.

A pair of finite sequences $(\{y_t\}_{t=T_1}^{T_2}, \{v_t\}_{t=T_1}^{T_2-1})$ is called a strategy of the second player if

$$\{y_t\}_{t=T_1}^{T_2} \subset Y, \ \{v_t\}_{t=T_1}^{T_2-1} \subset V \quad (6.4)$$

and for all integers $t = T_1, \dots, T_2 - 1$,

$$v_t \in V(y_t), \ y_{t+1} \in F_Y(y_t, v_t). \quad (6.5)$$

We denote by $\mathrm{Str}_2(T_1, T_2)$ the set of all the strategies of the second player.

Let $z_1, z_2 \in X$ and $\xi_1, \xi_2 \in Y$. We consider the game

$$\Gamma(z_1, z_2, \xi_1, \xi_2, T_1, T_2).$$

For this game the strategies of the first player are all the strategies

$$(\{x_t\}_{t=T_1}^{T_2}, \{u_t\}_{t=T_1}^{T_2-1}) \in \mathrm{Str}_1(T_1, T_2)$$

satisfying $x_{T_i} = z_i$, $i = 1, 2$, the strategies of the second player are all the strategies $(\{y_t\}_{t=T_1}^{T_2}, \{v_t\}_{t=T_1}^{T_2-1}) \in \mathrm{Str}_2(T_1, T_2)$ satisfying $y_{T_i} = \xi_i$, $i = 1, 2$ and the objective function for the first player associated with the strategies

$$(\{x_t\}_{t=T_1}^{T_2}, \{u_t\}_{t=T_1}^{T_2-1}) \in \mathrm{Str}_1(T_1, T_2), \ (\{y_t\}_{t=T_1}^{T_2}, \{v_t\}_{t=T_1}^{T_2-1}) \in \mathrm{Str}_2(T_1, T_2)$$

is given by $\sum_{t=T_1}^{T_2-1} f(x_t, u_t, y_t, v_t)$.

We also consider the game $\Gamma(z_1, \xi_1, T_1, T_2)$. For this game the strategies of the first player are all the strategies $(\{x_t\}_{t=T_1}^{T_2}, \{u_t\}_{t=T_1}^{T_2-1}) \in \mathrm{Str}_1(T_1, T_2)$

satisfying $x_{T_1} = z_1$, the strategies of the second player are all the strategies $(\{y_t\}_{t=T_1}^{T_2}, \{v_t\}_{t=T_1}^{T_2-1}) \in \mathrm{Str}_2(T_1, T_2)$ satisfying $y_{T_1} = \xi_1$, and the objective function for the first player associated with the strategies

$$(\{x_t\}_{t=T_1}^{T_2}, \{u_t\}_{t=T_1}^{T_2-1}) \in \mathrm{Str}_1(T_1, T_2),\ (\{y_t\}_{t=T_1}^{T_2}, \{v_t\}_{t=T_1}^{T_2-1}) \in \mathrm{Str}_2(T_1, T_2)$$

is given by $\sum_{t=T_1}^{T_2-1} f(x_t, u_t, y_t, v_t)$.

Let T be an integer. A pair of sequences $(\{x_t\}_{t=T}^{\infty}, \{u_t\}_{t=T}^{\infty})$ is called a strategy of the first player if $\{x_t\}_{t=T}^{\infty} \subset X$, $\{u_t\}_{t=T}^{\infty} \subset U$ and for all integers $t \geq T$,

$$u_t \in U(x_t),\ x_{t+1} \in F_X(x_t, u_t).$$

We denote by $\mathrm{Str}_1(T, \infty)$ the set of all the strategies $(\{x_t\}_{t=T}^{\infty}, \{u_t\}_{t=T}^{\infty})$ of the first player.

A pair of sequences $(\{y_t\}_{t=T}^{\infty}, \{v_t\}_{t=T}^{\infty})$ is called a strategy of the second player if $\{y_t\}_{t=T}^{\infty} \subset Y$, $\{v_t\}_{t=T}^{\infty} \subset V$ and for all integers $t \geq T$,

$$v_t \in V(y_t),\ y_{t+1} \in F_Y(y_t, v_t).$$

We denote by $\mathrm{Str}_2(T, \infty)$ the set of all the strategies $(\{y_t\}_{t=T}^{\infty}, \{v_t\}_{t=T}^{\infty})$ of the second player.

In this chapter we study turnpike properties of approximate solutions of the two-player zero-sum games defined above.

Suppose that

$$x^* \in X,\ u^* \in U,\ y^* \in Y,\ v^* \in V, \tag{6.6}$$

$$u^* \in U(x^*),\ x^* \in F_X(x^*, u^*), \tag{6.7}$$

$$v^* \in V(y^*),\ y^* \in F_Y(y^*, v^*) \tag{6.8}$$

and that the following assumption holds:

(A1)

the function $f(\cdot, \cdot, y^*, v^*) : \{(x, u) \in X \times U : u \in U(x)\} \to R^1$ is lower semicontinuous and it is continuous at (x^*, u^*);

the function $f(x^*, u^*, \cdot, \cdot) : \{(y, v) \in Y \times V : v \in V(y)\} \to R^1$ is upper semicontinuous and it is continuous at (y^*, v^*).

We also suppose that there exists a constant $c_* > 0$ such that the following assumption holds:

(A2)

for each integer $T \geq 1$ and each strategy $(\{x_t\}_{t=0}^{T}, \{u_t\}_{t=0}^{T-1}) \in \mathrm{Str}_1(0, T)$,

$$\sum_{t=0}^{T-1} f(x_t, u_t, y^*, v^*) \geq T f(x^*, u^*, y^*, v^*) - c_*;$$

for each integer $T \geq 1$ and each strategy $(\{y_t\}_{t=0}^{T}, \{v_t\}_{t=0}^{T-1}) \in \mathrm{Str}_2(0, T)$,

$$\sum_{t=0}^{T-1} f(x^*, u^*, y_t, v_t) \leq T f(x^*, u^*, y^*, v^*) + c_*.$$

The next proposition easily follows from assumption (A2).

Proposition 6.1. *1. Assume that a strategy* $(\{x_t\}_{t=0}^{\infty}, \{u_t\}_{t=0}^{\infty}) \in Str_1(0,\infty)$. *Then either*

$$\sup\{|\sum_{t=0}^{T-1} f(x_t, u_t, y^*, v^*) - Tf(x^*, u^*, y^*, v^*)| : T = 1, 2, \dots\} < \infty$$

or

$$\lim_{T\to\infty} [\sum_{t=0}^{T-1} f(x_t, u_t, y^*, v^*) - Tf(x^*, u^*, y^*, v^*)] = \infty.$$

2. Assume that a strategy $(\{y_t\}_{t=0}^{\infty}, \{v_t\}_{t=0}^{\infty}) \in Str_2(0,\infty)$. *Then either*

$$\sup\{|\sum_{t=0}^{T-1} f(x^*, u^*, y_t, v_t) - Tf(x^*, u^*, y^*, v^*)| : T = 1, 2, \dots\} < \infty$$

or

$$\lim_{T\to\infty} [\sum_{t=0}^{T-1} f(x^*, u^*, y_t, v_t) - Tf(x^*, u^*, y^*, v^*)] = -\infty.$$

In this chapter we suppose that the following two assumptions hold.

(A3) There exists a real positive constant $\bar{r}$ such that for every pair of points $x, x' \in X$ which satisfies

$$\rho_X(x, x^*),\ \rho_X(x', x^*) \le \bar{r}$$

there exists a point $u \in U(x)$ such that $x' \in F_X(x, u)$ and for every pair of points $y, y' \in Y$ which satisfies

$$\rho_Y(y, y^*),\ \rho_Y(y', y^*) \le \bar{r}$$

there exists a point $v \in V(y)$ such that $y' \in F_Y(y, v)$. Moreover, for every positive number ϵ there exists a positive number δ such that:

for every pair of points $x, x' \in X$ which satisfies

$$\rho_X(x, x^*),\ \rho_X(x', x^*) \le \delta$$

there exists a point $u \in U(x)$ such that $x' \in F_X(x, u)$ and $\rho_U(u, u^*) \le \epsilon$;

for every pair of points $y, y' \in Y$ which satisfies

$$\rho_Y(y, y^*),\ \rho_Y(y', y^*) \le \delta$$

there exists a point $v \in V(y)$ such that $y' \in F_Y(y, v)$ and $\rho_V(v, v^*) \le \epsilon$;

(A4)

for every strategy $(\{x_t\}_{t=0}^{\infty}, \{u_t\}_{t=0}^{\infty}) \in \mathrm{Str}_1(0,\infty)$ which satisfies

$$\sup\{|\sum_{t=0}^{T-1} f(x_t, u_t, y^*, v^*) - Tf(x^*, u^*, y^*, v^*)| : T = 1, 2, \dots\} < \infty \quad (6.9)$$

the equality $\lim_{t\to\infty} \rho_X(x_t, x^*) = 0$ is valid;

for every strategy $(\{y_t\}_{t=0}^{\infty}, \{v_t\}_{t=0}^{\infty}) \in \mathrm{Str}_2(0, \infty)$ which satisfies

$$\sup\{|\sum_{t=0}^{T-1} f(x^*, u^*, y_t, v_t) - Tf(x^*, u^*, y^*, v^*)| : T = 1, 2, \dots\} < \infty \quad (6.10)$$

the equality $\lim_{t \to \infty} \rho_Y(y_t, y^*) = 0$ is valid.

It should be mentioned that analogous assumptions (A2)–(A4) are usually posed in infinite horizon optimal control when a problem possesses a turnpike property and the turnpike is a singleton. In particular, these assumptions follow from standard convexity–concavity assumptions [67, 78]. Examples of games satisfying (A1)–(A4) are considered in Sect. 6.9.

It is easy to see that assumption (A3) is valid if the following assumption holds:

(A3)′ there exists a positive number $\bar{r}$ and mappings

$$G_X : \{x \subset X : \rho_X(x, x^*) \le \bar{r}\} \times \{x \in X : \rho_X(x, x^*) \le \bar{r}\} \to U$$

and

$$G_Y : \{y \in Y : \rho_Y(y, y^*) \le \bar{r}\} \times \{y \in Y : \rho_Y(y, y^*) \le \bar{r}\} \to V$$

such that:

$$G_X(x^*, x^*) = u^*, \ G_Y(y^*, y^*) = v^*;$$

for every pair of points $x, x' \in X$ which satisfies

$$\rho_X(x, x^*), \ \rho_X(x', x^*) \le \bar{r}$$

the inclusions

$$G_X(x, x') \in U(x), \ x' \in F_X(x, G_X(x, x'))$$

hold;

for every pair of points $y, y' \in Y$ which satisfies

$$\rho_Y(y, y^*), \ \rho_Y(y', y^*) \le \bar{r}$$

the inclusions

$$G_Y(y, y') \in V(y), \ y' \in F_Y(y, G_Y(y, y'))$$

hold;

the mapping G_X is continuous at the point (x^*, x^*) and the mapping G_Y is continuous at the point (y^*, y^*).

A point $z \in X$ is called good if there exists a strategy $(\{x_t\}_{t=0}^{\infty}, \{u_t\}_{t=0}^{\infty}) \in \mathrm{Str}_1(0, \infty)$ satisfying (6.9) and the equality $x_0 = z$. The point $z \in X$ is called (M)-good, where M is a positive number, if there exists a strategy $(\{x_t\}_{t=0}^{\infty}, \{u_t\}_{t=0}^{\infty}) \in \mathrm{Str}_1(0, \infty)$ such that $x_0 = z$ and

$$\sum_{t=0}^{T-1} f(x_t, u_t, y^*, v^*) - Tf(x^*, u^*, y^*, v^*) \leq M$$

for all natural numbers T.

A point $\xi \in Y$ is called good if there exists a strategy $(\{y_t\}_{t=0}^{\infty}, \{v_t\}_{t=0}^{\infty}) \in \mathrm{Str}_2(0, \infty)$ satisfying (6.10) and the equality $y_0 = \xi$. The point $\xi \in Y$ is called (M)-good, where M is a positive number, if there exists a strategy $(\{y_t\}_{t=0}^{\infty}, \{v_t\}_{t=0}^{\infty}) \in \mathrm{Str}_2(0, \infty)$ such that $y_0 = \xi$ and

$$\sum_{t=0}^{T-1} f(x^*, u^*, y_t, v_t) - Tf(x^*, u^*, y^*, v^*) \geq -M$$

for all natural numbers T.

Let M be a positive number. The set of all (M)-good points of X is denoted by X_M and the set of all (M)-good points of Y is denoted by Y_M.

Let $L \geq 1$ be an integer. Denote by $X^{(L)}$ the collection of all points $x \in X$ for which there exists a strategy $(\{x_t\}_{t=0}^{L}, \{u_t\}_{t=0}^{L-1}) \in \mathrm{Str}_1(0, L)$ such that $x_0 = x$ and $x_L = x^*$ and denote by $Y^{(L)}$ the collection of all points $y \in Y$ for which there exists a strategy $(\{y_t\}_{t=0}^{L}, \{v_t\}_{t=0}^{L-1}) \in \mathrm{Str}_2(0, L)$ such that $y_0 = y$ and $y_L = y^*$.

Denote by $\bar{X}^{(L)}$ the collection of all points $x \in X$ for which that there exists

$$(\{x_t\}_{t=0}^{L}, \{u_t\}_{t=0}^{L-1}) \in \mathrm{Str}_1(0, L)$$

such that $x_0 = x^*$ and $x_L = x$ and denote by $\bar{Y}^{(L)}$ the collection of all points $y \in Y$ for which that there exists a strategy $(\{y_t\}_{t=0}^{L}, \{v_t\}_{t=0}^{L-1}) \in \mathrm{Str}_2(0, L)$ such that $y_0 = y^*$ and $y_L = y$.

It is easy to see that the following result holds.

Proposition 6.2. *Let $L \geq 1$ be an integer. Then $X^{(L)} \subset X_{2L\|f\|}$ and $Y^{(L)} \subset Y_{2L\|f\|}$.*

For the proof of the next results see Lemmas 6.12 and 6.3.

Proposition 6.3. *Let M be a positive number. Then there exists an integer $L \geq 1$ such that $X_M \subset X^{(L)}$ and $Y_M \subset Y^{(L)}$.*

6.2 Main Results

Let integers T_1, T_2 satisfy $T_2 > T_1$, $M \geq 0$, $(\{x_t\}_{t=T_1}^{T_2}, \{u_t\}_{t=T_1}^{T_2-1}) \in \mathrm{Str}_1(T_1, T_2)$ and $(\{y_t\}_{t=T_1}^{T_2}, \{v_t\}_{t=T_1}^{T_2-1}) \in \mathrm{Str}_2(T_1, T_2)$. be a pair of strategies. This pair of strategies is called (M)-good if the following properties hold:

for every $(\{x'_t\}_{t=T_1}^{T_2}, \{u'_t\}_{t=T_1}^{T_2-1}) \in \mathrm{Str}_1(T_1, T_2)$ satisfying $x_{T_i} = x'_{T_i}$, $i = 1, 2$,

$$\sum_{t=T_1}^{T_2-1} f(x'_t, u'_t, y_t, v_t) \geq \sum_{t=T_1}^{T_2-1} f(x_t, u_t, y_t, v_t) - M; \tag{6.11}$$

for every $(\{y'_t\}_{t=T_1}^{T_2}, \{v'_t\}_{t=T_1}^{T_2-1}) \in \mathrm{Str}_2(T_1, T_2)$ satisfying $y_{T_i} = y'_{T_i}$, $i = 1, 2$,

$$\sum_{t=T_1}^{T_2-1} f(x_t, u_t, y'_t, v'_t) \le \sum_{t=T_1}^{T_2-1} f(x_t, u_t, y_t, v_t) + M. \tag{6.12}$$

The pair of strategies $(\{x_t\}_{t=T_1}^{T_2}, \{u_t\}_{t=T_1}^{T_2-1})$ and $(\{y_t\}_{t=T_1}^{T_2}, \{v_t\}_{t=T_1}^{T_2-1})$ is called $(M, *)$-good if the following properties hold:

for every $(\{x'_t\}_{t=T_1}^{T_2}, \{u'_t\}_{t=T_1}^{T_2-1}) \in \mathrm{Str}_1(T_1, T_2)$ satisfying $x_{T_1} = x'_{T_1}$ inequality (6.11) is valid;

for every $(\{y'_t\}_{t=T_1}^{T_2}, \{v'_t\}_{t=T_1}^{T_2-1}) \in \mathrm{Str}_2(T_1, T_2)$ satisfying $y_{T_1} = y'_{T_1}$ inequality (6.12) is valid.

In Sect. 6.3 we prove the following two results which establish the existence of good pairs of strategies.

Theorem 6.4. *Let $L \ge 1$ be an integer. Then for every natural number $T > 2L$, every point $z_1 \in X^{(L)}$, every point $z_2 \in \bar{X}^{(L)}$, every point $\xi_1 \in Y^{(L)}$, and every point $\xi_2 \in \bar{Y}^{(L)}$ there exists $(4L\|f\| + c_*)$-good pair of strategies $(\{x_t\}_{t=0}^{T}, \{u_t\}_{t=0}^{T-1}) \in Str_1(0, T)$ and $(\{y_t\}_{t=0}^{T}, \{v_t\}_{t=0}^{T-1}) \in Str_2(0, T)$ such that*

$$x_0 = z_1,\ x_T = z_2,\ y_0 = \xi_1,\ y_T = \xi_2,$$

$$x_t = x^*,\ y_t = y^* \text{ for all } t = L, \dots, T - L,$$

$$u_t = u^*,\ v_t = v^* \text{ for all } t = L, \dots, T - L - 1.$$

Theorem 6.5. *Let $L \ge 1$ be an integer. Then for every natural number $T > L$, every point $z \in X^{(L)}$, and every point $\xi \in Y^{(L)}$ there exists $(2L\|f\| + c_*, *)$-good pair of strategies $(\{x_t\}_{t=0}^{T}, \{u_t\}_{t=0}^{T-1}) \in Str_1(0, T)$ and $(\{y_t\}_{t=0}^{T}, \{v_t\}_{t=0}^{T-1}) \in Str_2(0, T)$ such that*

$$x_0 = z,\ y_0 = \xi,$$

$$x_t = x^*,\ y_t = y^* \text{ for all } t = L, \dots, T,$$

$$u_t = u^*,\ v_t = v^* \text{ for all } t = L, \dots, T - 1.$$

In Sect. 6.4 we prove the following result.

Theorem 6.6. *Let $T \ge 1$ be an integer and let*

$$x_t = x^*,\ y_t = y^* \text{ for all } t = 0, \dots, T,$$

$$u_t = u^*,\ v_t = v^* \text{ for all } t = 0, \dots, T - 1.$$

Then $(\{x_t\}_{t=0}^{T}, \{u_t\}_{t=0}^{T-1}) \in Str_1(0, T)$, $(\{y_t\}_{t=0}^{T}, \{v_t\}_{t=0}^{T-1}) \in Str_2(0, T)$ is (0)-good pair of strategies.

The proof of the following result is given in Sect. 6.5.

Theorem 6.7. *Suppose that the following assumption holds:*

(A^*) for every positive number ϵ there exists a positive number δ such that for every $(x, u, y, v) \in X \times U \times Y \times V$ satisfying

$$u \in U(x),\ v \in V(y),\ \rho_Y(y, y^*) \le \delta,\ \rho_V(v, v^*) \le \delta$$

the inequality

$$|f(x, u, y, v) - f(x, u, y^*, v^*)| \le \epsilon$$

is valid and such that for every $(x, u, y, v) \in X \times U \times Y \times V$ satisfying

$$u \in U(x),\ v \in V(y),\ \rho_X(x, x^*) \le \delta,\ \rho_U(u, u^*) \le \delta$$

the inequality

$$|f(x, u, y, v) - f(x^*, u^*, y, v)| \le \epsilon$$

is valid.

Let ϵ be a positive number. Then there exists a positive number δ such that for every natural number $T > 2$, every pair of points $z_0,\ z_1 \in X$, and every pair of points $\xi_0,\ \xi_1 \in Y$ which satisfy

$$\rho_X(z_i, x^*) \le \delta,\ i = 0, 1 \text{ and } \rho_Y(\xi_i, y^*) \le \delta,\ i = 0, 1$$

there exists an (ϵ)-good pair of strategies

$$(\{x_t\}_{t=0}^T, \{u_t\}_{t=0}^{T-1}) \in Str_1(0, T),\ (\{y_t\}_{t=0}^T, \{v_t\}_{t=0}^{T-1}) \in Str_2(0, T)$$

such that

$$\begin{gathered} x_0 = z_0,\ x_t = x^* \text{ for all } t = 1, \dots, T-1,\ x_T = z_1, \\ u_t = u^* \text{ for all } t = 1, \dots, T-2, \\ y_0 = \xi_0,\ y_t = y^* \text{ for all } t = 1, \dots, T-2,\ y_T = \xi_1, \\ v_t = v^* \text{ for all } t = 1, \dots, T-2, \\ \rho_U(u_0, u^*) \le \epsilon,\ \rho_U(u_{T-1}, u^*) \le \epsilon,\ \rho_V(v_0, v^*) \le \epsilon,\ \rho_V(v_{T-1}, v^*) \le \epsilon. \end{gathered}$$

In Sect. 6.7 we prove the following two theorems which are our main turnpike results in this chapter. Section 6.6 contains auxiliary results needed for these theorems.

Theorem 6.8. *Let $M, \epsilon > 0$ and L_0 be a natural number. Then there exist natural numbers l, Q such that for each integer $T > Ql$ and each (M)-good pair of strategies $(\{x_t\}_{t=0}^T, \{u_t\}_{t=0}^{T-1}) \in Str_1(0, T)$ and $(\{y_t\}_{t=0}^T, \{v_t\}_{t=0}^{T-1}) \in Str_2(0, T)$ such that*

$$x_0 \in X^{(L_0)},\ x_T \in \bar{X}^{(L_0)},\ y_0 \in Y^{(L_0)},\ y_T \in \bar{Y}^{(L_0)}$$

there exist a natural number $q \le Q$ and sequences of integers

$$\{a_i\}_{i=1}^q,\ \{b_i\}_{i=1}^q \subset [0, T]$$

such that

$$0 \le b_i - a_i \le l,\ i = 1, \dots, q,$$

$$\rho_X(x_t, x^*) \le \epsilon,\ \rho_Y(y_t, y^*) \le \epsilon$$

for all integers $t \in [0, T] \setminus \cup_{i=1}^{q} [a_i, b_i]$.

Theorem 6.9. *Let* $M, \epsilon > 0$ *and* L_0 *be a natural number. Then there exist natural numbers* l, Q *such that for each integer* $T > Ql$ *and each* $(M, *)$*-good pair of strategies* $(\{x_t\}_{t=0}^{T}, \{u_t\}_{t=0}^{T-1}) \in Str_1(0, T)$ *and* $(\{y_t\}_{t=0}^{T}, \{v_t\}_{t=0}^{T-1}) \in Str_2(0, T)$ *such that*

$$x_0 \in X^{(L_0)},\ y_0 \in Y^{(L_0)}$$

there exist a natural number $q \le Q$ *and sequences of integers*

$$\{a_i\}_{i=1}^{q},\ \{b_i\}_{i=1}^{q} \subset [0, T]$$

such that

$$0 \le b_i - a_i \le l,\ i = 1, \dots, q,$$

$$\rho_X(x_t, x^*) \le \epsilon,\ \rho_Y(y_t, y^*) \le \epsilon$$

for all integers $t \in [0, T] \setminus \cup_{i=1}^{q} [a_i, b_i]$.

Theorem 6.10. *Let assumption* (A^*) *of Theorem 6.7 hold,* $\epsilon > 0$ *and* L_0 *be a natural number. Then there exist a natural number* $L > L_0$ *and* $\delta > 0$ *such that for each integer* $T > 2L + 1$ *and each* (δ)*-good pair of strategies* $(\{x_t\}_{t=0}^{T}, \{u_t\}_{t=0}^{T-1}) \in Str_1(0, T)$ *and* $(\{y_t\}_{t=0}^{T}, \{v_t\}_{t=0}^{T-1}) \in Str_2(0, T)$ *satisfying*

$$x_0 \in X^{(L_0)},\ x_T \in \bar{X}^{(L_0)},\ y_0 \in Y^{(L_0)},\ y_T \in \bar{Y}^{(L_0)}$$

there exist integers

$$\tau_1 \in [0, L],\ \tau_2 \in [T - L, T]$$

such that

$$\rho_X(x_t, x^*) \le \epsilon,\ \rho_Y(y_t, y^*) \le \epsilon$$

for all integers $t \in [\tau_1, \tau_2]$. *Moreover, if* $\rho_X(x_0, x^*) \le \delta$, $\rho_Y(y_0, y^*) \le \delta$, *then* $\tau_1 = 0$ *and if* $\rho_X(x_T, x^*) \le \delta$, $\rho_Y(y_T, y^*) \le \delta$, *then* $\tau_2 = T$.

Theorem 6.11. *Let assumption* (A^*) *of Theorem 6.7 hold,* $\epsilon > 0$ *and* L_0 *be a natural number. Then there exist a natural number* $L > L_0$ *and* $\delta > 0$ *such that for each integer* $T > 2L + 1$ *and each* $(\delta, *)$*-good pair of strategies* $(\{x_t\}_{t=0}^{T}, \{u_t\}_{t=0}^{T-1}) \in Str_1(0, T)$ *and* $(\{y_t\}_{t=0}^{T}, \{v_t\}_{t=0}^{T-1}) \in Str_2(0, T)$ *satisfying*

$$x_0 \in X^{(L_0)},\ y_0 \in Y^{(L_0)}$$

there exist integers

$$\tau_1 \in [0, L],\ \tau_2 \in [T - L, T]$$

such that

$$\rho_X(x_t, x^*) \le \epsilon,\ \rho_Y(y_t, y^*) \le \epsilon$$

for all integers $t \in [\tau_1, \tau_2]$. *Moreover, if* $\rho_X(x_0, x^*) \le \delta$, $\rho_Y(y_0, y^*) \le \delta$, *then* $\tau_1 = 0$ *and if* $\rho_X(x_T, x^*) \le \delta$, $\rho_Y(y_T, y^*) \le \delta$, *then* $\tau_2 = T$.

The proofs of Theorems 6.10 and 6.11 are given in Sect. 6.8. Section 6.9 contains examples.

The results of this section were obtained in [88].

6.3 Proofs of Theorems 6.4 and 6.5

Proof of Theorem 6.4

Let an integer $T > 2L$ and let

$$z_1 \in X^{(L)},\ z_2 \in \bar{X}^{(L)},\ \xi_1 \in Y^{(L)},\ \xi_2 \in \bar{Y}^{(L)}. \tag{6.13}$$

In view of the definitions of $X^{(L)}$ and $Y^{(L)}$ and (6.13) there exist

$$(\{x_t\}_{t=0}^{L}, \{u_t\}_{t=0}^{L-1}) \in \mathrm{Str}_1(0, L) \tag{6.14}$$

such that

$$x_0 = z_1,\ x_L = x^* \tag{6.15}$$

and

$$(\{y_t\}_{t=0}^{L}, \{v_t\}_{t=0}^{L-1}) \in \mathrm{Str}_2(0, L) \tag{6.16}$$

which satisfies

$$y_0 = \xi_1,\ y_L = y^*. \tag{6.17}$$

Define

$$x_t = x^* \text{ for all } t = L+1, \dots, T-L, \tag{6.18}$$

$$u_t = u^* \text{ for all } t = L, \dots, T-L-1, \tag{6.19}$$

$$y_t = y^* \text{ for all } t = L+1, \dots, T-L, \tag{6.20}$$

$$v_t = v^* \text{ for all } t = L, \dots, T-L-1. \tag{6.21}$$

In view of the definitions of $\bar{X}^{(L)}$ and $\bar{Y}^{(L)}$, (6.13), (6.18), and (6.20) there exist

$$(\{x_t\}_{t=T-L}^{T}, \{u_t\}_{t=T-L}^{T-1}) \in \mathrm{Str}_1(T-L, T) \tag{6.22}$$

such that

$$x_T = z_2 \tag{6.23}$$

and

$$(\{y_t\}_{t=T-L}^{T}, \{v_t\}_{t=T-L}^{T-1}) \in \mathrm{Str}_2(T-L, T) \tag{6.24}$$

which satisfies

$$y_T = \xi_2. \tag{6.25}$$

It follows from (6.6), (6.8), (6.14), (6.16), (6.18), (6.21), (6.22), and (6.24) that

$$(\{x_t\}_{t=0}^{T}, \{u_t\}_{t=0}^{T-1}) \in \mathrm{Str}_1(0, T),\ (\{y_t\}_{t=0}^{T}, \{v_t\}_{t=0}^{T-1}) \in \mathrm{Str}_2(0, T).$$

By (6.15), (6.17), (6.23), and (6.25), we have

$$x_0 = z_1, \; x_T = z_2, \; y_0 = \xi_1, \; y_T = \xi_2. \tag{6.26}$$

We claim that the pair of strategies $(\{x_t\}_{t=0}^T, \{u_t\}_{t=0}^{T-1}), (\{y_t\}_{t=0}^T, \{v_t\}_{t=0}^{T-1})$ is $(4L\|f\| + c_*)$-good.

Assume that

$$(\{x'_t\}_{t=0}^T, \{u'_t\}_{t=0}^{T-1}) \in \mathrm{Str}_1(0, T)$$

and that

$$x'_0 = x_0, \; x'_T = x_T.$$

In view of (6.15), (6.18), (6.19), and (A2), we have

$$\sum_{t=0}^{T-1} f(x'_t, u'_t, y_t, v_t) - \sum_{t=0}^{T-1} f(x_t, u_t, y_t, v_t)$$

$$= \sum_{t=0}^{L-1} f(x'_t, u'_t, y_t, v_t) - \sum_{t=0}^{L-1} f(x_t, u_t, y_t, v_t)$$

$$+ \sum_{t=L}^{T-L-1} f(x'_t, u'_t, y_t, v_t) - (T - 2L) f(x^*, u^*, y^*, v^*)$$

$$+ \sum_{t=T-L}^{T-1} f(x'_t, u'_t, y_t, v_t) - \sum_{t=T-L}^{T-1} f(x_t, u_t, y_t, v_t) \geq -4L\|f\| - c_*.$$

Assume that

$$(\{y'_t\}_{t=0}^T, \{v'_t\}_{t=0}^{T-1}) \in \mathrm{Str}_2(0, T)$$

and that

$$y'_0 = y_0, \; y'_T = y_T.$$

In view of (6.17), (6.20), (6.21), and (A2), we have

$$\sum_{t=0}^{T-1} f(x_t, u_t, y'_t, v'_t) - \sum_{t=0}^{T-1} f(x_t, u_t, y_t, v_t)$$

$$= \sum_{t=0}^{L-1} f(x_t, u_t, y'_t, v'_t) - \sum_{t=0}^{L-1} f(x_t, u_t, y_t, v_t)$$

$$+ \sum_{t=L}^{T-L-1} f(x_t, u_t, y'_t, v'_t) - (T - 2L) f(x^*, u^*, y^*, v^*)$$

$$+ \sum_{t=T-L}^{T-1} f(x_t, u_t, y'_t, v'_t) - \sum_{t=T-L}^{T-1} f(x_t, u_t, y_t, v_t) \leq 4L\|f\| + c_*.$$

This completes the proof of Theorem 6.4. □

Proof of Theorem 6.5

Let an integer $T > L$ and let

$$z \in X^{(L)},\ \xi \in Y^{(L)}. \tag{6.27}$$

In view of the definitions of $X^{(L)}$ and $Y^{(L)}$ and (6.27) there exist

$$(\{x_t\}_{t=0}^{L}, \{u_t\}_{t=0}^{L-1}) \in \mathrm{Str}_1(0, L) \tag{6.28}$$

such that

$$x_0 = z,\ x_L = x^* \tag{6.29}$$

and

$$(\{y_t\}_{t=0}^{L}, \{v_t\}_{t=0}^{L-1}) \in \mathrm{Str}_2(0, L) \tag{6.30}$$

which satisfies

$$y_0 = \xi,\ y_L = y^*. \tag{6.31}$$

Define

$$x_t = x^* \text{ for all } t = L+1, \dots, T, \tag{6.32}$$

$$u_t = u^* \text{ for all } t = L, \dots, T-1, \tag{6.33}$$

$$y_t = y^* \text{ for all } t = L+1, \dots, T, \tag{6.34}$$

$$v_t = v^* \text{ for all } t = L, \dots, T-1. \tag{6.35}$$

It follows from (6.6), (6.8), and (6.28)–(6.35) that

$$(\{x_t\}_{t=0}^{T}, \{u_t\}_{t=0}^{T-1}) \in \mathrm{Str}_1(0, T),\ (\{y_t\}_{t=0}^{T}, \{v_t\}_{t=0}^{T-1}) \in \mathrm{Str}_2(0, T),$$

$$x_0 = z,\ y_0 = \xi.$$

We claim that the pair of strategies $(\{x_t\}_{t=0}^{T}, \{u_t\}_{t=0}^{T-1})$, $(\{y_t\}_{t=0}^{T}, \{v_t\}_{t=0}^{T-1})$ is $(2L\|f\| + c_*, *)$-good.

Assume that

$$(\{x'_t\}_{t=0}^{T}, \{u'_t\}_{t=0}^{T-1}) \in \mathrm{Str}_1(0, T)$$

and that

$$x'_0 = x_0.$$

In view of (6.29), (6.32), (6.33), and (A2), we have

$$\sum_{t=0}^{T-1} f(x'_t, u'_t, y_t, v_t) - \sum_{t=0}^{T-1} f(x_t, u_t, y_t, v_t)$$

$$= \sum_{t=0}^{L-1} f(x'_t, u'_t, y_t, v_t) - \sum_{t=0}^{L-1} f(x_t, u_t, y_t, v_t)$$

$$+ \sum_{t=L}^{T-1} f(x'_t, u'_t, y_t, v_t) - (T-L) f(x^*, u^*, y^*, v^*) \geq -2L\|f\| - c_*.$$

Assume that

$$(\{y'_t\}_{t=0}^T, \{v'_t\}_{t=0}^{T-1}) \in \mathrm{Str}_2(0,T)$$

and that

$$y'_0 = y_0.$$

It follows from (6.31), (6.34), (6.35), and (A2) that

$$\sum_{t=0}^{T-1} f(x_t, u_t, y'_t, v'_t) - \sum_{t=0}^{T-1} f(x_t, u_t, y_t, v_t)$$

$$= \sum_{t=0}^{L-1} f(x_t, u_t, y'_t, v'_t) - \sum_{t=0}^{L-1} f(x_t, u_t, y_t, v_t)$$

$$+ \sum_{t=L}^{T-1} f(x_t, u_t, y'_t, v'_t) - (T-L) f(x^*, u^*, y^*, v^*) \leq 2L\|f\| + c_*.$$

This completes the proof of Theorem 6.5. □

6.4 Proof of Theorem 6.6

Let

$$(\{x'_t\}_{t=0}^T, \{u'_t\}_{t=0}^{T-1}) \in \mathrm{Str}_1(0,T)$$

satisfy

$$x'_0 = x^*, \ x'_T = x^*.$$

There exist a point $x'_t \in X$ for all integers $t \geq T+1$ and a point $u'_t \in U$ for all integers $t \geq T$ such that for all nonnegative integers t, we have

$$x'_{t+T} = x'_t, \ u'_{t+T} = u'_t.$$

It is clear that

$$(\{x'_t\}_{t=0}^\infty, \{u'_t\}_{t=0}^\infty) \in \mathrm{Str}_1(0,\infty).$$

In view of (A2), for every natural number k,

$$-c_* \leq \sum_{t=0}^{kT-1} f(x'_t, u'_t, y^*, v^*) - kTf(x^*, u^*, y^*, v^*)$$

$$= k(\sum_{t=0}^{T-1} f(x'_t, u'_t, y^*, v^*) - Tf(x^*, u^*, y^*, v^*)).$$

It follows from the relation above that

$$\sum_{t=0}^{T-1} f(x'_t, u'_t, y_t, v_t) \geq \sum_{t=0}^{T-1} f(x_t, u_t, y_t, v_t).$$

Analogously we can show that for every

$$(\{y'_t\}_{t=0}^{T}, \{v'_t\}_{t=0}^{T-1}) \in \mathrm{Str}_2(0,T)$$

which satisfies

$$y'_0 = y^*, \; y'_T = y^*$$

the inequality

$$\sum_{t=0}^{T-1} f(x_t, u_t, y'_t, v'_t) \le \sum_{t=0}^{T-1} f(x_t, u_t, y_t, v_t)$$

holds. This completes the proof of Theorem 6.6. □

6.5 Proof of Theorem 6.7

Assumptions (A^*) implies that there exists a number $\epsilon_0 \in (0, \epsilon/32)$ such that:

(i) for every $(x,u,y,v) \in X \times U \times Y \times V$ which satisfies

$$u \in U(x), \; v \in V(y), \; \rho_X(x,x^*) \le \epsilon_0, \; \rho_U(u,u^*) \le \epsilon_0$$

the inequality

$$|f(x,u,y,v) - f(x^*,u^*,y,v)| \le \epsilon/64$$

holds;

(ii) for every $(x,u,y,v) \in X \times U \times Y \times V$ which satisfies

$$u \in U(x), \; v \in V(y), \; \rho_Y(y,y^*) \le \epsilon_0, \; \rho_V(v,v^*) \le \epsilon_0$$

the inequality

$$|f(x,u,y,v) - f(x,u,y^*,v^*)| \le \epsilon/64$$

holds.

Assumption (A3) implies that there exists a number $\delta \in (0, \epsilon_0)$ such that the following properties hold:

(iii) for every pair of points $x, x' \in X$ which satisfies

$$\rho_X(x,x^*), \; \rho_X(x',x^*) \le \delta$$

there exists a point $u \in U(x)$ such that $x' \in F_X(x,u)$ and $\rho_U(u,u^*) \le \epsilon_0$;

(iv) for every pair of point $y, y' \in Y$ which satisfies

$$\rho_Y(y,y^*), \; \rho_Y(y',y^*) \le \delta$$

there exists a point $v \in V(y)$ such that $y' \in F_Y(y,v)$ and $\rho_V(v,v^*) \le \epsilon_0$.

Let an integer $T > 2$ and let points

$$z_0, \; z_1 \in X, \; \xi_0, \; \xi_1 \in Y \tag{6.36}$$

satisfy

$$\rho_X(z_i, x^*) \le \delta,\ i = 0, 1 \text{ and } \rho_Y(\xi_i, y^*) \le \delta,\ i = 0, 1. \tag{6.37}$$

Define

$$x_0 = z_0,\ x_t = x^* \text{ for all } t = 1, \dots, T-1,\ x_T = z_1,$$
$$y_0 = \xi_0,\ y_t = y^* \text{ for all } t = 1, \dots, T-1,\ y_T = \xi_1,$$
$$u_t = u^*,\ v_t = v^* \text{ for all } t = 1, \dots, T-2. \tag{6.38}$$

In view of (6.37), (6.38), properties (iii) and (iv), there exist

$$u_0,\ u_{T-1} \in U,\ v_0,\ v_{T-1} \in V$$

such that

$$u_0 \in U(x_0),\ x^* \in F_X(x_0, u_0),\ \rho_U(u_0, u^*) \le \epsilon_0, \tag{6.39}$$
$$u_{T-1} \in U(x^*),\ x_T \in F_X(x_{T-1}, u_{T-1}),\ \rho_U(u_{T-1}, u^*) \le \epsilon_0, \tag{6.40}$$
$$v_0 \in V(y_0),\ y^* \in F_Y(y_0, v_0),\ \rho_V(y_0, v^*) \le \epsilon_0, \tag{6.41}$$
$$v_{T-1} \in V(y^*),\ y_T \in F_Y(y_{T-1}, v_{T-1}),\ \rho_V(v_{T-1}, v^*) \le \epsilon_0. \tag{6.42}$$

In view of (6.36)–(6.42), we have

$$(\{x_t\}_{t=0}^T, \{u_t\}_{t=0}^{T-1}) \in \mathrm{Str}_1(0, T),\ (\{y_t\}_{t=0}^T, \{v_t\}_{t=0}^{T-1}) \in \mathrm{Str}_2(0, T),$$

$$x_0 = z_0,\ x_T = z_1,\ y_0 = \xi_0,\ y_T = \xi_1.$$

In order to complete the proof of the theorem it is sufficient to show that the pair of strategies

$$(\{x_t\}_{t=0}^T, \{u_t\}_{t=0}^{T-1}) \in \mathrm{Str}_1(0, T),\ (\{y_t\}_{t=0}^T, \{v_t\}_{t=0}^{T-1}) \in \mathrm{Str}_2(0, T),$$

is (ϵ)-good.

Assume that

$$(\{x'_t\}_{t=0}^T, \{u'_t\}_{t=0}^{T-1}) \in \mathrm{Str}_1(0, T),\ x'_0 = z_0,\ x'_T = z_1. \tag{6.43}$$

In view of (6.37), (6.43), properties (iii) and (iv), there exist $u'_{-1},\ u'_T \in U$ such that

$$u'_{-1} \in U(x^*),\ x_0 \in F_X(x^*, u'_{-1}),\ \rho_U(u^*, u'_{-1}) \le \epsilon_0, \tag{6.44}$$

$$u'_T \in U(z_1),\ x^* \in F_X(z_1, u'_T),\ \rho_U(u_T, u^*) \le \epsilon_0. \tag{6.45}$$

Define

$$x'_{-1} = x^*,\ x'_{T+1} = x^*. \tag{6.46}$$

It follows from (6.43) to (6.46) that

$$(\{x'_t\}_{t=-1}^{T+1}, \{u'_t\}_{t=-1}^T) \in \mathrm{Str}_1(-1, T+1). \tag{6.47}$$

In view of (6.45), (6.47), and Theorem 6.6, we have

$$\sum_{t=-1}^{T} f(x'_t, u'_t, y^*, v^*) \geq (T+2) f(x^*, u^*, y^*, v^*). \tag{6.48}$$

Relations (6.38) and (6.48) imply that

$$\sum_{t=0}^{T-1} f(x'_t, u'_t, y_t, v_t) = f(x'_0, u'_0, y_0, v_0)$$

$$+ \sum_{t=1}^{T-2} f(x'_t, u'_t, y^*, v^*) + f(x'_{T-1}, u'_{T-1}, y_{T-1}, v_{T-1})$$

$$= \sum_{t=-1}^{T} f(x'_t, u'_t, y^*, v^*) - \sum_{t=-1}^{0} f(x'_t, u'_t, y^*, v^*)$$

$$- \sum_{t=T-1}^{T} f(x'_t, u'_t, y^*, v^*) + f(x'_0, u'_0, y_0, v_0) + f(x'_{T-1}, u'_{T-1}, y_{T-1}, v_{T-1})$$

$$\geq T f(x^*, u^*, y^*, v^*) + (f(x^*, u^*, y^*, v^*) - f(x'_{-1}, u'_{-1}, y^*, v^*))$$

$$+ (f(x^*, u^*, y^*, v^*) - f(x'_T, u'_T, y^*, v^*))$$

$$+ (f(x'_0, u'_0, y_0, v_0) - f(x'_0, u'_0, y^*, v^*))$$

$$+ f(x'_{T-1}, u'_{T-1}, y_{T-1}, v_{T-1}) - f(x'_{T-1}, u'_{T-1}, y^*, v^*). \tag{6.49}$$

It follows from (6.37), (6.38), (6.41)–(6.46), properties (i) and (ii) that

$$|f(x^*, u^*, y^*, v^*) - f(x'_{-1}, u'_{-1}, y^*, v^*)| \leq \epsilon/64,$$

$$|f(x^*, u^*, y^*, v^*) - f(x'_T, u'_T, y^*, v^*)| \leq \epsilon/64,$$

$$|f(x'_0, u'_0, y_0, v_0) - f(x'_0, u'_0, y^*, v^*)| \leq \epsilon/64,$$

$$|f(x'_{T-1}, u'_{T-1}, y_{T-1}, v_{T-1}) - f(x'_{T-1}, u'_{T-1}, y^*, v^*)| \leq \epsilon/64. \tag{6.50}$$

In view of (6.49) and (6.50),

$$\sum_{t=0}^{T-1} f(x'_t, u'_t, y_t, v_t) \geq T f(x^*, u^*, y^*, v^*) - \epsilon/16. \tag{6.51}$$

In view of (6.38), we have

$$\sum_{t=0}^{T-1} f(x_t, u_t, y_t, v_t) = f(x_0, u_0, y_0, v_0) + (T-2) f(x^*, u^*, y^*, v^*)$$

$$+ f(x_{T-1}, u_{T-1}, y_{T-1}, v_{T-1}). \tag{6.52}$$

By (6.37)–(6.42) and property (i),

$$|f(x_0, u_0, y_0, v_0) - f(x^*, u^*, y_0, v_0)| \le \epsilon/64,$$
$$|f(x^*, u^*, y^*, v^*) - f(x^*, u^*, y_0, v_0)| \le \epsilon/64,$$
$$|f(x_{T-1}, u_{T-1}, y_{T-1}, v_{T-1}) - f(x^*, u^*, y_{T-1}, v_{T-1})| \le \epsilon/64,$$
$$|f(x^*, u^*, y_{T-1}, v_{T-1}) - f(x^*, u^*, y^*, v^*)| \le \epsilon/64. \tag{6.53}$$

In view of (6.52) and (6.53),

$$\sum_{t=0}^{T-1} f(x_t, u_t, y_t, v_t) \le Tf(x^*, u^*, y^*, v^*) + \epsilon/16. \tag{6.54}$$

By (6.51) and (6.54), we have

$$\sum_{t=0}^{T-1} f(x'_t, u'_t, y_t, v_t) \ge \sum_{t=0}^{T-1} f(x_t, u_t, y_t, v_t) - \epsilon/8.$$

Analogously we can show that for every

$$(\{y'_t\}_{t=0}^{T}, \{v'_t\}_{t=0}^{T-1}) \in \mathrm{Str}_2(0, T)$$

which satisfies

$$y'_0 = y_0,\ y'_T = y_T$$

the inequality

$$\sum_{t=0}^{T-1} f(x_t, u_t, y'_t, v'_t) \le \sum_{t=0}^{T-1} f(x_t, u_t, y_t, v_t) + \epsilon/8$$

holds. This completes the proof of Theorem 6.7. □

6.6 Auxiliary Results

We suppose that the sum over the empty set is zero.

Lemma 6.12. *Let M_0, ϵ be positive numbers and let $L_0 \ge 1$ be an integer. Then there exists an integer $L_1 > L_0$ such that for every integer $T \ge L_1$, every strategy $(\{x_t\}_{t=0}^{T}, \{u_t\}_{t=0}^{T-1}) \in Str_1(0, T)$ satisfying*

$$\sum_{t=0}^{T-1} f(x_t, u_t, y^*, v^*) \le Tf(x^*, u^*, y^*, v^*) + M_0$$

and every integer $S \in [0, T - L_1]$ there exists an integer $S_1 \in [S, S + L_1 - L_0]$ such that

$$\rho_X(x_t, x^*) \le \epsilon,\ t = S_1, \dots, S_1 + L_0.$$

Proof. Assume the contrary. Then for every natural number $k > L_0$ there exist a natural number $T_k \geq k$, a strategy

$$(\{x_t^{(k)}\}_{t=0}^{T_k}, \{u_t^{(k)}\}_{t=0}^{T_k-1}) \in \mathrm{Str}_1(0, T_k) \tag{6.55}$$

which satisfies

$$\sum_{t=0}^{T_k-1} f(x_t^{(k)}, u_t^{(k)}, y^*, v^*) \leq T_k f(x^*, u^*, y^*, v^*) + M_0 \tag{6.56}$$

and an integer

$$S_k \in [0, T_k - k] \tag{6.57}$$

such that the following property holds:

(P1) for every integer $\tau \in [S_k, S_k + k - L_0]$ the inequality

$$\max\{\rho_X(x_t, x^*) : t = \tau, \dots, \tau + L_0\} > \epsilon \tag{6.58}$$

holds.

Let an integer $k > L_0$ and let integers τ_1, τ_2 satisfy

$$0 \leq \tau_1 < \tau_2 \leq T_k.$$

In view of (6.156) and (A2),

$$\sum_{t=\tau_1}^{\tau_2-1} f(x_t^{(k)}, u_t^{(k)}, y^*, v^*) = \sum_{t=0}^{T_k-1} f(x_t^{(k)}, u_t^{(k)}, y^*, v^*)$$

$$-\sum\{f(x_t^{(k)}, u_t^{(k)}, y^*, v^*) : t \in \{0, \dots, T_k - 1\} \text{ and } t < \tau_1\}$$

$$-\sum\{f(x_t^{(k)}, u_t^{(k)}, y^*, v^*) : t \in \{0, \dots, T_k - 1\} \text{ and } t \geq \tau_2\}$$

$$\leq T_k f(x^*, u^*, y^*, v^*) + M_0 - \tau_1 f(x^*, u^*, y^*, v^*) + c_*$$

$$-(T_k - \tau_2) f(x^*, u^*, y^*, v^*) + c_* = (\tau_2 - \tau_1) f(x^*, u^*, y^*, v^*) + M_0 + 2c_*.$$

Hence for every integer $k > L_0$ and every pair of integers τ_1, τ_2 satisfying

$$0 \leq \tau_1 < \tau_2 \leq T_k$$

we have

$$\sum_{t=\tau_1}^{\tau_2-1} f(x_t^{(k)}, u_t^{(k)}, y^*, v^*) \leq (\tau_2 - \tau_1) f(x^*, u^*, y^*, v^*) + M_0 + 2c_*. \tag{6.59}$$

Let $k > L_0$ be an integer. Define

$$\tilde{x}_t^{(k)} = x_{t+S_k}^{(k)}, \ t = 0, \dots, k, \tag{6.60}$$

$$\tilde{u}_t^{(k)} = u_{t+S_k}^{(k)},\ t = 0, \dots, k-1. \tag{6.61}$$

In view of (6.55), (6.57), (6.60), and (6.61),

$$(\{\tilde{x}_t^{(k)}\}_{t=0}^{k}, \{\tilde{u}_t^{(k)}\}_{t=0}^{k-1}) \in \mathrm{Str}_1(0, k). \tag{6.62}$$

It follows from (6.59), (6.60), and (6.61) that for every integer $k > L_0$ and every pair of integers τ_1, τ_2 which satisfy

$$0 \le \tau_1 < \tau_2 \le k$$

we have

$$\sum_{t=\tau_1}^{\tau_2 - 1} f(\tilde{x}_t^{(k)}, \tilde{u}_t^{(k)}, y^*, v^*) \le (\tau_2 - \tau_1) f(x^*, u^*, y^*, v^*) + M_0 + 2c_*. \tag{6.63}$$

In view of (6.57), (6.60), (6.61), and (P1), the following property holds:

(P2) for every integer $k > L_0$ and every integer $\tau \in [0, k - L_0]$, we have

$$\max\{\rho_X(\tilde{x}_t^{(k)}, x^*) : \ t = \tau, \dots, \tau + L_0\} > \epsilon.$$

Using diagonalization process, compactness of (X, ρ_X) and (U, ρ_U), and extracting subsequences if necessary we obtain that there exists a strictly increasing sequence of natural numbers $\{k_j\}_{j=1}^{\infty}$ such that $k_1 > L_0$ and that for every nonnegative integer t there exist

$$x_t = \lim_{j \to \infty} \tilde{x}_t^{(k_j)},\ u_t = \lim_{j \to \infty} \tilde{u}_t^{(k_j)}. \tag{6.64}$$

Since the sets $\{(x, u) \in X \times U : \ u \in U(x)\}$ and

$$\{(x, u, x') \in X \times U \times X : \ u \in U(x),\ x' \in F_X(x, u)\}$$

are closed it follows from (6.64) that

$$(\{x_t\}_{t=0}^{\infty}, \{u_t\}_{t=0}^{\infty}) \in \mathrm{Str}_1(0, \infty).$$

In view of (6.63), (6.64), and (A1), for every pair of integers τ_1, τ_2 which satisfy $0 \le \tau_1 < \tau_2$, we have

$$\sum_{t=\tau_1}^{\tau_2 - 1} f(x_t, u_t, y^*, v^*) \le (\tau_2 - \tau_1) f(x^*, u^*, y^*, v^*) + M_0 + 2c_*. \tag{6.65}$$

It follows from (6.65), (A2), and (A4) that

$$\lim_{t \to \infty} \rho_X(x_t, x^*) = 0.$$

There exists a natural number integer t_0 such that

$$\rho_X(x_t, x^*) < \epsilon/4 \text{ for all integers } t \geq t_0.$$

Combined with (6.64) the relation above implies that there exists a natural number j_0 such that for all natural numbers $j \geq j_0$, we have

$$\rho_X(\tilde{x}_t^{(k_j)}, x^*) \leq \epsilon/2,\ t = t_0, \dots, t_0 + L_0.$$

This contradicts (P2). The contradiction we have reached proves Lemma 6.12.
□

Analogously to Lemma 6.12 we can prove the next auxiliary result.

Lemma 6.13. *Let M_0, ϵ be positive numbers and $L_0 \geq 1$ be an integer. Then there exists an integer $L_1 > L_0$ such that for every integer $T \geq L_1$, every strategy*

$$(\{y_t\}_{t=0}^{T}, \{v_t\}_{t=0}^{T-1}) \in Str_2(0, T)$$

satisfying

$$\sum_{t=0}^{T-1} f(x^*, u^*, y_t, v_t) \geq Tf(x^*, u^*, y^*, v^*) - M_0$$

and every integer $S \in [0, T - L_1]$ there exists an integer $S_1 \in [S, S + L_1 - L_0]$ for which

$$\rho_Y(y_t, y^*) \leq \epsilon,\ t = S_1, \dots, S_1 + L_0.$$

We suppose that supremum over the empty set is $-\infty$ and that infimum over empty set is ∞.

For every pair of integers $T_2 > T_1$, every pair of points $z_1, z_2 \in X$, and every pair of points $\xi_1, \xi_2 \in Y$ define

$$\sigma_X(T_1, T_2, z_1, z_2) = \inf\{ \sum_{t=T_1}^{T_2-1} f(x_t, u_t, y^*, v^*) :$$

$$(\{x_t\}_{t=T_1}^{T_2}, \{u_t\}_{t=T_1}^{T_2-1}) \in \mathrm{Str}_1(T_1, T_2),\ x_{T_1} = z_1,\ x_{T_2} = z_2\}, \tag{6.66}$$

$$\sigma_Y(T_1, T_2, \xi_1, \xi_2) = \sup\{ \sum_{t=T_1}^{T_2-1} f(x^*, u^*, y_t, v_t) :$$

$$(\{y_t\}_{t=T_1}^{T_2}, \{v_t\}_{t=T_1}^{T_2-1}) \in \mathrm{Str}_2(T_1, T_2),\ y_{T_1} = \xi_1,\ y_{T_2} = \xi_2\}. \tag{6.67}$$

Lemma 6.14. *Let ϵ be a positive number. Then there exists a positive number δ such that for every pair of integers $T_2 > T_1$ and every pair of points $z_1, z_2 \in X$ which satisfies $\rho_X(z_i, x^*) \leq \delta,\ i = 1, 2$ the inequality*

$$\sigma_X(T_1, T_2, z_1, z_2) \leq (T_2 - T_1)f(x^*, u^*, y^*, v^*) + \epsilon \tag{6.68}$$

holds.

Proof. Assumption (A1) implies that there exists a number $\epsilon_0 \in (0, \epsilon)$ such that for every point $x \in X$ and every point $u \in U(x)$ which satisfy

$$\rho_X(x, x^*) \leq \epsilon_0,\ \rho_U(u, u^*) \leq \epsilon_0,$$

we have

$$|f(x, u, y^*, v^*) - f(x^*, u^*, y^*, v^*)| \leq \epsilon/8. \tag{6.69}$$

Assumption (A3) implies that there exists a number $\delta \in (0, \epsilon_0)$ such that the following property holds:

(P3) for every pair of points $x, x' \in X$ which satisfy

$$\rho_X(x, x^*),\ \rho_X(x', x^*) \leq \delta$$

there exists a point $u \in U(x)$ such that $x' \in F_X(x, u)$ and $\rho_U(u, u^*) \leq \epsilon_0$.

Let integers $T_2 > T_1$ and let

$$z_1, z_2 \in X,\ \rho_X(z_i, x^*) \leq \delta,\ i = 1, 2. \tag{6.70}$$

We claim that (6.68) is true. There are two cases:

$$T_2 = T_1 + 1; \tag{6.71}$$

$$T_2 > T_1 + 1. \tag{6.72}$$

Assume that (6.71) is valid. In view of (6.70) and property (P3), there exists

$$u_1 \in U(z_1) \tag{6.73}$$

for which

$$z_2 \in F_X(z_1, u_1),\ \rho_U(u_1, u^*) \leq \epsilon_0. \tag{6.74}$$

Put

$$x_{T_1} = z_1,\ x_{T_1+1} = z_2,\ u_{T_1} = u_1. \tag{6.75}$$

In view of (6.73)–(6.75),

$$(\{x_t\}_{t=T_1}^{T_2}, \{u_t\}_{t=T_1}^{T_2-1}) \in \mathrm{Str}_1(T_1, T_1 + 1),$$

$$\sigma_X(T_1, T_2, z_1, z_2) \leq f(x_{T_1}, u_{T_1}, y^*, v^*). \tag{6.76}$$

By (6.64), (6.69), (6.70), and (6.74)–(6.76), we have

$$f(x_{T_1}, u_{T_1}, y^*, v^*) = f(z_1, u_1, y^*, v^*) \leq f(x^*, u^*, y^*, v^*) + \epsilon,$$

$$\sigma_X(T_1, T_2, z_1, z_2) \leq (T_2 - T_1) f(x^*, u^*, y^*, v^*) + \epsilon.$$

Assume that (6.72) is true. Define

$$x_{T_1} = z_1,\ x_t = x^* \text{ for all } t = T_1 + 1, \ldots, T_2 - 1,\ x_{T_2} = z_2,$$

$$u_t = u^* \text{ for all integers } t \text{ satisfying } T_1 + 1 \leq t < T_2 - 1. \tag{6.77}$$

It follows from (6.70), (6.77), and property (P3) that there exist $u_{T_1}, u_{T_2-1} \in U$ such that

$$u_{T_1} \in U(x_{T_1}),\ \rho_U(u_{T_1}, u^*) \le \epsilon_0,\ x^* \in F_X(x_{T_1}, u_{T_1}), \tag{6.78}$$

$$u_{T_2-1} \in U(x_{T_2-1}),\ \rho_U(u_{T_2-1}, u^*) \le \epsilon_0,\ x_{T_2} \in F_X(x_{T_2-1}, u_{T_2-1}). \tag{6.79}$$

In view of (6.69), (6.70), and (6.77)–(6.79), we have

$$f(x_{T_1}, u_{T_1}, y^*, v^*) \le f(x^*, u^*, y^*, v^*) + \epsilon/8,$$

$$f(x_{T_2-1}, u_{T_2-1}, y^*, v^*) \le f(x^*, u^*, y^*, v^*) + \epsilon/8. \tag{6.80}$$

It follows from (6.70) and (6.77)–(6.79) that

$$(\{x_t\}_{t=T_1}^{T_2}, \{u_t\}_{t=T_1}^{T_2-1}) \in \mathrm{Str}_1(T_1, T_2), \tag{6.81}$$

Relations (6.77), (6.80), and (6.81) imply that

$$\sigma_X(T_1, T_2, z_1, z_2) \le \sum_{t=T_1}^{T_2-1} f(x_t, u_t, y^*, v^*) \le (T_2 - T_1) f(x^*, u^*, y^*, v^*) + \epsilon/4.$$

Lemma 6.14 is proved. □

Analogously to Lemma 6.14 we can prove the next auxiliary result.

Lemma 6.15. *Let ϵ be a positive number. Then there exists a positive number δ such that for every pair of integers $T_2 > T_1$ and every pair of points $\xi_1, \xi_2 \in Y$ which satisfies $\rho_Y(\xi_i, y^*) \le \delta,\ i = 1, 2$ the inequality*

$$\sigma_Y(T_1, T_2, \xi_1, \xi_2) \ge (T_2 - T_1) f(x^*, u^*, y^*, v^*) - \epsilon$$

holds.

Lemma 6.16. *Let $\epsilon \in (0, 1)$. Then there exists a positive number δ such that for every natural number T and every strategy*

$$(\{x_t\}_{t=0}^{T}, \{u_t\}_{t=0}^{T-1}) \in Str_1(0, T)$$

which satisfies

$$\rho_X(x_i, x^*) \le \delta,\ i = 0, T$$

and

$$\sum_{t=0}^{T-1} f(x_t, u_t, y^*, v^*) \le \sigma_X(0, T, x_0, x_T) + \delta$$

the following inequality is valid:

$$\rho_X(x_t, x^*) \le \epsilon,\ t = 0, \ldots, T.$$

Proof. Assumption (A1) implies that for every natural number k there exists a number

$$\epsilon_k \in (0, 4^{-k}\epsilon) \tag{6.82}$$

such that the following property holds:

(i) for every point $x \in X$ and every point $u \in U(x)$ which satisfy

$$\rho_X(x, x^*) \le \epsilon_k, \; \rho_U(u, u^*) \le \epsilon_k$$

the inequality

$$|f(x, u, y^*, v^*) - f(x^*, u^*, y^*, v^*)| \le 4^{-k}$$

is true.

Assumption (A3) and Lemma 6.14 imply that there exists a number

$$\delta_k \in (0, \epsilon_k) \tag{6.83}$$

such that the following properties hold:

(ii) for every pair of points $x, x' \in X$ which satisfies

$$\rho_X(x, x^*) \le \delta_k, \; \rho_X(x', x^*) \le \delta_k$$

there exists a point $u \in U(x)$ such that $x' \in F_X(x, u)$ and $\rho_U(u, u^*) \le \epsilon_k$;

(iii) for every pair of point $x, x' \in X$ which satisfies

$$\rho_X(x, x^*) \le \delta_k, \; \rho_X(x', x^*) \le \delta_k$$

and every natural number T we have

$$\sigma_X(0, T, x, x') \le Tf(x^*, u^*, y^*, v^*) + 4^{-k}.$$

Assume that the lemma does not hold. Then for every natural number k there exist a natural number $T_k > 1$ and a strategy

$$(\{x_t^{(k)}\}_{t=1}^{T_k}, \{u_t^{(k)}\}_{t=1}^{T_k-1}) \in \mathrm{Str}_1(1, T_k) \tag{6.84}$$

such that

$$\rho_X(x_1^{(k)}, x^*) \le \delta_k, \; \rho_X(x_{T_k}^{(k)}, x^*) \le \delta_k, \tag{6.85}$$

$$\sum_{t=1}^{T_k-1} f(x_t^{(k)}, u_t^{(k)}, y^*, v^*) \le \sigma_X(1, T_k, x_1^{(k)}, x_{T_k}^{(k)}) + \delta_k, \tag{6.86}$$

$$\max\{\rho_X(x_t^{(k)}, x^*) : \; t = 1, \dots, T_k\} > \epsilon. \tag{6.87}$$

Let k be an integer. Put

$$x_0^{(k)} = x^*, \; x_{T_k+1}^{(k)} = x^*. \tag{6.88}$$

In view of (6.85), (6.88), and property (ii), there exists a pair of points

$$u_0^{(k)} \in U(x^*), \; u_{T_k}^{(k)} \in U(x_{T_k}^{(k)}) \tag{6.89}$$

such that

$$x_1^{(k)} \in F_X(x^*, u_0^{(k)}),\ x^* \in F_X(x_{T_k}^{(k)}, u_{T_k}^{(k)}), \tag{6.90}$$

$$\rho_U(u_0^{(k)}, u^*) \le \epsilon_k,\ \rho_U(u_{T_k}^{(k)}, u^*) \le \epsilon_k. \tag{6.91}$$

It follows from (6.84) and (6.88)–(6.90) that

$$(\{x_t^{(k)}\}_{t=0}^{T_k+1}, \{u_t^{(k)}\}_{t=0}^{T_k}) \in \mathrm{Str}_1(0, T_k+1). \tag{6.92}$$

By (6.85), (6.86), and property (iii), we have

$$\sum_{t=1}^{T_k-1} f(x_t^{(k)}, u_t^{(k)}, y^*, v^*) \le (T_k - 1) f(x^*, u^*, y^*, v^*) + \delta_k + 4^{-k}. \tag{6.93}$$

Relations (6.83), (6.85), (6.88), (6.91), and property (i) imply that

$$f(x_0^{(k)}, u_0^{(k)}, y^*, v^*),\ f(x_{T_k}^{(k)}, u_{T_k}^{(k)}, y^*, v^*) \le f(x^*, u^*, y^*, v^*) + 4^{-k}. \tag{6.94}$$

It follows from (6.82), (6.83), (6.93), and (6.94) that

$$\sum_{t=0}^{T_k} f(x_t^{(k)}, u_t^{(k)}, y^*, v^*) \le (T_k + 1) f(x^*, u^*, y^*, v^*) + 4 \cdot 4^{-k}. \tag{6.95}$$

For all natural numbers k define

$$S_k = \sum_{i=1}^{k} (T_i + 1). \tag{6.96}$$

In view of (6.88) and (6.92), there exists

$$(\{\tilde{x}_t\}_{t=0}^{\infty}, \{\tilde{u}_t\}_{t=0}^{\infty}) \in \mathrm{Str}_1(0, \infty)$$

such that

$$\tilde{x}_t = x_t^{(1)},\ t = 0, \dots, S_1 \tag{6.97}$$

and that for all natural numbers k, we have

$$\tilde{x}_{t+S_k} = x_t^{(k+1)},\ t = 0, \dots, S_{k+1} - S_k,$$

$$\tilde{u}_{t+S_k} = u_t^{(k+1)},\ t = 0, \dots, S_{k+1} - S_k - 1. \tag{6.98}$$

It follows from (6.95), (6.97), and (6.98) that

$$\liminf_{T \to \infty} [\sum_{t=0}^{T-1} f(\tilde{x}_t, \tilde{u}_t, y^*, v^*) - T f(x^*, u^*, y^*, v^*)] < 4 \sum_{k=0}^{\infty} 4^{-k}. \tag{6.99}$$

By (6.99) and assumptions (A2) and (A4),

$$\lim_{t\to\infty} \rho_X(\tilde{x}_t, x^*) = 0.$$

The relation above implies that there exists an integer $\tau_1 \geq 1$ such that for all natural numbers $t \geq \tau_1$, we have

$$\rho_X(\tilde{x}_t, x^*) \leq \epsilon/2. \tag{6.100}$$

There exists a natural number k_0 such that $S_{k_0} > \tau_1$. Combined with relations (6.98) and (6.100) this implies that for all natural numbers $k > k_0$ and all natural numbers $t \in \{1, \dots, T_k\}$, the inequality $\rho_X(x_t^{(k)}, x^*) \leq \epsilon/2$ is valid. This contradicts (6.87). The contradiction we have reached proves Lemma 6.16.

Analogously to Lemma 6.16 we can prove the following result.

Lemma 6.17. *Let $\epsilon \in (0, 1)$. Then there exists a positive number δ such that for every natural number T and every strategy*

$$(\{y_t\}_{t=0}^T, \{v_t\}_{t=0}^{T-1}) \in Str_2(0, T)$$

which satisfies

$$\rho_Y(y_i, y^*) \leq \delta,\ i = 0, T$$

and

$$\sum_{t=0}^{T-1} f(x^*, u^*, y_t, v_t) \geq \sigma_Y(0, T, y_0, y_T) - \delta$$

the following inequality is valid:

$$\rho_Y(y_t, y^*) \leq \epsilon,\ t = 0, \dots, T.$$

Lemma 6.18. *Let M, ϵ be positive numbers. Then there exist natural numbers l_1, Q_1 such that for every natural number $T > Q_1 l_1$ and every strategy*

$$(\{x_t\}_{t=0}^T, \{u_t\}_{t=0}^{T-1}) \in Str_1(0, T)$$

which satisfies

$$\sum_{t=0}^{T-1} f(x_t, u_t, y^*, v^*) \leq T f(x^*, u^*, y^*, v^*) + M \tag{6.101}$$

there exist a natural number $q \leq Q_1$ and sequences of integers

$$\{a_i\}_{i=1}^q,\ \{b_i\}_{i=1}^q \subset [0, T]$$

such that

$$0 \leq b_i - a_i \leq l_1,\ i = 1, \dots, q, \tag{6.102}$$

$$\rho_X(x_t, x^*) \leq \epsilon \text{ for all integers } t \in [0, T] \setminus \cup_{i=1}^q [a_i, b_i]. \tag{6.103}$$

Proof. Lemma 6.16 implies that there exists a number $\delta \in (0, \epsilon)$ such that the following property holds:

(i) for every natural number T and every strategy

$$(\{x_t\}_{t=0}^{T}, \{u_t\}_{t=0}^{T-1}) \in \mathrm{Str}_1(0, T)$$

which satisfies

$$\rho_X(x_i, x^*) \le \delta,\ i = 0, T$$

and

$$\sum_{t=0}^{T-1} f(x_t, u_t, y^*, v^*) \le \sigma_X(0, T, x_0, x_T) + \delta,$$

we have

$$\rho_X(x_t, x^*) \le \epsilon,\ t = 0, \dots, T.$$

Lemma 6.12 implies that there exists an integer $L_1 > 1$ such that the following property holds:

(ii) for every natural number $T \ge L_1$, every strategy

$$(\{x_t\}_{t=0}^{T}, \{u_t\}_{t=0}^{T-1}) \in \mathrm{Str}_1(0, T)$$

which satisfies

$$\sum_{t=0}^{T-1} f(x_t, u_t, y^*, v^*) \le T f(x^*, u^*, y^*, v^*) + M$$

and every integer $S \in [0, T - L_1]$ there exists an integer $S_1 \in [S, S + L_1 - 1]$ for which

$$\rho_X(x_t, x^*) \le \delta,\ t = S_1,\ S_1 + 1.$$

Fix integers

$$Q_1 > 6 + 3\delta^{-1}(M + c_*) \text{ and } l_1 \ge 4L_1 + 4. \tag{6.104}$$

Assume that an integer $T > Q_1 l_1$ and that a strategy

$$(\{x_t\}_{t=0}^{T}, \{u_t\}_{t=0}^{T-1}) \in \mathrm{Str}_1(0, T)$$

satisfies (6.101). Assumption (A2) and (6.101) imply that

$$\sum_{t=0}^{T-1} f(x_t, u_t, y^*, v^*) \le \sigma_X(0, T, x_0, x_T) + c_* + M. \tag{6.105}$$

Set

$$t_0 = 0. \tag{6.106}$$

By induction we define a finite strictly increasing sequence of integers t_i, $i = 0, 1, \dots, q$ where $q \ge 1$ is an integer such that

$$t_q = T$$

and the following properties hold:

(iii) if an integer i satisfies $0 \le i < q-1$, then

$$\sum_{t=t_i}^{t_{i+1}-1} f(x_t, u_t, y^*, v^*) > \sigma_X(t_i, t_{i+1}, x_{t_i}, x_{t_{i+1}}) + \delta; \tag{6.107}$$

(iv) if an integer i satisfies $0 \le i \le q-1$ and (6.107) holds, then $t_{i+1} > t_i+1$ and

$$\sum_{t=t_i}^{t_{i+1}-2} f(x_t, u_t, y^*, v^*) \le \sigma_X(t_i, t_{i+1}-1, x_{t_i}, x_{t_{i+1}-1}) + \delta. \tag{6.108}$$

Assume that $k \ge 0$ is an integer and we have defined a finite strictly increasing sequence of integers $t_i \in [0, T-1]$, $i = 0, \dots, k$ such that the following property holds:

(v) if an integer i satisfies $0 \le i \le k-1$, then (6.107) and (6.108) are true.

There are two cases:

$$\sum_{t=t_k}^{T-1} f(x_t, u_t, y^*, v^*) \le \sigma_X(t_k, T, x_{t_k}, x_T) + \delta; \tag{6.109}$$

$$\sum_{t=t_k}^{T-1} f(x_t, u_t, y^*, v^*) > \sigma_X(t_k, T, x_{t_k}, x_T) + \delta. \tag{6.110}$$

If (6.109) is valid, then we set $q = k+1$, $t_q = T$, the construction is completed and the properties (iii) and (iv) hold.

Assume that (6.110) is true and define

$$t_{k+1} = \min\{\tau \in \{t_k+1, \dots, T\} :$$

$$\sum_{t=t_k}^{\tau-1} f(x_t, u_t, y^*, v^*) > \sigma_X(t_k, \tau, x_{t_k}, x_\tau) + \delta\}. \tag{6.111}$$

If $t_{k+1} = T$, then we set $q = k+1$, the construction is completed and the properties (iii) and (iv) hold. If $t_{k+1} < T$, then the assumption made for k also holds for $k+1$.

It is clear that the construction will be completed after a finite number of steps and let t_q be the last element of the sequence. Evidently, $t_q = T$ and the properties (iii) and (iv) hold. In view of (6.105) and (iii),

$$c_* + M \ge \sum_{t=0}^{T-1} f(x_t, u_t, y^*, v^*) - \sigma_X(0, T, x_0, x_T)$$

$$\geq \sum \{ \sum_{t=t_i}^{t_{i+1}-1} f(x_t, u_t, y^*, v^*) - \sigma_X(t_i, t_{i+1}, x_{t_i}, x_{t_{i+1}}) : \; i \text{ is an integer and}$$

$$0 \leq i < q-1\} \geq \delta(q-2),$$

$$q \leq 2 + \delta^{-1}(c_* + M). \tag{6.112}$$

Define

$$A = \{i \in \{0, \dots, q-1\} : \; t_{i+1} - t_i \geq 4L_1 + 4\}. \tag{6.113}$$

Let

$$i \in A. \tag{6.114}$$

In view of (6.101), (6.104), (6.113), (6.114), and property (ii), there exist integers

$$S_{i,1} \in [t_i, t_i + L_1], \; S_{i,2} \in [t_{i+1} - L_1, t_{i+1} - 1] \tag{6.115}$$

such that

$$\rho_X(x_{S_{i,1}}, x^*) \leq \delta, \; \rho_X(x_{S_{i,2}-1}, x^*) \leq \delta. \tag{6.116}$$

It follows from (6.113) to (6.115) that

$$S_{i,2} - 1 - S_{i,1} \geq t_{i+1} - L_1 - t_i - L_1 \geq 2L_1 + 4. \tag{6.117}$$

Relations (6.115) and (6.117) imply that

$$\sum_{t=S_{i,1}}^{S_{i,2}-2} f(x_t, u_t, y^*, v^*) \leq \sigma_X(S_{i,1}, S_{i,2} - 1, x_{S_{i,1}}, x_{S_{i,2}-1}) + \delta. \tag{6.118}$$

By (6.116)–(6.118) and the property (i), we have

$$\rho_X(x_t, x^*) \leq \epsilon, \; t = S_{i,1}, \dots, S_{i,2} - 1.$$

Combined with (6.115) the relation above implies that

$$\rho_X(x_t, x^*) \leq \epsilon, \; t = t_i + L_1, \dots, t_{i+1} - L_1 - 1. \tag{6.119}$$

In view of (6.119),

$$\{t \in \{0, \dots, T\} : \; \rho_X(x_t, x^*) > \epsilon\}$$

$$\subset \cup \{\{t_i, \dots, t_{i+1}\} : \; i \in \{0, \dots, q-1\} \setminus A\}$$

$$\cup \{\{t_i, \dots, t_i + L_1 - 1\} \cup \{t_{i+1} - L_1, \dots, t_{i+1}\} : \; i \in A\}. \tag{6.120}$$

It is clear that the right-hand side of (6.120) is a union of a finite collection of the sets of the type $\{a, \dots, b\}$ where integers $a, b \in [0, T]$, and its cardinality does not exceed

$$3q \leq 6 + 3\delta^{-1}(c_* + M) < Q$$

(see (6.112), (6.113)) and the maximal difference $b - a$ does not exceed $4L_1 + 4 \leq l_1$ (see (6.110)). This completes the proof of Lemma 6.18. □

Analogously we can prove the next auxiliary result.

Lemma 6.19. *Let M, ϵ be positive numbers. Then there exist natural numbers l_1, Q_1 such that for every natural number $T > Q_1 l_1$ and every strategy*

$$(\{y_t\}_{t=0}^{T}, \{v_t\}_{t=0}^{T-1}) \in Str_2(0,T)$$

which satisfies

$$\sum_{t=0}^{T-1} f(x^*, u^*, y_t, v_t) \geq Tf(x^*, u^*, y^*, v^*) - M$$

there exist a natural number $q \leq Q_1$ and finite sequences of integers

$$\{a_i\}_{i=1}^{q}, \ \{b_i\}_{i=1}^{q} \subset [0,T]$$

such that

$$0 \leq b_i - a_i \leq l_1, \ i = 1, \dots, q,$$

$$\rho_Y(y_t, y^*) \leq \epsilon \text{ for all integers } t \in [0,T] \setminus \cup_{i=1}^{q} [a_i, b_i].$$

Lemma 6.20. *Let M be a positive number, $L_0 \geq 1$ be an integer, $T > 2L_0+1$ be a natural number and*

$$(\{x_t\}_{t=0}^{T}, \{u_t\}_{t=0}^{T-1}) \in Str_1(0,T) \tag{6.121}$$

and

$$(\{y_t\}_{t=0}^{T}, \{v_t\}_{t=0}^{T-1}) \in Str_2(0,T) \tag{6.122}$$

be an (M)-good pair of strategies such that

$$x_0 \in X^{(L_0)}, \ x_T \in \bar{X}^{(L_0)}, \ y_0 \in Y^{(L_0)}, \ y_T \in \bar{Y}^{(L_0)}. \tag{6.123}$$

Then

$$\sum_{t=0}^{T-1} f(x_t, u_t, y^*, v^*) \leq Tf(x^*, u^*, y^*, v^*) + c_* + 2M + 8L_0\|f\|, \tag{6.124}$$

$$\sum_{t=0}^{T-1} f(x^*, u^*, y_t, v_t) \geq Tf(x^*, u^*, y^*, v^*) - c_* - 2M - 8L_0\|f\|. \tag{6.125}$$

Proof. In view of (6.123) there exist

$$(\{\tilde{x}_t\}_{t=0}^{T}, \{\tilde{u}_t\}_{t=0}^{T-1}) \in \mathrm{Str}_1(0,T),$$

$$(\{\tilde{y}_t\}_{t=0}^{T}, \{\tilde{v}_t\}_{t=0}^{T-1}) \in \mathrm{Str}_2(0,T) \tag{6.126}$$

such that

$$\tilde{x}_0 = x_0, \ \tilde{x}_{L_0} = x^*,$$

$$\tilde{x}_t = x^*,\ \tilde{u}_t = u^*,\ t = L_0, \dots, T - L_0 - 1,\ \tilde{x}_{T-L_0} = x^*,\ \tilde{x}_T = x_T,$$
$$\tilde{y}_0 = y_0,\ \tilde{y}_{L_0} = y^*,$$
$$\tilde{y}_t = y^*,\ \tilde{v}_t = v^*,\ t = L_0, \dots, T - L_0 - 1,\ \tilde{y}_{T-L_0} = y^*,\ \tilde{y}_T = y_T. \quad (6.127)$$

Assumption (A2) implies that

$$\sum_{t=0}^{T-1} f(x_t, u_t, y^*, v^*) \geq T f(x^*, u^*, y^*, v^*) - c_*,$$
$$\sum_{t=0}^{T-1} f(x^*, u^*, y_t, v_t) \leq T f(x^*, u^*, y^*, v^*) + c_*. \quad (6.128)$$

Since the pair of strategies $(\{x_t\}_{t=0}^T, \{u_t\}_{t=0}^{T-1})$ and $(\{y_t\}_{t=0}^T, \{v_t\}_{t=0}^{T-1})$ is an (M)-good it follows from (6.127) that

$$-M - 4L_0\|f\| + \sum_{t=0}^{T-1} f(x_t, u_t, y^*, v^*) \leq -2L_0\|f\| - M + \sum_{t=L_0}^{T-L_0-1} f(x_t, u_t, y^*, v^*)$$
$$\leq \sum_{t=0}^{L_0-1} f(x_t, u_t, \tilde{y}_t, \tilde{v}_t) + \sum_{t=L_0}^{T-L_0-1} f(x_t, u_t, y^*, v^*) + \sum_{t=T-L_0}^{T-1} f(x_t, u_t, \tilde{y}, \tilde{v}_t) - M$$
$$= -M + \sum_{t=0}^{T-1} f(x_t, u_t, \tilde{y}_t, \tilde{v}_t) \leq \sum_{t=0}^{T-1} f(x_t, u_t, y_t, v_t) \leq M + \sum_{t=0}^{T-1} f(\tilde{x}_t, \tilde{u}_t, y_t, v_t)$$
$$= M + \sum_{t=0}^{L_0-1} f(\tilde{x}_t, \tilde{u}_t, y_t, v_t) + \sum_{t=L_0}^{T-L_0-1} f(x^*, u^*, y_t, v_t) + \sum_{t=T-L_0}^{T-1} f(\tilde{x}_t, \tilde{u}_t, y_t, v_t)$$
$$\leq M + 2L_0\|f\| + \sum_{t=L_0}^{T-L_0-1} f(x^*, u^*, y_t, v_t)$$
$$\leq M + 4L_0\|f\| + \sum_{t=0}^{T-1} f(x^*, u^*, y_t, v_t).$$

Assumption (A2) and the relation above imply that

$$-M - 4L_0\|f\| - c_* + T f(x^*, u^*, y^*, v^*) \leq -M - 4L_0\|f\| + \sum_{t=0}^{T-1} f(x_t, u_t, y^*, v^*)$$
$$\leq \sum_{t=0}^{T-1} f(x_t, u_t, y_t, v_t) \leq M + 4L_0\|f\| + \sum_{t=0}^{T-1} f(x^*, u^*, y_t, v_t)$$
$$\leq M + 4L_0\|f\| + c_* + T f(x^*, u^*, y^*, v^*).$$

This implies that

$$\sum_{t=0}^{T-1} f(x_t, u_t, y^*, v^*) \le Tf(x^*, u^*, y^*, v^*) + c_* + 2M + 8L_0\|f\|,$$

$$\sum_{t=0}^{T-1} f(x^*, u^*, y_t, v_t) \ge Tf(x^*, u^*, y^*, v^*) - c_* - 2M - 8L_0\|f\|.$$

This completes the proof of Lemma 6.20. □

Lemma 6.21. *Let M be a positive number, $L_0 \ge 1$ be an integer, $T > L_0 + 1$ be an integer and*

$$(\{x_t\}_{t=0}^{T}, \{u_t\}_{t=0}^{T-1}) \in Str_1(0, T),\ (\{y_t\}_{t=0}^{T}, \{v_t\}_{t=0}^{T-1}) \in Str_2(0, T)$$

be an $(M, *)$-good pair of strategies such that

$$x_0 \in X^{(L_0)},\ y_0 \in Y^{(L_0)}. \tag{6.129}$$

Then

$$\sum_{t=0}^{T-1} f(x_t, u_t, y^*, v^*) \le Tf(x^*, u^*, y^*, v^*) + c_* + 2M + 4L_0\|f\|,$$

$$\sum_{t=0}^{T-1} f(x^*, u^*, y_t, v_t) \ge Tf(x^*, u^*, y^*, v^*) - c_* - 2M - 4L_0\|f\|.$$

Proof. In view of (6.129), there exist

$$(\{\tilde{x}_t\}_{t=0}^{T}, \{\tilde{u}_t\}_{t=0}^{T-1}) \in \mathrm{Str}_1(0, T),\ (\{\tilde{y}_t\}_{t=0}^{T}, \{\tilde{v}_t\}_{t=0}^{T-1}) \in \mathrm{Str}_2(0, T)$$

such that

$$\tilde{x}_0 = x_0,\ \tilde{x}_t = x^*,\ t = L_0, \dots, T,\ \tilde{u}_t = u^*,\ t = L_0, \dots, T-1,$$

$$\tilde{y}_0 = y_0,\ \tilde{y}_t = y^*,\ t = L_0, \dots, T,\ \tilde{v}_t = v^*,\ t = L_0, \dots, T-1. \tag{6.130}$$

Assumption (A2) implies that

$$\sum_{t=0}^{T-1} f(x_t, u_t, y^*, v^*) \ge Tf(x^*, u^*, y^*, v^*) - c_*,$$

$$\sum_{t=0}^{T-1} f(x^*, u^*, y_t, v_t) \le Tf(x^*, u^*, y^*, v^*) + c_*. \tag{6.131}$$

Since the pair of strategies $(\{x_t\}_{t=0}^{T}, \{u_t\}_{t=0}^{T-1})$ and $(\{y_t\}_{t=0}^{T}, \{v_t\}_{t=0}^{T-1})$ is an $(M, *)$-good it follows from (6.130) that

$$-M - 2L_0\|f\| + \sum_{t=0}^{T-1} f(x_t, u_t, y^*, v^*) \le -M + \sum_{t=0}^{L_0-1} f(x_t, u_t, \tilde{y}_t, \tilde{v}_t)$$

$$+ \sum_{t=L_0}^{T-1} f(x_t, u_t, y^*, v^*) = \sum_{t=0}^{T-1} f(x_t, u_t, \tilde{y}_t, \tilde{v}_t) - M \le \sum_{t=0}^{T-1} f(x_t, u_t, y_t, v_t)$$

$$\le M + \sum_{t=0}^{T-1} f(\tilde{x}_t, \tilde{u}_t, y_t, v_t) = M + \sum_{t=0}^{L_0-1} f(\tilde{x}_t, \tilde{u}_t, y_t, v_t) + \sum_{t=L_0}^{T-1} f(x^*, u^*, y_t, v_t)$$

$$\le M + 2L_0\|f\| + \sum_{t=0}^{T-1} f(x^*, u^*, y_t, v_t).$$

It follows from (6.131) and the relation above that

$$-M - 2L_0\|f\| - c_* + Tf(x^*, u^*, y^*, v^*) \le -M - 2L_0\|f\| + \sum_{t=0}^{T-1} f(x_t, u_t, y^*, v^*)$$

$$\le \sum_{t=0}^{T-1} f(x_t, u_t, y_t, v_t) \le M + 2L_0\|f\| + \sum_{t=0}^{T-1} f(x^*, u^*, y_t, v_t)$$

$$\le M + 2L_0\|f\| + c_* + Tf(x^*, u^*, y^*, v^*).$$

This implies that

$$\sum_{t=0}^{T-1} f(x_t, u_t, y^*, v^*) \le Tf(x^*, u^*, y^*, v^*) + c_* + 2M + 4L_0\|f\|,$$

$$\sum_{t=0}^{T-1} f(x^*, u^*, y_t, v_t) \ge Tf(x^*, u^*, y^*, v^*) - c_* - 2M - 4L_0\|f\|.$$

This completes the proof of Lemma 6.21. □

6.7 Proofs of Theorems 6.8 and 6.9

We prove Theorems 6.9 and 6.9 simultaneously. Lemma 6.18 implies that there exist natural numbers l_1, Q_1 such that the following property holds:

(i) for every natural number $T > Q_1 l_1$ and every strategy

$$(\{x_t\}_{t=0}^{T}, \{u_t\}_{t=0}^{T-1}) \in \mathrm{Str}_1(0, T)$$

which satisfies

$$\sum_{t=0}^{T-1} f(x_t, u_t, y^*, v^*) \le Tf(x^*, u^*, y^*, v^*) + 2M + c_* + 8L_0\|f\|$$

there exist a natural number $q_1 \leq Q_1$ and finite sequences of integers $\{a_{i,1}\}_{i=1}^{q_1}$, $\{b_{i,1}\}_{i=1}^{q_1} \subset [0, T]$ such that

$$0 \leq b_{i,1} - a_{i,1} \leq l_1, \ i = 1, \dots, q_1, \tag{6.132}$$

$$\rho_X(x_t, x^*) \leq \epsilon \text{ for all integers } t \in [0, T] \setminus \cup_{i=1}^{q_1} [a_{i,1}, b_{i,1}]. \tag{6.133}$$

Lemma 6.19 implies that there exist natural numbers l_2, Q_2 such that the following property holds:

(ii) for every natural number $T > Q_2 l_2$ and every strategy

$$(\{y_t\}_{t=0}^T, \{v_t\}_{t=0}^{T-1}) \in \mathrm{Str}_2(0, T)$$

which satisfies

$$\sum_{t=0}^{T-1} f(x^*, u^*, y_t, v_t) \geq T f(x^*, u^*, y^*, v^*) - c_* - 2M - 8L_0 \|f\|$$

there exist a natural number $q_2 \leq Q_2$ and finite sequences of integers $\{a_{i,2}\}_{i=1}^{q_2}$, $\{b_{i,2}\}_{i=1}^{q_2} \subset [0, T]$ such that

$$0 \leq b_{i,2} - a_{i,2} \leq l_2, \ i = 1, \dots, q_2, \tag{6.134}$$

$$\rho_Y(y_t, y^*) \leq \epsilon \text{ for all integers } t \in [0, T] \setminus \cup_{i=1}^{q_2} [a_{i,2}, b_{i,2}]. \tag{6.135}$$

Put

$$Q = Q_1 + Q_2, \ l = \max\{l_1, l_2, 2L_0 + 2\}. \tag{6.136}$$

Let an integer $T > QL$ and let

$$(\{x_t\}_{t=0}^T, \{u_t\}_{t=0}^{T-1}) \in \mathrm{Str}_1(0, T), \ (\{y_t\}_{t=0}^T, \{v_t\}_{t=0}^{T-1}) \in \mathrm{Str}_2(0, T) \tag{6.137}$$

be an (M)-good pair of strategies in the case of Theorem 6.8 and be an $(M, *)$-good pair of strategies in the case of Theorem 6.9 such that

$$x_0 \in X^{(L_0)}, \ y_0 \in Y^{(L_0)} \tag{6.138}$$

and in the case of Theorem 6.8

$$x_T \in \bar{X}^{(L_0)}, \ y_T \in \bar{Y}^{(L_0)}. \tag{6.139}$$

It follows from (6.136) to (6.139) and Lemmas 6.20 and 6.21 that

$$\sum_{t=0}^{T-1} f(x_t, u_t, y^*, v^*) \leq T f(x^*, u^*, y^*, v^*) + c_* + 2M + 8L_0 \|f\|, \tag{6.140}$$

$$\sum_{t=0}^{T-1} f(x^*, u^*, y_t, v_t) \geq T f(x^*, u^*, y^*, v^*) - c_* - 2M - 8L_0 \|f\|. \tag{6.141}$$

By property (i), (6.136), (6.137), and (6.140), there exist a natural number $q_1 \leq Q_1$ and finite sequences of integers $\{a_{i,1}\}_{i=1}^{q_1}$, $\{b_{i,1}\}_{i=1}^{q_1} \subset [0, T]$ such that (6.132) and (6.133) are valid. In view of the property (ii), (6.136), (6.137), and (6.141), there exist a natural number $q_2 \leq Q_2$ and finite sequences of integers $\{a_{i,2}\}_{i=1}^{q_2}$, $\{b_{i,2}\}_{i=1}^{q_2} \subset [0, T]$ such that (6.134) and (6.135) are true. This completes the proofs of Theorems 6.8 and 6.9. □

6.8 Proofs of Theorems 6.10 and 6.11

Lemma 6.22. *Suppose that assumption* (A^*) *from Theorem 6.7 holds and let* ϵ *be a positive number. Then there exists a positive number* δ *such that for every natural number* $T > 2$ *and every* (δ)*-good pair of strategies*

$$(\{x_t\}_{t=0}^{T}, \{u_t\}_{t=0}^{T-1}) \in Str_1(0,T),$$

$$(\{y_t\}_{t=0}^{T}, \{v_t\}_{t=0}^{T-1}) \in Str_2(0,T)$$

which satisfies

$$\rho_X(x_i, x^*) \le \delta, \ i = 0, T, \ \rho_Y(y_i, y^*) \le \delta, \ i = 0, T \tag{6.142}$$

the following inequality holds:

$$\rho_X(x_t, x^*) \le \epsilon, \ t = 0, \dots, T, \ \rho_Y(y_t, y^*) \le \epsilon, \ t = 0, \dots, T. \tag{6.143}$$

Proof. Lemmas 6.16 and 6.17 imply that there exists a number $\delta_0 \in (0, \epsilon/4)$ such that:

(i) for every natural number T and every strategy

$$(\{x_t\}_{t=0}^{T}, \{u_t\}_{t=0}^{T-1}) \in \mathrm{Str}_1(0,T)$$

which satisfies

$$\rho_X(x_i, x^*) \le \delta_0, \ i = 0, T$$

and

$$\sum_{t=0}^{T-1} f(x_t, u_t, y^*, v^*) \le \sigma_X(0, T, x_0, x_T) + \delta_0$$

we have

$$\rho_X(x_t, x^*) \le \epsilon, \ t = 0, \dots, T;$$

(ii) for every natural number T and every strategy

$$(\{y_t\}_{t=0}^{T}, \{v_t\}_{t=0}^{T-1}) \in \mathrm{Str}_2(0,T)$$

which satisfies

$$\rho_Y(y_i, y^*) \le \delta_0, \ i = 0, T$$

and

$$\sum_{t=0}^{T-1} f(x^*, u^*, y_t, v_t) \ge \sigma_Y(0, T, y_0, y_T) - \delta_0$$

we have

$$\rho_Y(y_t, y^*) \le \epsilon, \ t = 0, \dots, T.$$

Assumption (A^*) implies that there exists a number $\delta_1 \in (0, \delta_0/32)$ such that the following properties hold:

(iii) for every $(x, u, y, v) \in X \times U \times Y \times V$ which satisfies

$$u \in U(x),\ v \in V(y),\ \rho_X(x, x^*) \le \delta_1,\ \rho_U(u, u^*) \le \delta_1$$

we have

$$|f(x, u, y, v) - f(x^*, u^*, y, v)| \le \delta_0/32;$$

(iv) for every $(x, u, y, v) \in X \times U \times Y \times V$ which satisfies

$$u \in U(x),\ v \in V(y),\ \rho_Y(y, y^*) \le \delta_1,\ \rho_V(v, v^*) \le \delta_1$$

we have

$$|f(x, u, y, v) - f(x, u, y^*, v^*)| \le \delta_0/32.$$

By Theorem 6.7 implies that there exists a number $\delta \in (0, \delta_1/2)$ such that the following property holds:

(v) for every natural number $T > 2$, every pair of points $z_0,\ z_1 \in X$, and every pair of points $\xi_0,\ \xi_1 \in Y$ which satisfies

$$\rho_X(z_i, x^*) \le \delta,\ i = 0, 1 \text{ and } \rho_Y(\xi_i, y^*) \le \delta,\ i = 0, 1$$

there exists a (δ_1)-good pair of strategies

$$(\{x_t\}_{t=0}^T, \{u_t\}_{t=0}^{T-1}) \in \mathrm{Str}_1(0, T) \text{ and } (\{y_t\}_{t=0}^T, \{v_t\}_{t=0}^{T-1}) \in \mathrm{Str}_2(0, T)$$

such that

$$x_0 = z_0,\ x_t = x^* \text{ for all } t = 1, \dots, T-1,\ x_T = z_1,$$

$$u_t = u^* \text{ for all } t = 1, \dots, T-2,$$

$$y_0 = \xi_0,\ y_t = y^* \text{ for all } t = 1, \dots, T-1,\ y_T = \xi_1,$$

$$v_t = v^* \text{ for all } t = 1, \dots, T-2,$$

$$\rho_U(u_0, u^*) \le \delta_1,\ \rho_U(u_{T-1}, u^*) \le \delta_1,\ \rho_V(v_0, v^*) \le \delta_1,\ \rho_U(v_{T-1}, v^*) \le \delta_1.$$

Let $T > 2$ be a natural number and let

$$(\{x_t\}_{t=0}^T, \{u_t\}_{t=0}^{T-1}) \in \mathrm{Str}_1(0, T),$$

$$(\{y_t\}_{t=0}^T, \{v_t\}_{t=0}^{T-1}) \in \mathrm{Str}_2(0, T)$$

be a (δ)-good pair of strategies which satisfies (6.142). In view of (6.142) and the property (v), there exists a (δ_1)-good pair of strategies

$$(\{\tilde{x}_t\}_{t=0}^T, \{\tilde{u}_t\}_{t=0}^{T-1}) \in \mathrm{Str}_1(0, T)$$

and

$$(\{\tilde{y}_t\}_{t=0}^T, \{\tilde{v}_t\}_{t=0}^{T-1}) \in \mathrm{Str}_2(0, T)$$

such that

$$\tilde{x}_i = x_i, \ i = 0, T, \ \tilde{y}_i = y_i, \ i = 0, T, \tag{6.144}$$

$$\tilde{x}_t = x^* \text{ and } \tilde{y}_t = y^* \text{ for all } t = 1, \dots, T-1, \tag{6.145}$$

$$\tilde{u}_t = u^* \text{ and } \tilde{v}_t = v^* \text{ for all } t = 1, \dots, T-2, \tag{6.146}$$

$$\rho_U(\tilde{u}_0, u^*) \le \delta_1, \ \rho_U(\tilde{u}_{T-1}, u^*) \le \delta_1, \tag{6.147}$$

$$\rho_V(\tilde{v}_0, v^*) \le \delta_1, \ \rho_U(\tilde{v}_{T-1}, v^*) \le \delta_1. \tag{6.148}$$

Since the pairs of strategies

$$(\{x_t\}_{t=0}^T, \{u_t\}_{t=0}^{T-1}) \in \mathrm{Str}_1(0, T),$$

$$(\{y_t\}_{t=0}^T, \{v_t\}_{t=0}^{T-1}) \in \mathrm{Str}_2(0, T)$$

and

$$(\{\tilde{x}_t\}_{t=0}^T, \{\tilde{u}_t\}_{t=0}^{T-1}) \in \mathrm{Str}_1(0, T),$$

$$(\{\tilde{y}_t\}_{t=0}^T, \{\tilde{v}_t\}_{t=0}^{T-1}) \in \mathrm{Str}_2(0, T)$$

are (δ_1)-good it follows from (6.144) that

$$\sum_{t=0}^{T-1} f(\tilde{x}_t, \tilde{u}_t, \tilde{y}_t, \tilde{v}_t) - 2\delta_1 \le \sum_{t=0}^{T-1} f(x_t, u_t, \tilde{y}_t, \tilde{v}_t) - \delta_1 \le \sum_{t=0}^{T-1} f(x_t, u_t, y_t, v_t)$$

$$\le \sum_{t=0}^{T-1} f(\tilde{x}_t, \tilde{u}_t, y_t, v_t) + \delta_1 \le \sum_{t=0}^{T-1} f(\tilde{x}_t, \tilde{u}_t, \tilde{y}_t, \tilde{v}_t) + 2\delta_1. \tag{6.149}$$

By (6.149), we have

$$\sum_{t=0}^{T-1} f(x_t, u_t, \tilde{y}_t, \tilde{v}_t) \le \sum_{t=0}^{T-1} f(\tilde{x}_t, \tilde{u}_t, \tilde{y}_t, \tilde{v}_t) + 3\delta_1, \tag{6.150}$$

$$\sum_{t=0}^{T-1} f(\tilde{x}_t, \tilde{u}_t, y_t, v_t) \ge \sum_{t=0}^{T-1} f(\tilde{x}_t, \tilde{u}_t, \tilde{y}_t, \tilde{v}_t) - 3\delta_1. \tag{6.151}$$

Let

$$(\{\widehat{x}_t\}_{t=0}^T, \{\widehat{u}_t\}_{t=0}^{T-1}) \in \mathrm{Str}_1(0, T), \ \widehat{x}_0 = x_0, \ \widehat{x}_T = x_T. \tag{6.152}$$

Since the pair of strategies

$$(\{\tilde{x}_t\}_{t=0}^T, \{\tilde{u}_t\}_{t=0}^{T-1}), \ (\{\tilde{y}_t\}_{t=0}^T, \{\tilde{v}_t\}_{t=0}^{T-1})$$

is (δ_1)-good relations (6.144), (6.150), and (6.152) imply that

$$\sum_{t=0}^{T-1} f(\widehat{x}_t, \widehat{u}_t, \tilde{y}_t, \tilde{v}_t) \ge -\delta_1$$

$$+\sum_{t=0}^{T-1} f(\tilde{x}_t, \tilde{u}_t, \tilde{y}_t, \tilde{v}_t) \geq -4\delta_1 + \sum_{t=0}^{T-1} f(x_t, u_t, \tilde{y}_t, \tilde{v}_t). \tag{6.153}$$

It follows from (6.142), (6.144)–(6.146), (6.148), and (6.153) that

$$\sum_{t=0}^{T-1} f(\widehat{x}_t, \widehat{u}_t, y^*, v^*) - \sum_{t=0}^{T-1} f(x_t, u_t, y^*, v^*)$$

$$= [\sum_{t=0}^{T-1} f(\widehat{x}_t, \widehat{u}_t, y^*, v^*) - \sum_{t=0}^{T-1} f(\widehat{x}_t, \widehat{u}_t, \tilde{y}_t, \tilde{v}_t)]$$

$$+[\sum_{t=0}^{T-1} f(\widehat{x}_t, \widehat{u}_t, \tilde{y}_t, \tilde{v}_t) - \sum_{t=0}^{T-1} f(x_t, u_t, \tilde{y}_t, \tilde{v}_t)]$$

$$+[\sum_{t=0}^{T-1} f(x_t, u_t, \tilde{y}_t, \tilde{v}_t) - \sum_{t=0}^{T-1} f(x_t, u_t, y^*, v^*)]$$

$$\geq [f(\widehat{x}_0, \widehat{u}_0, y^*, v^*) - f(\widehat{x}_0, \widehat{u}_0, y_0, \tilde{v}_0) + f(\widehat{x}_{T-1}, \widehat{u}_{T-1}, y^*, v^*)$$
$$-f(\widehat{x}_{T-1}, \widehat{u}_{T-1}, y^*, \tilde{v}_{T-1})] - 4\delta_1$$
$$+[f(x_0, u_0, \tilde{y}_0, \tilde{v}_0) - f(x_0, u_0, y^*, v^*) + f(x_{T-1}, u_{T-1}, y^*, \tilde{v}_{T-1})$$
$$-f(x_{T-1}, u_{T-1}, y^*, v^*)] \geq -\delta_0/16 - \delta_0/8 - \delta_0/16 = -\delta_0/4,$$

$$\sum_{t=0}^{T-1} f(x_t, u_t, y^*, v^*) \leq \sigma_X(0, T, x_0, x_T) + \delta_0/4. \tag{6.154}$$

Analogously we can show that

$$\sum_{t=0}^{T-1} f(x^*, u^*, y_t, v_t) \geq \sigma_Y(0, T, y_0, y_T) - \delta_0/4. \tag{6.155}$$

In view of (6.142), (6.154), (6.155), and the properties (i) and (ii), we have

$$\rho_X(x_t, x^*) \leq \epsilon \text{ and } \rho_Y(y_t, y^*) \leq \epsilon \text{ for all } t = 0, \dots, T.$$

This completes the proof of Lemma 6.22. □

We complete the proofs of Theorems 6.10 and 6.11 simultaneously. Lemma 6.22 implies that there exists a positive number $\delta < \min\{\epsilon, 1/4\}$ such that:

(vi) for every natural number $T > 2$ and every (δ)-good pair of strategies

$$(\{x_t\}_{t=0}^T, \{u_t\}_{t=0}^{T-1}) \in \mathrm{Str}_1(0, T),$$
$$(\{y_t\}_{t=0}^T, \{v_t\}_{t=0}^{T-1}) \in \mathrm{Str}_2(0, T)$$

which satisfies

$$\rho_X(x_i, x^*) \leq \delta,\ i = 0, T,\ \rho_Y(y_i, y^*) \leq \delta,\ i = 0, T$$

we have

$$\rho_X(x_t, x^*) \le \epsilon,\ t = 0, \dots, T,\ \rho_Y(y_t, y^*) \le \epsilon,\ t = 0, \dots, T.$$

Theorems 6.8 and 6.9 imply that there exists a natural number $L > L_0$ such that the following property holds:

(vii) for every natural number $T > 2L$ and every (1)-good pair of strategies

$$(\{x_t\}_{t=0}^{T},\ \{u_t\}_{t=0}^{T-1}) \in \mathrm{Str}_1(0, T)$$

and

$$(\{y_t\}_{t=0}^{T}, \{v_t\}_{t=0}^{T-1}) \in \mathrm{Str}_2(0, T)$$

which satisfies

$$x_0 \in X^{(L_0)},\ x_T \in \bar{X}^{(L_0)},\ y_0 \in Y^{(L_0)},\ y_T \in \bar{Y}^{(L_0)}$$

there exist integers

$$\tau_1 \in [0, L],\ \tau_2 \in [T - L, T] \tag{6.156}$$

for which

$$\rho_X(x_{\tau_i}, x^*) \le \delta,\ \rho_Y(y_{\tau_i}, y^*) \le \delta,\ i = 1, 2; \tag{6.157}$$

(viii) for every natural number $T > 2L$ and every $(1, *)$-good pair of strategies

$$(\{x_t\}_{t=0}^{T},\ \{u_t\}_{t=0}^{T-1}) \in \mathrm{Str}_1(0, T)$$

and

$$(\{y_t\}_{t=0}^{T}, \{v_t\}_{t=0}^{T-1}) \in \mathrm{Str}_2(0, T)$$

which satisfies

$$x_0 \in X^{(L_0)},\ y_0 \in Y^{(L_0)}$$

there exist integers τ_1, τ_2 satisfying (6.156) and (6.157).

Assume that an integer $T > 2L + 1$,

$$(\{x_t\}_{t=0}^{T}, \{u_t\}_{t=0}^{T-1}) \in \mathrm{Str}_1(0, T),\ (\{y_t\}_{t=0}^{T}, \{v_t\}_{t=0}^{T-1}) \in \mathrm{Str}_2(0, T), \tag{6.158}$$

$$x_0 \in X^{(L_0)},\ y_0 \in Y^{(L_0)}.$$

In the case of Theorem 6.10 we suppose that

$$x_T \in \bar{X}^{(L_0)},\ y_T \in \bar{Y}^{(L_0)}$$

and that the pair (6.158) is (δ)-good and in the case of Theorem 6.11 we suppose that the pair (6.158) is $(\delta, *)$-good.

Properties (vii) and (viii) imply that there exist integers $\tau_1 \in [0, L]$, $\tau_2 \in [T - L, T]$ such that (6.157) is valid. If

$$\rho_X(x_0, x^*) \le \delta,\ \rho_Y(y_0, y^*) \le \delta,$$

then we set $\tau_1 = 0$ and if

$$\rho_X(x_T, x^*) \le \delta,\ \rho_Y(y_T, y^*) \le \delta,$$

then we set $\tau_2 = T$. It is clear that the pair of strategies

$$(\{x_t\}_{t=\tau_1}^{\tau_2}, \{u_t\}_{t=\tau_1}^{\tau_2-1}),\ (\{y_t\}_{t=\tau_1}^{\tau_2}, \{v_t\}_{t=\tau_1}^{\tau_2-1})$$

is (δ)-good. Combined with (6.157) and property (vi) this implies that

$$\rho_X(x_t, x^*) \le \epsilon,\ \rho_Y(y_t, y^*) \le \epsilon,\ t = \tau_1, \dots, \tau_2.$$

This completes the proof of Theorems 6.10 and 6.11. □

6.9 Examples

Example 6.23. Let (X, ρ_X), (Y, ρ_Y), (U, ρ_U), and (V, ρ_V) be compact metric spaces. For every point $x \in X$, let $U(x)$ be a subset of U which can be empty and for every point $y \in Y$, let $V(y)$ be a subset of V which can also be empty. We suppose that the sets

$$\{(x, u) \in X \times U :\ u \in U(x)\} \text{ and } \{(y, v) \in Y \times V :\ v \in V(y)\}$$

are closed in the product topologies.

Let

$$F_X :\ \{(x, u) \in X \times U :\ u \in U(x)\} \to X,$$
$$F_Y :\ \{(y, v) \in Y \times V :\ v \in V(y)\} \to Y$$

be continuous mappings. Then the sets

$$\{(x, u, x') \in X \times U \times X :\ u \in U(x),\ x' = F_X(x, u)\},$$

$$\{(y, v, y') \in Y \times V \times Y :\ v \in V(y),\ y' = F_Y(y, v)\}$$

are closed in the product topologies.

Suppose that

$$f :\ \{(x, u, y, v) \in X \times U \times Y \times V :\ u \in U(x),\ v \in V(y)\} \to R^1$$

is a bounded function. Suppose that

$$x^* \in X,\ u^* \in U,\ y^* \in Y,\ v^* \in V, \tag{6.159}$$

$$u^* \in U(x^*),\ x^* = F_X(x^*, u^*), \tag{6.160}$$
$$v^* \in V(y^*),\ y^* = F_Y(y^*, v^*) \tag{6.161}$$

and that assumption (A3) holds.

Let $L_1 : \{(x, u) \in X \times U : u \in U(x)\} \to [0, \infty)$ be a lower semicontinuous function which is continuous at (x^*, u^*) and such that for all points $x \in X$ and all points $u \in U(x)$,

$$L_1(x, u) = 0 \text{ if and only if } x = x^*, \ u = u^* \tag{6.162}$$

and let $L_2 : \{(y, v) \in Y \times V : v \in V(y)\} \to [0, \infty)$ be a lower semicontinuous function which is continuous at (y^*, v^*) and such that for all points $y \in Y$ and all points $v \in V(y)$,

$$L_2(y, v) = 0 \text{ if and only if } y = y^*, \ v = v^*. \tag{6.163}$$

Let $\pi_1 : X \to R^1$ and $\pi_2 : Y \to R^1$ be continuous functions and suppose that for every point $x \in X$ and every point $u \in U(x)$,

$$f(x, u, y^*, v^*) = f(x^*, u^*, y^*, v^*) + L_1(x, u) + \pi_1(x) - \pi_1(F_X(x, u)) \tag{6.164}$$

and that for every point $y \in Y$ and every point $v \in V(y)$,

$$f(x^*, u^*, y, v) = f(x^*, u^*, y^*, v^*) - L_2(y, v) + \pi_2(y) - \pi_2(F_Y(y, v)). \tag{6.165}$$

It is clear that assumption (A1) holds. Using (6.164) and (6.165) it is easy to show that assumption (A2) holds.

We claim that assumption (A4) holds. Let

$$(\{x_t\}_{t=0}^{\infty}, \{u_t\}_{t=0}^{\infty}) \in \mathrm{Str}_1(0, \infty)$$

satisfy (6.9). In view of (6.164), we have $\sum_{t=0}^{\infty} L_1(x_t, u_t) < \infty$. Combined with (6.162) this implies that

$$\lim_{t\to\infty} x_t = x^*.$$

Let

$$(\{y_t\}_{t=0}^{\infty}, \{v_t\}_{t=0}^{\infty}) \in \mathrm{Str}_2(0, \infty)$$

satisfy (6.10). In view of (6.165), we have $\sum_{t=0}^{\infty} L_2(y_t, v_t) < \infty$. Combined with (6.163) this implies that $\lim_{t\to\infty} y_t = y^*$. Hence assumption (A4) holds. Therefore all the assumptions of Sect. 6.1 hold and our main results (Theorems 6.8 and 6.9) hold.

Example 6.24. Our second example is a particular case of Example 6.23 with $(U, \rho_U) = (X, \rho_X)$, $(V, \rho_V) = (Y, \rho_Y)$,

$$F_X(x, u) = u, \ x \in X, \ u \in U(x) \text{ and } F_Y(y, v) = v, \ y \in Y, \ v \in V(y),$$

$$x^* = u^*, \ y^* = v^*.$$

It is not difficult to see that Assumption (A3) is equivalent to the following assumption:

there exists a constant $\bar{r} > 0$ such that for each $x, x' \in X$ satisfying

$$\rho_X(x, x^*),\ \rho_X(x', x^*) \le \bar{r}$$

the inclusion $x' \in U(x)$ holds and that for each $y, y' \in Y$ satisfying

$$\rho_Y(y, y^*),\ \rho_Y(y', y^*) \le \bar{r}$$

the inclusion $y' \in V(y)$ holds.

Example 6.25. [78] Let $X \subset R^{m_1}$ and $Y \subset R^{m_2}$ be nonempty convex compact sets with the metrics induced by the Euclidean norms, $(U, \rho_U) = (X, \rho_X)$, $(V, \rho_V) = (Y, \rho_Y)$, $U(x) = X$, $x \in X$, $V(y) = Y$, $y \in Y$,

$$F_X(x, u) = u,\ (x, u) \in X \times X, \text{ and } F_Y(y, v) = v,\ (y, v) \in Y \times Y.$$

Suppose that a continuous function $f : X \times X \times Y \times Y \to R^1$ satisfies the following conditions:

for every $(y_1, y_2) \in Y \times Y$ the function $(x_1, x_2) \to f(x_1, x_2, y_1, y_2)$, $(x_1, x_2) \in X \times X$ is convex;

for every $(x_1, x_2) \in X \times X$ the function $(y_1, y_2) \to f(x_1, x_2, y_1, y_2)$, $(y_1, y_2) \in Y \times Y$ is concave.

There exists (see Sect. 2 of [78]) $(x^*, y^*) \in X \times Y$ such that

$$\sup_{y \in Y} f(x^*, x^*, y, y) = f(x^*, x^*, y^*, y^*) = \inf_{x \in X} f(x, x, y^* y^*). \tag{6.166}$$

It is clear that assumption (A1) holds. In view of Proposition 2.1 of [78], assumption (A2) holds. Clearly, assumption (A3) holds.

We suppose that the following assumption of [78] holds:

for every point $x \in X \setminus \{x^*\}$ and every point $x' \in X$, we have

$$f(2^{-1}(x^* + x), 2^{-1}(x^* + x'), y^*, y^*) < 2^{-1} f(x^*, x^*, y^*, y^*) + 2^{-1} f(x, x', y^*, y^*);$$

for every point $y \in Y \setminus \{y^*\}$ and every point $y' \in Y$, we have

$$f(x^*, x^*, 2^{-1}(y^* + y), 2^{-1}(y' + y^*)) > 2^{-1} f(x^*, x^*, y^*, y^*) + 2^{-1} f(x^*, x^*, y, y').$$

It is easy to see that this property implies assumption (A4). Thus all the assumptions of Sect. 6.1 hold and Theorems 6.8 and 6.9 hold.

Example 6.26. Our fourth example is a particular case of Example 6.23 with $(U, \rho_U) = (X, \rho_X)$, $(V, \rho_V) = (Y, \rho_Y)$. Let $X \subset R^{m_1}$ and $Y \subset R^{m_2}$ be nonempty compact sets with the metrics induced by the Euclidean norms. We consider the mappings $x \to U(x)$, $x \in X$, $y \to V(y)$, $y \in Y$ and the mappings F_X, F_Y, f introduced in Example 6.23. We also assume that there exists $(x^*, u^*, y^*, v^*) \in X \times U \times Y \times V$ satisfying (6.159)–(6.161). In Example 6.23 we assume that assumption (A3) holds. Here instead of this we make another assumption which implies assumption (A3).

For every integer $k \geq 1$, every positive number r, and every point $x \in R^k$, define

$$B_k(x, r) = \{x' \in R^k : \|x - x'\| \leq r\}.$$

We assume that there exists a number $r_0 \in (0, 1)$ such that for every point

$$x \in X \cap B_{m_1}(x^*, r_0)$$

the inclusion

$$B_{m_1}(u^*, r_0) \subset U(x)$$

holds, for every point

$$y \in Y \cap B_{m_2}(y^*, r_0)$$

the inclusion

$$B_{m_2}(v^*, r_0) \subset V(y)$$

holds and there exist a number $r_1 \in (0, 1)$ and functions

$$F_{X,0} : X \cap B_{m_1}(x^*, r_1) \to R^{m_1},$$

$$F_{Y,0} : Y \cap B_{m_2}(y^*, r_1) \to R^{m_2}$$

such that for every point

$$x \in X \cap B_{m_1}(x^*, r_1)$$

and every point $u \in X \cap B_{m_1}(u^*, r_1)$ the equality

$$F_X(x, u) = F_{X,0}(x) + u$$

holds and for every point

$$y \in Y \cap B_{m_2}(y^*, r_1)$$

and every point

$$v \in Y \cap B_{m_2}(v^*, r_1)$$

the equality

$$F_Y(y, v) = F_{Y,0}(y) + v$$

is true. It is not difficult to see that the assumption we made implies assumption (A3).

Example 6.27. Let $X, Y, U, V = [0, 1]$ with the usual Euclidean metric. For every pair of numbers $x, y \in [0, 1]$ define

$$U(x) = [0, x^{1/2}],\ V(y) = [0, y^{1/3}].$$

For every number $x \in [0, 1]$, every number $u \in U(x)$, every number $y \in [0, 1]$, and every number $v \in V(y)$ define

$$F_X(x, u) = x^{1/2} - u,\ F_Y(y, v) = y^{1/3} - v.$$

Set

$$x^* = 2^{-1},\ u^* = 2^{-1/2} - 2^{-1},\ y^* = 2^{-1}, v^* = 2^{-1/3} - 2^{-1}.$$

It is clear that $x^*, u^*, y^*, v^* \in [0, 1]$,

$$u^* \in U(x^*),\ x^* \in F_X(x^*, u^*),$$

$$v^* \in V(y^*),\ y^* \in F_Y(y^*, v^*).$$

For every number $x \in [0, 1]$, every number $u \in U(x)$, every number $y \in [0, 1]$, and every number $v \in V(y)$ define

$$f(x, u, y, v) = [(2x - 1)^2 + (2u + 1 - 2^{1/2})^2](x + u + 1)$$
$$-[(2y - 1)^2 + (2v + 1 - 2^{2/3})^2](y^2 + v^2 + 1)$$
$$+x - x^{1/2} + u + y - y^{1/3} + v.$$

It is easy to see that this example is a particular of Example 6.231 with

$$L_1(x, u) = [(2x - 1)^2 + (2u + 1 - 2^{1/2})^2](x + u + 1),\ x \in [0, 1],\ u \in U(x),$$

$$L_2(y, v) = [(2y - 1)^2 + (2v + 1 - 2^{2/3})^2](y^2 + v^2 + 1),\ y \in [0, 1],\ v \in V(y),$$

$$\pi_i(x) = x,\ x \in [0, 1],\ i = 1, 2.$$

7

Nonautonomous Games with Convexity–Concavity Assumptions

In this chapter we study turnpike properties of approximate solutions for a class of dynamic discrete-time two-player zero-sum games with convexity–concavity assumptions. These properties describe the structure of approximate solutions which is independent of the length of the interval, for all sufficiently large intervals. We also study the existence of equilibria over an infinite horizon employing a version of the overtaking optimality criterion.

7.1 Preliminaries

Let M, N be nonempty sets and let $f : M \times N \to R^1$. Define

$$f^a(x) = \sup_{y \in N} f(x, y),\ x \in M, \quad f^b(y) = \inf_{x \in M} f(x, y),\ y \in N,$$

$$v_f^a = \inf_{x \in M} \sup_{y \in N} f(x, y), \quad v_f^b = \sup_{y \in N} \inf_{x \in M} f(x, y).$$

It is clear that

$$v_f^b \le v_f^a.$$

The following result holds (see Chap. 6, Sect. 2, Proposition 1 of [7]).

Proposition 7.1. *Let $f : M \times N \to R^1$, $\bar{x} \in M$, $\bar{y} \in N$. Then*

$$\sup_{y \in N} f(\bar{x}, y) = f(\bar{x}, \bar{y}) = \inf_{x \in M} f(x, \bar{y}) \tag{7.1}$$

if and only if

$$v_f^a = v_f^b,\ \sup_{y \in N} f(\bar{x}, y) = v_f^a,\ \inf_{x \in M} f(x, \bar{y}) = v_f^b.$$

A.J. Zaslavski, *Discrete-Time Optimal Control and Games on Large Intervals*, Springer Optimization and Its Applications 119,
DOI 10.1007/978-3-319-52932-5_7

Let $f : M \times N \to R^1$. If $(\bar{x}, \bar{y}) \in M \times N$ satisfies (7.1), then it is called a saddle point (for f). The following result is valid (see Chap. 6, Sect. 2, Theorem 8 of [7]).

Proposition 7.2. *Let $M \subset R^m$, $N \subset R^n$ be convex compact sets and let $f : M \times N \to R^1$ be a continuous function. Assume that for each $y \in N$ the function $x \to f(x, y)$, $x \in M$ is convex and for each $x \in M$ the function $y \to f(x, y)$, $y \in N$ is concave. Then there exists a saddle point for f.*

Denote by $|\cdot|$ the Euclidean norm in R^m. Let $X \subset R^{m_1}$ and $Y \subset R^{m_2}$ be nonempty convex compact sets.

Denote by $C(X \times X \times Y \times Y)$ the set of all continuous functions $f : X \times X \times Y \times Y \to R^1$. For every function $f \in C(X \times X \times Y \times Y)$ put

$$\|f\| = \max\{|f(x_1, x_2, y_1, y_2)| : x_1, x_2 \in X, \ y_1, y_2 \in Y\}.$$

It is clear that $(C(X \times X \times Y \times Y), \|\cdot\|)$ is a Banach space. For every pair of functions $f, g \in C(X \times X \times Y \times Y)$ define

$$\rho_c(f, g) = \|f - g\|.$$

It is easy to see that $(C(X \times X \times Y \times Y), \rho_c)$ is a complete metric space.

Denote by $\mathcal{M}$ the set of all continuous functions $f : X \times X \times Y \times Y \to R^1$ such that:

for each $(y_1, y_2) \in Y \times Y$ the function $(x_1, x_2) \to f(x_1, x_2, y_1, y_2)$, $(x_1, x_2) \in X \times X$ is convex;

for each $(x_1, x_2) \in X \times X$ the function $(y_1, y_2) \to f(x_1, x_2, y_1, y_2)$, $(y_1, y_2) \in Y \times Y$ is concave.

It is not difficult to see that $\mathcal{M}$ is a closed subset of $(C(X \times X \times Y \times Y), \rho_c)$. We consider the space $\mathcal{M}$ equipped with the complete metric ρ_c.

Let $\{f_i\}_{i=0}^{\infty} \subset C(X \times X \times Y \times Y)$ be given. We associate with this sequence a dynamic discrete-time two-player zero-sum game and study its optimal solutions. Namely, given an integer $n \geq 1$ we consider a discrete-time two-player zero-sum game over the interval $[0, n]$. For this game $\{\{x_i\}_{i=0}^{n} : x_i \in X, \ i = 0, \dots n\}$ is the set of strategies for the first player, $\{\{y_i\}_{i=0}^{n} : y_i \in Y, \ i = 0, \dots n\}$ is the set of strategies for the second player, and the cost for the first player associated with the strategies $\{x_i\}_{i=0}^{n}$, $\{y_i\}_{i=0}^{n}$ is given by $\sum_{i=0}^{n-1} f_i(x_i, x_{i+1}, y_i, y_{i+1})$.

Let a sequence $\{f_i\}_{i=0}^{\infty} \subset C(X \times X \times Y \times Y)$ be given, M be a nonnegative number and let $n_1 < n_2$ be nonnegative integers. A pair of finite sequences $\{\bar{x}_i\}_{i=n_1}^{n_2} \subset X$, $\{\bar{y}_i\}_{i=n_1}^{n_2} \subset Y$ is called $(\{f_i\}_{i=0}^{\infty}, M)$-good if the following properties hold:

for every finite sequence $\{x_i\}_{i=n_1}^{n_2} \subset X$ which satisfies

$$x_{n_1} = \bar{x}_{n_1}, \ x_{n_2} = \bar{x}_{n_2},$$

we have

$$M + \sum_{i=n_1}^{n_2-1} f_i(x_i, x_{i+1}, \bar{y}_i, \bar{y}_{i+1}) \geq \sum_{i=n_1}^{n_2-1} f_i(\bar{x}_i, \bar{x}_{i+1}, \bar{y}_i, \bar{y}_{i+1}); \tag{7.2}$$

for every finite sequence $\{y_i\}_{i=n_1}^{n_2} \subset Y$ which satisfies

$$y_{n_1} = \bar{y}_{n_1},\ y_{n_2} = \bar{y}_{n_2},$$

we have

$$M + \sum_{i=n_1}^{n_2-1} f_i(\bar{x}_i, \bar{x}_{i+1}, \bar{y}_i, \bar{y}_{i+1}) \geq \sum_{i=n_1}^{n_2-1} f_i(\bar{x}_i, \bar{x}_{i+1}, y_i, y_{i+1}). \tag{7.3}$$

If a pair of sequences $\{\bar{x}_i\}_{i=n_1}^{n_2} \subset X$, $\{\bar{y}_i\}_{i=n_1}^{n_2} \subset Y$ is $(\{f_i\}_{i=0}^{\infty}, 0)$-good, then it is called $(\{f_i\}_{i=0}^{\infty})$-optimal.

A pair of sequences $\{x_i\}_{i=0}^{\infty} \subset X$, $\{y_i\}_{i=0}^{\infty} \subset Y$ is called $(\{f_i\}_{i=0}^{\infty})$-locally optimal if for each integer $n \geq 2$ the pair of sequences $\{x_i\}_{i=0}^{n}$, $\{y_i\}_{i=0}^{n}$ is $(\{f_i\}_{i=0}^{\infty})$-optimal.

Proposition 7.2 implies the following result.

Proposition 7.3. *Let $\{g_i\}_{i=0}^{\infty} \subset \mathcal{M}$. Then for every pair of points $x, x' \in X$, every pair of points $y, y' \in Y$ and every natural number n there exists a $(\{g_i\}_{i=0}^{\infty})$-optimal pair of sequences $\{x_i\}_{i=0}^{n} \subset X$, $\{y_i\}_{i=0}^{n} \subset Y$ such that $x_0 = x$, $x_n = x'$, $y_0 = y$, $y_n = y'$.*

We study the existence of equilibria over an infinite horizon and employ the following version of the overtaking optimality criterion.

Let $\{f_i\}_{i=0}^{\infty} \subset C(X \times X \times Y \times Y)$. A pair of sequences $\{\bar{x}_i\}_{i=0}^{\infty} \subset X$, $\{\bar{y}_i\}_{i=0}^{\infty} \subset Y$ is called $(\{f_i\}_{i=0}^{\infty})$-overtaking optimal if the following properties hold:

for each sequence $\{x_i\}_{i=0}^{\infty} \subset X$ satisfying $x_0 = \bar{x}_0$,

$$\limsup_{T\to\infty} [\sum_{i=0}^{T-1} f_i(\bar{x}_i, \bar{x}_{i+1}, \bar{y}_i, \bar{y}_{i+1}) - \sum_{i=0}^{T-1} f_i(x_i, x_{i+1}, \bar{y}_i, \bar{y}_{i+1})] \leq 0; \tag{7.4}$$

for each sequence $\{y_i\}_{i=0}^{\infty} \subset Y$ satisfying $y_0 = \bar{y}_0$

$$\limsup_{T\to\infty} [\sum_{i=0}^{T-1} f_i(\bar{x}_i, \bar{x}_{i+1}, y_i, y_{i+1}) - \sum_{i=0}^{T-1} f_i(\bar{x}_i, \bar{x}_{i+1}, \bar{y}_i, \bar{y}_{i+1})] \leq 0. \tag{7.5}$$

In the next section we prove the following result obtained in [80].

Theorem 7.4. *Let $\{f_i\}_{i=0}^{\infty} \subset \mathcal{M}$. Then for each $x \in X$ and each $y \in Y$ there exists an $(\{f_i\}_{i=0}^{\infty})$-locally optimal pair of sequences $\{x_i\}_{i=0}^{\infty} \subset X$, $\{y_i\}_{i=0}^{\infty} \subset Y$ such that $x_0 = x$, $y_0 = y$.*

7.2 Proof of Theorem 7.4

Let $\{f_i\}_{i=0}^{\infty} \in \mathcal{M}$ be given. The next result easily follows from Proposition 7.2.

Proposition 7.5. *Let $x \in X$, $y \in Y$ and let an integer $n \geq 2$. Then there exists an $(\{f_i\}_{i=0}^{\infty})$-optimal pair of sequences $\{x_i\}_{i=0}^{n} \subset X$, $\{y_i\}_{i=0}^{n} \subset Y$ such that $x_0 = x$ and $y_0 = y$.*

Proposition 7.6. *Let $n \geq 2$ be a natural number and let*

$$(\{x_i^{(k)}\}_{i=0}^{n}, \{y_i^{(k)}\}_{i=0}^{n}) \subset X \times Y, \quad k = 1, 2, \dots \tag{7.6}$$

be a sequence of $(\{f_i\}_{i=0}^{\infty})$-optimal pairs. Assume that

$$\lim_{k \to \infty} x_i^{(k)} = x_i, \quad \lim_{k \to \infty} y_i^{(k)} = y_i, \quad i = 0, 1, 2, \dots, n. \tag{7.7}$$

Then the pair of sequences $(\{x_i\}_{i=0}^{n}, \{y_i\}_{i=0}^{n})$ is $(\{f_i\}_{i=0}^{\infty})$-optimal.

Proof. Let

$$\{u_i\}_{i=0}^{n} \subset X, \quad u_0 = x_0, u_n = x_n. \tag{7.8}$$

We claim that

$$\sum_{i=0}^{n-1} f_i(x_i, x_{i+1}, y_i, y_{i+1}) \leq \sum_{i=0}^{n-1} f_i(u_i, u_{i+1}, y_i, y_{i+1}). \tag{7.9}$$

Assume the contrary. Then there exists a positive number ϵ for which

$$\sum_{i=0}^{n-1} f_i(x_i, x_{i+1}, y_i, y_{i+1}) > \sum_{i=0}^{n-1} f_i(u_i, u_{i+1}, y_i, y_{i+1}) + 8\epsilon. \tag{7.10}$$

There exists a positive number $\delta < \epsilon$ such that

$$|f_j(z_1, z_2, \xi_1, \xi_2) - f_j(\bar{z}_1, \bar{z}_2, \bar{\xi}_1, \bar{\xi}_2)| \leq \epsilon(8n)^{-1} \tag{7.11}$$

for each $z_1, z_2, \bar{z}_1, \bar{z}_2 \in X$, $\xi_1, \xi_2, \bar{\xi}_1, \bar{\xi}_2 \in Y$ which satisfies

$$|z_i - \bar{z}_i|, |\xi_i - \bar{\xi}_i| \leq \delta, \quad i = 1, 2$$

and every integer $j \in \{0, \dots, n-1\}$. In view of (7.7), there exists a natural number q such that

$$|x_i - x_i^{(q)}|, |y_i - y_i^{(q)}| \leq \delta, \quad i = 0, \dots n. \tag{7.12}$$

Define a finite sequence $\{u_i^{(q)}\}_{i=0}^{n} \subset X$ as follows:

$$u_0^{(q)} = x_0^{(q)}, \ u_n^{(q)} = x_n^{(q)}, \quad u_i^{(q)} = u_i, \ i = 1, \dots n-1. \tag{7.13}$$

Since the pair of sequences $(\{x_i^{(q)}\}_{i=0}^n, \{y_i^{(q)}\}_{i=0}^n)$ is $(\{f_i\}_{i=0}^\infty)$-optimal relation (7.13) implies that

$$\sum_{i=0}^{n-1} f_i(x_i^{(q)}, x_{i+1}^{(q)}, y_i^{(q)}, y_{i+1}^{(q)}) \le \sum_{i=0}^{n-1} f_i(u_i^{(q)}, u_{i+1}^{(q)}, y_i^{(q)}, y_{i+1}^{(q)}). \tag{7.14}$$

It follows from the definition of δ (see (7.11)), (7.12), (7.13), and (7.8) that for all integers $i = 0, \dots, n-1$ the following inequalities are valid:

$$|f_i(x_i^{(q)}, x_{i+1}^{(q)}, y_i^{(q)}, y_{i+1}^{(q)}) - f_i(x_i, x_{i+1}, y_i, y_{i+1})| \le (8n)^{-1}\epsilon,$$

$$|f_i(u_i^{(q)}, u_{i+1}^{(q)}, y_i^{(q)}, y_{i+1}^{(q)}) - f_i(u_i, u_{i+1}, y_i, y_{i+1})| \le (8n)^{-1}\epsilon. \tag{7.15}$$

In view of (7.10) and (7.15), we have

$$\sum_{i=0}^{n-1} f_i(x_i^{(q)}, x_{i+1}^{(q)}, y_i^{(q)}, y_{i+1}^{(q)}) - \sum_{i=0}^{n-1} f_i(u_i^{(q)}, u_{i+1}^{(q)}, y_i^{(q)}, y_{i+1}^{(q)}) > \epsilon.$$

This is contradictory to (7.14). The contradiction we have reached proves (7.9). Analogously we can show that for every finite sequence $\{u_i\}_{i=0}^n \subset Y$ which satisfies

$$u_0 = y_0, \ u_n = y_n$$

the

$$\sum_{i=0}^{n-1} f_i(x_i, x_{i+1}, y_i, y_{i+1}) \ge \sum_{i=0}^{n-1} f_i(x_i, x_{i+1}, u_i, u_{i+1})$$

is valid. Proposition 7.6 is proved. □

Completion of the Proof of Theorem 7.4. Let $x \in X$ and $y \in Y$ be given. Proposition 7.5 implies that for every natural number $n \ge 2$ there exists an $(\{f_i\}_{i=0}^\infty)$-optimal pair of sequences

$$\{x_i^{(n)}\}_{i=0}^n \subset X, \ \{y_i^{(n)}\}_{i=0}^n \subset Y$$

satisfying

$$x_0^{(n)} = x, \ y_0^{(n)} = y.$$

There exist a pair of sequences

$$\{x_i\}_{i=0}^\infty \subset X, \ \{y_i\}_{i=0}^\infty \subset Y$$

and a strictly increasing sequence of natural numbers $\{n_k\}_{k=1}^\infty$ such that for every nonnegative integer i we have

$$x_i^{(n_k)} \to x_i, \quad y_i^{(n_k)} \to y_i \text{ as } k \to \infty.$$

Proposition 7.6 implies that the pair of sequences

$$\{x_i\}_{i=0}^\infty \subset X, \ \{y_i\}_{i=0}^\infty \subset Y$$

is $(\{f_i\}_{i=0}^\infty)$-locally optimal. This completes the proof of Theorem 7.4. □

7.3 A Space of Objective Functions

Denote by $\mathcal{A}$ the set of all sequences of functions $\{f_i\}_{i=0}^{\infty} \subset \mathcal{M}$ such that

$$\sup\{\|f_i\| : i = 0, 1, \dots\} < \infty \tag{7.16}$$

and that the following property holds:

(P1) for every positive number ϵ there exists a positive number $\delta(\epsilon)$ such that for every nonnegative integer i, each $z_1, z_2, \bar{z}_1, \bar{z}_2 \in X$ and each $\xi_1, \xi_2, \bar{\xi}_1, \bar{\xi}_2 \in Y$ which satisfies

$$|z_i - \bar{z}_i| \leq \delta(\epsilon),\ |\xi_i - \bar{\xi}_i| \leq \delta(\epsilon),\ i = 1, 2$$

we have

$$|f_i(z_1, z_2, \xi_1, \xi_2) - f_i(\bar{z}_1, \bar{z}_2, \bar{\xi}_1, \bar{\xi}_2)| \leq \epsilon.$$

For every pair of sequences $\{f_i\}_{i=0}^{\infty}, \{g_i\}_{i=0}^{\infty} \in \mathcal{A}$ define

$$\rho_{\mathcal{A}}(\{f_i\}_{i=0}^{\infty}, \{g_i\}_{i=0}^{\infty}) = \sup\{\|f_i - g_i\| : i = 0, 1, \dots\}. \tag{7.17}$$

Clearly, $\rho_{\mathcal{A}}$ is a metric on $\mathcal{A}$ and the metric space $(\mathcal{A}, \rho_{\mathcal{A}})$ is complete.

Denote by $\mathcal{A}_*$ the set of all sequences of $\{f_i\}_{i=0}^{\infty} \in \mathcal{A}$ such that for every positive number ϵ there exists a positive number $\delta(\epsilon)$ such that the following property holds:

(P2)

for each nonnegative integer i, each $z_1, z_2, \bar{z}_1, \bar{z}_2 \in X$ which satisfy

$$\max\{|z_1 - \bar{z}_1|, |z_2 - \bar{z}_2|\} \geq \epsilon$$

and each pair of points $\xi_1, \xi_2 \in Y$, we have

$$\begin{aligned} &2^{-1} f_i(z_1, z_2, \xi_1, \xi_2) + 2^{-1} f_i(\bar{z}_1, \bar{z}_2, \xi_1, \xi_2) - \delta(\epsilon) \\ &\quad \geq f_i(2^{-1}(z_1 + \bar{z}_1), 2^{-1}(z_2 + \bar{z}_2), \xi_1, \xi_2); \end{aligned} \tag{7.18}$$

for each nonnegative integer i, each pair of points $z_1, z_2 \in X$, and each $\xi_1, \xi_2, \bar{\xi}_1, \bar{\xi}_2 \in Y$ which satisfy

$$\max\{|\xi_1 - \bar{\xi}_1|, |\xi_2 - \bar{\xi}_2|\} \geq \epsilon,$$

we have

$$\begin{aligned} &2^{-1} f_i(z_1, z_2, \xi_1, \xi_2) + 2^{-1} f_i(z_1, z_2, \bar{\xi}_1, \bar{\xi}_2) + \delta(\epsilon) \\ &\quad \leq f_i(z_1, z_2, 2^{-1}(\xi_1 + \bar{\xi}_1), 2^{-1}(\xi_2 + \bar{\xi}_2)). \end{aligned} \tag{7.19}$$

We show that most elements of $\mathcal{A}$ belong to $\mathcal{A}_*$. For this goal we use the following porosity notion [57, 74].

Let (Z, d) be a complete metric space. We denote by $B(y, r)$ the closed ball of center $y \in Z$ and radius $r > 0$. A subset $E \subset Z$ is called porous (with respect to the metric d) if there exists $\alpha \in (0,1)$ and $r_0 > 0$ such that for each $r \in (0, r_0]$ and each $y \in Z$ there is $z \in Z$ for which

$$B(z, \alpha r) \subset B(y, r) \setminus E.$$

A subset of the space Z is called σ-porous (with respect to d) if it is a countable union of porous subsets of Z.

The following result was obtained in [87].

Theorem 7.7. *$\mathcal{A} \setminus \mathcal{A}_*$ is a σ-porous subset of $\mathcal{A}$ with respect to d.*

Proof. For every integer $n \geq 1$ denote by $\mathcal{A}_n$ the set of all sequences of functions $\{f_i\}_{i=0}^{\infty} \in \mathcal{A}$ such that the following property holds:

(P3) there is a positive number $\delta = \delta(\{f_i\}_{i=0}^{\infty}, n)$ such that:

for each nonnegative integer i, each $z_1, z_2, \bar{z}_1, \bar{z}_2 \in X$ which satisfy

$$\max\{|z_1 - \bar{z}_1|, |z_2 - \bar{z}_2|\} \geq 1/n$$

and each pair of points $\xi_1, \xi_2 \in Y$ the inequality

$$2^{-1} f_i(z_1, z_2, \xi_1, \xi_2) + 2^{-1} f_i(\bar{z}_1, \bar{z}_2, \xi_1, \xi_2) - \delta$$

$$\geq f_i(2^{-1}(z_1 + \bar{z}_1), 2^{-1}(z_2 + \bar{z}_2), \xi_1, \xi_2)$$

is valid;

for each nonnegative integer i, each pair of points $z_1, z_2 \in X$, and each $\xi_1, \xi_2, \bar{\xi}_1, \bar{\xi}_2 \in Y$ which satisfy

$$\max\{|\xi_1 - \bar{\xi}_1|, |\xi_2 - \bar{\xi}_2|\} \geq 1/n$$

the inequality

$$2^{-1} f_i(z_1, z_2, \xi_1, \xi_2) + 2^{-1} f_i(z_1, z_2, \bar{\xi}_1, \bar{\xi}_2) + \delta$$

$$\leq f_i(z_1, z_2, 2^{-1}(\xi_1 + \bar{\xi}_1), 2^{-1}(\xi_2 + \bar{\xi}_2))$$

is valid.

It is clear that

$$\mathcal{A}_* = \cap_{n=1}^{\infty} \mathcal{A}_n. \tag{7.20}$$

In view of (7.20), it is sufficient to show that the set $\mathcal{A} \setminus \mathcal{A}_n$ is porous. Let $n \geq 1$ be an integer. Put

$$D_0 = \sup\{|x| : x \in X\} + \sup\{|y| : y \in Y\}. \tag{7.21}$$

Set

$$\alpha = 512^{-1} n^{-2} (D_0^2 + 1)^{-1}. \tag{7.22}$$

Assume that

$$\{f_i\}_{i=0}^{\infty} \in \mathcal{A},\ r \in (0, 1]. \tag{7.23}$$

Set

$$\gamma = 16^{-1}r(D_0^2+1)^{-1}. \tag{7.24}$$

For every nonnegative integer i define

$$g_i(x_1,x_2,y_1,y_2)$$
$$= f_i(x_1,x_2,y_1,y_2) + \gamma|x_1|^2 + \gamma|x_2|^2 - \gamma|y_1|^2 - \gamma|y_2|^2, \tag{7.25}$$
$$x_1,x_2 \in X,\ y_1,y_2 \in Y.$$

It is clear that $\{g_i\}_{i=0}^{\infty} \in \mathcal{A}$. In view of (7.17), (7.21), (7.25), and (7.24), we have

$$\rho_{\mathcal{A}}(\{f_i\}_{i=0}^{\infty}, \{g_i\}_{i=0}^{\infty}) \le 4\gamma D_0^2 \le r/4. \tag{7.26}$$

Let $\{h_i\}_{i=0}^{\infty} \in \mathcal{A}$ satisfy

$$\rho_{\mathcal{A}}(\{h_i\}_{i=0}^{\infty}, \{g_i\}_{i=0}^{\infty}) \le \alpha r. \tag{7.27}$$

It follows from (7.27), (7.26), and (7.22) that

$$\rho_{\mathcal{A}}(\{h_i\}_{i=0}^{\infty}, \{f_i\}_{i=0}^{\infty})$$
$$\le \rho_{\mathcal{A}}(\{h_i\}_{i=0}^{\infty}, \{g_i\}_{i=0}^{\infty}) + \rho_{\mathcal{A}}(\{g_i\}_{i=0}^{\infty}, \{f_i\}_{i=0}^{\infty}) \le r/2. \tag{7.28}$$

We show that $\{h_i\}_{i=0}^{\infty} \in \mathcal{A}_n$. Let i be a nonnegative integer. Assume that

$$z_1, z_2, \bar{z}_1, \bar{z}_2 \in X,\ \xi_1, \xi_2 \in Y$$

satisfy

$$\max\{|z_1 - \bar{z}_1|,\ |z_2 - \bar{z}_2|\} \ge 1/n. \tag{7.29}$$

In view of (7.17), (7.22), (7.24), (7.25), (7.27), and (7.29), we have

$$2^{-1}h_i(z_1,z_2,\xi_1,\xi_2) + 2^{-1}h_i(\bar{z}_1,\bar{z}_2,\xi_1,\xi_2)$$
$$-h_i(2^{-1}(z_1+\bar{z}_1), 2^{-1}(z_2+\bar{z}_2), \xi_1, \xi_2)$$
$$\ge -3\alpha r + 2^{-1}g_i(z_1,z_2,\xi_1,\xi_2)$$
$$+2^{-1}g_i(\bar{z}_1,\bar{z}_2,\xi_1,\xi_2) - g_i(2^{-1}(z_1+\bar{z}_1), 2^{-1}(z_2+\bar{z}_2), \xi_1, \xi_2)$$
$$\ge -3\alpha r + \gamma[2^{-1}(|z_1|^2+|z_2|^2) + 2^{-1}(|\bar{z}_1|^2+|\bar{z}_2|^2)$$
$$-4^{-1}|z_1+\bar{z}_1|^2 - 4^{-1}|z_2+\bar{z}_2|^2]$$
$$= -3\alpha r + \gamma[4^{-1}|z_1-\bar{z}_1|^2 + 4^{-1}|z_2-\bar{z}_2|^2] \ge -3\alpha r + 4^{-1}\gamma n^{-2}$$
$$= -3\alpha r + 64^{-1}n^{-2}(D_0^2+1)^{-1}r \ge 128^{-1}n^{-2}(D_0^2+1)^{-1}r. \tag{7.30}$$

Assume that

$$z_1, z_2 \in X,\ \xi_1, \xi_2, \bar{\xi}_1, \bar{\xi}_2 \in Y$$

satisfy

$$\max\{|\xi_1 - \bar{\xi}_1|,\ |\xi_2 - \bar{\xi}_2|\} \ge 1/n. \tag{7.31}$$

It follows from (7.17), (7.22), (7.24), (7.25), (7.27), and (7.31) that

$$\begin{aligned}
&h_i(z_1, z_2, 2^{-1}(\xi_1 + \bar{\xi}_1), 2^{-1}(\xi_2 + \bar{\xi}_2)) \\
&-2^{-1}h_i(z_1, z_2, \xi_1, \xi_2) + 2^{-1}h_i(z_1, z_2, \bar{\xi}_1, \bar{\xi}_2) \\
&\geq -3\alpha r + g_i(z_1, z_2, 2^{-1}(\xi_1 + \bar{\xi}_1), 2^{-1}(\xi_2 + \bar{\xi}_2)) \\
&-2^{-1}g_i(z_1, z_2, \xi_1, \xi_2) - 2^{-1}g_i(z_1, z_2, \bar{\xi}_1, \bar{\xi}_2) \\
&\geq -3\alpha r + \gamma[2^{-1}|\xi_1|^2 + 2^{-1}|\xi_2|^2 + 2^{-1}|\bar{\xi}_1|^2 \\
&+2^{-1}|\bar{\xi}_2|^2 - 4^{-1}|\xi_1 + \bar{\xi}_1|^2 - 4^{-1}|\xi_2 + \bar{\xi}_2|^2] \\
&= -3\alpha r + \gamma[4^{-1}|\xi_1 - \bar{\xi}_1|^2 + 4^{-1}|\xi_2 - \bar{\xi}_2|^2] \\
&\geq -3\alpha r + 4^{-1}\gamma n^{-2} \\
&= -3\alpha r + 64^{-1}n^{-2}(D_0^2 + 1)^{-1}r \geq 128^{-1}n^{-2}(D_0^2 + 1)^{-1}r. \qquad (7.32)
\end{aligned}$$

Thus we have shown that (7.29) implies (7.30) and (7.31) implies (7.32), Thus $\{h_i\}_{i=0}^{\infty} \in \mathcal{A}_n$.

We have shown that if $\{h_i\}_{i=0}^{\infty} \in \mathcal{A}$ satisfies (7.27), then $\{h_i\}_{i=0}^{\infty} \in \mathcal{A}_n$ and (7.28) is valid. Hence the set $\mathcal{A} \setminus \mathcal{A}_n$ is porous. Theorem 7.7 is proved. □

7.4 Turnpike Results

Let $\{f_i\}_{i=0}^{\infty} \in \mathcal{A}$ and let $\{\bar{x}_i\}_{i=0}^{\infty} \subset X$, $\{\bar{y}_i\}_{i=0}^{\infty} \subset Y$ be an $(\{f_i\}_{i=0}^{\infty})$-locally optimal pair of sequences. Its existence follows from Theorem 7.4.

We suppose that the following assumptions hold:

(A1) for every positive number ϵ there exists a positive number δ such that for every nonnegative integer i, every point $x \in X$ which satisfies $|x - \bar{x}_i| \geq \epsilon$ and every point $x' \in X$ the inequality

$$\begin{aligned}
&f_i(2^{-1}(x + \bar{x}_i), 2^{-1}(x' + \bar{x}_{i+1}), \bar{y}_i, \bar{y}_{i+1}) + \delta \\
&\leq 2^{-1}f_i(\bar{x}_i, \bar{x}_{i+1}, \bar{y}_i, \bar{y}_{i+1}) + 2^{-1}f_i(x, x', \bar{y}_i, \bar{y}_{i+1})
\end{aligned}$$

is valid;

(A2) for every positive number ϵ there exists a positive number δ such that for every nonnegative integer i, every point $y \in Y$ which satisfies $|y - \bar{y}_i| \geq \epsilon$ and every point $y' \in Y$ the inequality

$$\begin{aligned}
&f_i(\bar{x}_i, \bar{x}_{i+1}, 2^{-1}(y + \bar{y}_i), 2^{-1}(y' + \bar{y}_{i+1})) - \delta \\
&\geq 2^{-1}f_i(\bar{x}_i, \bar{x}_{i+1}, \bar{y}_i, \bar{y}_{i+1}) + 2^{-1}f_i(\bar{x}_i, \bar{x}_{i+1}, y, y')
\end{aligned}$$

is valid.

(It should be mentioned that every sequence of functions $\{f_i\}_{i=0}^{\infty} \in \mathcal{A}_*$ satisfies assumptions (A1) and (A2).)

The following theorem is our first main result which will be proved in Sect. 7.6.

Theorem 7.8. *Let $\epsilon \in (0,1)$. Then there exist a number $\delta \in (0,\epsilon)$ and a natural number $L \geq 4$ such that for every sequence of functions $\{g_i\}_{i=0}^{\infty} \subset C(X \times X \times Y \times Y)$ which satisfies*

$$\|g_i - f_i\| \leq \delta,\ i = 0, 1, \dots,$$

every pair of nonnegative integers T_1, $T_2 \geq T_1 + 2L$ and every pair of finite sequences

$$\{x_i\}_{i=T_1}^{T_2} \subset X,\ \{y_i\}_{i=T_1}^{T_2} \subset Y$$

such that the pair

$$\{x_i\}_{i=j}^{j+2L},\ \{y_i\}_{i=j}^{j+2L}$$

is $(\{g_i\}_{i=0}^{\infty}, \delta)$-good for every integer $j \in [T_1, T_2 - 2L]$, the inequalities

$$|x_i - \bar{x}_i|,\ |y_i - \bar{y}_i| \leq \epsilon \tag{7.33}$$

are valid for all integers $i \in [T_1 + L, T_2 - L]$. Moreover, if

$$|x_{T_1} - \bar{x}_{T_1}|,\ |y_{T_1} - \bar{y}_{T_1}| \leq \delta,$$

then (7.33) is true for all integers $i \in [T_1, T_2 - L]$, and if

$$|x_{T_2} - \bar{x}_{T_2}|,\ |y_{T_2} - \bar{y}_{T_2}| \leq \delta,$$

then (7.33) holds for all integers $i \in [T_1 + L, T_2]$.

The next theorem is proved in Sect. 7.7.

Theorem 7.9. *For every point $x \in X$ and every point $y \in Y$ there exists an $(\{f_i\}_{i=0}^{\infty})$-overtaking optimal pair of sequences $\{x_i\}_{i=0}^{\infty} \subset X$, $\{y_i\}_{i=0}^{\infty} \subset Y$ such that $x_0 = x$ and $y_0 = y$.*

.

The next result easily follows from Theorem 7.8.

Theorem 7.10. *Let $\epsilon \in (0,1)$. Then there exist a natural number $L \geq 4$, a positive number $\delta < \epsilon$, and a number $\lambda \in (0,1)$ such that for every sequence of functions $\{g_i\}_{i=0}^{\infty} \subset C(X \times X \times Y \times Y)$ which satisfies*

$$\|g_i - f_i\| \leq \delta,\ i = 0, 1, \dots,$$

every sequence $\{\alpha_i\}_{i=0}^{\infty} \subset (0,1]$ such that

$$\alpha_i \alpha_j^{-1} \geq \lambda$$

for every pair of nonnegative integers i, j which satisfies $|i - j| \leq 2L$, every pair of integers $T_1 \geq 0$, $T_2 \geq T_1 + 2L$ and every pair of finite sequences

$$\{x_i\}_{i=T_1}^{T_2} \subset X,\ \{y_i\}_{i=T_1}^{T_2} \subset Y$$

for which the pair $\{x_i\}_{i=j}^{j+2L}$, $\{y_i\}_{i=j}^{j+2L}$ *is* $(\{\alpha_i g_i\}_{i=0}^{\infty})$*-optimal for every integer* $j \in [T_1, T_2 - 2L]$, *the inequalities*

$$|x_i - \bar{x}_i|, \; |y_i - \bar{y}_i| \le \epsilon \tag{7.34}$$

are valid for all integers $i \in [T_1 + L, T_2 - L]$. *Moreover, if*

$$|x_{T_1} - \bar{x}_{T_1}|, \; |y_{T_1} - \bar{y}_{T_1}| \le \delta,$$

then (7.34) is valid for all integers $i \in [T_1, T_2 - L]$, *and if*

$$|x_{T_2} - \bar{x}_{T_2}|, \; |y_{T_2} - \bar{y}_{T_2}| \le \delta,$$

then (7.34) is true for all integers $i \in [T_1 + L, T_2]$.

The next result easily follows from Theorem 7.10.

Theorem 7.11. *Let* $\epsilon \in (0,1)$. *Then there exist a natural number* $L \ge 4$, *a number* $\delta \in (0, \epsilon)$, *and a number* $\lambda \in (0,1)$ *such that for every sequence of functions* $\{g_i\}_{i=0}^{\infty} \subset C(X \times X \times Y \times Y)$ *which satisfies*

$$\|g_i - f_i\| \le \delta, \; i = 0, 1, \dots,$$

every sequence $\{\alpha_i\}_{i=0}^{\infty} \subset (0,1]$ *such that*

$$\alpha_i \alpha_j^{-1} \ge \lambda$$

for every pair of nonnegative integers i, j *which satisfies* $|i - j| \le 2L$ *and every* $(\{\alpha_i g_i\}_{i=0}^{\infty})$*-overtaking optimal pair of sequences*

$$\{x_i\}_{i=0}^{\infty} \subset X, \; \{y_i\}_{i=0}^{\infty} \subset Y$$

the inequalities

$$|x_i - \bar{x}_i|, \; |y_i - \bar{y}_i| \le \epsilon \tag{7.35}$$

are valid for all integers $i \ge L$. *Moreover, if*

$$|x_0 - \bar{x}_0|, \; |y_0 - \bar{y}_0| \le \delta,$$

then (7.35) is true for all nonnegative integers i

7.5 Auxiliary Results for Theorem 7.8

Lemma 7.12. *Let* $\epsilon \in (0,1)$. *Then there exists a number* $\delta \in (0, \epsilon)$ *such that for every pair of nonnegative integers* T_1, $T_2 \ge T_1 + 2$ *and every* $(\{f_i\}_{i=0}^{\infty}, \delta)$*-good pair of finite sequences*

$$\{x_i\}_{i=T_1}^{T_2} \subset X, \; \{y_i\}_{i=T_1}^{T_2} \subset Y$$

which satisfies

$$x_{T_1} = \bar{x}_{T_1},\ x_{T_2} = \bar{x}_{T_2},\quad y_{T_1} = \bar{y}_{T_1},\ y_{T_2} = \bar{y}_{T_2} \tag{7.36}$$

the following relations are valid:

$$|x_i - \bar{x}_i|,\ |y_i - \bar{y}_i| \le \epsilon,\quad i = T_1, \dots T_2. \tag{7.37}$$

Proof. Assumptions (A1) and (A2) imply there exists $\gamma > 0$ such that the following properties hold:

for every nonnegative integer i, every point $x \in X$ which satisfies $|x - \bar{x}_i| \ge \epsilon$ and every point $x' \in X$, we have

$$f_i(2^{-1}(\bar{x}_i + x), 2^{-1}(\bar{x}_{i+1} + x'), \bar{y}_i, \bar{y}_{i+1}) + \gamma$$

$$\le 2^{-1} f_i(\bar{x}_i, \bar{x}_{i+1}, \bar{y}_i, \bar{y}_{i+1}) + 2^{-1} f_i(x, x', \bar{y}_i, \bar{y}_{i+1}); \tag{7.38}$$

for every nonnegative integer i, every point $y \in Y$ which satisfies $|y - \bar{y}_i| \ge \epsilon$ and every point $y' \in Y$, we have

$$f_i(\bar{x}_i, \bar{x}_{i+1}, 2^{-1}(\bar{y}_i + y), 2^{-1}(y' + \bar{y}_{i+1})) - \gamma$$

$$\ge 2^{-1} f_i(\bar{x}_i, \bar{x}_{i+1}, \bar{y}_i, \bar{y}_{i+1}) + 2^{-1} f_i(\bar{x}_i, \bar{x}_{i+1}, y, y'). \tag{7.39}$$

Fix $\delta > 0$ such that

$$\delta < \gamma/4,\ \delta < 8^{-1}\epsilon. \tag{7.40}$$

Assume that integers $T_1 \ge 0$, $T_2 \ge T_1 + 2$, $\{x_i\}_{i=T_1}^{T_2} \subset X$, $\{y_i\}_{i=T_1}^{T_2} \subset Y$ is an $(\{f_i\}_{i=0}^{\infty}, \delta)$-good pair of sequences and that relation (7.36) holds. Since

$$\{\bar{x}_i\}_{i=0}^{\infty}, \{\bar{y}_i\}_{i=0}^{\infty}$$

is an $(\{f_i\}_{i=0}^{\infty})$-locally optimal pair of sequences and

$$\{x_i\}_{i=T_1}^{T_2}, \{y_i\}_{i=T_1}^{T_2}$$

is an $(\{f_i\}_{i=0}^{\infty}, \delta)$-good pair of sequences relation (7.36) implies that

$$\sum_{i=T_1}^{T_2-1} f_i(\bar{x}_i, \bar{x}_{i+1}, \bar{y}_i, \bar{y}_{i+1}) \le \sum_{i=T_1}^{T_2-1} f_i(x_i, x_{i+1}, \bar{y}_i, \bar{y}_{i+1})$$

$$\le \sum_{i=T_1}^{T_2-1} f_i(x_i, x_{i+1}, y_i, y_{i+1}) + \delta$$

$$\le \sum_{i=T_1}^{T_2-1} f_i(\bar{x}_i, \bar{x}_{i+1}, y_i, y_{i+1}) + 2\delta \le \sum_{i=T_1}^{T_2-1} f_i(\bar{x}_i, \bar{x}_{i+1}, \bar{y}_i, \bar{y}_{i+1}) + 2\delta. \tag{7.41}$$

In view of (7.41), we have

$$|\sum_{i=T_1}^{T_2-1} f_i(x_i, x_{i+1}, y_i, y_{i+1}) - \sum_{i=T_1}^{T_2-1} f_i(\bar{x}_i, \bar{x}_{i+1}, \bar{y}_i, \bar{y}_{i+1})| \le \delta, \quad (7.42)$$

$$\sum_{i=T_1}^{T_2-1} f_i(x_i, x_{i+1}, \bar{y}_i, \bar{y}_{i+1}) - \sum_{i=T_1}^{T_2-1} f_i(\bar{x}_i, \bar{x}_{i+1}, \bar{y}_i, \bar{y}_{i+1}) \in [0, 2\delta], \quad (7.43)$$

$$\sum_{i=T_1}^{T_2-1} f_i(\bar{x}_i, \bar{x}_{i+1}, y_i, y_{i+1}) - \sum_{i=T_1}^{T_2-1} f_i(\bar{x}_i, \bar{x}_{i+1}, \bar{y}_i, \bar{y}_{i+1}) \in [-2\delta, 0]. \quad (7.44)$$

Define

$$\widehat{x}_i = 2^{-1}(x_i + \bar{x}_i),\ \widehat{y}_i = 2^{-1}(y_i + \bar{y}_i),\ i = T_1, \dots, T_2. \quad (7.45)$$

Since $\{x_i\}_{i=0}^{\infty}, \{\bar{y}_i\}_{i=0}^{\infty}$ is an $(\{f_i\}_{i=0}^{\infty})$-locally optimal pair of sequences relations (7.36) and (7.45) imply that

$$\sum_{i=T_1}^{T_2-1} f_i(\widehat{x}_i, \widehat{x}_{i+1}, \bar{y}_i, \bar{y}_{i+1}) \ge \sum_{i=T_1}^{T_2-1} f_i(\bar{x}_i, \bar{x}_{i+1}, \bar{y}_i, \bar{y}_{i+1})$$

$$\ge \sum_{i=T_1}^{T_2-1} f_i(\bar{x}_i, \bar{x}_{i+1}, \widehat{y}_i, \widehat{y}_{i+1}). \quad (7.46)$$

In order to complete the proof of the lemma it is sufficient to show that (7.37) is valid.

Assume the contrary. Then it follows from (7.36) that there exists an integer $k \in [T_1 + 1, T_2 - 1]$ for which

$$\max\{|x_k - \bar{x}_k|,\ |y_k - \bar{y}_k|\} > \epsilon. \quad (7.47)$$

In view of (7.45) for all integers $i = T_1, \dots, T_2 - 1$, we have

$$f_i(\widehat{x}_i, \widehat{x}_{i+1}, \bar{y}_i, \bar{y}_{i+1}) \le 2^{-1} f_i(x_i, x_{i+1}, \bar{y}_i, \bar{y}_{i+1}) + 2^{-1} f_i(\bar{x}_i, \bar{x}_{i+1}, \bar{y}_i, \bar{y}_{i+1}), \quad (7.48)$$

$$f_i(\bar{x}_i, \bar{x}_{i+1}, \widehat{y}_i, \widehat{y}_{i+1}) \ge 2^{-1} f_i(\bar{x}_i, \bar{x}_{i+1}, y_i, y_{i+1}) + 2^{-1} f_i(\bar{x}_i, \bar{x}_{i+1}, \bar{y}_i, \bar{y}_{i+1}). \quad (7.49)$$

It follows from (7.47), (7.45), (7.38), (7.39), (7.48), and (7.49) that

$$2^{-1} f_k(x_k, x_{k+1}, \bar{y}_k, \bar{y}_{k+1}) + 2^{-1} f_k(\bar{x}_k, \bar{x}_{k+1}, \bar{y}_k, \bar{y}_{k+1})$$

$$- f_k(\widehat{x}_k, \widehat{x}_{k+1}, \bar{y}_k, \bar{y}_{k+1}) + f_k(\bar{x}_k, \bar{x}_{k+1}, \widehat{y}_k, \widehat{y}_{k+1})$$

$$- 2^{-1} f_k(\bar{x}_k, \bar{x}_{k+1}, y_k, y_{k+1}) - 2^{-1} f_k(\bar{x}_k, \bar{x}_{k+1}, \bar{y}_k, \bar{y}_{k+1}) \ge \gamma. \quad (7.50)$$

In view of (7.43), (7.46), and (7.48)–(7.50), we have

$$\gamma \le \sum_{i=T_1}^{T_2-1} [2^{-1} f_i(x_i, x_{i+1}, \bar{y}_i, \bar{y}_{i+1})$$

$$+2^{-1} f_i(\bar{x}_i, \bar{x}_{i+1}, \bar{y}_i, \bar{y}_{i+1}) - f_i(\widehat{x}_i, \widehat{x}_{i+1}, \bar{y}_i, \bar{y}_{i+1})]$$

$$+ \sum_{i=T_1}^{T_2-1} [f_i(\bar{x}_i, \bar{x}_{i+1}, \widehat{y}_i, \widehat{y}_{i+1}) - 2^{-1} f_i(\bar{x}_i, \bar{x}_{i+1}, y_i, y_{i+1}) - 2^{-1} f_i(\bar{x}_i, \bar{x}_i, \bar{y}_i, \bar{y}_{i+1})]$$

$$= \sum_{i=T_1}^{T_2-1} [f_i(\bar{x}_i, \bar{x}_{i+1}, \widehat{y}_i, \widehat{y}_{i+1}) - f_i(\widehat{x}_i, \widehat{x}_{i+1}, \bar{y}_i, \bar{y}_{i+1})$$

$$+2^{-1} f_i(x_i, x_{i+1}, \bar{y}_i, \bar{y}_{i+1}) - 2^{-1} f_i(\bar{x}_i, \bar{x}_{i+1}, y_i, y_{i+1})]$$

$$\le 2^{-1} \sum_{i=T_1}^{T_2-1} [f_i(x_i, x_{i+1}, \bar{y}_i, \bar{y}_{i+1}) - f_i(\bar{x}_i, \bar{x}_{i+1}, y_i, y_{i+1})] \le \delta.$$

This contradicts (7.40). The contradiction we have reached proves (7.37). This completes the proof of Lemma 7.12. □

Define

$$D_1 = \sup\{|f_i(z_1, z_2, \xi_1, \xi_2)| : i \in \{0, 1, \dots\},\ z_1, z_2 \in X,\ \xi_1, \xi_2 \in Y\}. \quad (7.51)$$

It is not difficult to see that the following result is true.

Lemma 7.13. *Let integers $T_1 \ge 0$, $T_2 \ge T_1 + 2$, $M > 0$ and let*

$$\{x_i\}_{i=T_1}^{T_2} \subset X,\ \{y_i\}_{i=T_1}^{T_2} \subset Y$$

be an $(\{f_i\}_{i=0}^{\infty}, M)$-good pair of sequences. Then the pair of sequences

$$\{x'_i\}_{i=T_1}^{T_2} \subset X,\ \{y'_i\}_{i=T_1}^{T_2} \subset Y$$

such that

$$x'_i = x_i,\ y'_i = y_i,\ i = T_1 + 1, \dots, T_2 - 1, \quad x'_{T_1} = \bar{x}_{T_1},\ x'_{T_2} = \bar{x}_{T_2},$$

$$y'_{T_1} = \bar{y}_{T_1}, y'_{T_2} = \bar{y}_{T_2}$$

is $(\{f_i\}_{i=0}^{\infty}, M + 8D_1)$-good.

Clearly, the following auxiliary result is easily deduced from property (P1).

Lemma 7.14. *Let $\epsilon > 0$. Then there exists a positive number δ such that for every pair of nonnegative integers T_1, $T_2 \geq T_1 + 2$ and for pairs of sequences*

$$\{x_i\}_{i=T_1}^{T_2} \subset X,\ \{y_i\}_{i=T_1}^{T_2} \subset Y$$

and

$$\{x'_i\}_{i=T_1}^{T_2} \subset X,\ \{y'_i\}_{i=T_1}^{T_2} \subset Y$$

satisfying

$$|x_j - x'_j|,\ |y_j - y'_j| \leq \delta,\ j = T_1, T_2, \quad x'_j = x_j,\ y'_j = y_j,\ j = T_1 + 1, \dots T_2 - 1 \tag{7.52}$$

the relation

$$|\sum_{i=T_1}^{T_2-1} [f_i(x_i, x_{i+1}, y_i, y_{i+1}) - f_i(x'_i, x'_{i+1}, y'_i, y'_{i+1})]| \leq \epsilon$$

is valid.

The next auxiliary result easily follows from Lemma 7.14.

Lemma 7.15. *Let ϵ be a positive number. Then there exists a positive number δ such that for every pair of nonnegative integers T_1, $T_2 \geq T_1 + 2$, every $(\{f_i\}_{i=0}^{\infty}, \epsilon)$-good pair of finite sequences*

$$\{x_i\}_{i=T_1}^{T_2} \subset X,\ \{y_i\}_{i=T_1}^{T_2} \subset Y$$

and every pair of finite sequences

$$\{x'_i\}_{i=T_1}^{T_2} \subset X,\ \{y'_i\}_{i=T_1}^{T_2} \subset Y$$

the following assertion is true:

if (7.52) holds, then the pair of finite sequences $\{x'_i\}_{i=T_1}^{T_2}$, $\{y'_i\}_{i=T_1}^{T_2}$ is $(\{f_i\}_{i=0}^{\infty}, 2\epsilon)$-good.

The next lemma easily follows from Lemmas 7.12 and 7.15.

Lemma 7.16. *Assume that a number $\epsilon \in (0, 1)$. Then there exists $\delta \in (0, \epsilon)$ such that for every pair of nonnegative integers T_1, $T_2 \geq T_1 + 2$ and every $(\{f_i\}_{i=0}^{\infty}, \delta)$-good pair of finite sequences*

$$\{x_i\}_{i=T_1}^{T_2} \subset X,\ \{y_i\}_{i=T_1}^{T_2} \subset Y$$

which satisfies

$$|x_i - \bar{x}_i|,\ |y_i - \bar{y}_i| \leq \delta,\ t = T_1,\ T_2$$

the following inequalities are true:

$$|x_i - \bar{x}_i|,\ |y_i - \bar{y}_i| \leq \epsilon, \quad i = T_1, \dots T_2.$$

Lemma 7.17. *Assume that $M > 0$, $\epsilon \in (0,1)$. Then there exists a natural number $n_0 \geq 4$ such that for every nonnegative integer T and every $(\{f_i\}_{i=0}^{\infty}, M)$-good pair of finite sequences $\{x_i\}_{i=T}^{T+n_0} \subset X$, $\{y_i\}_{i=T}^{T+n_0} \subset Y$ satisfying*

$$x_T = \bar{x}_T,\ x_{T+n_0} = \bar{x}_{T+n_0},\quad y_T = \bar{y}_T,\ y_{T+n_0} = \bar{y}_{T+n_0} \tag{7.53}$$

there exists a integer $j \in \{T+1, \dots T+n_0-1\}$ such that

$$|x_j - \bar{x}_j|,\ |y_j - \bar{y}_j| \leq \epsilon. \tag{7.54}$$

Proof. Assumptions (A1) and (A2) imply that there exists $\gamma > 0$ such that:

for every nonnegative integer i, every point $x \in X$ which satisfies $|x - \bar{x}_i| \geq \epsilon$, and every point $x' \in X$, we have

$$f_i(2^{-1}(x + \bar{x}_i), 2^{-1}(x' + \bar{x}_{i+1}), \bar{y}_i, \bar{y}_{i+1}) + \gamma$$

$$\leq 2^{-1} f_i(\bar{x}_i, \bar{x}_{i+1}, \bar{y}_i, \bar{y}_{i+1}) + 2^{-1} f_i(x, x', \bar{y}_i, \bar{y}_{i+1}); \tag{7.55}$$

for every nonnegative integer i, every point $y \in Y$ which satisfies $|y - \bar{y}_i| \geq \epsilon$, and every point $y' \in Y$, we have

$$f_i(\bar{x}_i, \bar{x}_{i+1}, 2^{-1}(y + \bar{y}_i), 2^{-1}(y' + \bar{y}_{i+1})) - \gamma$$

$$\geq 2^{-1} f_i(\bar{x}_i, \bar{x}_{i+1}, \bar{y}_i, \bar{y}_{i+1}) + 2^{-1} f_i(\bar{x}_i, \bar{x}_{i+1}, y, y'). \tag{7.56}$$

Fix a natural number

$$n_0 > 8 + 2(\gamma)^{-1} M. \tag{7.57}$$

Assume that an integer $T \geq 0$,

$$\{x_i\}_{i=T}^{T+n_0} \subset X,\ \{y_i\}_{i=T}^{T+n_0} \subset Y$$

is an $(\{f_i\}_{i=0}^{\infty}, M)$-good pair of finite sequences satisfying relation (7.53). We claim that there exists an integer $j \in \{T+1, \dots T+n_0-1\}$ such that (7.54) is true.

Assume the contrary. Then

$$\max\{|x_j - \bar{x}_j|,\ |y_j - \bar{y}_j|\} > \epsilon,\ j = T+1, \dots, T+n_0-1. \tag{7.58}$$

Since $\{\bar{x}_i\}_{i=0}^{\infty}, \{\bar{y}_i\}_{i=0}^{\infty}$ is an $(\{f_i\}_{i=0}^{\infty})$-locally optimal pair of sequences and

$$\{x_i\}_{i=T}^{T+n_0}, \{y_i\}_{i=T}^{T+n_0}$$

is an $(\{f_i\}_{i=0}^{\infty}, M)$-good pair of sequences relation (7.53) implies that

$$\sum_{i=T}^{T+n_0-1} f_i(\bar{x}_i, \bar{x}_{i+1}, \bar{y}_i, \bar{y}_{i+1}) \leq \sum_{i=T}^{T+n_0-1} f_i(x_i, x_{i+1}, \bar{y}_i, y_{i+1})$$

$$\leq \sum_{i=T}^{T+n_0-1} f_i(x_i, x_{i+1}, y_i, y_{i+1}) + M$$

$$\leq \sum_{i=T}^{T+n_0-1} f_i(\bar{x}_i, \bar{x}_{i+1}, y_i, y_{i+1}) + 2M \leq \sum_{i=T}^{T+n_0-1} f_i(\bar{x}_i, \bar{x}_{i+1}, \bar{y}_i, \bar{y}_{i+1}) + 2M. \tag{7.59}$$

In view of (7.59), we have

$$|\sum_{i=T}^{T+n_0-1} f_i(\bar{x}_i, \bar{x}_{i+1}, \bar{y}_i, \bar{y}_{i+1}) - \sum_{i=T}^{T+n_0-1} f_i(x_i, x_{i+1}, y_i, y_{i+1})| \leq M, \tag{7.60}$$

$$\sum_{i=T}^{T+n_0-1} f_i(x_i, x_{i+1}, \bar{y}_i, \bar{y}_{i+1}) - \sum_{i=T}^{T+n_0-1} f_i(\bar{x}_i, \bar{x}_{i+1}, \bar{y}_i, \bar{y}_{i+1}) \in [0, 2M], \tag{7.61}$$

$$\sum_{i=T}^{T+n_0-1} f_i(\bar{x}_i, \bar{x}_{i+1}, y_i, y_{i+1}) - \sum_{i=T}^{T+n_0-1} f_i(\bar{x}_i, \bar{x}_{i+1}, \bar{y}_i, \bar{y}_{i+1}) \in [-2M, 0]. \tag{7.62}$$

Define

$$x'_i = 2^{-1}(x_i + \bar{x}_i),\ y'_i = 2^{-1}(y_i + \bar{y}_i),\ i = T, \dots, T + n_0. \tag{7.63}$$

In view of (7.63), (7.53) and since $\{\bar{x}_i\}_{i=0}^{\infty}, \{\bar{y}_i\}_{i=0}^{\infty}$ is an $(\{f_i\}_{i=0}^{\infty})$-locally optimal pair of sequences we have

$$\sum_{i=T}^{T+n_0-1} f_i(x'_i, x'_{i+1}, \bar{y}_i, \bar{y}_{i+1}) \geq \sum_{i=T}^{T+n_0-1} f_i(\bar{x}_i, \bar{x}_{i+1}, \bar{y}_i, \bar{y}_{i+1})$$

$$\geq \sum_{i=T}^{T+n_0-1} f_i(\bar{x}_i, \bar{x}_{i+1}, y'_i, y'_{i+1}). \tag{7.64}$$

By (7.63), for all integers $i = T, \dots, T + n_0 - 1$,

$$f_i(x'_i, x'_{i+1}, \bar{y}_i, \bar{y}_{i+1}) \leq 2^{-1} f_i(x_i, x_{i+1}, \bar{y}_i, \bar{y}_{i+1}) + 2^{-1} f_i(\bar{x}_i, \bar{x}_{i+1}, \bar{y}_i, \bar{y}_{i+1}), \tag{7.65}$$

$$f_i(\bar{x}_i, \bar{x}_{i+1}, y'_i, y'_{i+1}) \geq 2^{-1} f_i(\bar{x}_i, \bar{x}_{i+1}, y_i, y_{i+1}) + 2^{-1} f_i(\bar{x}_i, \bar{x}_{i+1}, \bar{y}_i, \bar{y}_{i+1}). \tag{7.66}$$

It follows from (7.63), (7.58), the choice of γ (see (7.55), (7.56)) that for all integers $i = T + 1, \dots, T + n_0 - 1$, at least one of the following inequalities is valid:

$$2^{-1} f_i(x_i, x_{i+1}, \bar{y}_i, \bar{y}_{i+1}) + 2^{-1} f_i(\bar{x}_i, \bar{x}_{i+1}, \bar{y}_i, \bar{y}_{i+1}) - f_i(x'_i, x'_{i+1}, \bar{y}_i, \bar{y}_{i+1}) \geq \gamma,$$

$$f_i(\bar{x}_i, \bar{x}_{i+1}, y'_i, y'_{i+1}) - 2^{-1} f_i(\bar{x}_i, \bar{x}_{i+1}, y_i, y_{i+1}) - 2^{-1} f_i(\bar{x}_i, \bar{x}_{i+1}, \bar{y}_i, \bar{y}_{i+1}) \geq \gamma.$$

Together with (7.65) and (7.66) this implies that for all integers $i = T, \dots, T + n_0 - 1$, we have

$$2^{-1} f_i(x_i, x_{i+1}, \bar{y}_i, \bar{y}_{i+1}) + 2^{-1} f_i(\bar{x}_i, \bar{x}_{i+1}, \bar{y}_i, \bar{y}_{i+1}) - f_i(x'_i, x'_{i+1}, \bar{y}_i, \bar{y}_{i+1})$$

$$+ f_i(\bar{x}_i, \bar{x}_{i+1}, y'_i, y'_{i+1}) - 2^{-1} f_i(\bar{x}_i, \bar{x}_{i+1}, y_i, y_{i+1}) - 2^{-1} f_i(\bar{x}_i, \bar{x}_{i+1}, \bar{y}_i, \bar{y}_{i+1}) \ge \gamma.$$

Together with (7.61), (7.62), and (7.64) this implies that

$$\gamma(n_0 - 1) \le \sum_{i=T}^{T+n_0-1} [2^{-1} f_i(x_i, x_{i+1}, \bar{y}_i, \bar{y}_{i+1})$$

$$+ 2^{-1} f_i(\bar{x}_i, \bar{x}_{i+1}, \bar{y}_i, \bar{y}_{i+1}) - f_i(x'_i, x'_{i+1}, \bar{y}_i, \bar{y}_{i+1})$$

$$+ f_i(\bar{x}_i, \bar{x}_{i+1}, y'_i, y'_{i+1}) - 2^{-1} f_i(\bar{x}_i, \bar{x}_{i+1}, y_i, y_{i+1}) - 2^{-1} f_i(\bar{x}_i, \bar{x}_{i+1}, \bar{y}_i, \bar{y}_{i+1})]$$

$$= \sum_{i=T}^{T+n_0-1} [f_i(\bar{x}_i, \bar{x}_{i+1}, y'_i, y'_{i+1}) - f_i(x'_i, x'_{i+1}, \bar{y}_i, \bar{y}_{i+1})$$

$$+ 2^{-1} f_i(x_i, x_{i+1}, \bar{y}_i, \bar{y}_{i+1}) - 2^{-1} f_i(\bar{x}_i, \bar{x}_{i+1}, y_i, y_{i+1})]$$

$$\le 2^{-1} \sum_{i=T}^{T+n_0-1} [f_i(x_i, x_{i+1}, \bar{y}_i, \bar{y}_{i+1}) - f_i(\bar{x}_i, \bar{x}_{i+1}, y_i, y_{i+1})]$$

$$\le 2^{-1} \sum_{i=T}^{T+n_0-1} f_i(\bar{x}_i, \bar{x}_{i+1}, \bar{y}_i, \bar{y}_{i+1}) + M - 2^{-1} \sum_{i=T}^{T+n_0-1} f_i(\bar{x}_i, \bar{x}_{i+1}, \bar{y}_i, \bar{y}_{i+1}) + M$$

and

$$\gamma(n_0 - 1) \le 2M.$$

This contradicts (7.57). The contradiction we have reached proves that there exists an integer $j \in \{T + 1, \dots, T + n_0 - 1\}$ for which relation (7.54) is valid. This completes the proof of Lemma 7.17. □

The next lemma easily follows from Lemmas 7.13 and 7.17.

Lemma 7.18. *Assume that $\epsilon \in (0, 1)$ and M is a positive number. Then there exists a natural number $n_0 \ge 4$ such that for every nonnegative integer T and every $(\{f_i\}_{i=0}^{\infty}, M)$-good pair of finite sequences*

$$\{x_i\}_{i=T}^{T+n_0} \subset X,\ \{y_i\}_{i=T}^{T+n_0} \subset Y$$

there exists an integer

$$j \in \{T + 1, \dots T + n_0 - 1\}$$

such that

$$|x_j - \bar{x}_j|,\ |y_j - \bar{y}_j| \le \epsilon.$$

The next auxiliary result easily follows from Lemma 7.18.

Lemma 7.19. *Assume that $\epsilon \in (0,1)$ and M is a positive number. Then there exist a natural number $n_0 \geq 4$ and a positive number δ such that for every sequence of functions $\{g_i\}_{i=0}^{\infty} \subset C(X \times X \times Y \times Y)$ which satisfies*

$$\|g_i - f_i\| \leq \delta,\ i = 0, 1, \dots,$$

every nonnegative integer T and every $(\{g_i\}_{i=0}^{\infty}, M)$-good pair of finite sequences

$$\{x_i\}_{i=T}^{T+n_0} \subset X,\ \{y_i\}_{i=T}^{T+n_0} \subset Y$$

there exists an integer

$$j \in \{T+1, \dots T+n_0-1\}$$

such that

$$|x_j - \bar{x}_j|,\ |y_j - \bar{y}_j| \leq \epsilon.$$

7.6 Proof of Theorem 7.8

Lemma 7.16 implies that there exists a number $\delta_0 \in (0, \epsilon)$ such that the following property holds:

(P4) for every pair of nonnegative integers T_1, $T_2 \geq T_1 + 2$ and every $(\{f_i\}_{i=0}^{\infty}, \delta_0)$-good pair of finite sequences

$$\{x_i\}_{i=T_1}^{T_2} \subset X,\ \{y_i\}_{i=T_1}^{T_2} \subset Y$$

which satisfies

$$|x_j - \bar{x}_j|,\ |y_j - \bar{y}_j| \leq \delta_0,\ j = T_1,\ T_2$$

the inequalities

$$|x_i - \bar{x}_i|,\ |y_i - \bar{y}_i| \leq \epsilon \tag{7.67}$$

are valid for all integers $i = T_1, \dots, T_2$.

Lemma 7.19 implies that there exists a natural number $n_0 \geq 4$ and a number $\delta_1 \in (0, \delta_0)$ such that the following property holds:

(P5) for every sequence of functions $\{g_i\}_{i=0}^{\infty} \subset C(X \times X \times Y \times Y)$ which satisfies

$$\|g_i - f_i\| \leq \delta_1,\ i = 0, 1, \dots,$$

every nonnegative integer T and every $(\{g_i\}_{i=0}^{\infty}, 8)$-good pair of finite sequences

$$\{x_i\}_{i=T}^{T+n_0} \subset X,\ \{y_i\}_{i=T}^{T+n_0} \subset Y$$

there exists an integer

$$j \in \{T+1, \dots T+n_0-1\}$$

such that

$$|x_j - \bar{x}_j|,\ |y_j - \bar{y}_j| \leq \delta_0.$$

Choose a natural number

$$L \geq 4n_0 \tag{7.68}$$

and a number

$$\delta \in (0, 8^{-1}L^{-1}\delta_1). \tag{7.69}$$

Assume that a sequence of functions $\{g_i\}_{i=0}^{\infty} \subset C(X \times X \times Y \times Y)$ satisfies

$$\|g_i - f_i\| \leq \delta,\ i = 0, 1, \ldots, \tag{7.70}$$

a pair of nonnegative integers T_1, $T_2 \geq T_1 + 2L$ and that a pair of finite sequences

$$\{x_i\}_{i=T_1}^{T_2} \subset X,\ \{y_i\}_{i=T_1}^{T_2} \subset Y$$

has the following property:

(P6) the pair $\{x_i\}_{i=j}^{j+2L}$, $\{y_i\}_{i=j}^{j+2L}$ is $(\{g_i\}_{i=0}^{\infty}, \delta)$-good pair for any integer $j \in [T_1, T_2 - 2L]$.

Let

$$s \in \{T_1 + L, \ldots, T_2 - L\} \tag{7.71}$$

be given. In view of property (P6), $\{x_i\}_{i=s-L}^{s+L}$, $\{y_i\}_{i=s-L}^{s+L}$ is a $(\{g_i\}_{i=0}^{\infty}, \delta)$-good pair. Together with (7.68), (7.70), and property (P5) applied to the pairs of sequences

$$\{x_i\}_{i=s-L}^{s-L+n_0},\ \{y_i\}_{i=s-L}^{s-L+n_0},;\{x_i\}_{i=s+L-n_0}^{s+L},\ \{y_i\}_{i=s+L-n_0}^{s+L}$$

this implies that there exist integers

$$j_1 \in \{s-L+1, \ldots, s-L+n_0-1\},\ j_2 \in \{s+L-n_0+1, \ldots, s+L-1\} \tag{7.72}$$

such that

$$|x_{j_1} - \bar{x}_{j_1}|,\ |y_{j_1} - \bar{y}_{j_1}|,\ |x_{j_2} - \bar{x}_{j_2}|,\ |y_{j_2} - \bar{y}_{j_2}| \leq \delta_0. \tag{7.73}$$

It is clear that $\{x_i\}_{i=j_1}^{j_2}$, $\{y_i\}_{i=j_1}^{j_2}$ is a $(\{g_i\}_{i=0}^{\infty}, \delta)$-good pair of sequences. Combined with (7.69), (7.70), and (7.72) this implies that $\{x_i\}_{i=j_1}^{j_2}$, $\{y_i\}_{i=j_1}^{j_2}$ is an $(\{f_i\}_{i=0}^{\infty}, \delta_0)$-good pair of sequences. Together with (7.73) and property (P4) this implies that

$$|x_i - \bar{x}_i|,\ |y_i - \bar{y}_i| \leq \epsilon \text{ for all integers } i \in \{j_1, \ldots, j_2\}$$

and in particular,

$$|x_s - \bar{x}_s|,\ |y_s - \bar{y}_s| \leq \epsilon \text{for all } s \in \{T_1 + L, \ldots, T_2 - L\}. \tag{7.74}$$

Moreover we have shown that the following property holds:

(P7) if $s \in \{T_1 + L, \ldots, T_2 - L\}$, then there exist integers j_1, j_2 for which (7.72) and (7.73) are true.

Assume that

$$|x_{T_1} - \bar{x}_{T_1}|,\ |y_{T_1} - \bar{y}_{T_1}| \le \delta. \tag{7.75}$$

In view of property (P7) applied with $S = T_1 + L$ there exists an integer

$$j_2 \in \{T_1 + 2L - n_0 + 1, \dots, , T_1 + 2L - 1\} \tag{7.76}$$

such that

$$|x_{j_2} - \bar{x}_{j_2}|,\ |y_{j_2} - \bar{y}_{j_2}| \le \delta_0. \tag{7.77}$$

Property (P6) and (7.76) imply that $\{x_i\}_{i=T_1}^{j_2}$, $\{y_i\}_{i=T_1}^{j_2}$ is a $(\{g_i\}_{i=0}^{\infty}, \delta)$-good pair of sequences. Combined with relations (7.76), (7.70), and (7.69) this implies that the pair of sequences $\{x_i\}_{i=T_1}^{j_2}$ and $\{y_i\}_{i=T_1}^{j_2}$ is $(\{f_i\}_{i=0}^{\infty}, \delta_0)$-good. Combined with property (P4) and relations (7.75)–(7.77) this implies that

$$|x_i - \bar{x}_i|,\ |y_i - \bar{y}_i| \le \epsilon \text{ for all integers } i = T_1, \dots, T_1 + 2L - n_0 + 1.$$

Together with (7.68) and (7.74) this implies that

$$|x_i - \bar{x}_i|,\ |y_i - \bar{y}_i| \le \epsilon \text{ for all integers } i = T_1, \dots, T_2 - L.$$

Assume that

$$|x_{T_2} - \bar{x}_{T_2}|,\ |y_{T_2} - \bar{y}_{T_2}| \le \delta. \tag{7.78}$$

Property (P7), applied with $S = T_2 - L$, implies that there exists an integer

$$j_1 \in \{T_2 - 2L + 1, \dots, , T_2 - 2L + n_0 - 1\} \tag{7.79}$$

such that

$$|x_{j_1} - \bar{x}_{j_1}|,\ |y_{j_1} - \bar{y}_{j_1}| \le \delta_0. \tag{7.80}$$

In view of (7.79) and property (P6), $\{x_i\}_{i=j_1}^{T_2}$ and $\{y_i\}_{i=j_1}^{T_2}$ is a $(\{g_i\}_{i=0}^{\infty}, \delta)$-good pair of sequences. Combined with relations (7.79), (7.70), and (7.69) this implies that the pair of sequences $\{x_i\}_{i=j_1}^{T_2}$ and $\{y_i\}_{i=j_1}^{T_2}$ is $(\{f_i\}_{i=0}^{\infty}, \delta_0)$-good. Together with property (P4) and relations (7.78)–(7.80) this implies that

$$|x_i - \bar{x}_i|,\ |y_i - \bar{y}_i| \le \epsilon \text{ for all integers } i = T_2 - 2L + n_0 - 1, \dots, T_2.$$

Combined with (7.68) and (7.74) this implies that

$$|x_i - \bar{x}_i|,\ |y_i - \bar{y}_i| \le \epsilon \text{ for all integers } i = T_1 + L, \dots, T_2.$$

This completes the proof of Theorem 7.8. □

7.7 Proof of Theorem 7.9

The following auxiliary result easily follows from Theorem 7.8.

Proposition 7.20. *Let $\epsilon \in (0,1)$. Then there exists an integer $n_1 \geq 1$ such that for every $(\{f_i\}_{i=0}^{\infty})$-locally optimal pair of sequences $\{x_i\}_{i=0}^{\infty} \subset X$, $\{y_i\}_{i=0}^{\infty} \subset Y$ and every natural number $i \geq n_1$ the inequalities*

$$|x_i - \bar{x}_i|, \ |y_i - \bar{y}_i| \leq \epsilon$$

hold.

It follows from Assumptions (A1) and (A2) and property (P1) that:

(P8) for every positive number ϵ there exists a number $\delta \in (0, \epsilon)$ such that for every nonnegative integer i, every pair of points $z_1, z_2 \in X$, every pair points $\xi_1, \xi_2 \in Y$ which satisfy

$$|z_1 - \bar{x}_i|, |z_2 - \bar{x}_{i+1}|, \ |\xi_1 - \bar{y}_i|, \ |\xi_2 - \bar{y}_{i+1}| \leq \delta$$

and each $z_1', z_2' \in X$, $\xi_1', \xi_2' \in Y$ if $\|z_1' - \bar{x}_i\| \geq \epsilon$, then

$$f_i(2^{-1}(z_1' + z_1), 2^{-1}(z_2' + z_2), \xi_1, \xi_2) + \delta$$
$$\leq 2^{-1} f_i(z_1, z_2, \xi_1, \xi_2) + 2^{-1} f_i(z_1', z_2', \xi_1, \xi_2) \tag{7.81}$$

and if $\|\xi_1' - \bar{y}_i\| \geq \epsilon$, then

$$f_i(z_1, z_2, 2^{-1}(\xi_1 + \xi_1'), 2^{-1}(\xi_2 + \xi_2')) - \delta$$
$$\geq 2^{-1} f_i(z_1, z_2, \xi_1, \xi_2) + 2^{-1} f_i(z_1, z_2, \xi_1', \xi_2'). \tag{7.82}$$

Let $x \in X$ and $y \in Y$ be given. In view of Theorem 7.4, there exists an $(\{f_i\}_{i=0}^{\infty})$-locally optimal pair of sequences

$$\{x_i^*\}_{i=0}^{\infty} \subset X, \ \{y_i^*\}_{i=0}^{\infty} \subset Y$$

which satisfies

$$x_0^* = x, \ y_0^* = y. \tag{7.83}$$

Proposition 7.20 implies that

$$\lim_{i \to \infty} |x_i^* - \bar{x}_i| = 0, \ \lim_{i \to \infty} |y_i^* - \bar{y}_i| = 0. \tag{7.84}$$

We claim that $\{x_i^*\}_{i=0}^{\infty}$, $\{y_i^*\}_{i=0}^{\infty}$ is $(\{f_i\}_{i=0}^{\infty})$-overtaking optimal pair. Recall (see (7.51)) that

$$D_1 = \sup\{|f_i(z_1, z_2, \xi_1, \xi_2)| : \ i \in \{0, 1, \dots\}, \ z_1, z_2 \in X, \ \xi_1, \xi_2 \in Y\}. \tag{7.85}$$

Lemma 7.21. *Let integers $T_1 \geq 0$, $T_2 \geq T_1 + 2$, $\{x_i\}_{i=T_1}^{T_2} \subset X$, $\{y_i\}_{i=T_1}^{T_2} \subset Y$. Then*

$$\sum_{i=T_1}^{T_2-1} f_i(x_i, x_{i+1}, y_i^*, y_{i+1}^*) \geq \sum_{i=T_1}^{T_2-1} f_i(x_i^*, x_{i+1}^*, y_i^*, y_{i+1}^*) - 4D_1,$$

$$\sum_{i=T_1}^{T_2-1} f_i(x_i^*, x_{i+1}^*, y_i, y_{i+1}) \leq \sum_{i=T_1}^{T_2-1} f_i(x_i^*, x_{i+1}^*, y_i^*, y_{i+1}^*) + 4D_1.$$

Proof. Define

$$x'_i = x^*_i,\ i = T_1, T_2,\ x'_i = x_i,\ i = T_1 + 1, \dots, T_2 - 1,$$

$$y'_i = y^*_i,\ i = T_1, T_2,\ y'_i = y_i,\ i = T_1 + 1, \dots, T_2 - 1. \tag{7.86}$$

Since $\{x^*_i\}_{i=0}^{\infty}$, $\{y^*_i\}_{i=0}^{\infty}$ is an $(\{f_i\}_{i=0}^{\infty})$-locally optimal pair of sequences relations (7.85) and (7.86) imply that

$$\sum_{i=T_1}^{T_2-1} f_i(x^*_i, x^*_{i+1}, y^*_i, y^*_{i+1}) \le \sum_{i=T_1}^{T_2-1} f_i(x'_i, x'_{i+1}, y^*_i, y^*_{i+1})$$

$$\le \sum_{i=T_1}^{T_2-1} f_i(x_i, x_{i+1}, y^*_i, y^*_{i+1}) + 4D_1,$$

$$\sum_{i=T_1}^{T_2-1} f_i(x^*_i, x^*_{i+1}, y^*_i, y^*_{i+1})$$

$$\ge \sum_{i=T_1}^{T_2-1} f_i(x^*_i, x^*_{i+1}, y'_i, y'_{i+1})$$

$$\ge \sum_{i=T_1}^{T_2-1} f_i(x^*_i, x^*_{i+1}, y_i, y_{i+1}).$$

This completes the proof of Lemma 7.21. □

Lemma 7.22. *Assume that* $\{x_i\}_{i=0}^{\infty} \subset X$ *satisfies*

$$\liminf_{T\to\infty}\Big[\sum_{i=0}^{T-1} f_i(x_i, x_{i+1}, y^*_i, y^*_{i+1}) - \sum_{i=0}^{T-1} f_i(x^*_i, x^*_{i+1}, y^*_i, y^*_{i+1})\Big] < \infty. \tag{7.87}$$

Then

$$\sup\Big\{\sum_{i=0}^{T-1} f_i(x_i, x_{i+1}, y^*_i, y^*_{i+1}) - \sum_{i=0}^{T-1} f_i(x^*_i, x^*_{i+1}, y^*_i, y^*_{i+1}) :\ T = 1, 2, \dots\Big\} < \infty.$$

Proof. Assume the contrary and let Q be an arbitrary positive number. Then there exists an integer $T_0 \ge 1$ for which

$$\sum_{i=0}^{T_0-1} f_i(x_i, x_{i+1}, y^*_i, y^*_{i+1}) - \sum_{i=0}^{T_0-1} f_i(x^*_i, x^*_{i+1}, y^*_i, y^*_{i+1}) > Q + 4D_1. \tag{7.88}$$

Lemma 7.21 and (7.88) imply that for every natural number $T > T_0 + 2$, we have

$$\sum_{i=0}^{T-1} f_i(x_i, x_{i+1}, y_i^*, y_{i+1}^*) - \sum_{i=0}^{T-1} f_i(x_i^*, x_{i+1}^*, y_i^*, y_{i+1}^*)$$

$$= \sum_{i=0}^{T_0-1} f_i(x_i, x_{i+1}, y_i^*, y_{i+1}^*) - \sum_{i=0}^{T_0-1} f_i(x_i^*, x_{i+1}^*, y_i^*, y_{i+1}^*)$$

$$+ \sum_{i=T_0}^{T-1} f_i(x_i, x_{i+1}, y_i^*, y_{i+1}^*) - \sum_{i=T_0}^{T-1} f_i(x_i^*, x_{i+1}^*, y_i^*, y_{i+1}^*) > Q + 4D_1 - 4D_1 = Q.$$

Hence for every natural number $T > T_0 + 2$, we have

$$\sum_{i=0}^{T-1} f_i(x_i, x_{i+1}, y_i^*, y_{i+1}^*) - \sum_{i=0}^{T-1} f_i(x_i^*, x_{i+1}^*, y_i^*, y_{i+1}^*) > Q$$

and

$$\liminf_{T\to\infty} [\sum_{i=0}^{T-1} f_i(x_i, x_{i+1}, y_i^*, y_{i+1}^*) - \sum_{i=0}^{T-1} f_i(x_i^*, x_{i+1}^*, y_i^*, y_{i+1}^*)] \geq Q.$$

Since Q is an arbitrary positive number this contradicts (7.87). The contradiction we have reached completes the proof of Lemma 7.22. □

Analogously we can prove the next auxiliary result.

Lemma 7.23. *Let* $\{y_i\}_{i=0}^{\infty} \subset Y$ *satisfy*

$$\limsup_{T\to\infty} [\sum_{i=0}^{T-1} f_i(x_i^*, x_{i+1}^*, y_i, y_{i+1}) - \sum_{i=0}^{T-1} f_i(x_i^*, x_{i+1}^*, y_i^*, y_{i+1}^*)] > -\infty. \quad (7.89)$$

Then

$$\inf\{\sum_{i=0}^{T-1} f_i(x_i^*, x_{i+1}^*, y_i, y_{i+1}) - \sum_{i=0}^{T-1} f_i(x_i^*, x_{i+1}^*, y_i^*, y_{i+1}^*) : \ T = 1, 2, \dots\} > -\infty.$$

Lemma 7.24. *Assume that* $\{x_i\}_{i=0}^{\infty} \subset X$ *satisfies*

$$\liminf_{T\to\infty} [\sum_{i=0}^{T-1} f_i(x_i, x_{i+1}, y_i^*, y_{i+1}^*) - \sum_{i=0}^{T-1} f_i(x_i^*, x_{i+1}^*, y_i^*, y_{i+1}^*)] < \infty. \quad (7.90)$$

Then

$$\lim_{i\to\infty} |x_i - \bar{x}_i| = 0. \quad (7.91)$$

Proof. Lemma 7.22 and (7.90) imply that there exists a positive number M_0 such that

$$\sum_{i=0}^{T-1} f_i(x_i, x_{i+1}, y_i^*, y_{i+1}^*) - \sum_{i=0}^{T-1} f_i(x_i^*, x_{i+1}^*, y_i^*, y_{i+1}^*)$$

$$< M_0 \text{ for all integers } T \geq 1. \tag{7.92}$$

Assume that (7.91) is not valid. Then there exists a number $\epsilon \in (0, 1)$ for which

$$\limsup_{i \to \infty} |x_i - \bar{x}_i| > 2\epsilon. \tag{7.93}$$

Let $\delta \in (0, \epsilon)$ be as guaranteed by property (P8). In view of (7.84) there exists an integer $i_0 \geq 1$ such that

$$|x_i^* - \bar{x}_i|, \ |y_i^* - \bar{y}_i| \leq \delta \text{ for all integers } i \geq i_0. \tag{7.94}$$

It follows from (7.93) that there exists a strictly increasing sequence of natural numbers $\{i_k\}_{k=1}^{\infty}$ such that

$$i_{k+1} - i_k \geq 10 \text{ for all integers } k \geq 0,$$

$$|x_{i_k} - \bar{x}_{i_k}| > 2\epsilon \text{ for all integers } k \geq 1. \tag{7.95}$$

For every nonnegative integer i set

$$x_i' = 2^{-1}(x_i + x_i^*). \tag{7.96}$$

It is clear that for all nonnegative integers i, we have

$$f_i(x_i', x_{i+1}', y_i^*, y_{i+1}^*)$$

$$\leq 2^{-1} f_i(x_i^*, x_{i+1}^*, y_i^*, y_{i+1}^*) + 2^{-1} f_i(x_i, x_{i+1}, y_i^*, y_{i+1}^*). \tag{7.97}$$

Property (P8), (7.94)–(7.96), and the choice of δ imply for all natural numbers k,

$$|x_{i_k}^* - \bar{x}_{i_k}|, \ |x_{i_k+1}^* - \bar{x}_{i_k+1}|, \ |y_{i_k}^* - \bar{y}_{i_k}|, \ |y_{i_k+1}^* - \bar{y}_{i_k+1}| \leq \delta$$

and

$$f_{i_k}(x_{i_k}', x_{i_k+1}', y_{i_k}^*, y_{i_k+1}^*)$$

$$= f_{i_k}(2^{-1}(x_{i_k}^* + x_{i_k}), 2^{-1}(x_{i_k+1}^* + x_{i_k+1}), y_{i_k}^*, y_{i_k+1}^*)$$

$$\leq 2^{-1} f_{i_k}(x_{i_k}^*, x_{i_k+1}^*, y_{i_k}^*, y_{i_k+1}^*) + 2^{-1} f_{i_k}(x_{i_k}, x_{i_k+1}, y_{i_k}^*, y_{i_k+1}^*) - \delta. \tag{7.98}$$

It follows from (7.97), (7.98), (7.94), and Lemma 7.21 that for all natural number k,

$$-4D_1 \le \sum_{i=0}^{i_k} f_i(x'_i, x'_{i+1}, y^*_i, y^*_{i+1}) - \sum_{i=0}^{i_k} f_i(x^*_i, x^*_{i+1}, y^*_i, y^*_{i+1})$$

$$\le -\sum_{i=0}^{i_k} f_i(x^*_i, x^*_{i+1}, y^*_i, y^*_{i+1})$$

$$+2^{-1}\sum_{i=0}^{i_k} f_i(x^*_i, x^*_{i+1}, y^*_i, y^*_{i+1}) + 2^{-1}\sum_{i=0}^{i_k} f_i(x_i, x_{i+1}, y^*_i, y^*_{i+1}) - k\delta$$

$$\le 2^{-1}\sum_{i=0}^{i_k} f_i(x_i, x_{i+1}, y^*_i, y^*_{i+1}) - 2^{-1}\sum_{i=0}^{i_k} f_i(x^*_i, x^*_{i+1}, y^*_i, y^*_{i+1}) - k\delta$$

$$\le 2^{-1}M_0 - k\delta \to -\infty \text{ as } k \to \infty.$$

The contradiction we have reached proves (7.91) and Lemma 7.24 itself. □

Lemma 7.25. *Assume that* $\{y_i\}_{i=0}^{\infty} \subset Y$ *satisfies*

$$\limsup_{T\to\infty}\left[\sum_{i=0}^{T-1} f_i(x^*_i, x^*_{i+1}, y_i, y_{i+1}) - \sum_{i=0}^{T-1} f_i(x^*_i, x^*_{i+1}, y^*_i, y^*_{i+1})\right] > -\infty. \quad (7.99)$$

Then

$$\lim_{i\to\infty} |y_i - \bar{y}_i| = 0. \quad (7.100)$$

Proof. It follows from (7.99) and Lemma 7.23 that there exists a positive number M_0 such that

$$\sum_{i=0}^{T-1} f_i(x^*_i, x^*_{i+1}, y_i, y_{i+1}) - \sum_{i=0}^{T-1} f_i(x^*_i, x^*_{i+1}, y^*_i, y^*_{i+1}) > -M_0$$

$$\text{for all integers } T \ge 1. \quad (7.101)$$

Assume that (7.100) is not valid. Then there exists a number $\epsilon \in (0, 1)$ for which

$$\limsup_{i\to\infty} |y_i - \bar{y}_i| > 2\epsilon. \quad (7.102)$$

Let $\delta \in (0, \epsilon)$ be as guaranteed by property (P8). In view of (7.84), there exists an integer $i_0 \ge 1$ such that

$$|x^*_i - \bar{x}_i|,\ |y^*_i - \bar{y}_i| \le \delta \text{ for all integers } i \ge i_0. \quad (7.103)$$

It follows from (7.102) that there exists a strictly increasing sequence of natural numbers $\{i_k\}_{k=1}^{\infty}$ such that

$$i_{k+1} - i_k \ge 10 \text{ for all integers } k \ge 0,$$

$$|y_{i_k} - \bar{y}_{i_k}| > 2\epsilon \text{ for all integers } k \ge 1. \quad (7.104)$$

For every nonnegative integer i set

$$y'_i = 2^{-1}(y_i + y^*_i). \tag{7.105}$$

It is clear that for all nonnegative integers i, we have

$$f_i(x^*_i, x^*_{i+1}, y'_i, y'_{i+1}) \geq 2^{-1} f_i(x^*_i, x^*_{i+1}, y_i, y_{i+1}) + 2^{-1} f_i(x^*_i, x^*_{i+1}, y^*_i, y^*_{i+1}). \tag{7.106}$$

It follows from (7.103)–(7.105), property (P8), and the choice of δ that for all natural numbers k,

$$|x^*_{i_k} - \bar{x}_{i_k}|,\ |x^*_{i_k+1} - \bar{x}_{i_k+1}|,\ |y^*_{i_k} - \bar{y}_{i_k}|,\ |y^*_{i_k+1} - \bar{y}_{i_k+1}| \leq \delta$$

and

$$f_{i_k}(x^*_{i_k}, x^*_{i_k+1}, y'_{i_k}, y'_{i_k+1}) - f_{i_k}(x^*_{i_k}, x^*_{i_k+1}, 2^{-1}(y^*_{i_k} + y_{i_k}), 2^{-1}(y^*_{i_k+1} + y_{i_k+1}))$$
$$\geq 2^{-1} f_{i_k}(x^*_{i_k}, x^*_{i_k+1}, y_{i_k}, y_{i_k+1}) + 2^{-1} f_{i_k}(x^*_{i_k}, x^*_{i_k+1}, y^*_{i_k}, y^*_{i_k+1}) + \delta. \tag{7.107}$$

Lemma 7.21, (7.106), (7.107), (7.104), and (7.101) imply that for all natural numbers k,

$$4D_1 \geq \sum_{i=0}^{i_k} f_i(x^*_i, x^*_{i+1}, y'_i, y'_{i+1}) - \sum_{i=0}^{i_k} f_i(x^*_i, x^*_{i+1}, y^*_i, y^*_{i+1})$$
$$\geq - \sum_{i=0}^{i_k} f_i(x^*_i, x^*_{i+1}, y^*_i, y^*_{i+1})$$
$$+2^{-1} \sum_{i=0}^{i_k} f_i(x^*_i, x^*_{i+1}, y_i, y_{i+1}) + 2^{-1} \sum_{i=0}^{i_k} f_i(x^*_i, x^*_{i+1}, y^*_i, y^*_{i+1}) + k\delta$$
$$= 2^{-1} \sum_{i=0}^{i_k} f_i(x^*_i, x^*_{i+1}, y_i, y_{i+1}) - 2^{-1} \sum_{i=0}^{i_k} f_i(x^*_i, x^*_{i+1}, y^*_i, y^*_{i+1}) + k\delta$$
$$\geq -2^{-1} M_0 + k\delta \to \infty \text{ as } k \to \infty.$$

The contradiction we have reached proves (7.100) and Lemma 7.25 itself. □

Completion of the proof of Theorem 7.9. Assume that $\{x_i\}_{i=0}^{\infty} \subset X$ and that

$$x_0 = x. \tag{7.108}$$

We claim that

$$\limsup_{T\to\infty} [\sum_{i=0}^{T-1} f_i(x^*_i, x^*_{i+1}, y^*_i, y^*_{i+1}) - \sum_{i=0}^{T-1} f_i(x_i, x_{i+1}, y^*_i, y^*_{i+1})] \leq 0. \tag{7.109}$$

Assume the contrary. Then there exists a number $\epsilon \in (0,1)$ such that

$$\limsup_{T\to\infty}[\sum_{i=0}^{T-1} f_i(x_i^*, x_{i+1}^*, y_i^*, y_{i+1}^*) - \sum_{i=0}^{T-1} f_i(x_i, x_{i+1}, y_i^*, y_{i+1}^*)] > 4\epsilon. \quad (7.110)$$

Lemma 7.24 and (7.110) imply that

$$\lim_{i\to\infty} |x_i - \bar{x}_i| = 0. \quad (7.111)$$

In view of property (P1), there exists a number $\delta \in (0, \epsilon)$ such that for every nonnegative integer i, each $z_1, z_2, \bar{z}_1, \bar{z}_2 \in X$ and each $\xi_1, \xi_2, \bar{\xi}_1, \bar{\xi}_2 \in Y$ which satisfy

$$|z_i - \bar{z}_i| \le \delta, \ |\xi_i - \bar{\xi}_i| \le \delta, \ i = 1, 2 \quad (7.112)$$

the following inequality holds:

$$|f_i(z_1, z_2, \xi_1, \xi_2) - f_i(\bar{z}_1, \bar{z}_2, \bar{\xi}_1, \bar{\xi}_2)| \le \epsilon/2. \quad (7.113)$$

It follows from (7.84) and (7.111) that there exists an integer $i_0 \ge 1$ such that for all natural numbers $i \ge i_0$, we have

$$|x_i - \bar{x}_i| \le \delta/2, \ |x_i^* - \bar{x}_i| \le \delta/2, \ |y_i^* - \bar{y}_i| \le \delta/2. \quad (7.114)$$

In view of (7.110), there exists a natural number $T_0 > i_0$ such that

$$\sum_{i=0}^{T_0-1} f_i(x_i^*, x_{i+1}^*, y_i^*, y_{i+1}^*) - \sum_{i=0}^{T_0-1} f_i(x_i, x_{i+1}, y_i^*, y_{i+1}^*) > 4\epsilon. \quad (7.115)$$

Set

$$z_i = x_i, \ i = 0, \dots, T_0, \ z_{T_0+1} = x_{T_0+1}^*. \quad (7.116)$$

It follows from (7.116), (7.108), (7.83), ($\{f_i\}_{i=0}^{\infty}$)-locally optimality of the pair

$$\{x_i^*\}_{i=0}^{\infty}, \{y_i^*\}_{i=0}^{\infty}$$

and (7.115) that

$$\sum_{i=0}^{T_0-1} f_i(x_i^*, x_{i+1}^*, y_i^*, y_{i+1}^*)$$

$$= \sum_{i=0}^{T_0} f_i(x_i^*, x_{i+1}^*, y_i^*, y_{i+1}^*) - f_{T_0}(x_{T_0}^*, x_{T_0+1}^*, y_{T_0}^*, y_{T_0+1}^*)$$

$$\le \sum_{i=0}^{T_0} f_i(z_i, z_{i+1}, y_i^*, y_{i+1}^*) - f_{T_0}(x_{T_0}^*, x_{T_0+1}^*, y_{T_0}^*, y_{T_0+1}^*)$$

$$= \sum_{i=0}^{T_0-1} f_i(x_i, x_{i+1}, y_i^*, y_{i+1}^*)$$

$$+f_{T_0}(x_{T_0}, x_{T_0+1}^*, y_{T_0}^*, y_{T_0+1}^*) - f_{T_0}(x_{T_0}^*, x_{T_0+1}^*, y_{T_0}^*, y_{T_0+1}^*)$$

$$< \sum_{i=0}^{T_0-1} f_i(x_i^*, x_{i+1}^*, y_i^*, y_{i+1}^*) - 4\epsilon + f_{T_0}(x_{T_0}, x_{T_0+1}^*, y_{T_0}^*, y_{T_0+1}^*)$$

$$-f_{T_0}(x_{T_0}^*, x_{T_0+1}^*, y_{T_0}^*, y_{T_0+1}^*)$$

and

$$4\epsilon \le f_{T_0}(x_{T_0}, x_{T_0+1}^*, y_{T_0}^*, y_{T_0+1}^*) - f_{T_0}(x_{T_0}^*, x_{T_0+1}^*, y_{T_0}^*, y_{T_0+1}^*). \tag{7.117}$$

In view of the inequality $T_0 > i_0$, the choice of δ (see (7.112), (7.113)) and (7.114), we have

$$|f_{T_0}(x_{T_0}, x_{T_0+1}^*, y_{T_0}^*, y_{T_0+1}^*) - f_{T_0}(x_{T_0}^*, x_{T_0+1}^*, y_{T_0}^*, y_{T_0+1}^*)| \le \epsilon/2.$$

This contradicts (7.117). The contradiction we have reached proves (7.109).

Analogously we can show that for every sequence $\{y_i\}_{i=0}^{\infty} \subset Y$ which satisfies $y_0 = y$,

$$\limsup_{T\to\infty} [\sum_{i=0}^{T-1} f_i(x_i^*, x_{i+1}^*, y_i, y_{i+1}) - \sum_{i=0}^{T-1} f_i(x_i^*, x_{i+1}^*, y_i^*, y_{i+1}^*)] \le 0.$$

This completes the proof of Theorem 7.9. □

8

Constrained Games with Convexity–Concavity Assumptions

In this chapter we study the existence and turnpike properties of approximate solutions for a class of dynamic constrained discrete-time two-player zero-sum games which satisfy convexity–concavity assumptions. We describe the structure of approximate solutions which is independent of the length of the interval, for all sufficiently large intervals and show that approximate solutions are determined mainly by the objective function, and are essentially independent of the choice of intervals and endpoint conditions. We show that the turnpike phenomenon is stable under small perturbations of objective functions and analyze the structure of approximate solutions in regions closed to the endpoints of domains.

8.1 Preliminaries

Let M, N be nonempty sets and let $h : M \times N \to R^1$. The triplet (h, M, N) describes the two-player zero-sum game, where M is the set of strategies of the first player, N is the set of strategies of the second player, and h and $-h$ are objective functions for the first and second player respectively.

Denote by $\langle \cdot, \cdot \rangle$ the inner product in the space R^m which induces the Euclidean norm $|\cdot|$. Let m_1, m_2 be natural numbers, $X_i \subset R^{m_i}$, $i = 1, 2$ be nonempty convex compact sets and $\Omega_i \subset X_i \times X_i$, $i = 1, 2$ be nonempty closed convex sets. For each bounded function $h : \Omega_1 \times \Omega_2 \to R^1$ set

$$\|h\| = \sup\{|h(z_1, z_2)| : \; z_i \in \Omega_i, \; i = 1, 2\}.$$

Denote by $\mathcal{M}$ the set of all continuous functions $f : \Omega_1 \times \Omega_2 \to R^1$ such that the following properties hold:

for each $(y_1, y_2) \in \Omega_2$, the function $(x_1, x_2) \to f(x_1, x_2, y_1, y_2)$, $(x_1, x_2) \in \Omega_1$ is convex;

A.J. Zaslavski, *Discrete-Time Optimal Control and Games on Large Intervals*, Springer Optimization and Its Applications 119,
DOI 10.1007/978-3-319-52932-5_8

for each $(x_1, x_2) \in \Omega_1$, the function $(y_1, y_2) \to f(x_1, x_2, y_1, y_2)$, $(y_1, y_2) \in \Omega_2$ is concave.

The set $\mathcal{M}$ is equipped with a metric $\rho : \mathcal{M} \times \mathcal{M} \to R^1$ defined by

$$\rho(f, g) = \sup\{|f(x_1, x_2, y_1, y_2) - g(x_1, x_2, y_1, y_2)| :$$
$$(x_1, x_2) \in \Omega_1,\ (y_1, y_2) \in \Omega_2\},\ f, g \in \mathcal{M}. \tag{8.1}$$

Clearly, $(\mathcal{M}, \rho)$ is a complete metric space. For $p = 1, 2$ set

$$\mathrm{diam}(X_p) = \sup\{|z_1 - z_2| :\ z_1, z_2 \in X_p\}. \tag{8.2}$$

Let $p \in \{1, 2\}$ and $T_2 > T_1$ be nonnegative integers. A sequence $\{x_t\}_{t=T_1}^{T_2} \subset X_p$ is called an (Ω_p)-program if $(x_t, x_{t+1}) \in \Omega_p$ for all integers $t \in \{T_1, \dots, T_2 - 1\}$. A sequence $\{x_t\}_{t=T_1}^{\infty} \subset X_p$ is called an (Ω_p)-program if $(x_t, x_{t+1}) \in \Omega_p$ for all integers $t \geq T_1$.

Proposition 8.1. *Let $p \in \{1, 2\}$. The following conditions are equivalent.*
(1) For each natural number n there exists an (Ω_p)-program $\{x_t\}_{t=0}^{n}$.
(2) $\Omega_p \cap \{(x, x) :\ x \in X_p\} \neq \emptyset$.

Proof. Clearly, condition (2) implies condition (1). Assume that condition (1) holds. For every natural number n, there exists an (Ω_p)-program $\{x_t^{(n)}\}_{t=0}^{n}$. Since the set Ω_p is convex, for every natural number n,

$$n^{-1}\Big(\sum_{i=0}^{n-1} x_i^{(n)}, \sum_{i=1}^{n} x_i^{(n)}\Big) = n^{-1}\sum_{i=0}^{n-1}(x_i^{(n)}, x_{i+1}^{(n)}) \in \Omega_p, \tag{8.3}$$

$$\Big|n^{-1}\sum_{i=0}^{n-1} x_i^{(n)} - n^{-1}\sum_{i=1}^{n} x_i^{(n)}\Big| \leq n^{-1}\mathrm{diam}(X_p). \tag{8.4}$$

There exists a strictly increasing sequence of natural numbers $\{n_k\}_{k=1}^{\infty}$ such that there exists

$$\lim_{k\to\infty} n_k^{-1}\sum_{i=0}^{n_k-1} x_i^{(n_k)}.$$

By (8.3) and (8.4),

$$\Big(\lim_{k\to\infty} n_k^{-1}\sum_{i=0}^{n_k-1} x_i^{(n_k)}, \lim_{k\to\infty} n_k^{-1}\sum_{i=0}^{n_k-1} x_i^{(n_k)}\Big) \in \Omega_p \cap \{(x, x) :\ x \in X_p\}.$$

Proposition 8.1 is proved. □

In this chapter we suppose that for every $p = 1, 2$,

$$\Omega_p \cap \{(x, x) :\ x \in X_p\} \neq \emptyset. \tag{8.5}$$

Given $f \in \mathcal{M}$ and an integer $n \geq 1$ we consider a discrete-time two-player zero-sum game over the interval $[0, n]$. For this game the set of all (Ω_1)-programs $\{x_i\}_{i=0}^n$ is the set of strategies for the first player, the set of all (Ω_2)-programs $\{y_i\}_{i=0}^n$ is the set of strategies for the second player, and the cost for the first player associated with the strategies $\{x_i\}_{i=0}^n$, $\{y_i\}_{i=0}^n$ is given by $\sum_{i=0}^{n-1} f(x_i, x_{i+1}, y_i, y_{i+1})$.

Let a function $f \in \mathcal{M}$ be given, M be a nonnegative number and let $n \geq 1$ be a natural number. A pair of an (Ω_1)-program $\{\bar{x}_i\}_{i=0}^n$ and an (Ω_2)-program $\{\bar{y}_i\}_{i=0}^n$ is called (f, M)-good if the following properties hold:

for every (Ω_1)-program $\{x_i\}_{i=0}^n$ which satisfies $x_i = \bar{x}_i$, $i = 0, n$ we have

$$M + \sum_{i=0}^{n-1} f(x_i, x_{i+1}, \bar{y}_i, \bar{y}_{i+1}) \geq \sum_{i=0}^{n-1} f(\bar{x}_i, \bar{x}_{i+1}, \bar{y}_i, \bar{y}_{i+1}); \tag{8.6}$$

for every (Ω_2)-program $\{y_i\}_{i=0}^n$ which satisfies $y_i = \bar{y}_i$, $i = 0, n$, we have

$$M + \sum_{i=0}^{n-1} f(\bar{x}_i, \bar{x}_{i+1}, \bar{y}_i, \bar{y}_{i+1}) \geq \sum_{i=0}^{n-1} f(\bar{x}_i, \bar{x}_{i+1}, y_i, y_{i+1}). \tag{8.7}$$

If the pair $\{\bar{x}_i\}_{i=0}^n$, $\{\bar{y}_i\}_{i=0}^n$ is $(f, 0)$-good, then it is called (f)-optimal.

We study the existence of equilibria over an infinite horizon and employ the following version of the overtaking optimality criterion.

Let $f \in \mathcal{M}$. A pair of an (Ω_1)-program $\{\bar{x}_i\}_{i=0}^\infty$ and an (Ω_2)-program $\{\bar{y}_i\}_{i=0}^\infty$ is called (f)-overtaking optimal if the following properties hold:

for each (Ω_1)-program $\{x_i\}_{i=0}^\infty$ satisfying $x_0 = \bar{x}_0$,

$$\limsup_{T \to \infty} [\sum_{i=0}^{T-1} f(\bar{x}_i, \bar{x}_{i+1}, \bar{y}_i, \bar{y}_{i+1}) - \sum_{i=0}^{T-1} f(x_i, x_{i+1}, \bar{y}_i, \bar{y}_{i+1})] \leq 0; \tag{8.8}$$

for each (Ω_2)-program $\{y_i\}_{i=0}^\infty$ satisfying $y_0 = \bar{y}_0$

$$\limsup_{T \to \infty} [\sum_{i=0}^{T-1} f(\bar{x}_i, \bar{x}_{i+1}, y_i, y_{i+1}) - \sum_{i=0}^{T-1} f(\bar{x}_i, \bar{x}_{i+1}, \bar{y}_i, \bar{y}_{i+1})] \leq 0. \tag{8.9}$$

A pair of an (Ω_1)-program $\{\bar{x}_i\}_{i=0}^\infty$ and an (Ω_2)-program $\{\bar{y}_i\}_{i=0}^\infty$ is called strongly (f)-overtaking optimal if (8.8) holds for each (Ω_1)-program $\{x_i\}_{i=0}^\infty$ and (8.9) is true for each (Ω_2)-program $\{y_i\}_{i=0}^\infty$.

Let M be a nonnegative number and let $n \geq 1$ be a natural number. A pair of an (Ω_1)-program $\{\bar{x}_i\}_{i=0}^n$ and an (Ω_2)-program $\{\bar{y}_i\}_{i=0}^n$ is called strongly (f, M)-good if (8.6) holds for every (Ω_1)-program $\{x_i\}_{i=0}^n$ which satisfies $x_n = \bar{x}_n$ and (8.7) holds for every (Ω_2)-program $\{y_i\}_{i=0}^n$ which satisfies $y_n = \bar{y}_n$.

If the pair $\{\bar{x}_i\}_{i=0}^n$, $\{\bar{y}_i\}_{i=0}^n$ is strongly $(f, 0)$-good, then it is called strongly (f)-optimal.

A pair of an (Ω_1)-program $\{x_i\}_{i=0}^\infty$ and an (Ω_2)-program $\{y_i\}_{i=0}^\infty$ is called (f)-optimal if for each integer $n \geq 2$ the pair $\{x_i\}_{i=n}^\infty$, $\{y_i\}_{i=0}^n$ is (f)-optimal.

For $p \in \{1, 2\}$, $x \in X_p$ and $r > 0$ set

$$B_p(x, r) = \{z \in R^{m_p} : |x - z| \le r\}, \; B_{X_p}(x, r) = B_p(x, r) \cap X_p. \tag{8.10}$$

Let $f \in \mathcal{M}$ and $p \in \{1, 2\}$. Set

$$\Delta_p = \{x \in X_p : (x, x) \in \Omega_p\}. \tag{8.11}$$

In view of (8.5),

$$\Delta_p \neq \emptyset. \tag{8.12}$$

Clearly, the set Δ_p is convex and closed. Define a function $\widehat{f} : \Delta_1 \times \Delta_2 \to R^1$ by

$$\widehat{f}(x, y) = f(x, x, y, y), \; (x, y) \in \Delta_1 \times \Delta_2. \tag{8.13}$$

By Proposition 7.2 there exists a saddle point

$$(x_f, y_f) \in \Delta_1 \times \Delta_2$$

for f such that

$$\sup\{\widehat{f}(x_f, y) : y \in \Delta_2\} = \widehat{f}(x_f, y_f) = \inf\{\widehat{f}(x, y_f) : x \in \Delta_1\}. \tag{8.14}$$

Set

$$\mu(f) = \widehat{f}(x_f, y_f). \tag{8.15}$$

Now we describe two-player zero-sum games considered in this chapter.

Let $n \ge 1$ be an integer and $\xi = (\xi_1, \xi_2, \xi_3, \xi_4) \in X_1 \times X_1 \times X_2 \times X_2$. Define

$$\Lambda_1(\xi, n) = \{(\Omega_1) - \text{programs } \{x_i\}_{i=0}^n : x_0 = \xi_1, \; x_n = \xi_2\}, \tag{8.16}$$

$$\Lambda_2(\xi, n) = \{(\Omega_2) - \text{programs } \{x_i\}_{i=0}^n : x_0 = \xi_3, \; x_n = \xi_4\} \tag{8.17}$$

(note that these sets can be empty),

$$f^{(\xi,n)}(\{x_0, \dots, x_n\}, \{y_0, \dots, y_n\}) = \sum_{i=0}^{n-1} f(x_i, x_{i+1}, y_i, y_{i+1}), \tag{8.18}$$

$$\{x_i\}_{i=0}^n \in \Lambda_1(\xi, n), \; \{y_i\}_{i=0}^n \in \Lambda_2(\xi, n).$$

Assume that

$$\Lambda_1(\xi, n) \neq \emptyset, \; \Lambda_2(\xi, n) \neq \emptyset.$$

Then $\Lambda_1(\xi, n)$, $\Lambda_2(\xi, n)$ are closed convex sets, for every $y \in \Lambda_2(\xi, n)$, the function $f^{(\xi,n)}(x, y)$, $x \in \Lambda_1(\xi, n)$ is convex and for every $x \in \Lambda_1(\xi, n)$, the function $f^{(\xi,n)}(x, y)$, $y \in \Lambda_2(\xi, n)$ is concave.

We consider the game for which $\Lambda_1(\xi, n)$ is the set of strategies of the first player, $\Lambda_2(\xi, n)$ is the set of strategies of the second player, and $f^{(\xi,n)}$ is the objective function of the first player.

By Proposition 7.2, there exist

$$x^{(\xi,n)} = \{x_0^{(\xi,n)}, \dots, x_n^{(\xi,n)}\} \in \Lambda_1(\xi,n),\ y^{(\xi,n)} = \{y_0^{(\xi,n)}, \dots, y_n^{(\xi,n)}\} \in \Lambda_2(\xi,n) \tag{8.19}$$

such that

$$\sup\{f^{(\xi,n)}(x^{(\xi,n)}, y) :\ y \in \Lambda_2(\xi,n)\} = f^{(\xi,n)}(x^{(\xi,n)}, y^{(\xi,n)})$$

$$= \inf\{f^{(\xi,n)}(x, y^{(\xi,n)}) : x \in \Lambda_1(\xi,n).\} \tag{8.20}$$

Let $n \geq 1$ be an integer and $\xi = (\xi_1, \xi_2) \in X_1 \times X_2$. Define

$$\Lambda_1(\xi,n) = \{(\Omega_1) - \text{programs } \{x_i\}_{i=0}^n :\ x_0 = \xi_1\},$$

$$\Lambda_2(\xi,n) = \{(\Omega_2) - \text{programs } \{y_i\}_{i=0}^n :\ y_0 = \xi_2\}$$

(note that these sets can be empty) and define $f^{(\xi,n)} : \Lambda_1(\xi,n) \times \Lambda_2(\xi,n) \to R^1$ by (8.18).

Assume that

$$\Lambda_1(\xi,n) \neq \emptyset,\ \Lambda_2(\xi,n) \neq \emptyset.$$

Then $\Lambda_1(\xi,n)$, $\Lambda_2(\xi,n)$ are closed convex sets, for every $y \in \Lambda_2(\xi,n)$, the function $f^{(\xi,n)}(x,y)$, $x \in \Lambda_1(\xi,n)$ is convex and for every $x \in \Lambda_1(\xi,n)$, the function $f^{(\xi,n)}(x,y)$, $y \in \Lambda_2(\xi,n)$ is concave.

We consider the game for which $\Lambda_1(\xi,n)$ is the set of strategies of the first player, $\Lambda_2(\xi,n)$ is the set of strategies of the second player, and $f^{(\xi,n)}$ is the objective function of the first player.

By Proposition 7.2, there exist

$$x^{(\xi,n)} = \{x_0^{(\xi,n)}, \dots, x_n^{(\xi,n)}\} \in \Lambda_1(\xi,n),\ y^{(\xi,n)} = \{y_0^{(\xi,n)}, \dots, y_n^{(\xi,n)}\} \in \Lambda_2(\xi,n) \tag{8.21}$$

such that (8.20) holds.

Let $n \geq 1$ be an integer. Define

$$\Lambda_1(n) = \{(\Omega_1) - \text{programs } \{x_i\}_{i=0}^n\},$$

$$\Lambda_2(n) = \{(\Omega_2) - \text{programs } \{y_i\}_{i=0}^n\}$$

and define

$$f^{(n)}(\{x_0, \dots, x_n\}, \{y_0, \dots, y_n\}) = \sum_{i=0}^{n-1} f(x_i, x_{i+1}, y_i, y_{i+1}),$$

$$\{x_i\}_{i=0}^n \in \Lambda_1(n),\ \{y_i\}_{i=0}^n \in \Lambda_2(n).$$

Clearly, $\Lambda_1(n)$, $\Lambda_2(n)$ are nonempty closed convex sets, for every $y \in \Lambda_2(n)$, the function $f^{(n)}(x,y)$, $x \in \Lambda_1(n)$ is convex and for every $x \in \Lambda_1(n)$, the function $f^{(n)}(x,y)$, $y \in \Lambda_2(n)$ is concave.

We consider the game for which $\Lambda_1(n)$ is the set of strategies of the first player, $\Lambda_2(n)$ is the set of strategies of the second player, and $f^{(n)}$ is the objective function of the first player.

By Proposition 7.2, there exist

$$x^{(n)} = \{x_0^{(n)}, \dots, x_n^{(n)}\} \in \Lambda_1(n), \; y^{(n)} = \{y_0^{(n)}, \dots, y_n^{(n)}\} \in \Lambda_2(n) \tag{8.22}$$

such that

$$\begin{aligned} \sup\{f^{(n)}(x^{(n)}, y) : \; y \in \Lambda_2(n)\} &= f^{(n)}(x^{(n)}, y^{(n)}) \\ &= \inf\{f^{(n)}(x, y^{(n)}) : x \in \Lambda_1(n).\} \end{aligned} \tag{8.23}$$

Let $T_2 > T_1 \geq 0$ be integers and $\xi = (\xi_1, \xi_2, \xi_3, \xi_4) \in X_1 \times X_1 \times X_2 \times X_2$. Define

$$\begin{aligned} \Lambda_1(\xi, T_1, T_2) &= \Lambda_1(\xi_1, \xi_2, \xi_3, \xi_4, T_1, T_2) \\ = \{(\Omega_1) - \text{programs } &\{x_t\}_{t=T_1}^{T_2} \; : \; x_{T_1} = \xi_1, \; x_{T_2} = \xi_2\}, \end{aligned} \tag{8.24}$$

$$\begin{aligned} \Lambda_2(\xi, T_1, T_2) &= \Lambda_2(\xi_1, \xi_2, \xi_3, \xi_4, T_1, T_2) \\ = \{(\Omega_2) - \text{programs } &\{x_t\}_{t=T_1}^{T_2} \; : \; x_{T_1} = \xi_3, \; x_{T_2} = \xi_4\}. \end{aligned} \tag{8.25}$$

Assume that

$$\Lambda_1(\xi_1, \xi_2, \xi_3, \xi_4, T_1, T_2) \neq \emptyset, \; \Lambda_2(\xi_1, \xi_2, \xi_3, \xi_4, T_1, T_2) \neq \emptyset$$

and that $F = \{f_t\}_{t=T_1}^{T_2-1} \subset \mathcal{M}$. Define

$$F_{T_1,T_2}^{(\xi)}(\{x_{T_1}, \dots, x_{T_2}\}, \{y_{T_1}, \dots, y_{T_2}\}) = \sum_{i=T_1}^{T_2-1} f_t(x_t, x_{t+1}, y_t, y_{t+1}), \tag{8.26}$$

$$\{x_t\}_{i=T_1}^{T_2} \in \Lambda_1(\xi, T_1, T_2), \; \{y_t\}_{t=T_1}^{T_2} \in \Lambda_2(\xi, T_1, T_2).$$

Clearly, $\Lambda_1(\xi, T_1, T_2)$, $\Lambda_2(\xi, T_1, T_2)$ are closed convex sets in $R^{(T_2-T_1+1)m_1}$, $R^{(T_2-T_1+1)m_2}$ respectively, for every $\{y_{T_1}, \dots, y_{T_2}\} \in \Lambda_2(\xi, T_1, T_2)$, the function $F_{T_1,T_2}^{(\xi)}(\{x_{T_1}, \dots, x_{T_2}\}, \{y_{T_1}, \dots, y_{T_2})\})$, $\{x_{T_1}, \dots, x_{T_2}\} \in \Lambda_1(\xi, T_1, T_2)$ is convex and for every $\{x_{T_1}, \dots, x_{T_2}\} \in \Lambda_1(\xi, T_1, T_2)$, the function

$$F_{T_1,T_2}^{(\xi)}(\{x_{T_1}, \dots, x_{T_2}\}, \{y_{T_1}, \dots, y_{T_2}\}), \; \{y_{T_1}, \dots, y_{T_2}\} \in \Lambda_2(\xi, T_1, T_2)$$

is concave.

We consider the game for which $\Lambda_1(\xi, T_1, T_2)$ is the set of strategies of the first player, $\Lambda_2(\xi, T_1, T_2)$ is the set of strategies of the second player, and $F_{T_1,T_2}^{(\xi)}$ is the objective function of the first player.

By Proposition 7.2, there exist

$$\{\widehat{x}_{T_1}, \dots, \widehat{x}_{T_2}\} \in \Lambda_1(\xi, T_1, T_2), \{\widehat{y}_{T_1}, \dots, \widehat{y}_{T_2}\} \in \Lambda_2(\xi, T_1, T_2) \tag{8.27}$$

such that

$$\sup\{\sum_{t=T_1}^{T_2-1} f_t(\widehat{x}_t, \widehat{x}_{t+1}, y_t, y_{t+1}) : \{y_{T_1}, \dots, y_{T_2}\} \in \Lambda_2(\xi, T_1, T_2)\}$$

$$= \sum_{t=T_1}^{T_2-1} f_t(\widehat{x}_t, \widehat{x}_{t+1}, \widehat{y}_t, \widehat{y}_{t+1})$$

$$= \inf\{\sum_{t=T_1}^{T_2-1} f_t(x_t, x_{t+1}, \widehat{y}_t, \widehat{y}_{t+1}) : \{x_{T_1}, \dots, x_{T_2}\} \in \Lambda_1(\xi, T_1, T_2)\}. \quad (8.28)$$

Let $M \geq 0$. A pair of strategies

$$\{\tilde{x}_t\}_{t=T_1}^{T_2} \in \Lambda_1(\xi, T_1, T_2), \ \{\tilde{y}_t\}_{t=T_1}^{T_2} \in \Lambda_2(\xi, T_1, T_2)$$

is called $(\{f_t\}_{t=T_1}^{T_2-1}, \xi, T_1, T_2, M)$-good if

$$\sum_{t=T_1}^{T_2-1} f_t(\tilde{x}_t, \tilde{x}_{t+1}, y_t, y_{t+1}) - M \leq \sum_{t=T_1}^{T_2-1} f_t(\tilde{x}_t, \tilde{x}_{t+1}, \tilde{y}_t, \tilde{y}_{t+1})$$

$$\leq \sum_{t=T_1}^{T_2-1} f_t(x_t, x_{t+1}, \tilde{y}_t, \tilde{y}_{t+1}) + M \quad (8.29)$$

for every $\{x_t\}_{t=T_1}^{T_2} \in \Lambda_1(\xi, T_1, T_2)$ and every $\{y_t\}_{t=T_1}^{T_2} \in \Lambda_2(\xi, T_1, T_2)$.

Let $T_2 > T_1 \geq 0$ be integers, $\xi = (\xi_1, \xi_2) \in X_1 \times X_2$ and $F=\{f_t\}_{t=T_1}^{T_2-1} \subset \mathcal{M}$. Define

$$\Lambda_1(\xi, T_1, T_2) = \Lambda_1(\xi_1, \xi_2, T_1, T_2) = \{(\Omega_1) - \text{programs } \{x_t\}_{t=T_1}^{T_2} : \ x_{T_1} = \xi_1\}, \quad (8.30)$$

$$\Lambda_2(\xi, T_1, T_2) = \Lambda_2(\xi_1, \xi_2, T_1, T_2) = \{(\Omega_2) - \text{programs } \{y_t\}_{t=T_1}^{T_2} : \ y_{T_1} = \xi_2\}. \quad (8.31)$$

Assume that

$$\Lambda_1(\xi, T_1, T_2) \neq \emptyset, \ \Lambda_2(\xi, T_1, T_2) \neq \emptyset$$

and define

$$F_{T_1,T_2}^{(\xi)}(\{x_{T_1}, \dots, x_{T_2}\}, \{y_{T_1}, \dots, y_{T_2}\}) = \sum_{t=T_1}^{T_2-1} f_t(x_t, x_{t+1}, y_t, y_{t+1}), \quad (8.32)$$

$$\{x_t\}_{t=T_1}^{T_2} \in \Lambda_1(\xi, T_1, T_2), \ \{y_t\}_{t=T_1}^{T_2} \in \Lambda_2(\xi, T_1, T_2).$$

Clearly, $\Lambda_1(\xi, T_1, T_2)$, $\Lambda_2(\xi, T_1, T_2)$ are closed convex sets in $R^{(T_2-T_1+1)m_1}$, $R^{(T_2-T_1+1)m_2}$ respectively, for every $\{y_{T_1}, \dots, y_{T_2}\} \in \Lambda_2(\xi, T_1, T_2)$, the function

$$F_{T_1,T_2}^{(\xi)}(\{x_{T_1}, \dots, x_{T_2}\}, \{y_{T_1}, \dots, y_{T_2}\}), \ \{x_{T_1}, \dots, x_{T_2}\} \in \Lambda_1(\xi, T_1, T_2)$$

is convex and for every $\{x_{T_1}, \dots, x_{T_2}\} \in \Lambda_1(\xi, T_1, T_2)$, the function

$$F^{(\xi)}_{T_1,T_2}(\{x_{T_1}, \dots, x_{T_2}\}, \{(y_{T_1}, \dots, y_{T_2}\}), \ \{y_{T_1}, \dots, y_{T_2}\} \in \Lambda_2(\xi, T_1, T_2)$$

is concave.

We consider the game for which $\Lambda_1(\xi, T_1, T_2)$ is the set of strategies of the first player, $\Lambda_2(\xi, T_1, T_2)$ is the set of strategies of the second player, and $F^{(\xi)}_{T_1,T_2}$ is the objective function of the first player.

By Proposition 7.2, there exist

$$\{\widehat{x}_t\}_{t=T_1}^{T_2} \in \Lambda_1(\xi, T_1, T_2), \{\widehat{y}_t\}_{t=T_1}^{T_2} \in \Lambda_2(\xi, T_1, T_2) \tag{8.33}$$

such that

$$\sup\{\sum_{t=T_1}^{T_2-1} f_t(\widehat{x}_t, \widehat{x}_{t+1}, y_t, y_{t+1}) : \ \{y_t\}_{t=T_1}^{T_2} \in \Lambda_2(\xi, T_1, T_2)\}$$

$$= \sum_{t=T_1}^{T_2-1} f_t(\widehat{x}_t, \widehat{x}_{t+1}, \widehat{y}_t, \widehat{y}_{t+1})$$

$$= \inf\{\sum_{t=T_1}^{T_2-1} f_t(x_t, x_{t+1}, \widehat{y}_t, \widehat{y}_{t+1}) : \ \{x_t\}_{t=T_1}^{T_2} \in \Lambda_1(\xi, T_1, T_2)\}. \tag{8.34}$$

Let $M \geq 0$. A pair of programs

$$\{\tilde{x}_t\}_{t=T_1}^{T_2} \in \Lambda_1(\xi, T_1, T_2), \ \{\tilde{y}_t\}_{t=T_1}^{T_2} \in \Lambda_2(\xi, T_1, T_2)$$

is called $(\{f_t\}_{t=T_1}^{T_2-1}, \xi, T_1, T_2, M)$-good if

$$\sum_{t=T_1}^{T_2-1} f_t(\tilde{x}_t, \tilde{x}_{t+1}, y_t, y_{t+1}) - M \leq \sum_{t=T_1}^{T_2-1} f_t(\tilde{x}_t, \tilde{x}_{t+1}, \tilde{y}_t, \tilde{y}_{t+1})$$

$$\leq \sum_{t=T_1}^{T_2-1} f_t(x_t, x_{t+1}, \tilde{y}_t, \tilde{y}_{t+1}) + M \tag{8.35}$$

for every $\{x_t\}_{t=T_1}^{T_2} \in \Lambda_1(\xi, T_1, T_2)$ and every $\{y_t\}_{t=T_1}^{T_2} \in \Lambda_2(\xi, T_1, T_2)$.

Let $T_2 > T_1 \geq 0$ be integers and $F = \{f_t\}_{t=T_1}^{T_2-1} \subset \mathcal{M}$. Define

$$\Lambda_1(T_1, T_2) = \{(\Omega_1) - \text{programs } \{x_t\}_{t=T_1}^{T_2}\},$$

$$\Lambda_2(T_1, T_2) = \{(\Omega_2) - \text{programs } \{y_t\}_{t=T_1}^{T_2}\} \tag{8.36}$$

and define

$$F_{T_1,T_2}(\{x_{T_1},\dots,x_{T_2}\},\{y_{T_1},\dots,y_{T_2}\}) = \sum_{t=T_1}^{T_2-1} f_t(x_t,x_{t+1},y_t,y_{t+1}), \quad (8.37)$$

$$\{x_t\}_{t=T_1}^{T_2} \in \Lambda_1(T_1,T_2),\ \{y_t\}_{t=T_1}^{T_2} \in \Lambda_2(T_1,T_2).$$

Clearly, $\Lambda_1(T_1,T_2)$, $\Lambda_2(T_1,T_2)$ are closed convex sets in $R^{(T_2-T_1+1)m_1}$, $R^{(T_2-T_1+1)m_2}$ respectively, for every $\{y_{T_1},\dots,y_{T_2}\} \in \Lambda_2(T_1,T_2)$, the function $F_{T_1,T_2}(\{x_{T_1},\dots,x_{T_2}\},\{y_{T_1},\dots,y_{T_2}\})$, $\{x_{T_1},\dots,x_{T_2}\} \in \Lambda_1(T_1,T_2)$ is convex and for every $\{x_{T_1},\dots,x_{T_2}\} \in \Lambda_1(T_1,T_2)$, the function

$$F_{T_1,T_2}(\{x_{T_1},\dots,x_{T_2}\},\{y_{T_1},\dots,y_{T_2}\}),\ \{y_{T_1},\dots,y_{T_2}\} \in \Lambda_2(T_1,T_2)$$

is concave.

We consider the game for which $\Lambda_1(T_1,T_2)$ is the set of strategies of the first player, $\Lambda_2(T_1,T_2)$ is the set of strategies of the second player, and F_{T_1,T_2} is the objective function of the first player.

By Proposition 7.2, there exist

$$\{\widehat{x}_t\}_{t=T_1}^{T_2} \in \Lambda_1(T_1,T_2), \{\widehat{y}_t\}_{t=T_1}^{T_2} \in \Lambda_2(T_1,T_2) \quad (8.38)$$

such that

$$\sup\{\sum_{t=T_1}^{T_2-1} f_t(\widehat{x}_t,\widehat{x}_{t+1},y_t,y_{t+1}) :\ \{y_t\}_{t=T_1}^{T_2} \in \Lambda_2(T_1,T_2)\}$$

$$= \sum_{t=T_1}^{T_2-1} f_t(\widehat{x}_t,\widehat{x}_{t+1},\widehat{y}_t,\widehat{y}_{t+1})$$

$$= \inf\{\sum_{t=T_1}^{T_2-1} f_t(x_t,x_{t+1},\widehat{y}_t,\widehat{y}_{t+1}) :\ \{x_t\}_{t=T_1}^{T_2} \in \Lambda_1(T_1,T_2)\}. \quad (8.39)$$

Let $M \ge 0$. A pair of sequences

$$\{\tilde{x}_t\}_{t=T_1}^{T_2} \in \Lambda_1(T_1,T_2),\ \{\tilde{y}_t\}_{t=T_1}^{T_2} \in \Lambda_2(T_1,T_2)$$

is called $(\{f_t\}_{t=T_1}^{T_2-1},T_1,T_2,M)$-good if

$$\sum_{t=T_1}^{T_2-1} f_t(\tilde{x}_t,\tilde{x}_{t+1},y_t,y_{t+1}) - M \le \sum_{t=T_1}^{T_2-1} f_t(\tilde{x}_t,\tilde{x}_{t+1},\tilde{y}_t,\tilde{y}_{t+1})$$

$$\le \sum_{t=T_1}^{T_2-1} f_t(x_t,x_{t+1},\tilde{y}_t,\tilde{y}_{t+1}) + M \quad (8.40)$$

for every $\{x_t\}_{t=T_1}^{T_2} \in \Lambda_1(T_1,T_2)$ and every $\{y_t\}_{t=T_1}^{T_2} \in \Lambda_2(T_1,T_2)$.

Let $T_2 > T_1 \geq 0$ be integers, $\xi = (\xi_1, \xi_2) \in X_1 \times X_2$ and

$$F = \{f_t\}_{t=T_1}^{T_2-1} \subset \mathcal{M}.$$

Define

$$\Lambda_1^-(\xi, T_1, T_2) = \Lambda_1^-(\xi_1, \xi_2, T_1, T_2) = \{(\Omega_1) - \text{programs } \{x_t\}_{t=T_1}^{T_2} : x_{T_2} = \xi_1\}, \tag{8.41}$$

$$\Lambda_2^-(\xi, T_1, T_2) = \Lambda_2^-(\xi_1, \xi_2, T_1, T_2) = \{(\Omega_2) - \text{programs } \{y_t\}_{t=T_1}^{T_2} : y_{T_2} = \xi_2\}. \tag{8.42}$$

Assume that

$$\Lambda_1^-(\xi, T_1, T_2) \neq \emptyset,\ \Lambda_2^-(\xi, T_1, T_2) \neq \emptyset$$

and define

$$F_{T_1,T_2}^{(\xi)}(\{x_{T_1}, \dots, x_{T_2}\}, \{y_{T_1}, \dots, y_{T_2}\}) = \sum_{t=T_1}^{T_2-1} f_t(x_t, x_{t+1}, y_t, y_{t+1}), \tag{8.43}$$

$$\{x_t\}_{t=T_1}^{T_2} \in \Lambda_1^-(\xi, T_1, T_2),\ \{y_t\}_{t=T_1}^{T_2} \in \Lambda_2^-(\xi, T_1, T_2).$$

Clearly, $\Lambda_1^-(\xi, T_1, T_2)$, $\Lambda_2^-(\xi, T_1, T_2)$ are closed convex sets in $R^{(T_2-T_1+1)m_1}$, $R^{(T_2-T_1+1)m_2}$ respectively, for every $\{y_{T_1}, \dots, y_{T_2}\} \in \Lambda_2^-(\xi, T_1, T_2)$, the function $F_{T_1,T_2}^{(\xi)}(\{x_{T_1}, \dots, x_{T_2}\}, \{y_{T_1}, \dots, y_{T_2}\})$, $\{x_{T_1}, \dots, x_{T_2}\} \in \Lambda_1^-(\xi, T_1, T_2)$ is convex and for every $\{x_{T_1}, \dots, x_{T_2}\} \in \Lambda_1^-(\xi, T_1, T_2)$, the function

$$F_{T_1,T_2}^{(\xi)}(\{x_{T_1}, \dots, x_{T_2}\}, \{y_{T_1}, \dots, y_{T_2}\}),\ \{y_{T_1}, \dots, y_{T_2}\} \in \Lambda_2^-(\xi, T_1, T_2)$$

is concave.

We consider the game for which $\Lambda_1^-(\xi, T_1, T_2)$ is the set of strategies of the first player, $\Lambda_2^-(\xi, T_1, T_2)$ is the set of strategies of the second player, and $F_{T_1,T_2}^{(\xi)}$ is the objective function of the first player.

By Proposition 7.2, there exist

$$\{\widehat{x}_t\}_{t=T_1}^{T_2} \in \Lambda_1^-(\xi, T_1, T_2), \{\widehat{y}_t\}_{t=T_1}^{T_2} \in \Lambda_2^-(\xi, T_1, T_2) \tag{8.44}$$

such that

$$\sup\{\sum_{t=T_1}^{T_2-1} f_t(\widehat{x}_t, \widehat{x}_{t+1}, y_t, y_{t+1}) : \{y_t\}_{t=T_1}^{T_2} \in \Lambda_2^-(\xi, T_1, T_2)\}$$

$$= \sum_{t=T_1}^{T_2-1} f_t(\widehat{x}_t, \widehat{x}_{t+1}, \widehat{y}_t, \widehat{y}_{t+1})$$

$$= \inf\{\sum_{t=T_1}^{T_2-1} f_t(x_t, x_{t+1}, \widehat{y}_t, \widehat{y}_{t+1}) : \{x_t\}_{t=T_1}^{T_2} \in \Lambda_1^-(\xi, T_1, T_2)\}. \tag{8.45}$$

Let $M \geq 0$. A pair of sequences

$$\{\tilde{x}_t\}_{t=T_1}^{T_2} \in \Lambda_1^-(\xi, T_1, T_2), \ \{\tilde{y}_t\}_{t=T_1}^{T_2} \in \Lambda_2^-(\xi, T_1, T_2)$$

is called $(\{f_t\}_{t=T_1}^{T_2-1}, \xi^-, T_1, T_2, M)$-good if

$$\sum_{t=T_1}^{T_2-1} f_t(\tilde{x}_t, \tilde{x}_{t+1}, y_t, y_{t+1}) - M \leq \sum_{t=T_1}^{T_2-1} f_t(\tilde{x}_t, \tilde{x}_{t+1}, \tilde{y}_t, \tilde{y}_{t+1})$$

$$\leq \sum_{t=T_1}^{T_2-1} f_t(x_t, x_{t+1}, \tilde{y}_t, \tilde{y}_{t+1}) + M \tag{8.46}$$

for every $\{x_t\}_{t=T_1}^{T_2} \in \Lambda_1^-(\xi, T_1, T_2)$ and every $\{y_t\}_{t=T_1}^{T_2} \in \Lambda_2^-(\xi, T_1, T_2)$.

Note that if $f \in \mathcal{M}$ and $f_t = f$, $t = T_1, \dots, T_2 - 1$, then in the definitions of good pairs of sequences given above, the notation $\{f_t\}_{t=T_1}^{T_2-1}$ is replaced by f.

Definition 8.2. Let $l \geq 0$ be an integer. Denote by $Y_1(l)$ the set of all $x \in X_1$ such that there exists an (Ω_1)-program $\{x_t\}_{t=0}^{l}$ such that

$$x_0 = x_f, \ x_l = x,$$

by $\bar{Y}_1(l)$ the set of all $x \in X_1$ such that there exists an (Ω_1)-program $\{x_t\}_{t=0}^{l}$ such that

$$x_0 = x, \ x_l = x_f,$$

by $Y_2(l)$ the set of all $y \in X_2$ such that there exists an (Ω_2)-program $\{y_t\}_{t=0}^{l}$ such that

$$y_0 = y_f, \ y_l = y$$

and by $\bar{Y}_2(l)$ the set of all $y \in X_2$ such that there exists an (Ω_2)-program $\{y_t\}_{t=0}^{l}$ such that $y_0 = y$, $y_l = y_f$.

8.2 Turnpike Properties

Let $f \in \mathcal{M}$. Then there exist

$$x_f \in \Delta_1, \ y_f \in \Delta_2 \tag{8.47}$$

such that

$$\sup\{\widehat{f}(x_f, y) : \ y \in \Delta_2\} = \widehat{f}(x_f, y_f) = \inf\{\widehat{f}(x, y_f) : \ x \in \Delta_1\} \tag{8.48}$$

(see (8.11)–(8.14)).

Proposition 8.3. *Suppose that $n \geq 2$ is an integer and that*

$$\bar{x}_i = x_f,\ \bar{y}_i = y_f,\ i = 0, \ldots, n.$$

Then the pair of sequences $\{\bar{x}_i\}_{i=0}^n$, $\{\bar{y}_i\}_{i=0}^n$ is (f)-optimal.

Proof. Assume that $\{x_i\}_{i=0}^n$ is an (Ω_1)-program and $\{y_i\}_{i=0}^n$ is an (Ω_2)-program satisfying

$$x_0, x_n = x_f,\ y_0, y_n = y_f. \tag{8.49}$$

By the convexity assumptions, (8.11), (8.48), and (8.49),

$$\sum_{i=0}^{n-1} f(x_i, x_{i+1}, \bar{y}_i, \bar{y}_{i+1}) = \sum_{i=0}^{n-1} f(x_i, x_{i+1}, y_f, y_f)$$

$$\geq nf(n^{-1}\sum_{i=0}^{n-1} x_i, n^{-1}\sum_{i=0}^{n-1} x_{i+1}, y_f, y_f) \geq nf(x_f, x_f, y_f, y_f),$$

$$\sum_{i=0}^{n-1} f(\bar{x}_i, \bar{x}_{i+1}, y_i, y_{i+1}) = \sum_{i=0}^{n-1} f(x_f, x_f, y_i, y_{i+1})$$

$$\leq nf(x_f, x_f, n^{-1}\sum_{i=0}^{n-1} y_i, n^{-1}\sum_{i=0}^{n-1} y_{i+1})$$

$$= nf(x_f, x_f, n^{-1}\sum_{i=0}^{n-1} y_i, n^{-1}\sum_{i=0}^{n-1} y_i)$$

$$\leq nf(x_f, x_f, y_f, y_f).$$

This completes the proof of Proposition 8.3. □

In this chapter we suppose that there exists $r_f \in (0, 1)$ such that for each $x, y \in B_1(x_f, r_f)$,

$$(x, y) \in \Omega_1, \tag{8.50}$$

for each $x, y \in B_2(y_f, r_f)$,

$$(x, y) \in \Omega_2 \tag{8.51}$$

and that the following two assumptions hold.

(A1) The function $f(\cdot, \cdot, y_f, y_f) : \Omega_1 \to R^1$ is strictly convex.

(A2) The function $f(x_f, x_f, \cdot, \cdot) : \Omega_2 \to R^1$ is strictly concave.

In view of (A1), (A2) and Example 2.11 of [84] there exist $l_1 \in R^{m_1}$ such that

$$f(x_1, x_2, y_f, y_f) \geq f(x_f, x_f, y_f, y_f) + \langle l_1, x_1 - x_2 \rangle \tag{8.52}$$

for all $(x_1, x_2) \in \Omega_1$ and $l_2 \in R^{m_2}$ such that

$$f(x_f, x_f, y_1, y_2) \leq f(x_f, x_f, y_f, y_f) + \langle l_2, y_1 - y_2 \rangle \tag{8.53}$$

for all $(y_1, y_2) \in \Omega_2$. Define

$$L_1(x_1, x_2) = -f(x_f, x_f, y_f, y_f) + \langle l_1, x_2 - x_1 \rangle + f(x_1, x_2, y_f, y_f), \ (x_1, x_2) \in \Omega_1. \tag{8.54}$$

Clearly, $L_1 : \Omega_1 \to R^1$ is a continuous strictly convex function such that for all $(x_1, x_2) \in \Omega_1$,

$$L_1(x_1, x_2) \geq 0,$$
$$L_1(x_1, x_2) = 0 \text{ if and only if } (x_1, x_2) = (x_f, x_f). \tag{8.55}$$

Define

$$L_2(y_1, y_2) = f(x_f, x_f, y_f, y_f) + \langle l_2, y_1 - y_2 \rangle - f(x_f, x_f, y_1, y_2), \ (y_1, y_2) \in \Omega_2. \tag{8.56}$$

Clearly, $L_2 : \Omega_2 \to R^1$ is a continuous strictly convex function such that for all $(y_1, y_2) \in \Omega_2$,

$$L_2(y_1, y_2) \geq 0,$$
$$L_2(y_1, y_2) = 0 \text{ if and only if } (y_1, y_2) = (y_f, y_f). \tag{8.57}$$

By (8.55) and (8.57), the following property holds:

(P1) for each $\epsilon > 0$ there exists $\delta > 0$ such that for every $(x_1, x_2) \in \Omega_1$ satisfying $L_1(x_1, x_2) \leq \delta$, we have

$$|x_i - x_f| \leq \epsilon, \ i = 1, 2$$

and for every $(y_1, y_2) \in \Omega_2$ satisfying $L_2(y_1, y_2) \leq \delta$, we have

$$|y_i - y_f| \leq \epsilon, \ i = 1, 2.$$

Proposition 8.4. *Let $\epsilon > 0$. Then there exists $\delta \in (0, r_f)$ such that for each integer $n \geq 2$ and each*

$$\xi_1, \xi_2 \in B_{X_1}(x_f, \delta), \ \eta_1, \eta_2 \in B_{X_2}(y_f, \delta), \tag{8.58}$$

a sequence $\{\bar{x}_i\}_{i=0}^n$ defined by

$$\bar{x}_0 = \xi_1, \ \bar{x}_i = x_f, \ i = 1, \dots, n-1, \ \bar{x}_n = \xi_2 \tag{8.59}$$

is an (Ω_1)-program and a sequence $\{\bar{y}_i\}_{i=0}^n$ defined by

$$\bar{y}_0 = \eta_1, \ \bar{y}_i = y_f, \ i = 1, \dots, n-1, \ \bar{y}_n = \eta_2 \tag{8.60}$$

is an (Ω_2)-program and the pair of sequences $\{\bar{x}_i\}_{i=0}^n$, $\{\bar{y}_i\}_{i=0}^n$ is (f, ϵ)-good.

Proof. There exists $\delta \in (0, r_f)$ such that for each $(u_1, v_1), (u_2, v_2) \in \Omega_1 \times \Omega_2$ satisfying

$$|u_1 - u_2|, \ |v_1 - v_2| \leq \delta$$

we have

$$|f(u_1, v_1) - f(u_2, v_2)| \leq \epsilon/8. \tag{8.61}$$

Let $n \geq 2$ be an integer, (8.58) hold and let sequences $\{\bar{x}_i\}_{i=0}^n$, $\{\bar{y}_i\}_{i=0}^n$ be defined by (8.59) and (8.60). Clearly, $\{\bar{x}_i\}_{i=0}^n$ is an (Ω_1)-program and $\{\bar{y}_i\}_{i=0}^n$ is an (Ω_2)-program. We show that the pair of sequences $\{\bar{x}_i\}_{i=0}^n$, $\{\bar{y}_i\}_{i=0}^n$ is (f, ϵ)-good.

Assume that $\{x_i\}_{i=0}^n$ is an (Ω_1)-program satisfying

$$x_0 = \xi_1, \; x_n = \xi_2. \tag{8.62}$$

Define

$$u_0 = x_f, \; u_i = x_{i-1}, \; i = 1, \dots, n+1, \; u_{n+2} = x_f. \tag{8.63}$$

By (8.58), (8.62), and (8.63), $\{u_i\}_{i=0}^{n+2}$ is an (Ω_1)-program. Proposition 8.3 implies that

$$\sum_{i=0}^{n+1} f(u_i, u_{i+1}, y_f, y_f) \geq (n+2) f(x_f, x_f, y_f, y_f). \tag{8.64}$$

It follows from (8.57), (8.58), the choice of δ and (8.61)–(8.64) that

$$(n+2) f(x_f, x_f, y_f, y_f)$$

$$\leq f(x_f, \xi_1, y_f, y_f) + \sum_{i=0}^{n-1} f(x_i, x_{i+1}, y_f, y_f) + f(\xi_2, x_f, y_f, y_f)$$

$$\leq \epsilon/4 + 2f(x_f, x_f, y_f, y_f) + \sum_{i=0}^{n-1} f(x_i, x_{i+1}, y_f, y_f),$$

$$n f(x_f, x_f, y_f, y_f) - \epsilon/4 \leq \sum_{i=0}^{n-1} f(x_i, x_{i+1}, y_f, y_f). \tag{8.65}$$

In view of the choice of δ, (8.57)–(8.61) and (8.65),

$$\sum_{i=0}^{n-1} f(\bar{x}_i, \bar{x}_{i+1}, \bar{y}_i, \bar{y}_{i+1}) \leq n f(x_f, x_f, y_f, y_f) + \epsilon/4$$

$$\leq \epsilon/2 + \sum_{i=0}^{n-1} f(x_i, x_{i+1}, y_f, y_f) \leq \sum_{i=0}^{n-1} f(x_i, x_{i+1}, \bar{y}_i, \bar{y}_{i+1}) + \epsilon. \tag{8.66}$$

Assume that $\{y_i\}_{i=0}^n$ is an (Ω_2)-program satisfying

$$y_0 = \eta_1, \; y_n = \eta_2. \tag{8.67}$$

Define

$$v_0 = y_f, \; v_i = y_{i-1}, \; i = 1, \dots, n+1, \; v_{n+2} = y_f. \tag{8.68}$$

By (8.58), (8.67), and (8.68), $\{v_i\}_{i=0}^{n+2}$ is an (Ω_2)-program. Proposition 8.3 and (8.68) imply that

$$\sum_{i=0}^{n+1} f(x_f, x_f, v_i, v_{i+1}) \le (n+2) f(x_f, x_f, y_f, y_f). \tag{8.69}$$

It follows from (8.58), the choice of δ and (8.61), (8.67)–(8.69) that

$$(n+2) f(x_f, x_f, y_f, y_f)$$

$$\ge f(x_f, x_f, y_f, \eta_1) + \sum_{i=0}^{n-1} f(x_f, x_f, y_i, y_{i+1}) + f(x_f, x_f, \eta_2, y_f)$$

$$\ge -\epsilon/4 + 2f(x_f, x_f, y_f, y_f) + \sum_{i=0}^{n-1} f(x_f, x_f, y_i, y_{i+1}),$$

$$n f(x_f, x_f, y_f, y_f) \ge -\epsilon/4 + \sum_{i=0}^{n-1} f(x_f, x_f, y_i, y_{i+1}). \tag{8.70}$$

In view of the choice of δ, (8.58)–(8.61) and (8.70),

$$\sum_{i=0}^{n-1} f(\bar{x}_i, \bar{x}_{i+1}, \bar{y}_i, \bar{y}_{i+1}) \ge n f(x_f, x_f, y_f, y_f) - \epsilon/4$$

$$\ge -\epsilon/2 + \sum_{i=0}^{n-1} f(x_f, x_f, y_i, y_{i+1}) \ge \sum_{i=0}^{n-1} f(\bar{x}_i, \bar{x}_{i+1}, y_i, y_{i+1}) - \epsilon.$$

Together with (8.66) this completes the proof of Proposition 8.4.

In Sect. 8.4 we prove the following turnpike result.

Theorem 8.5. *Let $M > 0$, $\epsilon \in (0, r_f)$ and l be a natural number. Then there exist a natural number $L > 2l$ and $\delta > 0$ such that for each integer $T \ge 2L$, each $f_t \in \mathcal{M}$, $t = 0, \dots, T-1$ satisfying*

$$\|f_t - f\| \le \delta,\ t = 0, \dots, T-1, \tag{8.71}$$

each (Ω_1)-program $\{x_t\}_{t=0}^T$ and each (Ω_2)-program $\{y_t\}_{t=0}^T$ such that at least one of the following conditions hold:

(a) $x_0 \in \bar{Y}_1(l)$, $y_0 \in \bar{Y}_2(l)$, $x_T \in Y_1(l)$, $y_T \in Y_2(l)$, the pair $\{x_t\}_{t=0}^T$, $\{y_t\}_{t=0}^T$ is $(\{f_t\}_{t=0}^{T-1}, x_0, x_T, y_0, y_T, 0, T, M)$-good;

(b) $x_0 \in \bar{Y}_1(l)$, $y_0 \in \bar{Y}_2(l)$, the pair $\{x_t\}_{t=0}^T$, $\{y_t\}_{t=0}^T$ is

$$(\{f_t\}_{t=0}^{T-1}, x_0, y_0, 0, T, M) - good;$$

(c) the pair $\{x_t\}_{t=0}^T$, $\{y_t\}_{t=0}^T$ is $(\{f_t\}_{t=0}^{T-1}, 0, T, M)$-good;

(d) $x_T \in Y_1(l)$, $y_T \in Y_2(l)$, *the pair* $\{x_t\}_{t=0}^T$, $\{y_t\}_{t=0}^T$ *is*

$$(\{f_t\}_{t=0}^{T-1}, (x_T, y_T)^-, 0, T, M) - good$$

and such that for each integer $S \in [0, T - L]$, *the pair* $\{x_t\}_{t=S}^{S+L}$, $\{y_t\}_{t=S}^{S+L}$ *is* $(\{f_t\}_{t=S}^{S+L-1}, x_S, x_{S+L}, y_S, y_{S+L}, S, S + L, \delta)$*-good, there exist integers*

$$\tau_1 \in [0, L] \text{ and } \tau_2 \in [T - L, T]$$

such that for all $t = \tau_1, \dots, \tau_2$,

$$|x_t - x_f| \le \epsilon, \ |y_t - y_f| \le \epsilon.$$

Moreover, if $|x_0 - x_f| \le \delta$, $|y_0 - y_f| \le \delta$, *then* $\tau_1 = 0$ *and if*

$$|x_T - x_f| \le \delta, \ |y_T - y_f| \le \delta,$$

then $\tau_2 = T$.

We denote by Card(A) the cardinality of a set A.

In Sect. 8.5 we prove the following turnpike result.

Theorem 8.6. *Let* $M > 0$, $\epsilon \in (0, r_f)$, *and* l *be a natural number. Then there exist a natural number* $L > 2l$ *and* $\delta > 0$ *such that for each integer* $T > L$, *each* $f_t \in \mathcal{M}$, $t = 0, \dots, T - 1$ *satisfying*

$$\|f_t - f\| \le \delta, \ t = 0, \dots, T - 1,$$

each (Ω_1)*-program* $\{x_t\}_{t=0}^T$ *and each* (Ω_2)*-program* $\{y_t\}_{t=0}^T$ *such that at least one of the following conditions hold:*

(a) $x_0 \in \bar{Y}_1(l)$, $y_0 \in \bar{Y}_2(l)$, $x_T \in Y_1(l)$, $y_T \in Y_2(l)$, *the pair* $\{x_t\}_{t=0}^T$, $\{y_t\}_{t=0}^T$ *is* $(\{f_t\}_{t=0}^{T-1}, x_0, x_T, y_0, y_T, 0, T, M)$*-good;*

(b) $x_0 \in \bar{Y}_1(l)$, $y_0 \in \bar{Y}_2(l)$, *the pair* $\{x_t\}_{t=0}^T$, $\{y_t\}_{t=0}^T$ *is*

$$(\{f_t\}_{t=0}^{T-1}, x_0, y_0, 0, T, M)\text{-good;}$$

(c) the pair $\{x_t\}_{t=0}^T$, $\{y_t\}_{t=0}^T$ *is* $(\{f_t\}_{t=0}^{T-1}, 0, T, M)$*-good;*

(d) $x_T \in Y_1(l)$, $y_T \in Y_2(l)$, *the pair* $\{x_t\}_{t=0}^T$, $\{y_t\}_{t=0}^T$ *is*

$$(\{f_t\}_{t=0}^{T-1}, (x_T, y_T)^-, 0, T, M) - good$$

the inequality

$$Card(\{t \in \{0, \dots, T\} : \max\{|x_t - x_f|, |y_t - y_f|\} \ge \epsilon\}) \le L$$

holds.

8.3 Auxiliary Results for Theorems 8.5 and 8.6

Lemma 8.7. *Let $\epsilon \in (0,1)$. Then there exists $\delta \in (0,\epsilon)$ such that for each integer $n \geq 2$ and each (f,δ)-good pair of an (Ω_1)-program $\{x_t\}_{t=0}^T$ and an (Ω_2)-program $\{y_t\}_{t=0}^T$ satisfying*

$$|x_0 - x_f| \leq \delta,\ |x_n - x_f| \leq \delta,\ |y_0 - y_f| \leq \delta,\ |y_n - y_f| \leq \delta,$$

the following inequalities hold:

$$|x_t - x_f| \leq \epsilon,\ |y_t - y_f| \leq \epsilon,\ t = 0,\dots,n.$$

Proof. In view of assumptions (A1), (A2) and the continuity of the function f, there exists $\gamma \in (0,\epsilon)$ such that the following property holds:

(P2) for each $(x,x') \in \Omega_1$ satisfying $|x - x_f| \geq \epsilon$,

$$2^{-1}f(x_f,x_f,y_f,y_f) + 2^{-1}f(x,x',y_f,y_f)$$
$$-f(2^{-1}(x+x_f),2^{-1}(x'+x_f),y_f,y_f) \geq \gamma$$

and for each $(y,y') \in \Omega_2$ satisfying $|y - y_f| \geq \epsilon$,

$$f(x_f,x_f,2^{-1}(y+y_f),2^{-1}(y'+y_f))$$
$$-2^{-1}f(x_f,x_f,y_f,y_f) - 2^{-1}f(x_f,x_f,y,y') \geq \gamma.$$

By property (P2) and the uniform continuity of the function f, there exists a positive number

$$\delta_0 < \min\{\epsilon,\gamma\}/4 \tag{8.72}$$

such that the following properties hold:

(P3) for each $(x,x') \in \Omega_1$ satisfying $|x - x_f| \geq \epsilon$, each $(u,u') \in \Omega_1$ satisfying

$$|u - x_f| \leq \delta_0,\ |u' - x_f| \leq \delta_0$$

and each $(y,y') \in \Omega_2$ satisfying

$$|y - y_f| \leq \delta_0,\ |y' - y_f| \leq \delta_0$$

we have

$$2^{-1}f(u,u',y,y') + 2^{-1}f(x,x',y,y')$$
$$-f(2^{-1}(x+u),2^{-1}(x'+u'),y,y') \geq \gamma/2;$$

(P4) for each $(y,y') \in \Omega_2$ satisfying $|y - y_f| \geq \epsilon$, each $(v,v') \in \Omega_2$ satisfying

$$|v - y_f| \leq \delta_0,\ |v' - y_f| \leq \delta_0$$

and each $(x,x') \in \Omega_1$ satisfying

$$|x - x_f| \leq \delta_0,\ |x' - x_f| \leq \delta_0$$

we have

$$f(x, x', 2^{-1}(y+v), 2^{-1}(y'+v'))$$
$$-2^{-1}f(x, x', v, v') - 2^{-1}f(x, x', y, y') \geq \gamma/2.$$

By Proposition 8.4, there exists a positive number

$$\delta < \min\{r_f, \gamma/16, \delta_0\} \tag{8.73}$$

such that the following property holds:

(P5) for each integer $n \geq 2$ if

$$x_0, x_n \in B_1(x_f, \delta),\ x_i = x_f,\ i = 1, \dots, n-1,$$

$$y_0, y_n \in B_2(y_f, \delta),\ y_i = y_f,\ i = 1, \dots, n-1,$$

then the sequence $\{x_i\}_{i=0}^n$ is an (Ω_1)-program, the sequence $\{y_i\}_{i=0}^n$ an (Ω_2)-program, and the pair of sequences $\{x_i\}_{i=0}^n$, $\{y_i\}_{i=0}^n$ is $(f, \gamma/16)$-good.

Assume that an integer $n \geq 2$, $\{x_t\}_{t=0}^n$ is an (Ω_1)-program, $\{y_t\}_{t=0}^n$ is an (Ω_2)-program, the pair of sequences $\{x_t\}_{t=0}^n$, $\{y_t\}_{t=0}^n$ is (f, δ)-good and that

$$|x_0 - x_f| \leq \delta,\ |x_n - x_f| \leq \delta,\ |y_0 - y_f| \leq \delta,\ |y_n - y_f| \leq \delta. \tag{8.74}$$

Define

$$\bar{x}_0 = x_0,\ \bar{x}_i = x_f,\ i = 1, \dots, n-1,\ \bar{x}_n = x_n,$$

$$\bar{y}_0 = y_0,\ \bar{y}_i = y_f,\ i = 1, \dots, n-1,\ \bar{y}_n = y_n. \tag{8.75}$$

Property (P5), (8.74) and (8.75) imply that $\{\bar{x}_i\}_{i=0}^n$ is an (Ω_1)-program, $\{\bar{y}_i\}_{i=0}^n$ is an (Ω_2)-program and that the pair of sequences $\{\bar{x}_i\}_{i=0}^n$, $\{\bar{y}_i\}_{i=0}^n$ is $(f, \gamma/16)$-good.

Therefore in view of (8.73) and (8.75),

$$\sum_{i=0}^{n-1} f(x_i, x_{i+1}, y_i, y_{i+1}) \leq \sum_{i=0}^{n-1} f(\bar{x}_i, \bar{x}_{i+1}, y_i, y_{i+1}) + 16^{-1}\gamma$$

$$\leq \sum_{i=0}^{n-1} f(\bar{x}_i, \bar{x}_{i+1}, \bar{y}_i, \bar{y}_{i+1}) + \gamma/8 \leq \sum_{i=0}^{n-1} f(x_i, x_{i+1}, \bar{y}_i, \bar{y}_{i+1}) + 3\gamma/16$$

$$\leq \sum_{i=0}^{n-1} f(x_i, x_{i+1}, y_i, y_{i+1}) + \gamma/4. \tag{8.76}$$

We show that for all $i = 0, \dots, n$,

$$|x_i - x_f| \leq \epsilon, |y_i - y_f| \leq \epsilon.$$

Assume the contrary. Then at least one of the following cases holds:

(a) $\max\{|x_i - x_f| : i = 1, \dots, n-1\} > \epsilon$;

(b) $\max\{|y_i - y_f| : i = 1, \dots, n-1\} > \epsilon$;

Assume that case (a) holds. Then there exists $j \in \{1, \dots, n-1\}$ such that

$$|x_j - x_f| > \epsilon. \tag{8.77}$$

We estimate

$$\sum_{i=0}^{n-1} f(2^{-1}\bar{x}_i + 2^{-1}x_i, 2^{-1}\bar{x}_{i+1} + 2^{-1}x_{i+1}, \bar{y}_i, \bar{y}_{i+1}) - \sum_{i=0}^{n-1} f(\bar{x}_i, \bar{x}_{i+1}, \bar{y}_i, \bar{y}_{i+1}).$$

It is clear that $\{2^{-1}\bar{x}_i + 2^{-1}x_i\}_{i=0}^n$ is an (Ω_1)-program. Since the pair of sequences $\{\bar{x}_i\}_{i=0}^n$, $\{\bar{y}_i\}_{i=0}^n$ is $(f, \gamma/16)$-good it follows from (8.75) that

$$\sum_{i=0}^{n-1} f(2^{-1}\bar{x}_i + 2^{-1}x_i, 2^{-1}\bar{x}_{i+1} + 2^{-1}x_{i+1}, \bar{y}_i, \bar{y}_{i+1})$$

$$- \sum_{i=0}^{n-1} f(\bar{x}_i, \bar{x}_{i+1}, \bar{y}_i, \bar{y}_{i+1}) \geq -\gamma/16. \tag{8.78}$$

By definition, for every $i = 0, \dots, n-1$,

$$f(2^{-1}\bar{x}_i + 2^{-1}x_i, 2^{-1}\bar{x}_{i+1} + 2^{-1}x_{i+1}, \bar{y}_i, \bar{y}_{i+1})$$

$$\leq 2^{-1}f(\bar{x}_i, \bar{x}_{i+1}, \bar{y}_i, \bar{y}_{i+1}) + 2^{-1}f(x_i, x_{i+1}, \bar{y}_i, \bar{y}_{i+1}). \tag{8.79}$$

Property (P3), (8.73)–(8.75), and (8.77) imply that

$$f(2^{-1}\bar{x}_j + 2^{-1}x_j, 2^{-1}\bar{x}_{j+1} + 2^{-1}x_{j+1}, \bar{y}_j, \bar{y}_{j+1})$$

$$\leq 2^{-1}f(\bar{x}_j, \bar{x}_{j+1}, \bar{y}_j, \bar{y}_{j+1}) + 2^{-1}f(x_j, x_{j+1}, \bar{y}_j, \bar{y}_{j+1}) - \gamma/2. \tag{8.80}$$

In view of (8.75), (8.76), (8.79), and (8.80),

$$\sum_{i=0}^{n-1} f(2^{-1}\bar{x}_i + 2^{-1}x_i, 2^{-1}\bar{x}_{i+1} + 2^{-1}x_{i+1}, \bar{y}_i, \bar{y}_{i+1}) - \sum_{i=0}^{n-1} f(\bar{x}_i, \bar{x}_{i+1}, \bar{y}_i, \bar{y}_{i+1})$$

$$\leq -\gamma/2 + 2^{-1}\sum_{i=0}^{n-1} f(\bar{x}_i, \bar{x}_{i+1}, \bar{y}_i, \bar{y}_{i+1}) + 2^{-1}\sum_{i=0}^{n-1} f(x_i, x_{i+1}, \bar{y}_i, \bar{y}_{i+1})$$

$$- \sum_{i=0}^{n-1} f(\bar{x}_i, \bar{x}_{i+1}, \bar{y}_i, \bar{y}_{i+1})$$

$$\leq -\gamma/2 + 2^{-1}\sum_{i=0}^{n-1} f(x_i, x_{i+1}, \bar{y}_i, \bar{y}_{i+1}) - 2^{-1}\sum_{i=0}^{n-1} f(\bar{x}_i, \bar{x}_{i+1}, \bar{y}_i, \bar{y}_{i+1})$$

$$\leq -\gamma/2 + \gamma/16 + \gamma/8 \leq -\gamma/4.$$

This contradicts (8.78). The contradiction we have reached proves that case (a) does not hold.

Assume that case (b) holds. Then there exists $j \in \{1, \dots, n-1\}$ such that

$$|y_j - y_f| > \epsilon. \tag{8.81}$$

We estimate

$$\sum_{i=0}^{n-1} f(\bar{x}_i, \bar{x}_{i+1}, 2^{-1}\bar{y}_i + 2^{-1}y_i, 2^{-1}\bar{y}_{i+1} + 2^{-1}y_{i+1}) - \sum_{i=0}^{n-1} f(\bar{x}_i, \bar{x}_{i+1}, \bar{y}_i, \bar{y}_{i+1}).$$

It is clear that $\{2^{-1}\bar{y}_i + 2^{-1}y_i\}_{i=0}^n$ is an (Ω_2)-program. Since the pair of sequences $\{\bar{x}_i\}_{i=0}^n$, $\{\bar{y}_i\}_{i=0}^n$ is $(f, \gamma/16)$-good it follows from (8.75) that

$$\sum_{i=0}^{n-1} f(\bar{x}_i, \bar{x}_{i+1}, 2^{-1}\bar{y}_i + 2^{-1}y_i, 2^{-1}\bar{y}_{i+1} + 2^{-1}y_{i+1})$$
$$- \sum_{i=0}^{n-1} f(\bar{x}_i, \bar{x}_{i+1}, \bar{y}_i, \bar{y}_{i+1}) \le \gamma/16. \tag{8.82}$$

By definition, for every $i = 0, \dots, n-1$,

$$f(\bar{x}_i, \bar{x}_{i+1}, 2^{-1}\bar{y}_i + 2^{-1}y_i, 2^{-1}\bar{y}_{i+1} + 2^{-1}y_{i+1})$$
$$\ge 2^{-1}f(\bar{x}_i, \bar{x}_{i+1}, \bar{y}_i, \bar{y}_{i+1}) + 2^{-1}f(\bar{x}_i, \bar{x}_{i+1}, y_i, y_{i+1}). \tag{8.83}$$

Property (P4), (8.73)–(8.75), and (8.81) imply that

$$f(\bar{x}_j, \bar{x}_{j+1}, 2^{-1}\bar{y}_j + 2^{-1}y_j, 2^{-1}\bar{y}_{j+1} + 2^{-1}y_{j+1})$$
$$- 2^{-1}f(\bar{x}_j, \bar{x}_{j+1}, \bar{y}_j, \bar{y}_{j+1}) - 2^{-1}f(\bar{x}_j, \bar{x}_{j+1}, y_j, y_{j+1}) \ge \gamma/2. \tag{8.84}$$

In view of (8.76), (8.83), and (8.84),

$$\sum_{i=0}^{n-1} f(\bar{x}_i, \bar{x}_{i+1}, 2^{-1}\bar{y}_i + 2^{-1}y_i, 2^{-1}\bar{y}_{i+1} + 2^{-1}y_{i+1}) - \sum_{i=0}^{n-1} f(\bar{x}_i, \bar{x}_{i+1}, \bar{y}_i, \bar{y}_{i+1})$$
$$\ge 2^{-1}\sum_{i=0}^{n-1} f(\bar{x}_i, \bar{x}_{i+1}, \bar{y}_i, \bar{y}_{i+1}) + 2^{-1}\sum_{i=0}^{n-1} f(\bar{x}_i, \bar{x}_{i+1}, y_i, y_{i+1}) + \gamma/2$$
$$- \sum_{i=0}^{n-1} f(\bar{x}_i, \bar{x}_{i+1}, \bar{y}_i, \bar{y}_{i+1})$$
$$\ge \gamma/2 + 2^{-1}\sum_{i=0}^{n-1} f(\bar{x}_i, \bar{x}_{i+1}, y_i, y_{i+1}) - 2^{-1}\sum_{i=0}^{n-1} f(\bar{x}_i, \bar{x}_{i+1}, \bar{y}_i, \bar{y}_{i+1})$$
$$\ge \gamma/2 - 3\gamma/16.$$

This contradicts (8.82). The contradiction we have reached proves Lemma 8.7. □

Corollary 8.8. *Let $\epsilon \in (0,1)$ and $L_0 \geq 1$ be an integer. Then there exists $\delta \in (0,\epsilon)$ such that for each integer $T \in \{1,\dots,L_0\}$, each $f_t \in \mathcal{M}$, $t = 0,\dots,T-1$ satisfying*

$$\|f_t - f\| \leq \delta,\ t = 0,\dots,T-1,$$

each (Ω_1)-program $\{x_t\}_{t=0}^T$ and each (Ω_2)-program $\{y_t\}_{t=0}^T$ such that

$$|x_0 - x_f| \leq \delta,\ |x_T - x_f| \leq \delta,\ |y_0 - y_f| \leq \delta,\ |y_T - y_f| \leq \delta,$$

and such that the pair $\{x_t\}_{t=0}^T, \{y_t\}_{t=0}^T$ is $(\{f_t\}_{t=0}^{T-1}, x_0, x_T, y_0, y_T, \delta)$-good the following inequalities hold:

$$|x_t - x_f| \leq \epsilon,\ |y_t - y_f| \leq \epsilon,\ t = 0,\dots,T.$$

Lemma 8.9. *Let $M > 0$, $\epsilon \in (0, r_f)$, and l be a natural number. Then there exist a natural number $L > 2l$ and $\delta > 0$ such that for each integer $T \geq L$, each $f_t \subset \mathcal{M}$, $t = 0,\dots,T-1$ satisfying*

$$\|f_t - f\| \leq \delta,\ t = 0,\dots,T-1, \tag{8.85}$$

each (Ω_1)-program $\{x_t\}_{t=0}^T$ and each (Ω_2)-program $\{y_t\}_{t=0}^T$ such that at least one of the following conditions holds:

(a) $x_0 \in \bar{Y}_1(l)$, $y_0 \in \bar{Y}_2(l)$, $x_T \in Y_1(l)$, $y_T \in Y_2(l)$, the pair $\{x_t\}_{t=0}^T$, $\{y_t\}_{t=0}^T$ is $(\{f_t\}_{t=0}^{T-1}, x_0, x_T, y_0, y_T, 0, T, M)$-good;

(b) $x_0 \in \bar{Y}_1(l)$, $y_0 \in \bar{Y}_2(l)$, the pair $\{x_t\}_{t=0}^T$, $\{y_t\}_{t=0}^T$ is

$$(\{f_t\}_{t=0}^{T-1}, x_0, y_0, 0, T, M)\text{-good;}$$

(c) the pair $\{x_t\}_{t=0}^T$, $\{y_t\}_{t=0}^T$ is $(\{f_t\}_{t=0}^{T-1}, 0, T, M)$-good;

(d) $x_T \in Y_1(l)$, $y_T \in Y_2(l)$, the pair $\{x_t\}_{t=0}^T$, $\{y_t\}_{t=0}^T$ is

$$(\{f_t\}_{t=0}^{T-1}, (x_T, y_T)^-, 0, T, M)\text{-good}$$

there exists an integer $j \in \{1,\dots,T-1\}$ such that

$$|x_j - x_f|,\ |y_j - y_f| \leq \epsilon. \tag{8.86}$$

Proof. By property (P1), there exists $\gamma \in (0,1)$ such that the following property holds:

(P6) for every $(\xi_1, \xi_2) \in \Omega_1$ satisfying $L_1(\xi_1, \xi_2) \leq \gamma$, we have

$$|\xi_1 - x_f|,\ |\xi_2 - x_f| \leq \epsilon/2;$$

for every $(\xi_1, \xi_2) \in \Omega_2$ satisfying $L_2(\xi_1, \xi_2) \leq \gamma$, we have

$$|\xi_1 - y_f|, |\xi_2 - y_f| \leq \epsilon/2.$$

Choose

$$\delta \in (0, \gamma/4)$$

and a natural number

$$L > 2(2M + 8l(\|f\| + 2))\gamma^{-1} + 4 + 2\gamma^{-1}(|l_1| + |l_2)(\mathrm{diam}(X_1) + \mathrm{diam}(X_2)). \tag{8.87}$$

Assume that an integer $T \geq L$, $f_t \in \mathcal{M}$, $t = 0, \ldots, T-1$ satisfy (8.85), an (Ω_1)-program $\{x_t\}_{t=0}^T$ and an (Ω_2)-program $\{y_t\}_{t=0}^T$ satisfy at least one of conditions (a)–(d). We show that there exists an integer $j \in \{1, \ldots, T-1\}$ such that (8.86) holds.

Assume the contrary. Then for every integer $i \in \{1, \ldots, T-1\}$,

$$\max\{|x_i - x_f|,\ |y_i - y_f|\} > \epsilon. \tag{8.88}$$

By Definition 8.2 and (8.87), there exist an (Ω_1)-program $\{u_t\}_{t=0}^T$ and an (Ω_2)-program $\{v_t\}_{t=0}^T$ such that

$$u_i = x_f,\ v_i = y_f \text{ for all integers } i \in [l, T-l], \tag{8.89}$$

in the case (a)

$$u_0 = x_0,\ u_T = x_T,\ v_0 = y_0,\ v_T = y_T, \tag{8.90}$$

in the case (b)

$$u_0 = x_0,\ v_0 = y_0, \tag{8.91}$$

in the case (d)

$$u_T = x_T,\ v_T = y_T. \tag{8.92}$$

By (a)–(d) and (8.89)–(8.92),

$$\begin{aligned} -M + \sum_{t=0}^{T-1} f_t(x_t, x_{t+1}, v_t, v_{t+1}) \\ \leq \sum_{t=0}^{T-1} f_t(x_t, x_{t+1}, y_t, y_{t+1}) \\ \leq \sum_{t=0}^{T-1} f_t(u_t, u_{t+1}, y_t, y_{t+1}) + M. \end{aligned} \tag{8.93}$$

It follows from (8.85) and (8.93) that

$$\begin{aligned} -M + \sum_{t=l}^{T-l-1} f_t(x_t, x_{t+1}, v_t, v_{t+1}) - 2l(\|f\| + \delta) \\ \leq -M + \sum_{t=0}^{T-1} f_t(x_t, x_{t+1}, v_t, v_{t+1}) \\ \leq \sum_{t=0}^{T-1} f_t(x_t, x_{t+1}, y_t, y_{t+1}) \leq \sum_{t=0}^{T-1} f_t(u_t, u_{t+1}, y_t, y_{t+1}) + M \end{aligned}$$

$$\leq \sum_{t=l}^{T-l-1} f_t(u_t, u_{t+1}, y_t, y_{t+1}) + 2l(\|f\| + \delta) + M. \quad (8.94)$$

In view of (8.85),

$$|\sum_{t=0}^{T-1} f_t(x_t, x_{t+1}, y_t, y_{t+1}) - \sum_{t=l}^{T-l-1} f_t(x_t, x_{t+1}, y_t, y_{t+1})| \leq 2l(\|f\| + \delta). \quad (8.95)$$

By (8.94) and (8.95),

$$\sum_{t=l}^{T-l-1} f_t(x_t, x_{t+1}, v_t, v_{t+1}) - M - 4l(\|f\| + 1)$$

$$\leq \sum_{t=l}^{T-l-1} f_t(x_t, x_{t+1}, y_t, y_{t+1})$$

$$\leq \sum_{t=l}^{T-l-1} f_t(u_t, u_{t+1}, y_t, y_{t+1}) + M + 4l(\|f\| + 1). \quad (8.96)$$

Conditions (a)–(d) imply that the pair $\{x_t\}_{t=l}^{T-l}$, $\{y_t\}_{t=l}^{T-l}$ is

$$(\{f_t\}_{t=l}^{T-l-1}, x_l, x_{T-l}, y_l, y_{T-l}, l, T-l, M) - \text{good}.$$

It follows from (8.85), (8.89), and (8.93) that

$$\sum_{t=l}^{T-l-1} f_t(x_t, x_{t+1}, y_t, y_{t+1}) - M - 4l(\|f\| + 1)$$

$$\leq \sum_{t=l}^{T-l-1} f_t(x_t, x_{t+1}, y_t, y_{t+1})$$

$$\leq \sum_{t=l}^{T-l-1} f_t(x_f, x_f, y_t, y_{t+1}) + M + 4l(\|f\| + 1). \quad (8.97)$$

By (8.54), (8.56), and (8.85),

$$\sum_{t=l}^{T-l-1} f_t(x_t, x_{t+1}, y_f, y_f) \geq \sum_{t=l}^{T-l-1} f(x_t, x_{t+1}, y_f, y_f) - \delta(T-2l)$$

$$= \sum_{t=l}^{T-l-1} L_1(x_t, x_{t+1}) + (T-2l) f(x_f, x_f, y_f, y_f)$$

$$+ \langle l_1, x_l - x_{T-l} \rangle - \delta(T-2l), \quad (8.98)$$

$$\sum_{t=l}^{T-l-1} f_t(x_f, x_f, y_t, y_{t+1}) \le \sum_{t=l}^{T-l-1} f(x_f, x_f, y_t, y_{t+1}) + \delta(T-2l)$$

$$= -\sum_{t=l}^{T-l-1} L_2(y_t, y_{t+1}) + (T-2l) f(x_f, x_f, y_f, y_f)$$

$$+ \langle l_2, y_l - y_{T-l}\rangle + \delta(T-2l). \tag{8.99}$$

In view of (8.97)–(8.99),

$$2M + 8l(\|f\|+1) \ge \sum_{t=l}^{T-l-1} f_t(x_t, x_{t+1}, y_f, y_f) - \sum_{t=l}^{T-l-1} f_t(x_f, x_f, y_t, y_{t+1})$$

$$\ge \sum_{t=l}^{T-l-1} L_1(x_t, x_{t+1}) + \sum_{t=l}^{T-l-1} L_2(y_t, y_{t+1})$$

$$-|l_1||x_l - x_{T-l}| - |l_2||y_l - y_{T-l}| - 2\delta(T-2l)$$

$$\ge \sum_{t=l}^{T-l-1} (L_1(x_t, x_{t+1}) + L_2(y_t, y_{t+1}))$$

$$- (\mathrm{diam}(X_1) + \mathrm{diam}(X_2))(|l_1| + |l_2|) - 2\delta(T-2l). \tag{8.100}$$

Property (P6) and (8.88) imply that for all $t = 0, \dots, T-1$,

$$L_1(x_t, x_{t+1}) + L_2(y_t, y_{t+1}) > \gamma.$$

Together with (8.100) this implies that

$$2M + 8l(\|f\|+1) + (\mathrm{diam}(X_1) + \mathrm{diam}(X_2))(|l_1| + |l_2|)$$

$$\ge (T-2l)\gamma - 2\delta(T-2l)$$

$$\ge 2^{-1}\gamma(T-2l) \ge 2^{-1}\gamma L - 2l,$$

$$L \le 2\gamma^{-1}(2M + 8l(\|f\|+2)) + (\mathrm{diam}(X_1) + \mathrm{diam}(X_2))(|l_1| + |l_2|).$$

This contradicts (8.87). The contradiction we have reached completes the proof of Lemma 8.9.

Lemma 8.10. *Let $M > 0$, $\epsilon \in (0, r_f)$, and l be a natural number. Then there exist a natural number $L > 2l$ and $\delta > 0$ such that for each integer $T \ge L$, each $f_t \in \mathcal{M}$, $t = 0, \dots, T-1$ satisfying*

$$\|f_t - f\| \le \delta,\ t = 0, \dots, T-1, \tag{8.101}$$

each (Ω_1)-program $\{x_t\}_{t=0}^T$ and each (Ω_2)-program $\{y_t\}_{t=0}^T$ such that at least one of the following conditions holds:

(a) $x_0 \in \bar{Y}_1(l)$, $y_0 \in \bar{Y}_2(l)$, $x_T \in Y_1(l)$, $y_T \in Y_2(l)$, *the pair* $\{x_t\}_{t=0}^T$, $\{y_t\}_{t=0}^T$ *is* $(\{f_t\}_{t=0}^{T-1}, x_0, x_T, y_0, y_T, 0, T, M)$*-good;*

(b) $x_0 \in \bar{Y}_1(l)$, $y_0 \in \bar{Y}_2(l)$, *the pair* $\{x_t\}_{t=0}^T$, $\{y_t\}_{t=0}^T$ *is*

$$(\{f_t\}_{t=0}^{T-1}, x_0, y_0, 0, T, M) - good;$$

(c) the pair $\{x_t\}_{t=0}^T$, $\{y_t\}_{t=0}^T$ *is* $(\{f_t\}_{t=0}^{T-1}, 0, T, M)$*-good;*

(d) $x_T \in Y_1(l)$, $y_T \in Y_2(l)$, *the pair* $\{x_t\}_{t=0}^T$, $\{y_t\}_{t=0}^T$ *is*

$$(\{f_t\}_{t=0}^{T-1}, (x_T, y_T)^-, 0, T, M) - good$$

there exists an integer $j \in \{1, \dots, L\}$ *such that*

$$|x_j - x_f|,\ |y_j - y_f| \le \epsilon. \tag{8.102}$$

Proof. By Lemma 8.9, there exist a natural number $L > 2l$ and $\delta > 0$ such that the following property holds:

(P7) for each integer $T \ge L$, each $f_t \in \mathcal{M}$, $t = 0, \dots, T-1$ satisfying (8.101), each (Ω_1)-program $\{x_t\}_{t=0}^T$ and each (Ω_2)-program $\{y_t\}_{t=0}^T$ such that at least one of conditions (a)–(d) holds there exists an integer $j \in \{1, \dots, T-1\}$ such that (8.102) is true.

Assume that an integer $T \ge L$, $f_t \in \mathcal{M}$, $t = 0, \dots, T-1$ satisfy (8.101), an (Ω_1)-program $\{x_t\}_{t=0}^T$ and an (Ω_2)-program $\{y_t\}_{t=0}^T$ satisfy at least one of conditions (a)–(d). In order to complete the proof it is sufficient to show that there exists $j \in \{1, \dots, L\}$ such that (8.102) holds.

Assume the contrary. Then

$$\max\{|x_i - x_f|,\ |y_i - y_f|\} > \epsilon,\ i = 1, \dots, L. \tag{8.103}$$

Property (P7) implies that there exists

$$j \in \{1, \dots, T-1\}$$

such that (8.102) holds. By (8.102) and (8.103),

$$j > L. \tag{8.104}$$

We may assume without loss of generality that

$$\max\{|x_i - x_f|,\ |y_i - y_f|\} > \epsilon,\ i = 1, \dots, j-1. \tag{8.105}$$

It is not difficult to see that the (Ω_1)-program $\{x_t\}_{t=0}^j$ and the (Ω_2)-program $\{y_t\}_{t=0}^j$ satisfy at least one of conditions (a)–(d). Combined with property (P7) and (8.104) this implies that there exists an integer

$$j_0 \in \{1, \dots, j-1\}$$

such that

$$|x_{j_0} - x_f|,\ |y_{j_0} - y_f| \le \epsilon.$$

This contradicts (8.105). The contradiction we have reached proves Lemma 8.10.

8.4 Proof of Theorem 8.5

By Lemma 8.7, there exists a positive number

$$\delta_0 < \min\{1, r_f, \epsilon\}$$

such that the following property holds:

(P8) for each integer $n \geq 1$ and each (f, δ_0)-good pair of an (Ω_1)-program $\{x_t\}_{t=0}^T$ and an (Ω_2)-program $\{y_t\}_{t=0}^T$ satisfying

$$|x_0 - x_f| \leq \delta_0,\ |x_n - x_f| \leq \delta_0,\ |y_0 - y_f| \leq \delta_0,\ |y_n - y_f| \leq \delta_0,$$

we have

$$|x_t - x_f| \leq \epsilon,\ |y_t - y_f| \leq \epsilon,\ t = 0, \dots, n.$$

By Lemma 8.10, there exist a natural number $L > 2l$ and $\delta_1 > 0$ such that the following property holds:

(P9) for each integer $T \geq L$, each $f_t \in \mathcal{M}$, $t = 0, \dots, T-1$ satisfying

$$\|f_t - f\| \leq \delta_1,\ t = 0, \dots, T-1,$$

each (Ω_1)-program $\{x_t\}_{t=0}^T$ and each (Ω_2)-program $\{y_t\}_{t=0}^T$ for which at least one of conditions (a)–(d) holds there exists an integer $j \in \{1, \dots, L\}$ such that

$$|x_j - x_f|,\ |y_j - y_f| \leq \delta_0. \tag{8.106}$$

Choose a positive number

$$\delta \leq \min\{\delta_0, \delta_1\}(2L+1)^{-1}. \tag{8.107}$$

Assume that an integer $T \geq 2L$, $f_t \in \mathcal{M}$, $t = 0, \dots, T-1$ satisfy (8.71), an (Ω_1)-program $\{x_t\}_{t=0}^T$ and an (Ω_2)-program $\{y_t\}_{t=0}^T$ satisfy at least one of conditions (a)–(d) and that for each integer $S \in [0, T-L]$ the pair $\{x_t\}_{t=S}^{S+L}$, $\{y_t\}_{t=S}^{S+L}$ is $(\{f_t\}_{t=S}^{S+L-1}, x_S, x_{S+L}, y_S, y_{S+L}, S, S+L, \delta)$-good. Property (P9), (8.71), and (8.107) imply that there exists an integer $S_0 \in \{0, \dots, L\}$ such that

$$|x_{S_0} - x_f| \leq \delta_0,\ |y_{S_0} - y_f| \leq \delta_0.$$

If

$$|x_0 - x_f| \leq \delta,\ |y_0 - y_f| \leq \delta,$$

then we set $S_0 = 0$. By induction, we continue to construct the sequence S_i, $i = 0, 1, \dots$. Assume that an integer $q \geq 0$ and that we have already defined nonnegative integers S_i, $i = 0, \dots, q$ such that for every integer i satisfying $0 \leq i < q$,

$$1 \leq S_{i+1} - S_i \leq L \tag{8.108}$$

and for all integers $i = 0, \dots, q$,

$$|x_{S_i} - x_f| \leq \delta_0,\ |y_{S_i} - y_f| \leq \delta_0. \tag{8.109}$$

Assume that $T - S_q \geq L$. By (8.71), (8.107), (8.109), and property (P9) applied to the pair of sequences $\{x_{t+S_q}\}_{t=0}^{T-S_q}$, $\{y_{t+S_q}\}_{t=0}^{T-S_q}$ which satisfies at least one of conditions (a), (b) there exists an integer $S_{q+1} \in \{S_q + 1, \dots, S_q + L\}$ such that

$$|x_{S_{q+1}} - x_f| \leq \delta_0, \ |y_{S_{q+1}} - y_f| \leq \delta_0.$$

Thus by induction we constructed the sequence of integers $\{S_i\}_{i=0}^{k}$ such that if

$$|x_0 - x_f| \leq \delta, \ |y_0 - y_f| \leq \delta,$$

then $S_0 = 0$,

$$0 \leq S_0 \leq L,$$

for every integer i satisfying $0 \leq i < k$,

$$1 \leq S_{i+1} - S_i \leq L, \tag{8.110}$$

for all integers $i = 0, \dots, k$,

$$|x_{S_i} - x_f| \leq \delta_0, \ |y_{S_i} - y_f| \leq \delta_0 \tag{8.111}$$

and

$$T \geq S_k > T - L.$$

If

$$|x_T - x_f| \leq \delta, \ |y_T - y_f| \leq \delta,$$

then we may assume that $S_k = T$.

Assume that an integer $i \in \{0, \dots, k-1\}$. By conditions (a)–(d), the pair of sequences $\{x_t\}_{t=S_i}^{S_{i+1}}$, $\{y_t\}_{t=S_i}^{S_{i+1}}$ is $(\{f_t\}_{t=S_i}^{S_{i+1}-1}, x_{S_i}, x_{S_{i+1}}, y_{S_i}, y_{S_{i+1}}, S_i, S_{i+1}, \delta)$-good. Together with (8.71), (8.107), and (8.110) this implies that the pair of sequences $\{x_t\}_{t=S_i}^{S_{i+1}}$, $\{y_t\}_{t=S_i}^{S_{i+1}}$ is $(f, \delta(1+2L))$-good and (f, δ_0)-good. Combined with (8.111) and property (P8) this implies that

$$|x_t - x_f|, \ |y_t - x_f| \leq \epsilon, \ t = S_i, \dots, S_{i+1}$$

for $i = 0, \dots, k-1$. This completes the proof of Theorem 8.5. □

8.5 Proof of Theorem 8.6

By Theorem 8.5 and Corollary 8.8, there exist a natural number $L_0 > 2l$ and $\delta_0 \in (0, \epsilon)$ such that the following property holds:

(P10) for each integer $T \geq 1$, each $f_t \in \mathcal{M}$, $t = 0, \dots, T-1$ satisfying

$$\|f_t - f\| \leq \delta_0, \ t = 0, \dots, T-1,$$

each (Ω_1)-program $\{x_t\}_{t=0}^T$ and each (Ω_2)-program $\{y_t\}_{t=0}^T$ such that

$$|x_0 - x_f|,\ |x_T - x_f|,\ |y_0 - y_f|,\ |y_T - y_f| \le \delta_0$$

and that the pair of sequences $\{x_t\}_{t=0}^T$, $\{y_t\}_{t=0}^T$ is

$$(\{f_t\}_{t=0}^{T-1}, x_0, x_T, y_0, y_T, 0, T, \delta_0)\text{-good}$$

we have

$$|x_t - x_f| \le \epsilon,\ |y_t - y_f| \le \epsilon,\ t = 0, \dots, T.$$

By Lemma 8.10, there exist a natural number $L_1 > 2l$ and $\delta \in (0, \delta_0)$ such that the following property holds:

(P11) for each integer $T \ge L_1$, each $f_t \in \mathcal{M}$, $t = 0, \dots, T-1$ satisfying

$$\|f_t - f\| \le \delta,\ t = 0, \dots, T-1,$$

each (Ω_1)-program $\{x_t\}_{t=0}^T$ and each (Ω_2)-program $\{y_t\}_{t=0}^T$ satisfying at least one of conditions (a)–(d), there exists an integer $j \in \{1, \dots, L_1\}$ such that

$$|x_j - x_f|,\ |y_j - y_f| \le \delta_0.$$

Fix a natural number

$$L > L_0 + L_1 + (L_1 + 2)(2 + 2\delta^{-1}M). \tag{8.112}$$

Assume that an integer $T > L$, $f_t \in \mathcal{M}$, $t = 0, \dots, T-1$ satisfy

$$\|f_t - f\| \le \delta,\ t = 0, \dots, T-1, \tag{8.113}$$

and an (Ω_1)-program $\{x_t\}_{t=0}^T$ and an (Ω_2)-program $\{y_t\}_{t=0}^T$ satisfy at least one of conditions (a)–(d). Property (P11), (8.112), and (8.113) imply that there exists an integer $S_0 \in \{0, \dots, L_1\}$ such that

$$|x_{S_0} - x_f| \le \delta_0,\ |y_{S_0} - y_f| \le \delta_0. \tag{8.114}$$

By induction, we continue to construct the sequence S_i, $i = 0, 1, \dots$.

Assume that an integer $q \ge 0$ and that we have already defined nonnegative integers S_i, $i = 0, \dots, q$ such that

$$0 \le S_0 \le L_1,$$

for every integer i satisfying $0 \le i < q$,

$$1 \le S_{i+1} - S_i \le L_1$$

and for all integers $i = 0, \dots, q$,

$$|x_{S_i} - x_f| \le \delta_0,\ |y_{S_i} - y_f| \le \delta_0. \tag{8.115}$$

Assume that $T - S_q \geq L_1$. By (8.113), (8.115), and property (P11) applied to the pair of sequences $\{x_{t+S_q}\}_{t=0}^{T-S_q}$, $\{y_{t+S_q}\}_{t=0}^{T-S_q}$ which satisfies at least one of conditions (a), (b) there exists an integer $S_{q+1} \in \{S_q + 1, \dots, S_q + L_1\}$ such that

$$|x_{S_{q+1}} - x_f| \leq \delta_0, \ |y_{S_{q+1}} - y_f| \leq \delta_0.$$

Thus by induction we constructed the sequence of integers $\{S_i\}_{i=0}^{k}$ such that

$$0 \leq S_0 \leq L_1, \ T \geq S_k > T - L_1, \tag{8.116}$$

for every integer i satisfying $0 \leq i < k$,

$$1 \leq S_{i+1} - S_i \leq L_1, \tag{8.117}$$

for all integers $i = 0, \dots, k$,

$$|x_{S_i} - x_f| \leq \delta_0, \ |y_{S_i} - y_f| \leq \delta_0. \tag{8.118}$$

Set

$$E_1 = \{i \in \{0, \dots, k-1\} : \text{ there is an } (\Omega_1) - \text{program}\{\xi_t\}_{t=S_i}^{S_{i+1}} \text{ such that}$$

$$\xi_p = x_p, \ p = S_i, S_{i+1},$$

$$\sum_{t=S_i}^{S_{i+1}-1} f_t(\xi_t, \xi_{t+1}, y_t, y_{t+1}) + \delta_0 < \sum_{t=S_i}^{S_{i+1}-1} f_t(x_t, x_{t+1}, y_t, y_{t+1})\} \tag{8.119}$$

and

$$E_2 = \{i \in \{0, \dots, k-1\} : \text{ there is an } (\Omega_2) - \text{program}\{\eta_t\}_{t=S_i}^{S_{i+1}} \text{ such that}$$

$$\eta_p = y_p, \ p = S_i, S_{i+1},$$

$$\sum_{t=S_i}^{S_{i+1}-1} f_t(x_t, x_{t+1}, \eta_t, \eta_{t+1}) - \delta_0 > \sum_{t=S_i}^{S_{i+1}-1} f_t(x_t, x_{t+1}, y_t, y_{t+1})\}. \tag{8.120}$$

By (8.119) and (8.120), for each $i \in \{0, \dots, k-1\} \setminus (E_1 \cup E_2)$, the pair of sequences $\{x_t\}_{t=S_i}^{S_{i+1}}$, $\{y_t\}_{t=S_i}^{S_{i+1}}$ is is $(\{f_t\}_{t=S_i}^{S_{i+1}-1}, x_{S_i}, x_{S_{i+1}}, y_{S_i}, y_{S_{i+1}}, S_i, S_{i+1}, \delta_0)$-good. Combined with (8.113), (8.118), and property (P10) this implies that the following property holds:

(P12) for each $i \in \{0, \dots, k-1\} \setminus (E_1 \cup E_2)$, we have

$$|x_t - x_f| \leq \epsilon, \ |y_t - y_f| \leq \epsilon, \ t = S_i, \dots, S_{i+1}.$$

It is not difficult to see that there exist an (Ω_1)-program $\{\xi_t\}_{t=0}^{T}$ and an (Ω_2)-program $\{\eta_t\}_{t=0}^{T}$ such that

$$\xi_0 = x_0, \ \xi_T = x_T, \ \xi_{S_i} = x_{S_i}, \ i = 0, \dots, q, \tag{8.121}$$

$$\xi_t = x_t, \ t \in \{0, \dots, T\} \setminus \cup\{\{S_i, \dots, S_{i+1}\} : \ i \in E_1\}, \tag{8.122}$$

for every $i \in E_1$,

$$\sum_{t=S_i}^{S_{i+1}-1} f_t(\xi_t, \xi_{t+1}, y_t, y_{t+1}) + \delta_0 < \sum_{t=S_i}^{S_{i+1}-1} f_t(x_t, x_{t+1}, y_t, y_{t+1}), \tag{8.123}$$

$$\eta_0 = y_0,\ \eta_T = y_T,\ \eta_{S_i} = y_{S_i},\ i = 0, \dots, q, \tag{8.124}$$

$$\eta_t = y_t,\ t \in \{0, \dots, T\} \setminus \cup\{\{S_i, \dots, S_{i+1}\} : \ i \in E_2\}, \tag{8.125}$$

for every $i \in E_2$,

$$\sum_{t=S_i}^{S_{i+1}-1} f_t(x_t, x_{t+1}, \eta_t, \eta_{t+1}) - \delta_0 > \sum_{t=S_i}^{S_{i+1}-1} f_t(x_t, x_{t+1}, y_t, y_{t+1}). \tag{8.126}$$

By conditions (a)–(d) and (8.121)–(8.123),

$$M \ge \sum_{t=0}^{T-1} f_t(x_t, x_{t+1}, y_t, y_{t+1}) - \sum_{t=0}^{T-1} f_t(\xi_t, \xi_{t+1}, y_t, y_{t+1})$$

$$\ge \sum_{i \in E_1} [\sum_{t=S_i}^{S_{i+1}-1} f_t(x_t, x_{t+1}, y_t, y_{t+1}) - \sum_{t=S_i}^{S_{i+1}-1} f_t(\xi_t, \xi_{t+1}, y_t, y_{t+1})]$$

$$\ge \delta_0 \mathrm{Card}(E_1),$$

$$\mathrm{Card}(E_1) \le \delta_0^{-1} M. \tag{8.127}$$

By conditions (a)–(d) and (8.124)–(8.126),

$$M \ge \sum_{t=0}^{T-1} f_t(x_t, x_{t+1}, \eta_t, \eta_{t+1}) - \sum_{t=0}^{T-1} f_t(x_t, x_{t+1}, y_t, y_{t+1})$$

$$\ge \sum_{i \in E_2} [\sum_{t=S_i}^{S_{i+1}-1} f_t(x_t, x_{t+1}, \eta_t, \eta_{t+1}) - \sum_{t=S_i}^{S_{i+1}-1} f_t(x_t, x_{t+1}, y_t, y_{t+1})]$$

$$\ge \delta_0 \mathrm{Card}(E_2),$$

$$\mathrm{Card}(E_2) \le \delta_0^{-1} M. \tag{8.128}$$

Property (P12) implies that

$$E := \{t \in \{0, \dots, T\} : \ \max\{|x_t - x_f|,\ |y_t - y_f|\} \ge \epsilon\}$$

$$\subset \{0, \dots, S_0\} \cup \{S_q, \dots, T\} \cup \{\{S_i, \dots, S_{i+1}\} : \ i \in E_1\}$$

$$\cup\{\{S_i, \dots, S_{i+1}\} : \ i \in E_2\}.$$

Together with (8.112), (8.116), (8.117), (8.127), and (8.128) this implies that

$$\mathrm{Card}(E) \le (L_1 + 1)(2 + \mathrm{Card}(E_1) + \mathrm{Card}(E_2))$$

$$\le (L_1 + 2)(2 + 2\delta_0^{-1} M) < L.$$

Theorem 8.6 is proved. □

8.6 Infinite Horizon Games

We prove the following results which establish the existence of overtaking optimal pairs of programs.

Theorem 8.11. *Let $l \geq 1$ be an integer and let $\{x_t\}_{t=0}^{\infty}$ be an (Ω_1)-program and $\{y_t\}_{t=0}^{\infty}$ be an (Ω_2)-program such that*

$$x_0 \in \bar{Y}_1(l), \ y_0 \in \bar{Y}_2(l),$$

the pair of sequences $\{x_t\}_{t=0}^{T}$, $\{y_t\}_{t=0}^{T}$ is (f)-optimal for all integers $T \geq 1$ and that

$$\lim_{t \to \infty} x_t = x_f, \ \lim_{t \to \infty} y_t = y_f.$$

Then the pair of sequences $\{x_t\}_{t=0}^{\infty}$, $\{y_t\}_{t=0}^{\infty}$ is (f)-overtaking optimal.

Theorem 8.12. *Let $\{x_t\}_{t=0}^{\infty}$ be an (Ω_1)-program and $\{y_t\}_{t=0}^{\infty}$ be an (Ω_2)-program such that*

$$\lim_{t \to \infty} x_t = x_f, \ \lim_{t \to \infty} y_t = y_f$$

and that for all integers $T \geq 1$ the pair of sequences $\{x_t\}_{t=0}^{T}$, $\{y_t\}_{t=0}^{T}$ is strongly (f)-optimal. Then the pair of sequences $\{x_t\}_{t=0}^{\infty}$, $\{y_t\}_{t=0}^{\infty}$ is strongly (f)-overtaking optimal.

Theorems 8.11 and 8.12 are proved in Sect. 8.7.

Theorem 8.13. *Let $\xi_0 \in \cup\{\bar{Y}_1(l) : l = 1, 2, \dots\}$ and $\xi_1 \in \cup\{\bar{Y}_2(l) : l = 1, 2, \dots\}$. Then there exist an (Ω_1)-program $\{x_t\}_{t=0}^{\infty}$ and an (Ω_2)-program $\{y_t\}_{t=0}^{\infty}$ such that*

$$x_0 = \xi_0, \ y_0 = \xi_1,$$

for all integers $T \geq 1$, the pair of sequences $\{x_t\}_{t=0}^{T}$, $\{y_t\}_{t=0}^{T}$ is (f)-optimal and that

$$\lim_{t \to \infty} x_t = x_f, \ \lim_{t \to \infty} y_t = y_f.$$

Theorem 8.14. *There exist an (Ω_1)-program $\{x_t\}_{t=0}^{\infty}$ and an (Ω_2)-program $\{y_t\}_{t=0}^{\infty}$ such that for all integers $T \geq 1$, the pair of sequences $\{x_t\}_{t=0}^{T}$, $\{y_t\}_{t=0}^{T}$ is strongly (f)-optimal and that*

$$\lim_{t \to \infty} x_t = x_f, \ \lim_{t \to \infty} y_t = y_f.$$

Theorems 8.13 and 8.14 are proved in Sect. 8.8.

Theorems 8.11 and 8.13 imply the following result.

Theorem 8.15. *Let $\xi_0 \in \cup\{\bar{Y}_1(l) : l = 1, 2, \dots\}$ and $\xi_1 \in \cup\{\bar{Y}_2(l) : l = 1, 2, \dots\}$. Then there exist an (Ω_1)-program $\{x_t\}_{t=0}^{\infty}$ and an (Ω_2)-program $\{y_t\}_{t=0}^{\infty}$ such that*

$$x_0 = \xi_0, \ y_0 = \xi_1,$$

the pair of sequences $\{x_t\}_{t=0}^{\infty}$, $\{y_t\}_{t=0}^{\infty}$ is (f)- overtaking optimal and that

$$\lim_{t \to \infty} x_t = x_f, \ \lim_{t \to \infty} y_t = y_f.$$

Theorems 8.12 and 8.14 imply the following result.

Theorem 8.16. *There exist an (Ω_1)-program $\{x_t\}_{t=0}^{\infty}$ and an (Ω_2)-program $\{y_t\}_{t=0}^{\infty}$ such that the pair of sequences $\{x_t\}_{t=0}^{\infty}$, $\{y_t\}_{t=0}^{\infty}$ is strongly (f)-overtaking optimal and that*

$$\lim_{t\to\infty} x_t = x_f, \ \lim_{t\to\infty} y_t = y_f.$$

For each compact metric space K denote by $C(K)$ the set of all continuous functions on K and for each $\phi \in C(K)$ set

$$\|\phi\| = \sup\{|\phi(z)| : \ z \in K\}.$$

Define functions $f^{(1)} : \Omega_1 \to R^1$, $f^{(2)} : \Omega_2 \to R^1$ by

$$f^{(1)}(x_1, x_2) = f(x_1, x_2, y_f, y_f), \ (x_1, x_2) \in \Omega_1,$$

$$f^{(2)}(y_1, y_2) = f(x_f, x_f, y_1, y_2), \ (y_1, y_2) \in \Omega_2.$$

By assumptions (A1), (A2), (8.11), (8.46), (8.47), (8.50), and (8.51), the assumptions made in Example 2.11 of [84] hold for the pairs $(f^{(2)}, \Omega_2)$ and $(-f^{(1)}, \Omega_1)$. This implies that assumptions (A1), (A2), (A3) of Sect. 2.2 are true for these pairs and the results of Chaps. 2–4 hold for the pairs $(f^{(2)}, \Omega_2)$ and $(-f^{(1)}, \Omega_1)$.

8.7 Proofs of Theorems 8.11 and 8.12

We prove Theorems 8.11 and 8.12 simultaneously. Let $\{u_t\}_{t=0}^{\infty}$ be an (Ω_1)-program. In the case of Theorem 8.11 we assume that

$$u_0 = x_0. \tag{8.129}$$

We show that

$$\limsup_{T\to\infty}[\sum_{t=0}^{T-1} f(x_t, x_{t+1}, y_t, y_{t+1}) - \sum_{t=0}^{T-1} f(u_t, u_{t+1}, y_t, y_{t+1})] \le 0. \tag{8.130}$$

Assume the contrary. Then there exist $\Gamma_0 > 0$ and a strictly increasing sequence of natural numbers $\{T_k\}_{k=1}^{\infty}$ such that for all natural numbers k,

$$\sum_{t=0}^{T_k-1} f(x_t, x_{t+1}, y_t, y_{t+1}) - \sum_{t=0}^{T_k-1} f(u_t, u_{t+1}, y_t, y_{t+1}) \ge \Gamma_0. \tag{8.131}$$

We show that

$$\lim_{j\to\infty} u_j = x_f.$$

The equality $\lim_{t\to\infty} y_t = y_f$ implies that

$$\lim_{t\to\infty} \|f^{(1)} - f(\cdot,\cdot,y_t,y_{t+1})\| = 0. \tag{8.132}$$

By Lemma 4.18, (8.132) and the equality $\lim_{t\to\infty} x_t = x_f$ there exist an integer $\tau_0 > 1$ and $M_0 > 0$ such that for each pair of integers $S_2 > S_1 \geq \tau_0$,

$$\sigma(\{-f(\cdot,\cdot,y_t,y_{t+1})\}_{t=S_1}^{S_2-1}, S_1, S_2) \leq M_0 - \sum_{t=S_1}^{S_2-1} f(x_t,x_{t+1},y_t,y_{t+1}) \tag{8.133}$$

and therefore in view of (8.133), for each pair of integers $S_2 > S_1 \geq \tau_0$,

$$-\sum_{t=S_1}^{S_2-1} f(u_t,u_{t+1},y_t,y_{t+1}) \leq M_0 - \sum_{t=S_1}^{S_2-1} f(x_t,x_{t+1},y_t,y_{t+1}). \tag{8.134}$$

Set

$$M_1 = 2\tau_0\|f\| + 2M_0 + 2. \tag{8.135}$$

Assume that an integer $S \geq \tau_0$ and

$$\sum_{t=0}^{S-1} f(x_t,x_{t+1},y_t,y_{t+1}) - \sum_{t=0}^{S-1} f(u_t,u_{t+1},y_t,y_{t+1}) \leq -M_0 - 1.$$

Together with (8.134) this implies that for every integer $T > S$,

$$\sum_{t=0}^{T-1} f(x_t,x_{t+1},y_t,y_{t+1}) - \sum_{t=0}^{T-1} f(u_t,u_{t+1},y_t,y_{t+1}) \leq -1.$$

This contradicts (8.131). The contradiction we have reached proves that for all integers $S \geq \tau_0$,

$$\sum_{t=0}^{S-1} f(x_t,x_{t+1},y_t,y_{t+1}) - \sum_{t=0}^{S-1} f(u_t,u_{t+1},y_t,y_{t+1}) \geq -M_0 - 1. \tag{8.136}$$

Assume that integers $S_1 > S_0 > \tau_0$ and that

$$\sum_{t=S_0}^{S_1-1} f(x_t,x_{t+1},y_t,y_{t+1}) - \sum_{t=S_0}^{S_1-1} f(u_t,u_{t+1},y_t,y_{t+1}) \leq -M_1. \tag{8.137}$$

By (8.134)–(8.137),

$$-M_0 - 1 \leq \sum_{t=0}^{S_1-1} f(x_t,x_{t+1},y_t,y_{t+1}) - \sum_{t=0}^{S_1-1} f(u_t,u_{t+1},y_t,y_{t+1})$$

$$= \sum_{t=0}^{\tau_0-1} f(x_t, x_{t+1}, y_t, y_{t+1}) - \sum_{t=0}^{\tau_0-1} f(u_t, u_{t+1}, y_t, y_{t+1})$$
$$+ \sum_{t=\tau_0}^{S_0-1} f(x_t, x_{t+1}, y_t, y_{t+1}) - \sum_{t=\tau_0}^{S_0-1} f(u_t, u_{t+1}, y_t, y_{t+1})$$
$$+ \sum_{t=S_0}^{S_1-1} f(x_t, x_{t+1}, y_t, y_{t+1}) - \sum_{t=S_0}^{S_1-1} f(u_t, u_{t+1}, y_t, y_{t+1})$$
$$\le 2\tau_0\|f\| + M_0 - M_1 \le -M_0 - 2.$$

The contradiction we have reached shows that for each pair of integers $S_1 > S_0 > \tau_0$,

$$\sum_{t=S_0}^{S_1-1} f(x_t, x_{t+1}, y_t, y_{t+1}) - \sum_{t=S_0}^{S_1-1} f(u_t, u_{t+1}, y_t, y_{t+1}) > -M_1. \tag{8.138}$$

By (8.133) and (8.138), for each pair of integers $S_1 > S_0 > \tau_0$,

$$-\sum_{t=S_0}^{S_1-1} f(u_t, u_{t+1}, y_t, y_{t+1}) > -M_1 - M_0 + \sigma(\{-f(\cdot,\cdot,y_t,y_{t+1})\}_{t=S_0}^{S_1-1}, S_0, S_1).$$

Together with assertion 1 of Theorem 2.5 and (8.132) this implies that for every $\epsilon > 0$, the set

$$\{t \in \{0, 1, \dots,\} : |u_t - x_f| > \epsilon\}$$

is finite. This implies that

$$\lim_{t\to\infty} u_t = x_f. \tag{8.139}$$

By the continuity of f, there exists $\epsilon \in (0, r_f)$ such that for each $\xi_1, \xi_2 \in X_1$ and each $\xi_3, \xi_4 \in X_2$ satisfying

$$|\xi_i - x_f| \le \epsilon,\ i = 1, 2,\ |\xi_i - y_f| \le \epsilon,\ i = 3, 4$$

we have

$$|f(\xi_1, \xi_2, \xi_3, \xi_4) - f(x_f, x_f, y_f, y_f)| \le \Gamma_0/8. \tag{8.140}$$

In view of (8.139) and the equalities

$$\lim_{t\to\infty} x_t = x_f,\ \lim_{t\to\infty} y_t = y_f,$$

there exists a natural number S_0 such that for each integer $t \ge S_0$,

$$|u_t - x_f| \le \epsilon,\ |x_t - x_f| \le \epsilon,\ |y_t - y_f| \le \epsilon. \tag{8.141}$$

There exists a natural number k such that

$$T_k > S_0 + 2. \tag{8.142}$$

Set

$$z_t = u_t,\ t = 0, \dots, T_k - 1,\ z_t = x_t \text{ for all integers } t \geq T_k. \tag{8.143}$$

In view of (8.141)–(8.143), $\{z_t\}_{t=0}^{T_k}$ is an (Ω_1)-program. Since the pair of sequences $\{x_t\}_{t=0}^{T_k}, \{y_t\}_{t=0}^{T_k}$ is (f)-optimal (strongly (f)-optimal, respectively) we have

$$\sum_{t=0}^{T_k-1} f(z_t, z_{t+1}, y_t, y_{t+1}) \geq \sum_{t=0}^{T_k-1} f(x_t, x_{t+1}, y_t, y_{t+1}).$$

It follows from the relation above, (8.131) and (8.140)–(8.143) that

$$0 \geq \sum_{t=0}^{T_k-1} f(x_t, x_{t+1}, y_t, y_{t+1}) - \sum_{t=0}^{T_k-1} f(z_t, z_{t+1}, y_t, y_{t+1})$$

$$= \sum_{t=0}^{T_k-1} f(x_t, x_{t+1}, y_t, y_{t+1}) - \sum_{t=0}^{T_k-1} f(u_t, u_{t+1}, y_t, y_{t+1})$$

$$+ f(u_{T_k-1}, u_{T_k}, y_{T_k-1}, y_{T_k}) - f(u_{T_k-1}, x_{T_k}, y_{T_k-1}, y_{T_k}) \geq \Gamma_0 - \Gamma_0/4.$$

The contradiction we have reached shows that (8.130) holds.

Let $\{v_t\}_{t=0}^{\infty}$ be an (Ω_2)-program. In the case of Theorem 8.11 we assume that

$$v_0 = y_0.$$

We show that

$$\liminf_{T\to\infty} [\sum_{t=0}^{T-1} f(x_t, x_{t+1}, y_t, y_{t+1}) - \sum_{t=0}^{T-1} f(x_t, x_{t+1}, v_t, v_{t+1})] \geq 0. \tag{8.144}$$

Assume the contrary. Then there exist $\Gamma_0 > 0$ and a strictly increasing sequence of natural numbers $\{T_k\}_{k=1}^{\infty}$ such that for all natural numbers k,

$$\sum_{t=0}^{T_k-1} f(x_t, x_{t+1}, y_t, y_{t+1}) - \sum_{t=0}^{T_k-1} f(x_t, x_{t+1}, v_t, v_{t+1}) \leq -\Gamma_0. \tag{8.145}$$

We show that

$$\lim_{t\to\infty} v_t = y_f.$$

The equality $\lim_{t\to\infty} x_t = x_f$ implies that

$$\lim_{t\to\infty} \|f^{(2)} - f(x_t, x_{t+1}, \cdot, \cdot)\| = 0. \tag{8.146}$$

By Lemma 4.18, (8.146) and (f)-optimality of the pair $\{x_t\}_{t=0}^T, \{y_t\}_{t=0}^T$ for all integers $T \geq 1$, there exist an integer $\tau_0 > 1$ and $M_0 > 0$ such that for each pair of integers $S_2 > S_1 \geq \tau_0$,

$$\sigma(\{f(x_t, x_{t+1}, \cdot, \cdot\}_{t=S_1}^{S_2-1}, S_1, S_2) \leq M_0 + \sum_{t=S_1}^{S_2-1} f(x_t, x_{t+1}, y_t, y_{t+1}) \tag{8.147}$$

and therefore in view of (8.147), for each pair of integers $S_2 > S_1 \geq \tau_0$,

$$\sum_{t=S_1}^{S_2-1} f(x_t, x_{t+1}, v_t, v_{t+1}) \leq M_0 + \sum_{t=S_1}^{S_2-1} f(x_t, x_{t+1}, y_t, y_{t+1}). \tag{8.148}$$

Set

$$M_1 = 2\tau_0\|f\| + 2M_0 + 2. \tag{8.149}$$

Assume that an integer $S \geq \tau_0$ and that

$$\sum_{t=0}^{S-1} f(x_t, x_{t+1}, y_t, y_{t+1}) - \sum_{t=0}^{S-1} f(x_t, x_{t+1}, v_t, v_{t+1}) \geq M_0 + 1.$$

Together with (8.148) this implies that for every integer $T > S$,

$$\sum_{t=0}^{T-1} f(x_t, x_{t+1}, y_t, y_{t+1}) - \sum_{t=0}^{T-1} f(x_t, x_{t+1}, v_t, v_{t+1}) \geq M_0 + 1 - M_0.$$

This contradicts (8.145). The contradiction we have reached proves that for all integers $S \geq \tau_0$,

$$\sum_{t=0}^{S-1} f(x_t, x_{t+1}, y_t, y_{t+1}) - \sum_{t=0}^{S-1} f(x_t, x_{t+1}, v_t, v_{t+1}) < M_0 + 1. \tag{8.150}$$

Assume that integers $S_1 > S_0 > \tau_0$ and that

$$\sum_{t=S_0}^{S_1-1} f(x_t, x_{t+1}, y_t, y_{t+1}) - \sum_{t=S_0}^{S_1-1} f(x_t, x_{t+1}, v_t, v_{t+1}) \geq M_1. \tag{8.151}$$

By (8.148)–(8.151),

$$M_0 + 1 > \sum_{t=0}^{S_1-1} f(x_t, x_{t+1}, y_t, y_{t+1}) - \sum_{t=0}^{S_1-1} f(x_t, x_{t+1}, v_t, v_{t+1})$$

$$= \sum_{t=0}^{\tau_0-1} f(x_t, x_{t+1}, y_t, y_{t+1}) - \sum_{t=0}^{\tau_0-1} f(x_t, x_{t+1}, v_t, v_{t+1})$$

$$+\sum_{t=\tau_0}^{S_0-1} f(x_t, x_{t+1}, y_t, y_{t+1}) - \sum_{t=\tau_0}^{S_0-1} f(x_t, x_{t+1}, v_t, v_{t+1})$$
$$+\sum_{t=S_0}^{S_1-1} f(x_t, x_{t+1}, y_t, y_{t+1}) - \sum_{t=S_0}^{S_1-1} f(x_t, x_{t+1}, v_t, v_{t+1})$$
$$\geq -2\tau_0\|f\| - M_0 + M_1 \geq M_0 + 2.$$

The contradiction we have reached shows that for each pair of integers $S_1 > S_0 > \tau_0$,

$$\sum_{t=S_0}^{S_1-1} f(x_t, x_{t+1}, y_t, y_{t+1}) - \sum_{t=S_0}^{S_1-1} f(x_t, x_{t+1}, v_t, v_{t+1}) < M_1. \tag{8.152}$$

By (8.147) and (8.152), for each pair of integers $S_1 > S_0 > \tau_0$,

$$\sum_{t=S_0}^{S_1-1} f(x_t, x_{t+1}, v_t, v_{t+1}) > -M_1 - M_0 + \sigma(\{f(x_t, x_{t+1}, \cdot, \cdot)\}_{t=S_0}^{S_1-1}, S_0, S_1).$$

Together with assertion 1 of Theorem 2.5 and (8.146) this implies that for every $\epsilon > 0$, the set

$$\{t \in \{0, 1, \ldots, \} : |v_t - y_f| > \epsilon\}$$

is finite. This implies that

$$\lim_{t\to\infty} v_t = y_f. \tag{8.153}$$

By the continuity of f, there exists $\epsilon \in (0, r_f)$ such that for each $\xi_1, \xi_2 \in X_1$ and each $\xi_3, \xi_4 \in X_2$ satisfying

$$|\xi_i - x_f| \leq \epsilon,\ i = 1, 2,\ |\xi_i - y_f| \leq \epsilon,\ i = 3, 4$$

we have

$$|f(\xi_1, \xi_2, \xi_3, \xi_4) - f(x_f, x_f, y_f, y_f)| \leq \Gamma_0/8. \tag{8.154}$$

In view of (8.153) and the equalities

$$\lim_{t\to\infty} x_t = x_f,\ \lim_{t\to\infty} y_t = y_f,$$

there exists a natural number S_0 such that for each integer $t \geq S_0$,

$$|v_t - y_f| \leq \epsilon,\ |x_t - x_f| \leq \epsilon,\ |y_t - y_f| \leq \epsilon. \tag{8.155}$$

There exists a natural number k such that

$$T_k > S_0 + 2. \tag{8.156}$$

Set

$$\xi_t = v_t,\ t = 0, \ldots, T_k - 1,\ \xi_t = y_t \text{ for all integers } t \geq T_k. \tag{8.157}$$

In view of (8.155)–(8.157), $\{\xi_t\}_{t=0}^{T_k}$ is an (Ω_2)-program. Since the pair of sequences $\{x_t\}_{t=0}^{T_k}, \{y_t\}_{t=0}^{T_k}$ is (f)-optimal (strongly (f)-optimal, respectively) we have

$$\sum_{t=0}^{T_k-1} f(x_t, x_{t+1}, \xi_t, \xi_{t+1}) \le \sum_{t=0}^{T_k-1} f(x_t, x_{t+1}, y_t, y_{t+1}).$$

It follows from the relation above, (8.145) and (8.155)–(8.157) that

$$0 \le \sum_{t=0}^{T_k-1} f(x_t, x_{t+1}, y_t, y_{t+1}) - \sum_{t=0}^{T_k-1} f(x_t, x_{t+1}, \xi_t, \xi_{t+1})$$

$$= \sum_{t=0}^{T_k-1} f(x_t, x_{t+1}, y_t, y_{t+1}) - \sum_{t=0}^{T_k-1} f(x_t, x_{t+1}, v_t, v_{t+1})$$

$$+ f(x_{T_k-1}, x_{T_k}, v_{T_k-1}, v_{T_k}) - f(x_{T_k-1}, x_{T_k}, v_{T_k-1}, y_{T_k}) \le -\Gamma_0 + \Gamma_0/4.$$

The contradiction we have reached shows that (8.144) holds. This completes the proof of Theorems 8.11 and 8.12. □

8.8 Proofs of Theorems 8.13 and 8.14

We prove Theorems 8.13 and 8.14 simultaneously. Let $\{T_k\}_{k=1}^{\infty}$ be a sequence of strictly increasing natural numbers. In the case of Theorem 8.13, $T_1 > l$. By Proposition 7.2, for each natural number k, there exist an (Ω_1)-program $\{x_t^{(k)}\}_{t=0}^{T_k}$ and an (Ω_2)-program $\{y_t^{(k)}\}_{t=0}^{T_k}$ such that in the case of Theorem 8.14 the pair $\{x_t^{(k)}\}_{t=0}^{T_k}, \{y_t^{(k)}\}_{t=0}^{T_k}$ is strongly (f)-optimal,

$$x_{T_k}^{(k)} = x_f,\ y_{T_k}^{(k)} = y_f \tag{8.158}$$

and in the case of Theorem 8.13,

$$x_0^{(k)} = \xi_0,\ x_{T_k}^{(k)} = x_f,\ y_0^{(k)} = \xi_1,\ y_{T_k}^{(k)} = y_f \tag{8.159}$$

and the pair $\{x_t^{(k)}\}_{t=0}^{T_k}, \{y_t^{(k)}\}_{t=0}^{T_k}$ is (f)-optimal.

Extracting a subsequence and re-indexing, if necessary, we may assume without loss of generality that for every integer $t \ge 0$ there exist

$$\bar{x}_t = \lim_{k\to\infty} x_t^{(k)},\ \bar{y}_t = \lim_{k\to\infty} y_t^{(k)}. \tag{8.160}$$

By Theorem 8.5, (8.158), and (8.159), for every $\epsilon > 0$, there exists a natural number $k(\epsilon)$ such that for every integer $k \ge 2k(\epsilon)$,

$$|x_t^{(k)} - x_f| \le \epsilon,\ |y_t^{(k)} - y_f| \le \epsilon$$

$$\text{for all integers } t \text{ satisfying } k(\epsilon) \le t \le T_k - k(\epsilon). \tag{8.161}$$

In view of (8.160) and (8.161), for every $\epsilon > 0$,

$$|\bar{x}_t - x_f| \le \epsilon,\ |\bar{y}_t - y_f| \le \epsilon \text{ for all integers } t \ge k(\epsilon). \tag{8.162}$$

It follows from (8.162) that

$$\lim_{t\to\infty} \bar{x}_t = x_f,\ \lim_{t\to\infty} \bar{y}_t = y_f. \tag{8.163}$$

Let $S_0 \ge 1$ be an integer and $\{x_t\}_{t=0}^{S_0}$ be an (Ω_1)-program such that

$$x_{S_0} = \bar{x}_{S_0}. \tag{8.164}$$

In the case of Theorem 8.13 we also assume that

$$x_0 = \bar{x}_0 \tag{8.165}$$

We show that

$$\sum_{t=0}^{S_0-1} f(x_t, x_{t+1}, \bar{y}_t, \bar{y}_{t+1}) \ge \sum_{t=0}^{S_0-1} f(\bar{x}_t, \bar{x}_{t+1}, \bar{y}_t, \bar{y}_{t+1}). \tag{8.166}$$

Assume the contrary. Then there exists $\Gamma_0 > 0$ such that

$$\sum_{t=0}^{S_0-1} f(x_t, x_{t+1}, \bar{y}_t, \bar{y}_{t+1}) < \sum_{t=0}^{S_0-1} f(\bar{x}_t, \bar{x}_{t+1}, \bar{y}_t, \bar{y}_{t+1}) - \Gamma_0. \tag{8.167}$$

By the continuity of f, there exists $\delta \in (0, r_f/4)$ such that for each $\xi_1, \xi_2 \in X_1$ and each $\xi_3, \xi_4 \in X_2$ satisfying

$$|\xi_i - x_f| \le \delta,\ i = 1, 2,\ |\xi_i - y_f| \le \delta,\ i = 3, 4$$

we have

$$|f(\xi_1, \xi_2, \xi_3, \xi_4) - f(x_f, x_f, y_f, y_f)| \le \Gamma_0/8. \tag{8.168}$$

In view of the choice of $k(\epsilon)$, $\epsilon > 0$ (see (8.161)) and (8.163), for every integer $k \ge 2k(\delta)$,

$$|x_t^{(k)} - x_f| \le \delta,\ |y_t^{(k)} - y_f| \le \delta \tag{8.169}$$

for all integers t satisfying $k(\delta) \le t \le T_k - k(\delta)$ and

$$|\bar{x}_t - x_f| \le \delta,\ |\bar{y}_t - y_f| \le \delta \tag{8.170}$$

for all integers $t \ge k(\delta)$.

Fix a natural number $k_1 > k(\delta)$ such that

$$T_{k_1} > 4T_{k(\delta)} + 4S_0. \tag{8.171}$$

Set

$$x_t = \bar{x}_t, \ t = S_0 + 1, \dots, T_{k_1}. \tag{8.172}$$

By the continuity of f, there exists $\delta_0 \in (0, \delta)$ such that for each $\xi_1, \xi_2, \bar{\xi}_1, \bar{\xi}_2 \in X_1$ and each $\eta_1, \eta_2, \bar{\eta}_1, \bar{\eta}_2 \in X_2$ satisfying

$$(\xi_1, \xi_2), \ (\bar{\xi}_1, \bar{\xi}_2) \in \Omega_1, \ (\eta_1, \eta_2), \ (\bar{\eta}_1, \bar{\eta}_2) \in \Omega_2,$$

$$|\xi_i - \bar{\xi}_i| \leq \delta_0, \ i = 1, 2, \ |\eta_i - \bar{\eta}_i| \leq \delta_0, \ i = 1, 2,$$

we have

$$|f(\xi_1, \xi_2, \eta_1, \eta_2) - f(\bar{\xi}_1, \bar{\xi}_2, \bar{\eta}_1, \bar{\eta}_2)| \leq \Gamma_0(8(T_{k_1} + 1))^{-1}. \tag{8.173}$$

In view of (8.160), there exists an integer $k_2 > k_1$ such that

$$|x_t^{(k_2)} - \bar{x}_t| \leq \delta_0, \ |y_t^{(k_2)} - \bar{y}_t| \leq \delta_0, \ t = 0, \dots, T_{k_1}. \tag{8.174}$$

Set

$$z_t = x_t, \ t = 0, \dots, T_{k_1} - 1, \ z_{T_{k_1}} = x_{T_{k_1}}^{(k_2)}. \tag{8.175}$$

It follows from (8.159), (8.164), (8.165), (8.169), (8.170), (8.172), (8.174), and (8.175) that $\{z_t\}_{t=0}^{T_{k_1}}$ is an (Ω_1)-program and that in the case of Theorem 8.13,

$$z_0 = x_0^{(k_2)}. \tag{8.176}$$

By the choice of the sequences $\{x_t^{(k_2)}\}_{t=0}^{T_{k_2}}$, $\{y_t^{(k_2)}\}_{t=0}^{T_{k_2}}$, (8.175) and (8.176),

$$\sum_{t=0}^{T_{k_1}-1} f(z_t, z_{t+1}, y_t^{(k_2)}, y_{t+1}^{(k_2)}) \geq \sum_{t=0}^{T_{k_1}-1} f(x_t^{(k_2)}, x_{t+1}^{(k_2)}, y_t^{(k_2)}, y_{t+1}^{(k_2)}). \tag{8.177}$$

It follows from (8.173) and (8.174) that

$$|\sum_{t=0}^{T_{k_1}-1} f(x_t^{(k_2)}, x_{t+1}^{(k_2)}, y_t^{(k_2)}, y_{t+1}^{(k_2)}) - \sum_{t=0}^{T_{k_1}-1} f(\bar{x}_t, \bar{x}_{t+1}, \bar{y}_t, \bar{y}_{t+1})| \leq 8^{-1}\Gamma_0, \tag{8.178}$$

$$|\sum_{t=0}^{T_{k_1}-1} f(z_t, z_{t+1}, y_t^{(k_2)}, y_{t+1}^{(k_2)}) - \sum_{t=0}^{T_{k_1}-1} f(z_t, z_{t+1}, \bar{y}_t, \bar{y}_{t+1})| \leq 8^{-1}\Gamma_0. \tag{8.179}$$

In view of (8.175),

$$|\sum_{t=0}^{T_{k_1}-1} f(z_t, z_{t+1}, \bar{y}_t, \bar{y}_{t+1}) - \sum_{t=0}^{T_{k_1}-1} f(x_t, x_{t+1}, \bar{y}_t, \bar{y}_{t+1})|$$

$$= |f(x_{T_{k_1}-1}, x_{T_{k_1}}^{(k_2)}, \bar{y}_{T_{k_1}-1}, \bar{y}_{T_{k_1}}) - f(x_{T_{k_1}-1}, x_{T_{k_1}}, \bar{y}_{T_{k_1}-1}, \bar{y}_{T_{k_1}})| \leq 8^{-1}\Gamma_0. \tag{8.180}$$

By (8.167), (8.172), and (8.177)–(8.180),

$$0 \leq \sum_{t=0}^{T_{k_1}-1} f(z_t, z_{t+1}, y_t^{(k_2)}, y_{t+1}^{(k_2)}) - \sum_{t=0}^{T_{k_1}-1} f(x_t^{(k_2)}, x_{t+1}^{(k_2)}, y_t^{(k_2)}, y_{t+1}^{(k_2)})$$

$$\leq \sum_{t=0}^{T_{k_1}-1} f(x_t, x_{t+1}, \bar{y}_t, \bar{y}_{t+1}) + 4^{-1}\Gamma_0$$

$$- \sum_{t=0}^{T_{k_1}-1} f(\bar{x}_t, \bar{x}_{t+1}, \bar{y}_t, \bar{y}_{t+1}) + 8^{-1}\Gamma_0 \leq -(5/8)\Gamma_0.$$

The contradiction we have reached shows that (8.166) holds for any integer $S_0 \geq 1$. Analogously we can show that for every integer $S_0 \geq 1$ and every (Ω_2)-program $\{y_t\}_{t=0}^{S_0}$ such that

$$y_{S_0} = \bar{y}_{S_0}$$

and in the case of Theorem 8.13

$$y_0 = \bar{y}_0$$

we have

$$\sum_{t=0}^{S_0-1} f(\bar{x}_t, \bar{x}_{t+1}, y_t, y_{t+1}) \leq \sum_{t=0}^{S_0-1} f(\bar{x}_t, \bar{x}_{t+1}, \bar{y}_t, \bar{y}_{t+1}).$$

This completes the proof of Theorems 8.13 and 8.14. □

8.9 Uniqueness of Overtaking Optimal Solutions

In this section we use the following assumptions.

(A3) For each $(y_1, y_2) \in \Omega_2$, the function $f(\cdot, \cdot, y_1, y_2) : \Omega_1 \to R^1$ is strictly convex.

(A4) For each $(x_1, x_2) \in \Omega_1$, the function $f(x_1, x_2, \cdot, \cdot) : \Omega_2 \to R^1$ is strictly concave.

Theorem 8.17. *Let assumptions (A3), (A4) hold and let* $\xi_1 \in \cup\{\bar{Y}_1(l) : l = 1, 2, \dots\}$ *and* $\xi_2 \in \cup\{\bar{Y}_2(l) : l = 1, 2, \dots\}$. *Then there exists a unique* (f)*-overtaking optimal pair of an* (Ω_1)*-program* $\{x_t\}_{t=0}^{\infty}$ *and an* (Ω_2)*-program* $\{y_t\}_{t=0}^{\infty}$ *such that*

$$x_0 = \xi_1, \; y_0 = \xi_2.$$

Theorem 8.18. *Let assumptions (A3), (A4) hold. Then there exist a unique strongly* (f)*-overtaking optimal pair of an* (Ω_1)*-program* $\{x_t\}_{t=0}^{\infty}$ *and an* (Ω_2)*-program* $\{y_t\}_{t=0}^{\infty}$.

Proof. We prove Theorems 8.17 and 8.18 simultaneously. In the case of Theorem 8.17, we assume that for $i = 1, 2$, $\{x_t^{(i)}\}_{t=0}^\infty$ is an (Ω_1)-program, $\{y_t^{(i)}\}_{t=0}^\infty$ is an (Ω_2)-program, the pair $\{x_t^{(i)}\}_{t=0}^\infty$, $\{y_t^{(i)}\}_{t=0}^\infty$ is (f)-overtaking optimal,

$$x_0^{(1)} = x_0^{(2)} \in \cup\{\bar{Y}_1(l) :\ l = 1, 2, \dots\},$$

$$y_0^{(1)} = y_0^{(2)} \in \cup\{\bar{Y}_1(l) :\ l = 1, 2, \dots\}.$$

In the case of Theorem 8.18, we assume that for $i = 1, 2$, $\{x_t^{(i)}\}_{t=0}^\infty$ is an (Ω_1)-program, $\{y_t^{(i)}\}_{t=0}^\infty$ is an (Ω_2)-program, the pair $\{x_t^{(i)}\}_{t=0}^\infty$, $\{y_t^{(i)}\}_{t=0}^\infty$ is strongly (f)-overtaking optimal. Then

$$\limsup_{T\to\infty}[\sum_{t=0}^{T-1} f(x_t^{(1)}, x_{t+1}^{(1)}, y_{t+1}^{(1)}, y_{t+1}^{(1)}) - \sum_{t=0}^{T-1} f(x_t^{(2)}, x_{t+1}^{(2)}, y_{t+1}^{(1)}, y_{t+1}^{(1)})] \le 0,$$

$$\limsup_{T\to\infty}[\sum_{t=0}^{T-1} f(x_t^{(2)}, x_{t+1}^{(2)}, y_{t+1}^{(2)}, y_{t+1}^{(2)}) - \sum_{t=0}^{T-1} f(x_t^{(1)}, x_{t+1}^{(1)}, y_{t+1}^{(2)}, y_{t+1}^{(2)})] \le 0,$$

$$\liminf_{T\to\infty}[\sum_{t=0}^{T-1} f(x_t^{(1)}, x_{t+1}^{(1)}, y_{t+1}^{(1)}, y_{t+1}^{(1)}) - \sum_{t=0}^{T-1} f(x_t^{(1)}, x_{t+1}^{(1)}, y_{t+1}^{(2)}, y_{t+1}^{(2)})] \ge 0,$$

$$\liminf_{T\to\infty}[\sum_{t=0}^{T-1} f(x_t^{(2)}, x_{t+1}^{(2)}, y_{t+1}^{(2)}, y_{t+1}^{(2)}) - \sum_{t=0}^{T-1} f(x_t^{(2)}, x_{t+1}^{(2)}, y_{t+1}^{(1)}, y_{t+1}^{(1)})] \ge 0.$$

These inequalities imply that

$$\lim_{T\to\infty}[\sum_{t=0}^{T-1} f(x_t^{(1)}, x_{t+1}^{(1)}, y_{t+1}^{(1)}, y_{t+1}^{(1)}) - \sum_{t=0}^{T-1} f(x_t^{(2)}, x_{t+1}^{(2)}, y_{t+1}^{(1)}, y_{t+1}^{(1)})] = 0, \quad (8.181)$$

$$\lim_{T\to\infty}[\sum_{t=0}^{T-1} f(x_t^{(2)}, x_{t+1}^{(2)}, y_{t+1}^{(2)}, y_{t+1}^{(2)}) - \sum_{t=0}^{T-1} f(x_t^{(1)}, x_{t+1}^{(1)}, y_{t+1}^{(2)}, y_{t+1}^{(2)})] = 0, \quad (8.182)$$

$$\lim_{T\to\infty}[\sum_{t=0}^{T-1} f(x_t^{(1)}, x_{t+1}^{(1)}, y_{t+1}^{(1)}, y_{t+1}^{(1)}) - \sum_{t=0}^{T-1} f(x_t^{(1)}, x_{t+1}^{(1)}, y_{t+1}^{(2)}, y_{t+1}^{(2)})] = 0, \quad (8.183)$$

$$\lim_{T\to\infty}[\sum_{t=0}^{T-1} f(x_t^{(2)}, x_{t+1}^{(2)}, y_{t+1}^{(2)}, y_{t+1}^{(2)}) - \sum_{t=0}^{T-1} f(x_t^{(2)}, x_{t+1}^{(2)}, y_{t+1}^{(1)}, y_{t+1}^{(1)})] = 0. \quad (8.184)$$

Set

$$\widehat{x}_t = 2^{-1}(x_t^{(1)} + x_t^{(2)}),\ \widehat{y}_t = 2^{-1}(y_t^{(1)} + y_t^{(2)}). \quad (8.185)$$

If an integer $j \ge 0$ satisfies $x_j^{(1)} \ne x_j^{(2)}$, then it follows from (A3), (8.181), and (8.185) that

$$\limsup_{T\to\infty}[\sum_{t=0}^{T-1} f(x_t^{(1)}, x_{t+1}^{(1)}, y_{t+1}^{(1)}, y_{t+1}^{(1)}) - \sum_{t=0}^{T-1} f(\widehat{x}_t, \widehat{x}_{t+1}, y_{t+1}^{(1)}, y_{t+1}^{(1)})]$$

$$> \limsup_{T\to\infty}[\sum_{t=0}^{T-1} f(x_t^{(1)}, x_{t+1}^{(1)}, y_{t+1}^{(1)}, y_{t+1}^{(1)}) - 2^{-1}\sum_{t=0}^{T-1} f(x_t^{(1)}, x_{t+1}^{(1)}, y_{t+1}^{(1)}, y_{t+1}^{(1)})$$

$$-2^{-1}\sum_{t=0}^{T-1} f(x_t^{(2)}, x_{t+1}^{(2)}, y_{t+1}^{(1)}, y_{t+1}^{(1)})] = 0,$$

a contradiction. Then $x_j^{(1)} = x_j^{(2)}$, $j = 0, 1, \dots$. If an integer $j \geq 0$ satisfies $y_j^{(1)} \neq y_j^{(2)}$, then it follows from (A4), (8.183), and (8.185) that

$$\liminf_{T\to\infty}[\sum_{t=0}^{T-1} f(x_t^{(1)}, x_{t+1}^{(1)}, y_{t+1}^{(1)}, y_{t+1}^{(1)}) - \sum_{t=0}^{T-1} f(x_t^{(1)}, x_{t+1}^{(1)}, \widehat{y}_t, \widehat{y}_{t+1})]$$

$$< \liminf_{T\to\infty}[\sum_{t=0}^{T-1} f(x_t^{(1)}, x_{t+1}^{(1)}, y_{t+1}^{(1)}, y_{t+1}^{(1)}) - 2^{-1}\sum_{t=0}^{T-1} f(x_t^{(1)}, x_{t+1}^{(1)}, y_{t+1}^{(1)}, y_{t+1}^{(1)})$$

$$-2^{-1}\sum_{t=0}^{T-1} f(x_t^{(1)}, x_{t+1}^{(1)}, y_t^{(2)}, y_{t+1}^{(2)})] = 0,$$

a contradiction. Then $y_j^{(1)} = y_j^{(2)}$, $j = 0, 1, \dots$. Theorems 8.17 and 8.18 are proved. □

8.10 Turnpike Properties for Games with Discounting

In Sect. 8.12 we prove the following turnpike result.

Theorem 8.19. *Let $\epsilon \in (0, r_f)$ and l be a natural number. Then there exist a natural number $L > 2l$ and $\delta \in (0, \epsilon)$, $\lambda \in (0, 1)$ such that for each integer $T > 2L$, each $f_t \in \mathcal{M}$, $t = 0, \dots, T-1$ satisfying*

$$\|f_t - f\| \leq \delta,\ t = 0, \dots, T-1,$$

each sequence $\{\alpha_t\}_{t=0}^{T-1} \subset (0, 1]$ such that for each integer $\tau \in [0, T-L]$

$$\alpha_i \alpha_j^{-1} \geq \lambda \text{ for all } i, j \in \{\tau, \dots, \tau + L\},$$

each (Ω_1)-program $\{x_t\}_{t=0}^T$ and each (Ω_2)-program $\{y_t\}_{t=0}^T$ such that at least one of the following conditions holds:

(a) $x_0 \in \bar{Y}_1(l)$, $y_0 \in \bar{Y}_2(l)$, $x_T \in Y_1(l)$, $y_T \in Y_2(l)$, the pair $\{x_t\}_{t=0}^T$, $\{y_t\}_{t=0}^T$ is $(\{\alpha_t f_t\}_{t=0}^{T-1}, x_0, x_T, y_0, y_T, 0, T, \delta \min\{\alpha_t :\ t = 0, \dots, T-1\})$-good;

(b) $x_0 \in \bar{Y}_1(l)$, $y_0 \in \bar{Y}_2(l)$, *the pair* $\{x_t\}_{t=0}^T$, $\{y_t\}_{t=0}^T$ *is*

$$(\{\alpha_t f_t\}_{t=0}^{T-1}, x_0, y_0, 0, T, \delta \min\{\alpha_t : t = 0, \dots, T-1\}) - good;$$

(c) the pair $\{x_t\}_{t=0}^T$, $\{y_t\}_{t=0}^T$ *is*

$$(\{\alpha_t f_t\}_{t=0}^{T-1}, 0, T, \delta \min\{\alpha_t : t = 0, \dots, T-1\}) - good;$$

(d) $x_T \in Y_1(l)$, $y_T \in Y_2(l)$, *the pair* $\{x_t\}_{t=0}^T$, $\{y_t\}_{t=0}^T$ *is*

$$(\{\alpha_t f_t\}_{t=0}^{T-1}, (x_T, y_T)^-, 0, T, \delta \min\{\alpha_t : t = 0, \dots, T-1\}) - good$$

there exist integers $\tau_1 \in [0, L]$ *and* $\tau_2 \in [T - L, T]$ *such that for all* $t = \tau_1, \dots, \tau_2$,

$$|x_t - x_f| \le \epsilon,\ |y_t - y_f| \le \epsilon.$$

Moreover, if

$$|x_0 - x_f| \le \delta,\ |y_0 - y_f| \le \delta,$$

then $\tau_1 = 0$ *and if*

$$|x_T - x_f| \le \delta,\ |y_T - y_f| \le \delta,$$

then $\tau_2 = T$.

8.11 Auxiliary Results for Theorem 8.19

Lemma 8.20. *Let* $\epsilon \in (0, r_f)$ *and* l *be a natural number. Then there exist a natural number* $L > 2l$ *and* $\delta \in (0, \epsilon)$, $\lambda \in (0, 1)$ *such that for each integer* $T \ge L$, *each* $f_t \in \mathcal{M}$, $t = 0, \dots, T-1$ *satisfying*

$$\|f_t - f\| \le \delta,\ t = 0, \dots, T-1, \tag{8.186}$$

each sequence $\{\alpha_t\}_{t=0}^{T-1} \subset (0, 1]$ *such that for each integer* $\tau \in [0, T - L]$

$$\alpha_i \alpha_j^{-1} \ge \lambda \text{ for all } i, j \in \{\tau, \dots, \tau + L\}, \tag{8.187}$$

each (Ω_1)*-program* $\{x_t\}_{t=0}^T$ *and each* (Ω_2)*-program* $\{y_t\}_{t=0}^T$ *such that at least one of the following conditions holds:*

(a) $x_0 \in \bar{Y}_1(l)$, $y_0 \in \bar{Y}_2(l)$, $x_T \in Y_1(l)$, $y_T \in Y_2(l)$, *the pair* $\{x_t\}_{t=0}^T$, $\{y_t\}_{t=0}^T$ *is* $(\{\alpha_t f_t\}_{t=0}^{T-1}, x_0, x_T, y_0, y_T, 0, T, \delta \min\{\alpha_t : t = 0, \dots, T-1\})$*-good;*

(b) $x_0 \in \bar{Y}_1(l)$, $y_0 \in \bar{Y}_2(l)$, *the pair* $\{x_t\}_{t=0}^T$, $\{y_t\}_{t=0}^T$ *is*

$$(\{\alpha_t f_t\}_{t=0}^{T-1}, x_0, y_0, 0, T, \delta \min\{\alpha_t : t = 0, \dots, T-1\}) - good;$$

(c) the pair $\{x_t\}_{t=0}^T$, $\{y_t\}_{t=0}^T$ *is* $(\{\alpha_t f_t\}_{t=0}^{T-1}, 0, T, \delta \min\{\alpha_t : t = 0, \dots, T-1\})$*-good;*

(d) $x_T \in Y_1(l)$, $y_T \in Y_2(l)$, *the pair* $\{x_t\}_{t=0}^T$, $\{y_t\}_{t=0}^T$ *is*

$$(\{\alpha_t f_t\}_{t=0}^{T-1}, (x_T, y_T)^-, 0, T, \delta \min\{\alpha_t : t = 0, \dots, T-1\}) - good$$

there exist an integer $j \in \{1, \dots, T-1\}$ *such that*

$$|x_j - x_f| \le \epsilon,\ |y_j - y_f| \le \epsilon. \tag{8.188}$$

Proof. By property (P1), there exists $\gamma \in (0, \epsilon)$ such that the following property holds:

(P13) for every $(\xi_1, \xi_2) \in \Omega_1$ satisfying $L_1(\xi_1, \xi_2) \le \gamma$, we have

$$|\xi_1 - x_f|, \ |\xi_2 - x_f| \le \epsilon/2;$$

for every $(\xi_1, \xi_2) \in \Omega_2$ satisfying $L_2(\xi_1, \xi_2) \le \gamma$, we have

$$|\xi_1 - y_f|, |\xi_2 - y_f| \le \epsilon/2.$$

Choose

$$\delta \in (0, \gamma/8), \tag{8.189}$$

a natural number L_0 such that

$$L_0\gamma > 4(|l_1| + |l_2|)(\mathrm{diam}(X_1) + \mathrm{diam}(X_2)) + 2l + 4, \tag{8.190}$$

a number λ such that

$$\lambda \in (2^{-1}, 1), \ (1 - \lambda)\lambda^{-1}(\|f\| + 1) \le \delta \tag{8.191}$$

and natural numbers

$$q_1 > 64(\|f\| + 1)\gamma^{-1}\lambda^{-2}, \ q_0 = 2q_1, \tag{8.192}$$

$$L = 2L_0q_0 + 4l. \tag{8.193}$$

Assume that an integer $T \ge L$, $f_t \in \mathcal{M}$, $t = 0, \dots, T - 1$ satisfy (8.186), a sequence $\{\alpha_t\}_{t=0}^{T-1} \subset (0, 1]$ satisfies (8.187) for each integer $\tau \in [0, T - L]$, an (Ω_1)-program $\{x_t\}_{t=0}^{T}$ and an (Ω_2)-program $\{y_t\}_{t=0}^{T}$ satisfy at least one of conditions (a)–(d). We show that there exists an integer $j \in \{1, \dots, T - 1\}$ such that (8.188) holds.

Assume the contrary. Then for every integer $i \in \{1, \dots, T - 1\}$,

$$\max\{|x_i - x_f|, \ |y_i - y_f|\} > \epsilon. \tag{8.194}$$

By Definition 8.2, (a)–(d), and (8.193), there exist an (Ω_1)-program $\{u_t\}_{t=0}^{T}$ and an (Ω_2)-program $\{v_t\}_{t=0}^{T}$ such that

$$u_i = x_f, \ v_i = y_f \text{ for all integers } i \in [l, T - l], \tag{8.195}$$

in the case (a)

$$u_0 = x_0, \ u_T = x_T, \ v_0 = y_0, \ v_T = y_T, \tag{8.196}$$

in the case (b)

$$u_0 = x_0, \ v_0 = y_0, \tag{8.197}$$

in the case (d)

$$u_T = x_T, \ v_T = y_T.$$

In view of (8.193),

$$T - 2l \geq L - 2l \geq 2L_0 q_0. \tag{8.198}$$

Relation (8.198) implies that there exists an integer $q \geq 2q_0$ such that

$$qL_0 \leq T - 2l \leq (q+1)L_0. \tag{8.199}$$

Set

$$\alpha_* = \min\{\alpha_t : t \in \{0, \ldots, T-1\}\} \tag{8.200}$$

By (a)–(d) and (8.195)–(8.197),

$$\begin{aligned} &-\delta\alpha_* + \sum_{t=0}^{T-1} \alpha_t f_t(x_t, x_{t+1}, v_t, v_{t+1}) \\ &\leq \sum_{t=0}^{T-1} \alpha_t f_t(x_t, x_{t+1}, y_t, y_{t+1}) \\ &\leq \sum_{t=0}^{T-1} \alpha_t f_t(u_t, u_{t+1}, y_t, y_{t+1}) + \delta\alpha_*. \end{aligned} \tag{8.201}$$

It follows from (8.186), (8.187), (8.193), (8.199), and (8.201) that

$$\begin{aligned} &-\delta\alpha_* + \sum_{t=l}^{l+qL_0-1} \alpha_t f_t(x_t, x_{t+1}, v_t, v_{t+1}) \\ &-(\|f\|+1)l\alpha_0\lambda^{-1} - (\|f\|+1)(L_0+l)\alpha_T\lambda^{-1} \\ &\leq \sum_{t=0}^{T-1} \alpha_t f_t(x_t, x_{t+1}, y_t, y_{t+1}) \\ &\leq \sum_{t=0}^{T-1} \alpha_t f_t(u_t, u_{t+1}, y_t, y_{t+1}) + \delta\alpha_* \\ &\leq \delta\alpha_* + \sum_{t=l}^{l+qL_0-1} \alpha_t f_t(u_t, u_{t+1}, y_t, y_{t+1}) \\ &+ (\|f\|+1)l\alpha_0\lambda^{-1} + (\|f\|+1)(L_0+l)\alpha_T\lambda^{-1}. \end{aligned} \tag{8.202}$$

By (8.186), (8.187), (8.193), and (8.199),

$$\begin{aligned} &\left|\sum_{t=0}^{T-1} \alpha_t f_t(x_t, x_{t+1}, y_t, y_{t+1}) - \sum_{t=l}^{l+qL_0-1} \alpha_t f_t(x_t, x_{t+1}, y_t, y_{t+1})\right| \\ &\leq (\|f\|+1)l\alpha_0\lambda^{-1} + (\|f\|+1)(L_0+l)\alpha_T\lambda^{-1}. \end{aligned} \tag{8.203}$$

In view of (8.202) and (8.203),

$$\sum_{t=l}^{l+qL_0-1} \alpha_t f_t(x_t, x_{t+1}, v_t, v_{t+1}) - \delta\alpha_* - 2(\|f\|+1)[l\alpha_0\lambda^{-1} + (L_0+l)\alpha_T\lambda^{-1}]$$

$$\leq \sum_{t=l}^{l+qL_0-1} \alpha_t f_t(x_t, x_{t+1}, y_t, y_{t+1})$$

$$\leq \delta\alpha_* + \sum_{t=l}^{l+qL_0-1} \alpha_t f_t(u_t, u_{t+1}, y_t, y_{t+1})$$

$$+ 2(\|f\|+1)[l\alpha_0\lambda^{-1} + (L_0+l)\alpha_T\lambda^{-1}]. \tag{8.204}$$

By (8.195) and (8.204),

$$\sum_{t=l}^{l+qL_0-1} \alpha_t f_t(x_t, x_{t+1}, y_f, y_f)$$

$$-\delta\alpha_* - 2(\|f\|+1)[l\alpha_0\lambda^{-1} + (L_0+l)\alpha_T\lambda^{-1}]$$

$$\leq \sum_{t=l}^{l+qL_0-1} \alpha_t f_t(x_t, x_{t+1}, y_t, y_{t+1})$$

$$\leq \delta\alpha_* + \sum_{t=l}^{l+qL_0-1} \alpha_t f_t(x_f, x_f, y_t, y_{t+1})$$

$$+ 2(\|f\|+1)[l\alpha_0\lambda^{-1} + (L_0+l)\alpha_T\lambda^{-1}]. \tag{8.205}$$

Fix an integer $k \in \{0, \ldots . q-1\}$. For each (Ω_1)-program $\{\xi_t\}_{t=l+kL_0}^{l+(k+1)L_0}$ and each (Ω_2)-program $\{\eta_t\}_{t=l+kL_0}^{l+(k+1)L_0}$ it follows from (8.186), (8.187), (8.191), and (8.193) that

$$|\sum_{t=l+kL_0}^{l+(k+1)L_0-1} \alpha_t f_t(\xi_t, \xi_{t+1}, \eta_t, \eta_{t+1})$$

$$-\alpha_{l+kL_0} \sum_{t=l+kL_0}^{l+(k+1)L_0-1} f_t(\xi_t, \xi_{t+1}, \eta_t, \eta_{t+1})|$$

$$= |\alpha_{l+kL_0} \sum_{t=l+kL_0}^{l+(k+1)L_0-1} \alpha_t \alpha_{l+kL_0}^{-1} f_t(\xi_t, \xi_{t+1}, \eta_t, \eta_{t+1})$$

$$-\alpha_{l+kL_0} \sum_{t=l+kL_0}^{l+(k+1)L_0-1} f_t(\xi_t, \xi_{t+1}, \eta_t, \eta_{t+1})|$$

$$\leq \alpha_{l+kL_0} L_0 \max\{\|\alpha_t \alpha_{l+kL_0}^{-1} f_t - f\| :$$

$$t \in \{l+kL_0, \ldots, l+(k+1)L_0-1\}\}$$

$$\leq \alpha_{l+kL_0} L_0 \max\{\|\alpha_t \alpha_{l+kL_0}^{-1} f_t - f_t\| + \|f_t - f\| :$$
$$t \in \{l + kL_0, \dots, l + (k+1)L_0 - 1\}\}$$
$$\leq \alpha_{l+kL_0} L_0(\delta + (1-\lambda)\lambda^{-1}(\|f\| + 1)) \leq 2\delta\alpha_{l+kL_0} L_0. \tag{8.206}$$

By (8.54), (8.56), and (8.206),

$$\sum_{t=l+kL_0}^{l+(k+1)L_0-1} \alpha_t f_t(x_t, x_{t+1}, y_f, y_f)$$
$$\geq \alpha_{l+kL_0} \sum_{t=l+kL_0}^{l+(k+1)L_0-1} f(x_t, x_{t+1}, y_f, y_f) - 2\delta L_0 \alpha_{l+kL_0}$$
$$= \alpha_{l+kL_0}(\sum_{t=l+kL_0}^{l+(k+1)L_0-1} L_1(x_t, x_{t+1}) + L_0 f(x_f, x_f, y_f, y_f)$$
$$+ \langle l_1, x_{l+kL_0} - x_{l+(k+1)L_0}\rangle) - 2\delta L_0 \alpha_{l+kL_0}, \tag{8.207}$$
$$\sum_{t=l+kL_0}^{l+(k+1)L_0-1} \alpha_t f_t(x_f, x_f, y_t, y_{t+1})$$
$$\leq \alpha_{l+kL_0} \sum_{t=l+kL_0}^{l+(k+1)L_0-1} f(x_f, x_f, y_t, y_{t+1}) + 2\delta L_0 \alpha_{l+kL_0}$$
$$= \alpha_{l+kL_0}(- \sum_{t=l+kL_0}^{l+(k+1)L_0-1} L_2(y_t, y_{t+1}) + L_0 f(x_f, x_f, y_f, y_f)$$
$$+ \langle l_2, y_{l+kL_0} - y_{l+(k+1)L_0}\rangle) + 2\delta L_0 \alpha_{l+kL_0}, \tag{8.208}$$

In view of (8.207) and (8.208),

$$\sum_{t=l+kL_0}^{l+(k+1)L_0-1} \alpha_t f_t(x_t, x_{t+1}, y_f, y_f) - \sum_{t=l+kL_0}^{l+(k+1)L_0-1} \alpha_t f_t(x_f, x_f, y_t, y_{t+1})$$
$$\geq \alpha_{l+kL_0}(\sum_{t=l+kL_0}^{l+(k+1)L_0-1} (L_1(x_t, x_{t+1}) + L_2(y_t, y_{t+1}))$$
$$- (|l_1| + |l_2|)(\mathrm{diam}(X_1) + \mathrm{diam}(X_2))) - 4\delta L_0 \alpha_{l+kL_0}. \tag{8.209}$$

Property (P13) and (8.194) imply that

$$L_1(x_t, x_{t+1}) + L_2(y_t, y_{t+1}) \geq \gamma \text{ for all } t = l + kL_0, \dots, l + (k+1)L_0 - 1. \tag{8.210}$$

It follows from (8.189), (8.190), (8.209), and (8.210) that

$$\sum_{t=l+kL_0}^{l+(k+1)L_0-1} \alpha_t f_t(x_t, x_{t+1}, y_f, y_f) - \sum_{t=l+kL_0}^{l+(k+1)L_0-1} \alpha_t f_t(x_f, x_f, y_t, y_{t+1})$$

$$\geq \alpha_{l+kL_0}(L_0\gamma - (|l_1| + |l_2|)(\mathrm{diam}(X_1) + \mathrm{diam}(X_2)) - 4\delta L_0)$$

$$\geq 4^{-1}\alpha_{l+kL_0} L_0\gamma,\ k = 0, \dots, q-1.$$

By the relation above, (8.187), (8.190), (8.192), (8.193), and (8.205),

$$8(\alpha_0 + \alpha_T)(\|f\| + 1)\lambda^{-1}(L_0 + l)$$

$$\geq 2\delta\alpha_* + 4(\|f\| + 1)[l\alpha_0\lambda^{-1} + (L_0 + l)\alpha_T\lambda^{-1}]$$

$$\geq \sum_{t=l}^{l+qL_0-1} \alpha_t f_t(x_t, x_{t+1}, y_f, y_f) - \sum_{t=l}^{l+qL_0-1} \alpha_t f_t(x_f, x_f, y_t, y_{t+1})$$

$$\geq 4^{-1}L_0\gamma \sum_{k=0}^{q-1} \alpha_{l+kL_0} \geq 4^{-1}L_0\gamma \sum_{k=0}^{q_1-1} \alpha_{l+kL_0} + 4^{-1}L_0\gamma \sum_{k=q-q_1}^{q-1} \alpha_{l+kL_0}$$

$$\geq 4^{-1}L_0\gamma\lambda\alpha_0 q_1 + 4^{-1}L_0\gamma\lambda\alpha_T q_1 \geq 4^{-1}L_0\gamma\lambda(\alpha_0 + \alpha_T)q_1,$$

$$q_1 \leq 64(\|f\| + 1)\gamma^{-1}\lambda^{-2}.$$

This contradicts (8.192). The contradiction we have reached proves Lemma 8.20. □

Lemma 8.21. *Let $\epsilon \in (0, r_f)$ and l be a natural number. Then there exist a natural number $L > 2l$, $\delta \in (0, \epsilon)$ and $\lambda \in (0, 1)$ such that for each integer $T \geq L$, each $f_t \in \mathcal{M}$, $t = 0, \dots, T-1$ satisfying*

$$\|f_t - f\| \leq \delta,\ t = 0, \dots, T-1, \tag{8.211}$$

each sequence $\{\alpha_t\}_{t=0}^{T-1} \subset (0, 1]$ such that for each integer $\tau \in [0, T-L]$

$$\alpha_i\alpha_j^{-1} \geq \lambda \text{ for all } i, j \in \{\tau, \dots, \tau + L\}, \tag{8.212}$$

each (Ω_1)-program $\{x_t\}_{t=0}^T$ and each (Ω_2)-program $\{y_t\}_{t=0}^T$ such that at least one of the following conditions holds:

(a) $x_0 \in \bar{Y}_1(l)$, $y_0 \in \bar{Y}_2(l)$, $x_T \in Y_1(l)$, $y_T \in Y_2(l)$, the pair $\{x_t\}_{t=0}^T$, $\{y_t\}_{t=0}^T$ is $(\{\alpha_t f_t\}_{t=0}^{T-1}, x_0, x_T, y_0, y_T, 0, T, \delta\min\{\alpha_t :\ t = 0, \dots, T-1\})$-good;

(b) $x_0 \in \bar{Y}_1(l)$, $y_0 \in \bar{Y}_2(l)$, the pair $\{x_t\}_{t=0}^T$, $\{y_t\}_{t=0}^T$ is

$$(\{\alpha_t f_t\}_{t=0}^{T-1}, x_0, y_0, 0, T, \delta\min\{\alpha_t :\ t = 0, \dots, T-1\}) - good;$$

(c) the pair $\{x_t\}_{t=0}^T$, $\{y_t\}_{t=0}^T$ is $(\{\alpha_t f_t\}_{t=0}^{T-1}, 0, T, \delta\min\{\alpha_t :\ t = 0, \dots, T-1\})$-good;

(d) $x_T \in Y_1(l)$, $y_T \in Y_2(l)$, *the pair* $\{x_t\}_{t=0}^T$, $\{y_t\}_{t=0}^T$ *is*

$$(\{\alpha_t f_t\}_{t=0}^{T-1}, (x_T, y_T)^-, 0, T, \delta \min\{\alpha_t : t = 0, \dots, T-1\}) - good$$

there exists an integer $j \in \{1, \dots, L\}$ *such that*

$$|x_j - x_f|, \ |y_j - y_f| \le \epsilon.$$

Proof. By Lemma 8.20, there exist a natural number $L > 2l$, $\delta \in (0, \epsilon)$ and $\lambda \in (0, 1)$ such that the following property holds:

(P14) for each integer $T \ge L$, each $f_t \in \mathcal{M}$, $t = 0, \dots, T-1$ satisfying (8.211), each sequence $\{\alpha_t\}_{t=0}^{T-1} \subset (0, 1]$ satisfying (8.212) for each integer $\tau \in [0, T-L]$, each (Ω_1)-program $\{x_t\}_{t=0}^T$ and each (Ω_2)-program $\{y_t\}_{t=0}^T$ such that at least one of conditions (a)–(d) holds there exists an integer $j \in \{1, \dots, T-1\}$ such that

$$|x_j - x_f|, \ |y_j - y_f| \le \epsilon. \tag{8.213}$$

Assume that an integer $T \ge L$, each $f_t \in \mathcal{M}$, $t = 0, \dots, T-1$ satisfy (8.211), a sequence $\{\alpha_t\}_{t=0}^{T-1} \subset (0, 1]$ satisfies (8.212) for each integer $\tau \in [0, T-L]$, an (Ω_1)-program $\{x_t\}_{t=0}^T$ and an (Ω_2)-program $\{y_t\}_{t=0}^T$ satisfy at least one of conditions (a)–(d). Property (P14) implies that there exists

$$j \in \{1, \dots, T-1\}$$

such that (8.213) holds. We may assume without loss of generality that

$$\max\{|x_t - x_f|, \ |y_t - y_f|\} > \epsilon \tag{8.214}$$

for every integer t satisfying $1 \le t < j$. In order to complete the proof it is sufficient to show that $j \le L$.

Assume the contrary. Then by property (P14) applied to the pair of sequences $\{x_t\}_{t=0}^j$ and $\{y_t\}_{t=0}^j$ satisfying at least one of conditions (a), (b), (d) there exists an integer

$$s \in \{1, \dots, j-1\}$$

such that

$$|x_s - x_f|, \ |y_s - y_f| \le \epsilon.$$

This contradicts (8.214). The contradiction we have reached proves Lemma 8.21.

8.12 Proof of Theorem 8.19

By Lemma 8.7, there exists a positive number $\delta_0 < \epsilon$ such that the following property holds:

(P15) for each integer $n \ge 1$ and each (f, δ_0)-good pair of an (Ω_1)-program $\{x_t\}_{t=0}^T$ and an (Ω_2)-program $\{y_t\}_{t=0}^T$ satisfying

$$|x_0 - x_f| \le \delta_0, \ |x_n - x_f| \le \delta_0, \ |y_0 - y_f| \le \delta_0, \ |y_n - y_f| \le \delta_0,$$

we have

$$|x_t - x_f| \le \epsilon,\ |y_t - y_f| \le \epsilon,\ t = 0, \dots, n.$$

By Lemma 8.21, there exist a natural number $L > 2l$, $\lambda_1 \in (0, 1)$ and $\delta_1 \in (0, \epsilon)$ such that the following property holds:

(P16) for each integer $T \ge L$, each $f_t \in \mathcal{M}$, $t = 0, \dots, T-1$ satisfying

$$\|f_t - f\| \le \delta_1,\ t = 0, \dots, T-1,$$

each sequence $\{\alpha_t\}_{t=0}^{T-1} \subset (0, 1]$ such that for each integer $\tau \in [0, T-L]$

$$\alpha_i \alpha_j^{-1} \ge \lambda \text{ for all } i, j \in \{\tau, \dots, \tau + L\},$$

each (Ω_1)-program $\{x_t\}_{t=0}^{T}$ and each (Ω_2)-program $\{y_t\}_{t=0}^{T}$ for which at least one of conditions (a)–(d) holds there exists an integer $j \in \{1, \dots, L\}$ such that

$$|x_j - x_f|,\ |y_j - y_f| \le \delta_0.$$

Choose a positive number

$$\delta \le 4^{-1} \min\{\delta_0, \delta_1\}(2L+1)^{-1} \tag{8.215}$$

and a number λ such that

$$\lambda \in (\lambda_1, 1),\ (\|f\| + 1)\lambda^{-1}(1 - \lambda) < \delta. \tag{8.216}$$

Assume that an integer $T > 2L$, $f_t \in \mathcal{M}$, $t = 0, \dots, T-1$ satisfy

$$\|f_t - f\| \le \delta,\ t = 0, \dots, T-1, \tag{8.217}$$

a sequence $\{\alpha_t\}_{t=0}^{T-1} \subset (0, 1]$ satisfies for each integer $\tau \in [0, T-L]$

$$\alpha_i \alpha_j^{-1} \ge \lambda \text{ for all } i, j \in \{\tau, \dots, \tau + L\}, \tag{8.218}$$

$\{x_t\}_{t=0}^{T}$ is an (Ω_1)-program, and an $\{y_t\}_{t=0}^{T}$ is an (Ω_2)-program and at least one of conditions (a)–(d) holds. Property (P16), (8.215), (8.217), and (8.218) imply that there exists an integer $S_0 \in \{0, \dots, L\}$ such that

$$|x_{S_0} - x_f| \le \delta_0,\ |y_{S_0} - y_f| \le \delta_0.$$

If

$$|x_0 - x_f| \le \delta,\ |y_0 - y_f| \le \delta,$$

then we set $S_0 = 0$.

Assume that an integer $k \ge 0$ and that we have already defined nonnegative integers S_i, $i = 0, \dots, k$ such that for every integer i satisfying $0 \le i < k$,

$$1 \le S_{i+1} - S_i \le L \tag{8.219}$$

and for all integers $i = 0, \dots, k$,

$$|x_{S_i} - x_f| \le \delta_0,\ |y_{S_i} - y_f| \le \delta_0. \tag{8.220}$$

Assume that $T - S_k \geq L$. By property (P16) applied to the pair of sequences $\{x_{t+S_k}\}_{t=0}^{T-S_k}$, $\{y_{t+S_k}\}_{t=0}^{T-S_k}$ which satisfies at least one of conditions (a), (b) there exists an integer $S_{k+1} \in \{S_k + 1, \dots, S_k + L\}$ such that

$$|x_{S_{k+1}} - x_f| \leq \delta_0, \ |y_{S_{k+1}} - y_f| \leq \delta_0.$$

Thus by induction we constructed the sequence of integers $\{S_i\}_{i=0}^{p}$ such that

$$0 \leq S_0 \leq L,$$

for every integer i satisfying $0 \leq i < p$, (8.219) is true, for all integers $i = 0, \dots, p$, (8.220) holds and

$$0 \leq T - S_p < L.$$

If

$$|x_T - x_f| \leq \delta_0, \ |y_T - y_f| \leq \delta_0,$$

then we may assume that $S_p = T$,

Assume that an integer $j \in \{0, \dots, p-1\}$. It is not difficult to see that in view of (a)–(d) the following property holds:

(e) the pair of sequences $\{x_t\}_{t=S_j}^{S_{j+1}}$, $\{y_t\}_{t=S_j}^{S_{j+1}}$ is

$$(\{\alpha_t f_t\}_{t=S_j}^{S_{j+1}-1}, x_{S_j}, x_{S_{j+1}}, y_{S_j}, y_{S_{j+1}}, S_j, S_{j+1},$$
$$\delta \min\{\alpha_t : \ t = 0, \dots, T-1\})\text{-good}.$$

By (8.217)–(8.219), for each (Ω_1)-program $\{\xi_t\}_{t=S_j}^{S_{j+1}}$ and each (Ω_2)-program $\{\eta_t\}_{t=S_j}^{S_{j+1}}$,

$$|\alpha_{S_j} \sum_{t=S_j}^{S_{j+1}-1} \alpha_{S_j}^{-1} \alpha_t f_t(\xi_t, \xi_{t+1}, \eta_t, \eta_{t+1}) - \alpha_{S_j} \sum_{t=S_j}^{S_{j+1}-1} f_t(\xi_t, \xi_{t+1}, \eta_t, \eta_{t+1})|$$

$$\leq \alpha_{S_j} \sum_{t=S_j}^{S_{j+1}-1} \|\alpha_t \alpha_{S_j}^{-1} f_t - f\|$$

$$\leq \alpha_{S_j} \sum_{t=S_j}^{S_{j+1}-1} (|\alpha_t \alpha_{S_j}^{-1} - 1| \|f_t\| + \|f_t - f\|)$$

$$\leq \alpha_{S_j} L(\delta + \lambda^{-1}(1-\lambda)(\|f\| + 1)) \leq 2\alpha_{S_j} \delta L. \tag{8.221}$$

It follows from (8.215), (8.221), and property (e) that the pair of sequences $\{x_t\}_{t=S_j}^{S_{j+1}}$, $\{y_t\}_{t=S_j}^{S_{j+1}}$ is $(f, x_{S_j}, x_{S_{j+1}}, y_{S_j}, y_{S_{j+1}}, S_j, S_{j+1}, \delta_0)$-good. Together with with (8.220) and property (P15) this implies that

$$|x_t - x_f|, \ |y_t - x_f| \leq \epsilon, \ t = S_j, \dots, S_{j+1}$$

for $j = 0, \dots, p-1$. This completes the proof of Theorem 8.19. □

8.13 Convergence of Overtaking Optimal Solutions for Games with Discounting

In this section we prove the following turnpike result.

Theorem 8.22. *Let $\epsilon \in (0, r_f)$ and l be a natural number. Then there exist a natural number $L > 2l$ and $\delta \in (0, \epsilon)$, $\lambda \in (0, 1)$ such that for each $f_t \in \mathcal{M}$, $t = 0, 1, \dots,$ satisfying*

$$\|f_t - f\| \le \delta, \; t = 0, 1, \dots,$$

each sequence $\{\alpha_t\}_{t=0}^{\infty} \subset (0, 1]$ such that for each integer $\tau \ge 0$,

$$\alpha_i \alpha_j^{-1} \ge \lambda \text{ for all } i, j \in \{\tau, \dots, \tau + L\},$$

each (Ω_1)-program $\{x_t\}_{t=0}^{\infty}$ and each (Ω_2)-program $\{y_t\}_{t=0}^{\infty}$ such that at least one of the following conditions hold:

(a) $x_0 \in \bar{Y}_1(l)$, $y_0 \in \bar{Y}_2(l)$, the pair $\{x_t\}_{t=0}^{T}$, $\{y_t\}_{t=0}^{T}$ is $(\{\alpha_t f_t\}_{t=0}^{\infty})$-overtaking optimal;

(b) the pair $\{x_t\}_{t=0}^{\infty}$, $\{y_t\}_{t=0}^{\infty}$ is strongly $(\{\alpha_t f_t\}_{t=0}^{\infty})$-overtaking optimal

$$|x_t - x_f| \le \epsilon, \; |y_t - y_f| \le \epsilon \tag{8.222}$$

for all integers $t \ge L$. If

$$|x_0 - x_f| \le \delta, \; |y_0 - y_f| \le \delta,$$

then (8.222) holds for all integers $t \ge 0$.

We begin with the following auxiliary result.

Lemma 8.23. *Let $\epsilon \in (0, r_f)$ and l be a natural number. Then there exist a natural number $L > 2l$ and $\delta \in (0, \epsilon)$, $\lambda \in (0, 1)$ such that for each $f_t \in \mathcal{M}$, $t = 0, 1, \dots,$ satisfying*

$$\|f_t - f\| \le \delta, \; t = 0, 1, \dots, \tag{8.223}$$

each sequence $\{\alpha_t\}_{t=0}^{\infty} \subset (0, 1]$ such that for each integer $\tau \ge 0$,

$$\alpha_i \alpha_j^{-1} \ge \lambda, \; i, j \in \{\tau, \dots, \tau + L\}, \tag{8.224}$$

each (Ω_1)-program $\{x_t\}_{t=0}^{\infty}$ and each (Ω_2)-program $\{y_t\}_{t=0}^{\infty}$ such that at least one of conditions (a), (b) of Theorem 8.22 holds there exists an integer $j \ge 1$ such that

$$|x_j - x_f| \le \epsilon, \; |y_j - y_f| \le \epsilon. \tag{8.225}$$

Proof. By property (P1), there exists $\gamma \in (0, \epsilon)$ such that the following properties hold:

(P17) for every $(\xi_1, \xi_2) \in \Omega_1$ satisfying $L_1(\xi_1, \xi_2) \le \gamma$, we have

$$|\xi_1 - x_f|,\ |\xi_2 - x_f| \le \epsilon/2;$$

(P18) for every $(\xi_1, \xi_2) \in \Omega_2$ satisfying $L_2(\xi_1, \xi_2) \le \gamma$, we have

$$|\xi_1 - y_f|, |\xi_2 - y_f| \le \epsilon/2.$$

Choose

$$\delta \in (0, \gamma/16), \tag{8.226}$$

a natural number $L_0 \ge 4 + 2l$ such that

$$L_0\gamma > 4(|l_1| + |l_2|)(\mathrm{diam}(X_1) + \mathrm{diam}(X_2)) \tag{8.227}$$

a number λ such that

$$\lambda \in (0,1),\ (1-\lambda)\lambda^{-1}(\|f\| + 1) \le \delta \tag{8.228}$$

and natural numbers

$$q_0 > 32(\|f\| + 1)\gamma^{-1}\lambda^{-2}, \tag{8.229}$$

$$L = 2L_0q_0 + 4l. \tag{8.230}$$

Assume that a sequence $f_t \in \mathcal{M}$, $t = 0, 1, \dots$, satisfy (8.223), a sequence $\{\alpha_t\}_{t=0}^{\infty} \subset (0, 1]$ satisfies (8.224) for each integer $\tau \ge 0$, an (Ω_1)-program $\{x_t\}_{t=0}^{\infty}$ and an (Ω_2)-program $\{y_t\}_{t=0}^{\infty}$ satisfy at least one of conditions (a), (b). In order to complete the proof of the lemma it is sufficient to show that there exists an integer $j \ge 1$ such that (8.225) holds.

Assume the contrary. Then

$$\max\{|x_t - x_f|,\ |y_t - y_f|\} > \epsilon \text{ for all integers } t \ge 1. \tag{8.231}$$

By (a), (b), there exist an (Ω_1)-program $\{u_t\}_{t=0}^{\infty}$ and an (Ω_2)-program $\{v_t\}_{t=0}^{\infty}$ such that

$$u_i = x_f,\ v_i = y_f \text{ for all integers } i \in [l, \infty), \tag{8.232}$$

in the case (a)

$$u_0 = x_0,\ \ v_0 = y_0. \tag{8.233}$$

By (a), (b), and (8.233),

$$\limsup_{T\to\infty}[\sum_{t=0}^{T-1} \alpha_t f_t(x_t, x_{t+1}, y_t, y_{t+1}) - \sum_{t=0}^{T-1} \alpha_t f_t(u_t, u_{t+1}, y_t, y_{t+1})] \le 0, \tag{8.234}$$

$$\liminf_{T\to\infty}[\sum_{t=0}^{T-1} \alpha_t f_t(x_t, x_{t+1}, y_t, y_{t+1}) - \sum_{t=0}^{T-1} \alpha_t f_t(x_t, x_{t+1}, v_t, v_{t+1})] \ge 0. \tag{8.235}$$

In view of (8.234) and (8.235),

$$\limsup_{T\to\infty}[\sum_{t=0}^{T-1}\alpha_t f_t(x_t,x_{t+1},v_t,v_{t+1})-\sum_{t=0}^{T-1}\alpha_t f_t(u_t,u_{t+1},y_t,y_{t+1})]\leq 0. \quad (8.236)$$

Let $q \geq q_0$ be an integer. It follows from (8.223), (8.224), and (8.230),

$$|\sum_{t=0}^{l-1}\alpha_t f_t(x_t,x_{t+1},v_t,v_{t+1})-\sum_{t=0}^{l-1}\alpha_t f_t(u_t,u_{t+1},y_t,y_{t+1})|$$
$$\leq 2l(\|f\|+1)\alpha_0\lambda^{-1}. \quad (8.237)$$

Let $k \in \{0,\dots.q-1\}$. For each (Ω_1)-program $\{\xi_t\}_{l+kL_0}^{l+(k+1)L_0}$ and each (Ω_2)-program $\{\eta_t\}_{l+kL_0}^{l+(k+1)L_0}$ it follows from (8.223), (8.224), (8.228), and (8.230) that

$$|\sum_{t=l+kL_0}^{l+(k+1)L_0-1}\alpha_t f_t(\xi_t,\xi_{t+1},\eta_t,\eta_{t+1})-\alpha_{l+kL_0}\sum_{t=l+kL_0}^{l+(k+1)L_0-1}f_t(\xi_t,\xi_{t+1},\eta_t,\eta_{t+1})|$$
$$=\alpha_{l+kL_0}|\sum_{t=l+kL_0}^{l+(k+1)L_0-1}\alpha_t\alpha_{l+kL_0}^{-1}f_t(\xi_t,\xi_{t+1},\eta_t,\eta_{t+1})$$
$$-\sum_{t=l+kL_0}^{l+(k+1)L_0-1}f_t(\xi_t,\xi_{t+1},\eta_t,\eta_{t+1})|$$
$$\leq \alpha_{l+kL_0}L_0\max\{\|\alpha_t\alpha_{l+kL_0}^{-1}f_t-f\|:$$
$$t\in\{l+kL_0,\dots,l+(k+1)L_0-1\}\}$$
$$\leq \alpha_{l+kL_0}L_0\max\{\|\alpha_t\alpha_{l+kL_0}^{-1}f_t-f_t\|+\|f_t-f\|:$$
$$t\in\{l+kL_0,\dots,l+(k+1)L_0-1\}\}$$
$$\leq \alpha_{l+kL_0}L_0(\delta+(1-\lambda)\lambda^{-1}(\|f\|+1))\leq 2\delta\alpha_{l+kL_0}L_0. \quad (8.238)$$

By (8.54), (8.56), (8.232), and (8.238),

$$\sum_{t=l+kL_0}^{l+(k+1)L_0-1}\alpha_t f_t(x_t,x_{t+1},v_t,v_{t+1})$$
$$=\sum_{t=l+kL_0}^{l+(k+1)L_0-1}\alpha_t f_t(x_t,x_{t+1},y_f,y_f)$$
$$\geq \alpha_{l+kL_0}\sum_{t=l+kL_0}^{l+(k+1)L_0-1}f_t(x_t,x_{t+1},y_f,y_f)-2\delta L_0\alpha_{l+kL_0}$$

$$= \alpha_{l+kL_0}\Big(\sum_{t=l+kL_0}^{l+(k+1)L_0-1} L_1(x_t, x_{t+1}) + L_0 f(x_f, x_f, y_f, y_f)$$

$$+ \langle l_1, x_{l+kL_0} - x_{l+(k+1)L_0}\rangle\Big) - 2\delta L_0 \alpha_{l+kL_0}, \tag{8.239}$$

$$\sum_{t=l+kL_0}^{l+(k+1)L_0-1} \alpha_t f_t(u_t, u_{t+1}, y_t, y_{t+1})$$

$$= \sum_{t=l+kL_0}^{l+(k+1)L_0-1} \alpha_t f_t(x_f, x_f, y_t, y_{t+1})$$

$$\le \alpha_{l+kL_0} \sum_{t=l+kL_0}^{l+(k+1)L_0-1} f_t(x_f, x_f, y_t, y_{t+1}) + 2\delta L_0 \alpha_{l+kL_0}$$

$$\le \alpha_{l+kL_0}\Big(-\sum_{t=l+kL_0}^{l+(k+1)L_0-1} L_2(y_t, y_{t+1}) + L_0 f(x_f, x_f, y_f, y_f)$$

$$+ \langle l_2, y_{l+kL_0} - y_{l+(k+1)L_0}\rangle\Big) + 2\delta L_0 \alpha_{l+kL_0}. \tag{8.240}$$

In view of (8.239) and (8.240),

$$\sum_{t=l+kL_0}^{l+(k+1)L_0-1} \alpha_t f_t(x_t, x_{t+1}, v_t, v_{t+1}) - \sum_{t=l+kL_0}^{l+(k+1)L_0-1} \alpha_t f_t(u_t, u_{t+1}, y_t, y_{t+1})$$

$$\ge \alpha_{l+kL_0}\Big(\sum_{t=l+kL_0}^{l+(k+1)L_0-1} (L_1(x_t, x_{t+1}) + L_2(y_t, y_{t+1}))$$

$$- (|l_1| + |l_2|)(\mathrm{diam}(X_1) + \mathrm{diam}(X_2))\Big) - 4\delta L_0 \alpha_{l+kL_0}. \tag{8.241}$$

Properties (P17) and (P18) and (8.231) imply that

$$L_1(x_t, x_{t+1}) + L_2(y_t, y_{t+1}) \ge \gamma \text{ for all } t = l+kL_0, \dots, l+(k+1)L_0-1. \tag{8.242}$$

It follows from (8.227), (8.241), and (8.242) that

$$\sum_{t=l+kL_0}^{l+(k+1)L_0-1} \alpha_t f_t(x_t, x_{t+1}, v_t, v_{t+1}) - \sum_{t=l+kL_0}^{l+(k+1)L_0-1} \alpha_t f_t(u_t, u_{t+1}, y_t, y_{t+1})$$

$$\ge \alpha_{l+kL_0}(L_0\gamma - (|l_1| + |l_2|)(\mathrm{diam}(X_1) + \mathrm{diam}(X_2))) - 4\delta L_0 \alpha_{l+kL_0}$$

$$\ge 2^{-1}\alpha_{l+kL_0} L_0 \gamma - 4\alpha_{l+kL_0} L_0 \delta \ge 4^{-1} \alpha_{l+kL_0} L_0 \gamma, \; k = 0, \dots, q-1. \tag{8.243}$$

By (8.224), (8.229), (8.230), (8.237), and (8.243),

$$\sum_{t=0}^{l+qL_0-1} \alpha_t f_t(x_t, x_{t+1}, v_t, v_{t+1}) - \sum_{t=0}^{l+qL_0-1} \alpha_t f_t(u_t, u_{t+1}, y_t, y_{t+1})$$

$$\geq -2l(\|f\|+1)\alpha_0\lambda^{-1} + \sum_{t=l}^{l+qL_0-1} \alpha_t f_t(x_t, x_{t+1}, v_t, v_{t+1})$$

$$- \sum_{t=l}^{l+qL_0-1} \alpha_t f_t(u_t, u_{t+1}, y_t, y_{t+1})$$

$$\geq -2l(\|f\|+1)\alpha_0\lambda^{-1} + 4^{-1}L_0\gamma \sum_{k=0}^{q-1} \alpha_{l+kL_0}$$

$$\geq -2l(\|f\|+1)\alpha_0\lambda^{-1} + 4^{-1}L_0\gamma \sum_{k=0}^{q_0-1} \alpha_{l+kL_0}$$

$$\geq -2l(\|f\|+1)\alpha_0\lambda^{-1} + 4^{-1}L_0\gamma q_0\alpha_0\lambda$$

$$\geq \alpha_0(4^{-1}L_0\gamma q_0\lambda - 2l(\|f\|+1)\lambda^{-1})$$

$$\geq \alpha_0(8(\|f\|+1)\lambda^{-1}L_0 - 2l(\|f\|+1)\lambda^{-1})$$

$$\geq 6\alpha_0(\|f\|+1)\lambda^{-1}L_0.$$

Since the relation above holds for every integer $q \geq q_0$ this contradicts (8.236). The contradiction we have reached proves Lemma 8.23. □

Applying by induction Lemma 8.23 we obtain the following result.

Lemma 8.24. *Let $\epsilon \in (0, r_f)$ and l be a natural number. Then there exist a natural number $L > 2l$, $\delta \in (0, \epsilon)$ and $\lambda \in (0,1)$ such that for each sequence $f_t \in \mathcal{M}$, $t = 0, 1, \dots$ satisfying (8.223), each sequence $\{\alpha_t\}_{t=0}^{\infty} \subset (0,1]$ such that (8.224) holds for each integer $\tau \geq 0$, each (Ω_1)-program $\{x_t\}_{t=0}^{\infty}$ and each (Ω_2)-program $\{y_t\}_{t=0}^{\infty}$ such that at least one of conditions (a), (b) of Theorem 8.22 holds there exists a strictly increasing sequence of nonnegative integers $\{S_i\}_{i=0}^{\infty}$ such that*

$$|x_{S_i} - x_f|, \ |y_{S_i} - y_f| \leq \epsilon$$

for all integers $i \geq 0$.

Theorem 8.22 follows from Theorem 8.14 and Lemma 8.24.

8.14 Existence of Overtaking Optimal Solutions for Games with Discounting

We suppose that the sum over empty set is zero. Let $r_1 \in (0, r_f/4)$. In this section we use the following properties.

(P19) For each $(\xi_1, \xi_2) \in \Omega_2$ satisfying $|\xi_i - y_f| \le r_1,\ i = 1, 2$ and each $(x_1, x_2) \in \Omega_1$ satisfying

$$\max\{|x_1 - x_f|,\ |x_2 - x_f|\} \ge r_f/4$$

we have

$$f(x_1, x_2, \xi_1, \xi_2) - f(x_f, x_f, \xi_1, \xi_2) + \langle l_1, x_2 - x_1\rangle \ge r_1.$$

(P20) For each $(\xi_1, \xi_2) \in \Omega_1$ satisfying $|\xi_i - x_f| \le r_1,\ i = 1, 2$ and each $(y_1, y_2) \in \Omega_1$ satisfying

$$\max\{|y_1 - y_f|,\ |y_2 - y_f|\} \ge r_f/4$$

we have

$$-f(\xi_1, \xi_2, y_1, y_2) + f(\xi_1, \xi_2, y_f, y_f) + \langle l_2, y_1 - y_2\rangle \ge r_1.$$

Let $\epsilon = r_1/4$, $l \ge 1$ be an integer and let $\delta \in (0, r_1/4)$, an integer $L > 2l$ and $\lambda \in (0, 1)$ be as guaranteed by Theorem 8.19.

Choose an integer $L_1 \ge 4L + 4$ such that

$$L_1 r_1 \ge 2(\mathrm{diam}(X_1) + \mathrm{diam}(X_2))(|l_1| + |l_2| + 1) \tag{8.244}$$

and $\lambda_0 \in (\lambda, 1)$ such that

$$\lambda_0 > 2^{-1},\ 64(\|f\| + 1)(1 - \lambda_0) < r_1 \tag{8.245}$$

and a positive number

$$\delta_0 < \min\{\delta, 1, 64^{-1} r_1\}. \tag{8.246}$$

Choose an integer $q_0 \ge 8$ such that

$$L_1 r_1 q_0 > 16(\|f\| + 1)L \tag{8.247}$$

and set

$$L_0 = 2q_0 L_1 + 2L + 4. \tag{8.248}$$

Let continuous functions $f_t \in \mathcal{M}$, $t = 0, 1, \dots$ satisfy

$$\|f_t - f\| \le \delta_0,\ t = 0, 1, \dots \tag{8.249}$$

and let a sequence $\{\alpha_t\}_{t=0}^{\infty} \subset (0, 1]$ satisfy

$$\lim_{t \to \infty} \alpha_t = 0, \tag{8.250}$$

$$\alpha_i \alpha_j^{-1} \ge \lambda_0 \text{ for each pair of integers } i, j \text{ satisfying } |i - j| \le L_0. \tag{8.251}$$

In this section we prove the following result.

Theorem 8.25. *Suppose that properties (P19) and (P20) hold. For each* $z = (z_1, z_2) \in \bar{Y}_1(l) \times \bar{Y}_2(l)$ *there exist an* (Ω_1)*-program* $\{x_t^{(z)}\}_{t=0}^{\infty}$ *and an* (Ω_2)*-program* $\{y_t^{(z)}\}_{t=0}^{\infty}$ *such that the following property holds.*

For each $\gamma > 0$ *there exists an integer* $n_0 \geq 1$ *such that for each integer* $T \geq n_0$ *and each* $z = (z_1, z_2) \in \bar{Y}_1(l) \times \bar{Y}_2(l)$ *the pair of programs* $\{x_t^{(z)}\}_{t=0}^{T}$, $\{y_t^{(z)}\}_{t=0}^{T}$ *is* $(\{\alpha_t f_t\}_{t=0}^{T-1}, z_1, z_2, 0, T, \gamma)$*-good.*

Let $z = (z_1, z_2) \in \bar{Y}_1(l) \times \bar{Y}_2(l)$. It is clear that the pair of programs $\{x_t^{(z)}\}_{t=0}^{\infty}$, $\{y_t^{(z)}\}_{t=0}^{\infty}$ is $(\{\alpha_t f_t\}_{t=0}^{\infty})$-overtaking optimal.

Note that by Proposition 7.2, for each integer $T \geq 1$ and each $z = (z_1, z_2) \in \bar{Y}_1(l) \times \bar{Y}_2(l)$, there exists $(\{\alpha_t f_t\}_{t=0}^{T-1}, z_1, z_2, 0, T, 0)$-good pair of an (Ω_1)-program $\{x_t\}_{t=0}^{T}$ and an (Ω_2)-program $\{y_t\}_{t=0}^{T}$ such that

$$x_0 = z_1,\ y_0 = z_2.$$

We begin with the following auxiliary result.

Lemma 8.26. *Let* $\gamma > 0$*. Then there exists an integer* $n_0 \geq 1$ *such that for each pair of integers* $T > S \geq n_0$ *and each* $(\{\alpha_t f_t\}_{t=0}^{T-1}, x_0, y_0, 0, T, 0)$*-good pair of an* (Ω_1)*-program* $\{x_t\}_{t=0}^{T}$ *and an* (Ω_2)*-program* $\{y_t\}_{t=0}^{T}$ *such that*

$$x_0 \in \bar{Y}_1(l),\ y_0 \in \bar{Y}_2(l) \tag{8.252}$$

the pair of programs $\{x_t\}_{t=0}^{S}$, $\{y_t\}_{t=0}^{S}$ *is* $(\{\alpha_t f_t\}_{t=0}^{S-1}, x_0, y_0, 0, S, \gamma)$*-good.*

Proof. By (8.250), there exists an integer

$$n_0 > 4L_0 + 4 \tag{8.253}$$

such that for all integers $t > n_0 - L_0 - 4$,

$$\alpha_t \leq 8^{-1}\gamma(8L_0 + 8)^{-1}(\|f\| + 1)^{-1}. \tag{8.254}$$

Assume that $T > S \geq n_0$ are integers and that an

$$(\{\alpha_t f_t\}_{t=0}^{T-1}, x_0, y_0, 0, T, 0) - \text{good pair}$$

of an (Ω_1)-program $\{x_t\}_{t=0}^{T}$ and of an (Ω_2)-program $\{y_t\}_{t=0}^{T}$ satisfies (8.252). By (8.246), (8.248), (8.249), (8.251), (8.254), the choice of δ, λ, L and Theorem 8.19, for all $t = L, \dots, T - L$,

$$|x_t - x_f| \leq r_1/4,\ |y_t - y_f| \leq r_1/4. \tag{8.255}$$

Assume that an (Ω_1)-program $\{\xi_t\}_{t=0}^{S}$ satisfies

$$\xi_0 = x_0. \tag{8.256}$$

We show that

$$\sum_{t=0}^{S-1} \alpha_t f_t(x_t, x_{t+1}, y_t, y_{t+1}) - \gamma \leq \sum_{t=0}^{S-1} \alpha_t f_t(\xi_t, \xi_{t+1}, y_t, y_{t+1}). \tag{8.257}$$

We may assume without loss of generality that

$$\sum_{t=0}^{S-1} \alpha_t f_t(\xi_t, \xi_{t+1}, y_t, y_{t+1}) \leq \sum_{t=0}^{S-1} \alpha_t f_t(\xi'_t, \xi'_{t+1}, y_t, y_{t+1}) \tag{8.258}$$

for every (Ω_1)-program $\{\xi'_t\}_{t=0}^{S}$ satisfying $\xi'_0 = x_0$. Set

$$A = \{j \in \{0, \dots, S\} : \ \xi_j \in \bar{Y}_1(l)\}. \tag{8.259}$$

In view of (8.252), $0 \in A$. Set

$$S_0 = \max\{j : \ j \in A\}. \tag{8.260}$$

We show that

$$S_0 \geq S - L_0. \tag{8.261}$$

Assume the contrary. Then

$$S_0 < S - L_0. \tag{8.262}$$

It follows from (8.248) and (8.262) that

$$S - L - 1 - (S_0 + L) > L_0 - 2L - 1 > 2q_0 L_1. \tag{8.263}$$

By (8.248), (8.259), (8.260), and (8.262), there exists an (Ω_1)-program $\{\xi_t^{(1)}\}_{t=0}^{S}$ such that

$$\xi_t^{(1)} = \xi_t, \ t = 0, \dots, S_0, \ \xi_t^{(1)} = x_f, \ t = S_0 + l, \dots, S. \tag{8.264}$$

Relations (8.256), (8.258), and (8.264) imply that

$$\sum_{t=S_0}^{S-1} \alpha_t f_t(\xi_t, \xi_{t+1}, y_t, y_{t+1}) \leq \sum_{t=S_0}^{S-1} \alpha_t f_t(\xi_t^{(1)}, \xi_{t+1}^{(1)}, y_t, y_{t+1}). \tag{8.265}$$

It follows from (8.249), (8.251), and (8.265),

$$\begin{aligned}
&\sum_{t=S_0+L}^{S-L-1} \alpha_t f_t(\xi_t, \xi_{t+1}, y_t, y_{t+1}) \\
&\leq \sum_{t=S_0+L}^{S-L-1} \alpha_t f_t(\xi_t^{(1)}, \xi_{t+1}^{(1)}, y_t, y_{t+1}) \\
&+ 2\alpha_{S_0} \lambda_0^{-1}(\|f\| + 1)L + 2\alpha_S \lambda_0^{-1}(\|f\| + 1)L.
\end{aligned} \tag{8.266}$$

Assume that integers τ_1, τ_2 satisfy

$$L_1 \le \tau_2 - \tau_1 \le 2L_1, \tag{8.267}$$

$$[\tau_1, \tau_2] \subset [S_0 + L, S - L]. \tag{8.268}$$

By (8.251) and (8.267),

$$|\sum_{t=\tau_1}^{\tau_2-1} \alpha_t f_t(\xi_t, \xi_{t+1}, y_t, y_{t+1}) - \alpha_{\tau_1} \sum_{t=\tau_1}^{\tau_2-1} f_t(\xi_t, \xi_{t+1}, y_t, y_{t+1})|$$

$$\le 2L_1(\|f\| + 1)\alpha_{\tau_1}(1 - \lambda_0)\lambda_0^{-1}, \tag{8.269}$$

$$|\sum_{t=\tau_1}^{\tau_2-1} \alpha_t f_t(\xi_t^{(1)}, \xi_{t+1}^{(1)}, y_t, y_{t+1}) - \alpha_{\tau_1} \sum_{t=\tau_1}^{\tau_2-1} f_t(\xi_t^{(1)}, \xi_{t+1}^{(1)}, y_t, y_{t+1})|$$

$$\le 2L_1(\|f\| + 1)\alpha_{\tau_1}(1 - \lambda_0)\lambda_0^{-1}. \tag{8.270}$$

In view of (8.249) and (8.267),

$$|\sum_{t=\tau_1}^{\tau_2-1} f_t(\xi_t, \xi_{t+1}, y_t, y_{t+1}) - \sum_{t=\tau_1}^{\tau_2-1} f(\xi_t, \xi_{t+1}, y_t, y_{t+1})| \le 2\delta_0 L_1, \tag{8.271}$$

$$|\sum_{t=\tau_1}^{\tau_2-1} f_t(\xi_t^{(1)}, \xi_{t+1}^{(1)}, y_t, y_{t+1}) - \sum_{t=\tau_1}^{\tau_2-1} f(\xi_t^{(1)}, \xi_{t+1}^{(1)}, y_t, y_{t+1})| \le 2\delta_0 L_1. \tag{8.272}$$

By (8.244), ((8.255), (8.259), (8.264), and (8.266)–(8.268),

$$\sum_{t=\tau_1}^{\tau_2-1} f(\xi_t, \xi_{t+1}, y_t, y_{t+1}) - \sum_{t=\tau_1}^{\tau_2-1} f(\xi_t^{(1)}, \xi_{t+1}^{(1)}, y_t, y_{t+1})$$

$$= \sum_{t=\tau_1}^{\tau_2-1} f(\xi_t, \xi_{t+1}, y_t, y_{t+1}) - \sum_{t=\tau_1}^{\tau_2-1} f(x_f, x_f, y_t, y_{t+1})$$

$$= \sum_{t=\tau_1}^{\tau_2-1} [f(\xi_t, \xi_{t+1}, y_t, y_{t+1}) - f(x_f, x_f, y_t, y_{t+1}) + \langle l_1, \xi_{t+1} - \xi_t \rangle]$$

$$+\langle l_1, \xi_{\tau_1} - \xi_{\tau_2} \rangle$$

$$\ge (\tau_2 - \tau_1)r_1 - \mathrm{diam}(X_1)|l_1| \ge L_1 r_1 - \mathrm{diam}(X_1)|l_1| \ge 2^{-1} L_1 r_1. \tag{8.273}$$

It follows from (8.245), (8.246), (8.269), (8.270), and (8.273) that

$$\sum_{t=\tau_1}^{\tau_2-1} \alpha_t f_t(\xi_t, \xi_{t+1}, y_t, y_{t+1}) - \sum_{t=\tau_1}^{\tau_2-1} \alpha_t f_t(\xi_t^{(1)}, \xi_{t+1}^{(1)}, y_t, y_{t+1})$$

$$\ge -4L_1(\|f\| + 1)\alpha_{\tau_1}(1 - \lambda_0)\lambda_0^{-1}$$

$$+\alpha_{\tau_1}(\sum_{t=\tau_1}^{\tau_2-1} f_t(\xi_t, \xi_{t+1}, y_t, y_{t+1}) - \sum_{t=\tau_1}^{\tau_2-1} f_t(\xi_t^{(1)}, \xi_{t+1}^{(1)}, y_t, y_{t+1}))$$

$$\geq -8L_1(\|f\| + 1)\alpha_{\tau_1}(1 - \lambda_0) - 4\delta_0 L_1 \alpha_{\tau_1}$$

$$+\alpha_{\tau_1}(\sum_{t=\tau_1}^{\tau_2-1} f(\xi_t, \xi_{t+1}, y_t, y_{t+1}) - \sum_{t=\tau_1}^{\tau_2-1} f(\xi_t^{(1)}, \xi_{t+1}^{(1)}, y_t, y_{t+1}))$$

$$\geq 2^{-1}L_1 r_1 \alpha_{\tau_1} - 4\delta_0 L_1 \alpha_{\tau_1} - 8L_1(\|f\| + 1)\alpha_{\tau_1}(1 - \lambda_0)$$

$$= \alpha_{\tau_1} L_1[2^{-1}r_1 - 4\delta_0 - 8(\|f\| + 1)(1 - \lambda_0)]$$

$$\geq \alpha_{\tau_1} L_1[4^{-1}r_1 - 8(\|f\| + 1)(1 - \lambda_0)] \geq 8^{-1}\alpha_{\tau_1} L_1 r_1.$$

Thus for each pair of integers τ_1, τ_2 satisfying (8.267) and (8.268),

$$\sum_{t=\tau_1}^{\tau_2-1} \alpha_t f_t(\xi_t, \xi_{t+1}, y_t, y_{t+1}) - \sum_{t=\tau_1}^{\tau_2-1} \alpha_t f_t(\xi_t^{(1)}, \xi_{t+1}^{(1)}, y_t, y_{t+1}) \geq 8^{-1}\alpha_{\tau_1} L_1 r_1. \tag{8.274}$$

In view of (8.263), there exists an integer $q > 2q_0$ such that

$$qL_1 \leq (S - L) - (S_0 + L) < (q + 1)L_1. \tag{8.275}$$

For all integer $i = 1, \dots, q$ set

$$S_i = S_0 + L + (i - 1)L_1, \tag{8.276}$$

$$S_{q+1} = S - L. \tag{8.277}$$

It follows from (8.267), (8.268), and (8.274)–(8.277) that for all integers $i = 1, \dots, q$,

$$\sum_{t=S_i}^{S_{i+1}-1} \alpha_t f_t(\xi_t, \xi_{t+1}, y_t, y_{t+1}) - \sum_{t=S_i}^{S_{i+1}-1} \alpha_t f_t(\xi_t^{(1)}, \xi_{t+1}^{(1)}, y_t, y_{t+1}) \geq 8^{-1}\alpha_{S_i} L_1 r_1$$

and

$$\sum_{t=S_0+L}^{S-L-1} \alpha_t f_t(\xi_t, \xi_{t+1}, y_t, y_{t+1}) - \sum_{t=S_0+L}^{S-L-1} \alpha_t f_t(\xi_t^{(1)}, \xi_{t+1}^{(1)}, y_t, y_{t+1})$$

$$\geq 8^{-1}L_1 r_1 \sum_{i=1}^{q} \alpha_{S_i}. \tag{8.278}$$

Relations (8.266) and (8.278) imply that

$$2(\alpha_{S_0} + \alpha_S)\lambda_0^{-1}L(\|f\| + 1) \geq 8^{-1}L_1 r_1 \sum_{i=1}^{q} \alpha_{S_i}$$

$$\geq 8^{-1}L_1r_1(\sum_{i=1}^{q}\alpha_{S_i} + \sum_{i=q-q_0+1}^{q}\alpha_{S_i}). \tag{8.279}$$

It follows that (8.248), (8.251), and (8.276) that for all $i = 1, \dots, q_0$,

$$\alpha_{S_i} \leq \lambda_0^{-1}\alpha_{S_0}$$

and for all $i = q - q_0 + 1, \dots, q$,

$$\alpha_{S_i} \leq \lambda_0^{-1}\alpha_S.$$

Together with (8.279) this implies that

$$2(\alpha_{S_0} + \alpha_S)\lambda_0^{-1}L(\|f\| + 1) \geq 8^{-1}L_1r_1q_0(\alpha_{S_0} + \alpha_S)\lambda_0^{-1},$$

$$16L(\|f\| + 1) \geq L_1r_1q_0.$$

This contradicts (8.247). The contradiction we have reached proves that

$$S_0 \geq S - L_0. \tag{8.280}$$

In view of (8.259) and (8.280),

$$\xi_{S_0} \in \bar{Y}_1(l).$$

It follows from (8.253), (8.255), (8.259), and (8.280) that there exists an (Ω_1)-program $\{\tilde{\xi}_t\}_{t=0}^T$ such that

$$\tilde{\xi}_t = \xi_t, \ t = 0, \dots, S_0,$$

$$\text{if } S_0 + l \leq T, \text{ then } \tilde{\xi}_{S_0+l} = x_f \tag{8.281}$$

and if $S_0 + l < T - L$, then for all integers t satisfying $S_0 + l < t \leq T$,

$$\tilde{\xi}_t = x_t. \tag{8.282}$$

By (8.256), (8.259), (8.260), (8.280), and the choice of $\{\tilde{\xi}_t\}_{t=0}^T$,

$$\sum_{t=0}^{S-1}\alpha_t f_t(\xi_t, \xi_{t+1}, y_t, y_{t+1}) - \sum_{t=0}^{S-1}\alpha_t f_t(x_t, x_{t+1}, y_t, y_{t+1})$$

$$\geq \sum_{t=0}^{S_0-1}\alpha_t f_t(\tilde{\xi}_t, \tilde{\xi}_{t+1}, y_t, y_{t+1})$$

$$-\sum_{t=0}^{S_0-1}\alpha_t f_t(x_t, x_{t+1}, y_t, y_{t+1}) - 2(\|f\| + 1)\alpha_{S_0}\lambda_0^{-1}$$

$$= \sum_{t=0}^{T-1} \alpha_t f_t(\tilde{\xi}_t, \tilde{\xi}_{t+1}, y_t, y_{t+1}) - \sum_{t=0}^{T-1} \alpha_t f_t(x_t, x_{t+1}, y_t, y_{t+1})$$

$$- \sum_{t=S_0}^{T-1} \alpha_t f_t(\tilde{\xi}_t, \tilde{\xi}_{t+1}, y_t, y_{t+1})$$

$$+ \sum_{t=S_0}^{T-1} \alpha_t f_t(x_t, x_{t+1}, y_t, y_{t+1}) - 2(\|f\| + 1)\alpha_{S_0}\lambda_0^{-1}$$

$$\geq \sum_{t=S_0}^{T-1} \alpha_t f_t(x_t, x_{t+1}, y_t, y_{t+1})$$

$$- \sum_{t=S_0}^{T-1} \alpha_t f_t(\tilde{\xi}_t, \tilde{\xi}_{t+1}, y_t, y_{t+1}) - 2(\|f\| + 1)\alpha_{S_0}\lambda_0^{-1}. \tag{8.283}$$

If $S_0 + l \geq T - L$, then $S_0 + l + L \geq T$ and in view of (8.248), (8.251), (8.254), and (8.283),

$$\sum_{t=0}^{S-1} \alpha_t f_t(\xi_t, \xi_{t+1}, y_t, y_{t+1}) - \sum_{t=0}^{S-1} \alpha_t f_t(x_t, x_{t+1}, y_t, y_{t+1})$$

$$\geq -2(\|f\| + 1) \sum_{T-l-L}^{T-1} \alpha_t - 2(\|f\| + 1)\alpha_{S_0}\lambda_0^{-1}$$

$$\geq -4(\|f\| + 1)\alpha_{T-1}\lambda_0^{-2}(l + L) \geq -\gamma.$$

Assume that $S_0 + l < T - L$. Then by (8.245), (8.251), (8.253), (8.254), (8.259), (8.260), (8.280), (8.282), and (8.283),

$$\sum_{t=0}^{S-1} \alpha_t f_t(\xi_t, \xi_{t+1}, y_t, y_{t+1}) - \sum_{t=0}^{S-1} \alpha_t f_t(x_t, x_{t+1}, y_t, y_{t+1})$$

$$\geq \sum_{t=S_0}^{T-1} \alpha_t f_t(x_t, x_{t+1}, y_t, y_{t+1})$$

$$- \sum_{t=S_0}^{T-1} \alpha_t f_t(\tilde{\xi}_t, \tilde{\xi}_{t+1}, y_t, y_{t+1}) - 2(\|f\| + 1)\alpha_{S_0}\lambda_0^{-1}$$

$$= \sum \{\alpha_t f_t(x_t, x_{t+1}, y_t, y_{t+1}) - \alpha_t f_t(\tilde{\xi}_t, \tilde{\xi}_{t+1}, y_t, y_{t+1}) :$$

$$\text{an integer } t \text{ satisfies } S_0 \leq t \leq S_0 + l\}$$

$$-2(\|f\| + 1)\alpha_{S_0}\lambda_0^{-1} \geq -4L(\|f\| + 1)\alpha_{S_0} \geq -\gamma.$$

Thus (8.257) holds in both cases.

Analogously we can show that for every (Ω_2)-program $\{\eta_t\}_{t=0}^S$ which satisfies $\eta_0 = y_0$ we have

$$\sum_{t=0}^{S-1} \alpha_t f_t(x_t, x_{t+1}, y_t, y_{t+1}) + \gamma \geq \sum_{t=0}^{S-1} \alpha_t f_t(x_t, x_{t+1}, \eta_t, \eta_{t+1}).$$

Lemma 8.26 is proved. □

Completion of the proof of Theorem 8.25.

Proof. Let $z_1 \in \bar{Y}_1(l)$, $z_2 \in \bar{Y}_2(l)$, $z = (z_1, z_2)$ and let $\{T_k\}$ be a strictly increasing sequence of natural numbers. For every natural number k there exists an $(\{\alpha_t f_t\}_{t=0}^{T_k-1}, z_1, z_2, 0, T_k, 0)$-good pair of an (Ω_1)-program $\{x_t^{(z,k)}\}_{t=0}^{T_k}$ and an (Ω_2)-program $\{y_t^{(z,k)}\}_{t=0}^{T_k}$ such that

$$x_0^{(z,k)} = z_1, \; y_0^{(z,k)} = z_2. \tag{8.284}$$

Extracting a subsequence and re-indexing, if necessary, we may assume without loss of generality that for every integer $t \geq 0$ there exist

$$x_t^{(z)} = \lim_{k \to \infty} x_t^{(z,k)}, \; y_t^{(z)} = \lim_{k \to \infty} y_t^{(z,k)}. \tag{8.285}$$

In view of (8.284) and (8.285), $\{x_t^{(z)}\}_{t=0}^{\infty}$ is an (Ω_1)-program, $\{y_t^{(z)}\}_{t=0}^{\infty}$ is an $(\Omega_2$-program,

$$x_0^{(z)} = z_1, \; y_0^{(z)} = z_2.$$

Let $\gamma > 0$ and let an integer $n_0 \geq 1$ be as guaranteed by Lemma 8.26.

Let $z_1 \in \bar{Y}_1(l)$, $z_2 \in \bar{Y}_2(l)$, $z = (z_1, z_2)$, $T \geq n_0$ be an integer, $\{\xi_t\}_{t=0}^T$ an (Ω_1)-program satisfying $\xi_0 = z_1$ and $\{\eta_t\}_{t=0}^T$ be an (Ω_2)-program satisfying $\eta_0 = z_2$. By Lemma 8.26 and the choice of n_0, for every integer $k \geq 1$ satisfying $T_k > T$,

$$\sum_{t=0}^{T-1} \alpha_t f_t(x_t^{(z,k)}, x_{t+1}^{(z,k)}, y_t^{(z,k)}, y_{t+1}^{(z,k)}) \leq \sum_{t=0}^{T-1} \alpha_t f_t(\xi_t, \xi_{t+1}, y_t^{(z,k)}, y_{t+1}^{(z,k)}) + \gamma, \tag{8.286}$$

$$\sum_{t=0}^{T-1} \alpha_t f_t(x_t^{(z,k)}, x_{t+1}^{(z,k)}, y_t^{(z,k)}, y_{t+1}^{(z,k)}) \geq \sum_{t=0}^{T-1} \alpha_t f_t(x_t^{(z,k)}, x_{t+1}^{(z,k)}, \eta_t, \eta_{t+1}) - \gamma, \tag{8.287}$$

In view of (8.285)–(8.287) and continuity of f_t, $t = 0, 1, \dots$,

$$\sum_{t=0}^{T-1} \alpha_t f_t(x_t^{(z)}, x_{t+1}^{(z)}, y_t^{(z)}, y_{t+1}^{(z)}) \leq \sum_{t=0}^{T-1} \alpha_t f_t(\xi_t, \xi_{t+1}, y_t^{(z)}, y_{t+1}^{(z)}) + \gamma,$$

$$\sum_{t=0}^{T-1} \alpha_t f_t(x_t^{(z)}, x_{t+1}^{(z)}, y_t^{(z)}, y_{t+1}^{(z)}) \geq \sum_{t=0}^{T-1} \alpha_t f_t(x_t^{(z)}, x_{t+1}^{(z)}, \eta_t, \eta_{t+1}) - \gamma.$$

Theorem 8.25 is proved. □

8.15 Structure of Optimal Solutions in Regions Close to the Endpoints

Let $g \in \mathcal{M}$. Define $g^0 : \Omega_2 \times \Omega_1 \to R^1$ by

$$g^0(y_1, y_2, x_1, x_2) = -g(x_1, x_2, y_1, y_2),\ (x_1, x_2) \in \Omega_1,\ (y_1, y_2) \in \Omega_2. \quad (8.288)$$

In view of (8.288), for each $(x_1, x_2) \in \Omega_1$, the function $g^0(\cdot, \cdot, x_1, x_2) : \Omega_2 \to R^1$ is convex, for each $(y_1, y_2) \in \Omega_2$, the function $g^0(y_1, y_2, \cdot, \cdot) : \Omega_1 \to R^1$ is concave and for all $x \in \Delta_1$ and all $y \in \Delta_2$,

$$g^0(y_f, y_f, x_f, x_f) = -g(x_f, x_f, y_f, y_f) \geq -g(x, x, y_f, y_f) = g^0(y_f, y_f, x, x), \quad (8.289)$$

$$g^0(y_f, y_f, x_f, x_f) = -g(x_f, x_f, y_f, y_f) \leq -g(x_f, x_f, y, y) = g^0(y, y, x_f, x_f). \quad (8.290)$$

Assume that $T_2 > T_1 \geq 0$ are integers and $\{f_t\}_{t=T_1}^{T_2-1} \subset \mathcal{M}$. For all $t = T_1, \dots, T_2 - 1$ set

$$f_t^0(y_1, y_2, x_1, x_2) = -f_t(x_1, x_2, y_1, y_2),\ (x_1, x_2) \in \Omega_1,\ (y_1, y_2) \in \Omega_2.$$

It is not difficult to see that the following result holds.

Proposition 8.27. *Let $T_2 > T_1 \geq 0$ be integers, $M \geq 0$, $\{f_t\}_{t=T_1}^{T_2-1} \subset \mathcal{M}$, $\{x_t\}_{t=T_1}^{T_2}$, $\{\xi_t\}_{t=T_1}^{T_2}$ be (Ω_1)-programs, $\{y_t\}_{t=T_1}^{T_2}$, $\{\eta_t\}_{t=T_1}^{T_2}$ be (Ω_2)-programs. Then the following assertions hold.*

1.

$$\sum_{t=T_1}^{T_2-1} f_t(x_t, x_{t+1}, y_t, y_{t+1}) \leq \sum_{t=T_1}^{T_2-1} f_t(\xi_t, \xi_{t+1}, y_t, y_{t+1}) + M$$

if and only if

$$\sum_{t=T_1}^{T_2-1} f_t^0(y_t, y_{t+1}, x_t, x_{t+1}) + M \geq \sum_{t=T_1}^{T_2-1} f_t^0(y_t, y_{t+1}, \xi_t, \xi_{t+1}).$$

2.

$$\sum_{t=T_1}^{T_2-1} f_t(x_t, x_{t+1}, y_t, y_{t+1}) + M \geq \sum_{t=T_1}^{T_2-1} f_t(x_t, x_t, \eta_t, \eta_{t+1})$$

if and only if

$$\sum_{t=T_1}^{T_2-1} f_t^0(y_t, y_{t+1}, x_t, x_{t+1}) \leq \sum_{t=T_1}^{T_2-1} f_t^0(\eta_t, \eta_{t+1}, x_t, x_{t+1}) + M.$$

It is clear that all the results obtained for the triplet (f, Ω_1, Ω_2) are valid for the triplet $(f^0, \Omega_2, \Omega_1)$.

Proposition 8.27 implies the following result.

Proposition 8.28. *Let $T_2 > T_1 \geq 0$ be integers, $M \geq 0$, $\{f_t\}_{t=T_1}^{T_2-1} \subset \mathcal{M}$, $\{x_t\}_{t=T_1}^{T_2}$ be (Ω_1)-program, $\{y_t\}_{t=T_1}^{T_2}$ be (Ω_2)-program. Then the following assertions hold.*

1. The pair $\{x_t\}_{t=T_1}^{T_2}$, $\{y_t\}_{t=T_1}^{T_2}$ is $(\{f_t\}_{t=T_1}^{T_2-1}, T_1, T_2, M)$-good if and only if the pair $\{y_t\}_{t=T_1}^{T_2}$, $\{x_t\}_{t=T_1}^{T_2}$ is $(\{f_t^0\}_{t=T_1}^{T_2-1}, T_1, T_2, M)$-good.

2. The pair $\{x_t\}_{t=T_1}^{T_2}$, $\{y_t\}_{t=T_1}^{T_2}$ is $(\{f_t\}_{t=T_1}^{T_2-1}, x_{T_1}, x_{T_2}, y_{T_1}, y_{T_2}, T_1, T_2, M)$-good if and only if the pair $\{y_t\}_{t=T_1}^{T_2}$, $\{x_t\}_{t=T_1}^{T_2}$ is

$$(\{f_t^0\}_{t=T_1}^{T_2-1}, y_{T_1}, y_{T_2}, x_{T_1}, x_{T_2}, T_1, T_2, M)\text{-good.}$$

3. The pair $\{x_t\}_{t=T_1}^{T_2}$, $\{y_t\}_{t=T_1}^{T_2}$ is $(\{f_t\}_{t=T_1}^{T_2-1}, x_{T_1}, y_{T_1}, T_1, T_2, M)$-good if and only if the pair $\{y_t\}_{t=T_1}^{T_2}$, $\{x_t\}_{t=T_1}^{T_2}$ is $(\{f_t^0\}_{t=T_1}^{T_2-1}, y_{T_1}, x_{T_1}, T_1, T_2, M)$-good.

4. The pair $\{x_t\}_{t=T_1}^{T_2}$, $\{y_t\}_{t=T_1}^{T_2}$ is $(\{f_t\}_{t=T_1}^{T_2-1}, (x_{T_2}, y_{T_2})^-, T_1, T_2, M)$-good if and only if the pair $\{y_t\}_{t=T_1}^{T_2}$, $\{x_t\}_{t=T_1}^{T_2}$ is $(\{f_t^0\}_{t=T_1}^{T_2-1}, (y_{T_2}, x_{T_2})^-, T_1, T_2, M)$-good.

We suppose that assumptions (A3) and (A4) introduced in Sect. 8.9 holds. By (A3), (A4), and Theorem 8.17, for each $\xi_1 \in \cup\{\bar{Y}_1(l) : l = 1, 2, \dots\}$ and each $\xi_2 \in \cup\{\bar{Y}_2(l) : l = 1, 2, \dots\}$ there exists a unique (f)-overtaking optimal pair of an (Ω_1)-program $\{x_t^{(\xi_1,\xi_2)}\}_{t=0}^\infty$ and an (Ω_2)-program $\{y_t^{(\xi_1,\xi_2)}\}_{t=0}^\infty$ such that

$$x_0^{(\xi_1,\xi_2)} = \xi_1,\ y_0^{(\xi_1,\xi_2)} = \xi_2. \tag{8.291}$$

By Theorem 8.15, for each $\xi_1 \in \cup\{\bar{Y}_1(l) : l = 1, 2, \dots\}$ and each $\xi_2 \in \cup\{\bar{Y}_2(l) : l = 1, 2, \dots\}$,

$$\lim_{t\to\infty} x_t^{(\xi_1,\xi_2)} = x_f,\ \lim_{t\to\infty} y_t^{(\xi_1,\xi_2)} = y_f. \tag{8.292}$$

Theorems 8.16 and 8.18 imply that there exists a unique strongly (f)-overtaking optimal pair of an (Ω_1)-program $\{x_t^f\}_{t=0}^\infty$ and an (Ω_2)-program $\{y_t^f\}_{t=0}^\infty$ and that

$$\lim_{t\to\infty} x_t^f = x_f,\ \lim_{t\to\infty} y_t^f = y_f. \tag{8.293}$$

For $i = 1, 2$ define

$$\bar{\Omega}_i = \{(x, y) \in X_i \times X_i : (y, x) \in \Omega_i\} \tag{8.294}$$

which is a nonempty closed convex subset of $X_i \times X_i$. Denote by $\mathcal{N}$ the set of all continuous functions $g : \bar{\Omega}_1 \times \bar{\Omega}_2 \to R^1$ such that:

for each $(y_1, y_2) \in \bar{\Omega}_2$ the function $(x_1, x_2) \to g(x_1, x_2, y_1, y_2)$, $(x_1, x_2) \in \bar{\Omega}_1$ is convex;

for each $(x_1, x_2) \in \bar{\Omega}_1$ the function $(y_1, y_2) \to g(x_1, x_2, y_1, y_2)$, $(y_1, y_2) \in \bar{\Omega}_2$ is concave.

We equip the set $\mathcal{N}$ with a metric $\rho_{\mathcal{N}} : \mathcal{N} \times \mathcal{N} \to R^1$ defined by

$$\begin{aligned}\rho_{\mathcal{N}}(h, g) = \sup\{&|g(x_1, x_2, y_1, y_2) - h(x_1, x_2, y_1, y_2)| : \\ &(x_1, x_2) \in \bar{\Omega}_1,\ (y_1, y_2) \in \bar{\Omega}_2\}.\end{aligned}$$

Clearly, $(\mathcal{N}, \rho_{\mathcal{N}})$ is a complete metric space.

For each $f \in \mathcal{M}$ define $\bar{f} \in \mathcal{N}$ by

$$\bar{f}(x_1, x_2, y_1, y_2) = f(x_2, x_1, y_2, y_1),\ (x_1, x_2) \in \bar{\Omega}_1,\ (y_1, y_2) \in \bar{\Omega}_2. \quad (8.295)$$

It is clear that $f \to \bar{f}$, $f \in \mathcal{M}$ is a linearly invertible isometry operator.

Let $T_2 > T_1 \geq 0$ be integers, $\{x_t\}_{t=T_1}^{T_2}$ be (Ω_1)-program, $\{y_t\}_{t=T_1}^{T_2}$ be (Ω_2)-program. Define $\{\bar{x}_t\}_{t=T_1}^{T_2} \subset X_1$, $\{\bar{y}_t\}_{t=T_1}^{T_2} \subset X_2$ by

$$\bar{x}_t = x_{T_2-t+T_1},\ \bar{y}_t = y_{T_2-t+T_1},\ t = T_1, \dots, T_2. \quad (8.296)$$

Clearly, $\{\bar{x}_t\}_{t=T_1}^{T_2}$ is an $(\bar{\Omega}_1)$-program, $\{\bar{y}_t\}_{t=T_1}^{T_2}$ is an $(\bar{\Omega}_2)$-program.

Assume that $\{f_t\}_{t=T_1}^{T_2-1} \subset \mathcal{M}$. By (8.295) and (8.296),

$$\sum_{t=T_1}^{T_2-1} \bar{f}_{T_2-t+T_1-1}(\bar{x}_t, \bar{x}_{t+1}, \bar{y}_t, \bar{y}_{t+1})$$

$$\sum_{t=T_1}^{T_2-1} f_{T_2-t+T_1-1}(x_{T_2-t+T_1-1}, x_{T_2-t+T_1}, y_{T_2-t+T_1-1}, y_{T_2-t+T_1})$$

$$= \sum_{t=T_1}^{T_2-1} f_t(x_t, x_{t+1}, y_t, y_{t+1}). \quad (8.297)$$

The next result easily follows from (8.297).

Proposition 8.29. *Let $T_2 > T_1 \geq 0$ be integers, $M \geq 0$, $\{f_t\}_{t=T_1}^{T_2-1} \subset \mathcal{M}$, $\{x_t^{(i)}\}_{t=T_1}^{T_2}$, $i = 1, 2$ be (Ω_1)-programs, $\{y_t^{(i)}\}_{t=T_1}^{T_2}$, $i = 1, 2$ be (Ω_2)-programs. Then*

$$\sum_{t=T_1}^{T_2-1} f_t(x_t^{(1)}, x_{t+1}^{(1)}, y_t^{(1)}, y_{t+1}^{(1)}) \leq \sum_{t=T_1}^{T_2-1} f_t(x_t^{(2)}, x_{t+1}^{(2)}, y_t^{(2)}, y_{t+1})^{(2)} + M$$

if and only if

$$\sum_{t=T_1}^{T_2-1} \bar{f}_{T_2-t+T_1-1}(\bar{x}_t^{(1)}, \bar{x}_{t+1}^{(1)}, \bar{y}_t^{(1)}, \bar{y}_{t+1}^{(1)})$$

$$\leq M + \sum_{t=T_1}^{T_2-1} \bar{f}_{T_2-t+T_1-1}(\bar{x}_t^{(2)}, \bar{x}_{t+1}^{(2)}, \bar{y}_t^{(1)}, \bar{y}_{t+1}^{(1)});$$

$$\sum_{t=T_1}^{T_2-1} f_t(x_t^{(1)}, x_{t+1}^{(1)}, y_t^{(1)}, y_{t+1}^{(1)}) + M \geq \sum_{t=T_1}^{T_2-1} f_t(x_t^{(1)}, x_{t+1}^{(1)}, y_t^{(2)}, y_{t+1}^{(2)})$$

if and only if

$$\sum_{t=T_1}^{T_2-1} \bar{f}_{T_2-t+T_1-1}(\bar{x}_t^{(1)}, \bar{x}_{t+1}^{(1)}, \bar{y}_t^{(1)}, \bar{y}_{t+1}^{(1)}) + M$$
$$\geq \sum_{t=T_1}^{T_2-1} f_{T_2-t+T_1-1}(\bar{x}_t^{(1)}, \bar{x}_{t+1}^{(1)}, \bar{y}_t^{(2)}, \bar{y}_{t+1}^{(2)}).$$

Proposition 8.29 implies the following result.

Proposition 8.30. *Let $T_2 > T_1 \geq 0$ be integers, $M \geq 0$, $\{f_t\}_{t=T_1}^{T_2-1} \subset \mathcal{M}$, $\{x_t\}_{t=T_1}^{T_2}$ be (Ω_1)-program, $\{y_t\}_{t=T_1}^{T_2}$ be (Ω_2)-program. Then $\{\bar{x}_t\}_{t=T_1}^{T_2}$ is an $(\bar{\Omega}_1)$-program, $\{\bar{y}_t\}_{t=T_1}^{T_2}$ is an $(\bar{\Omega}_2)$-program and the following assertions hold:*

the pair $\{x_t\}_{t=T_1}^{T_2}$, $\{y_t\}_{t=T_1}^{T_2}$ is $(\{f_t\}_{t=T_1}^{T_2-1}, x_{T_1}, x_{T_2}, y_{T_1}, y_{T_2}, T_1, T_2, M)$-good if and only if the pair $\{\bar{x}_t\}_{t=T_1}^{T_2}$, $\{\bar{y}_t\}_{t=T_1}^{T_2}$ is

$$(\{f_{T_2-t+T_1-1}\}_{t=T_1}^{T_2-1}, \bar{x}_{T_1}, \bar{x}_{T_2}, \bar{y}_{T_1}, \bar{y}_{T_2}, T_1, T_2, M)\text{-good;}$$

the pair $\{x_t\}_{t=T_1}^{T_2}$, $\{y_t\}_{t=T_1}^{T_2}$ is $(\{f_t\}_{t=T_1}^{T_2-1}, T_1, T_2, M)$-good if and only if the pair $\{\bar{x}_t\}_{t=T_1}^{T_2}$, $\{\bar{y}_t\}_{t=T_1}^{T_2}$ is $(\{f_{T_2-t+T_1-1}\}_{t=T_1}^{T_2-1}, T_1, T_2, M)$-good;

the pair $\{x_t\}_{t=T_1}^{T_2}$, $\{y_t\}_{t=T_1}^{T_2}$ is $(\{f_t\}_{t=T_1}^{T_2-1}, x_{T_1}, y_{T_1}, T_1, T_2, M)$-good if and only if the pair $\{\bar{x}_t\}_{t=T_1}^{T_2}$, $\{\bar{y}_t\}_{t=T_1}^{T_2}$ is

$$(\{f_{T_2-t+T_1-1}\}_{t=T_1}^{T_2-1}, (\bar{x}_{T_2}, \bar{y}_{T_2})^-, T_1, T_2, M)\text{-good;}$$

the pair $\{x_t\}_{t=T_1}^{T_2}$, $\{y_t\}_{t=T_1}^{T_2}$ is $(\{f_t\}_{t=T_1}^{T_2-1}, (x_{T_2}, y_{T_2})^-, T_1, T_2, M)$-good if and only if the pair $\{\bar{x}_t\}_{t=T_1}^{T_2}$, $\{\bar{y}_t\}_{t=T_1}^{T_2}$ is $(\{f_{T_2-t+T_1-1}\}_{t=T_1}^{T_2-1}, \bar{x}_{T_1}, \bar{y}_{T_1}, T_1, T_2, M)$-good.

Since the triplet $(\bar{f}, \bar{\Omega}_1, \bar{\Omega}_2)$ satisfies the same assumptions as the triplet (f, Ω_1, Ω_2) all the results obtained for the triplet (f, Ω_1, Ω_2) are also true for the triplet $(\bar{f}, \bar{\Omega}_1, \bar{\Omega}_2)$ In particular, for each $\xi_1 \in \cup\{Y_1(l) : l = 1, 2, \dots\}$ and each $\xi_2 \in \cup\{Y_2(l) : l = 1, 2, \dots\}$ there exists a unique $(\bar{f})$-overtaking optimal pair of an $(\bar{\Omega}_1)$-program $\{\bar{x}_t^{(\xi_1,\xi_2)}\}_{t=0}^\infty$ and an $(\bar{\Omega}_2)$-program $\{\bar{y}_t^{(\xi_1,\xi_2)}\}_{t=0}^\infty$ such that

$$\bar{x}_0^{(\xi_1,\xi_2)} = \xi_1, \ \bar{y}_0^{(\xi_1,\xi_2)} = \xi_2$$

and there exists a unique strongly $(\bar{f})$-overtaking optimal pair of an $(\bar{\Omega}_1)$-program $\{\bar{x}_t^f\}_{t=0}^\infty$ and an $(\bar{\Omega}_2)$-program $\{\bar{y}_t^f\}_{t=0}^\infty$.

In this chapter we prove the following results.

Theorem 8.31. *Let $\epsilon \in (0, 1)$ and $L, l \geq 1$ be integers. Then there exist a natural number $L_0 > L$ and $\delta > 0$ such that for each integer $T \geq L_0$, each $f_t \in \mathcal{M}$, $t = 0, \dots, T-1$ satisfying*

$$\|f_t - f\| \leq \delta, \ t = 0, \dots, T-1,$$

each (Ω_1)*-program* $\{x_t\}_{t=0}^T$ *and each* (Ω_2)*-program* $\{y_t\}_{t=0}^T$ *such that*

$$x_0 \in \bar{Y}_1(l),\ y_0 \in \bar{Y}_2(l)$$

and the pair $\{x_t\}_{t=0}^T$, $\{y_t\}_{t=0}^T$ *is* $(\{f_t\}_{t=0}^{T-1}, x_0, y_0, 0, T, \delta)$*-good, for all* $t = 0, \dots, L$,

$$|x_{T-t} - \bar{x}_t^f| \le \epsilon,\ |y_{T-t} - \bar{y}_t^f| \le \epsilon.$$

Theorem 8.32. *Let* $\epsilon \in (0,1)$ *and* $L \ge 1$ *be an integer. Then there exist a natural number* $L_0 > L$ *and* $\delta > 0$ *such that for each integer* $T \ge L_0$*, each* $f_t \in \mathcal{M}$, $t = 0, \dots, T-1$ *satisfying*

$$\|f_t - f\| \le \delta,\ t = 0, \dots, T-1,$$

each (Ω_1)*-program* $\{x_t\}_{t=0}^T$ *and each* (Ω_2)*-program* $\{y_t\}_{t=0}^T$ *such that the pair* $\{x_t\}_{t=0}^T$, $\{y_t\}_{t=0}^T$ *is* $(\{f_t\}_{t=0}^{T-1}, 0, T, \delta)$*-good, for all* $t = 0, \dots, L$,

$$|x_t - x_t^f| \le \epsilon,\ |y_t - y_t^f| \le \epsilon,$$

$$|x_{T-t} - \bar{x}_t^f| \le \epsilon,\ |y_{T-t} - \bar{y}_t^f| \le \epsilon.$$

Theorem 8.33. *Let* $\epsilon \in (0,1)$ *and* $L, l \ge 1$ *be integers. Then there exist a natural number* $L_0 > L$ *and* $\delta > 0$ *such that for each integer* $T \ge L_0$*, each* $f_t \in \mathcal{M}$, $t = 0, \dots, T-1$ *satisfying*

$$\|f_t - f\| \le \delta,\ t = 0, \dots, T-1,$$

each (Ω_1)*-program* $\{x_t\}_{t=0}^T$ *and each* (Ω_2)*-program* $\{y_t\}_{t=0}^T$ *such that*

$$x_0 \in \bar{Y}_1(l),\ y_0 \in \bar{Y}_2(l), x_T \in Y_1(l),\ y_T \in Y_2(l)$$

and the pair $\{x_t\}_{t=0}^T$, $\{y_t\}_{t=0}^T$ *is* $(\{f_t\}_{t=0}^{T-1}, x_0, x_T, y_0, y_T, 0, T, \delta)$*-good, for all* $t = 0, \dots, L$,

$$|x_t - x_t^{(x_0,y_0)}| \le \epsilon,\ |y_t - y_t^{(x_0,y_0)}| \le \epsilon,$$

$$|x_{T-t} - \bar{x}_t^{(x_T,y_T)}| \le \epsilon,\ |y_{T-t} - \bar{y}_t^{(x_T,y_T)}| \le \epsilon.$$

8.16 Auxiliary Results for Theorems 8.31–8.33

Denote by $\mathrm{Card}(E)$ the cardinality of a set E. If $E = \emptyset$, then $\mathrm{Card}(E) = 0$.

It is easy to see that the following lemma holds.

Lemma 8.34. *Let* $\epsilon > 0$*. Then there exists* $\delta \in (0, r_f)$ *such that for each pair of integers* $T_2 > T_1 \ge 0$ *satisfying* $T_2 > T_1 + 1$*, each* (Ω_1)*-program* $\{x_t\}_{t=T_1}^{T_2}$ *and each* (Ω_2)*-program* $\{y_t\}_{t=T_1}^{T_2}$ *satisfying*

$$|x_{T_2} - x_f|,\ |x_{T_2-1} - x_f| \le \delta,$$

$$|y_{T_2} - y_f|,\ |y_{T_2-1} - y_f| \le \delta,$$

each $f_t \in \mathcal{M}$, $t = T_1, \dots, T_2 - 1$ *satisfying*

$$\|f_t - f\| \le \delta,\ t = T_1, \dots, T_2 - 1$$

and sequences $\{\xi_t\}_{t=T_1}^{T_2-1} \subset X_1$, $\{\eta_t\}_{t=T_1}^{T_2-1} \subset X_2$ *such that*

$$\xi_t = x_t,\ t = T_1, \dots, T_2 - 1,\ |\xi_{T_2} - x_{T_2}| \le \delta,$$

$$\eta_t = y_t,\ t = T_1, \dots, T_2 - 1,\ |\eta_{T_2} - y_{T_2}| \le \delta$$

the following assertions hold.

1. *If* $M \ge 0$ *and the pair* $\{x_t\}_{t=T_1}^{T_2}$, $\{y_t\}_{t=T_1}^{T_2}$ *is*

$$(\{f_t\}_{t=T_1}^{T_2-1}, x_{T_1}, x_{T_2}, y_{T_1}, y_{T_2}, T_1, T_2, M)\text{-good},$$

then the pair $\{\xi_t\}_{t=T_1}^{T_2}$, $\{\eta_t\}_{t=T_1}^{T_2}$ *is*

$$(\{f_t\}_{t=T_1}^{T_2-1}, \xi_{T_1}, \xi_{T_2}, \eta_{T_1}, \eta_{T_2}, T_1, T_2, M + \epsilon)\text{-good}.$$

2. *If* $M \ge 0$ *and the pair* $\{x_t\}_{t=T_1}^{T_2}$, $\{y_t\}_{t=T_1}^{T_2}$ *is*

$$(\{f_t\}_{t=T_1}^{T_2-1}, (x_{T_2}, y_{T_2})^-, T_1, T_2, M)\text{-good},$$

then the pair $\{\xi_t\}_{t=T_1}^{T_2}$, $\{\eta_t\}_{t=T_1}^{T_2}$ *is*

$$(\{f_t\}_{t=T_1}^{T_2-1}, (\xi_{T_2}, \eta_{T_2})^-, T_1, T_2, M + \epsilon)\text{-good}.$$

Lemma 8.35. *Let* $\epsilon > 0$ *and* $l \ge 1$ *be an integer. Then there exist a natural number* $L_1 > 4$ *and* $\delta > 0$ *such that for each integer* $T \ge L_1$, *each* (Ω_1)-*program* $\{x_t\}_{t=0}^{T}$, *each* (Ω_2)-*program* $\{y_t\}_{t=0}^{T}$ *satisfying*

$$|x_T - x_f| \le \delta,\ |y_T - y_f| \le \delta$$

and each $f_t \in \mathcal{M}$, $t = 0, \dots, T - 1$ *satisfying*

$$\|f_t - f\| \le \delta T^{-1},\ t = 0, \dots, T - 1,$$

the following assertions hold:

1. *If the pair* $\{x_t\}_{t=0}^{T}$, $\{y_t\}_{t=0}^{T}$ *is*

$$(\{f_t\}_{t=0}^{T-1}, x_0, x_T, y_0, y_T, 0, T, \delta)\text{-good}$$

and

$$x_0 \in \bar{Y}_1(l),\ y_0 \in \bar{Y}_2(l)$$

then for all $t = 0, \dots, T$,

$$|x_t - x_t^{(x_0, y_0)}| \le \epsilon,\ |y_t - y_t^{(x_0, y_0)}| \le \epsilon.$$

2. *If the pair* $\{x_t\}_{t=0}^{T}$, $\{y_t\}_{t=0}^{T}$ *is*

$$(\{f_t\}_{t=0}^{T-1}, (x_T, y_T)^-, 0, T, \delta)\text{-good},$$

then for all $t = 0, \dots, T$,

$$|x_t - x_t^f| \le \epsilon,\ |y_t - y_t^f| \le \epsilon.$$

Proof. By (A3), (A4), and the uniform continuity of f there exists a positive number

$$\delta_0 < \min\{\epsilon/2, 4^{-1}\}$$

such that the following properties hold:

(P21) for each (u_1, u_2), $(v_1, v_2) \in \Omega_1$ an each $(z_1, z_2) \in \Omega_2$ satisfying

$$\max\{|u_1 - v_1|\ |u_2 - v_2|\} \geq \epsilon$$

we have

$$f(2^{-1}(u_1 + v_1), 2^{-1}(u_2 + v_2), z_1, z_2)$$
$$\leq 2^{-1} f(u_1, u_2, , z_1, z_2) + 2^{-1} f(v_1, v_2, z_1, z_2) - 4\delta_0;$$

(P22) for each (u_1, u_2), $(v_1, v_2) \in \Omega_2$ an each $(z_1, z_2) \in \Omega_1$ satisfying

$$\max\{|u_1 - v_1|\ |u_2 - v_2|\} \geq \epsilon$$

we have

$$f(z_1, z_2, 2^{-1}(u_1 + v_1), 2^{-1}(u_2 + v_2))$$
$$\geq 2^{-1} f(z_1, z_2, u_1, u_2,) + 2^{-1} f(z_1, z_2, v_1, v_2) + 4\delta_0.$$

By Lemma 8.34, there exists a positive number

$$\delta_1 < \min\{\delta_0/4, r_f/4)$$

such that the following property holds:

(P23) for each pair of integers $T_2 > T_1 \geq 0$ satisfying $T_2 > T_1 + 1$, each $f_t \in \mathcal{M}$, $t = T_1, \dots, T_2 - 1$ satisfying

$$\|f_t - f\| \leq \delta_1,\ t = T_1, \dots, T_2 - 1,$$

each (Ω_1)-program $\{x_t\}_{t=T_1}^{T_2}$ and each (Ω_2)-program $\{y_t\}_{t=T_1}^{T_2}$ satisfying

$$|x_{T_2} - x_f|,\ |x_{T_2-1} - x_f| \leq 2\delta_1,$$

$$|y_{T_2} - y_f|,\ |y_{T_2-1} - y_f| \leq 2\delta_1,$$

and sequences $\{\xi_t\}_{t=T_1}^{T_2-1} \subset X_1$, $\{\eta_t\}_{t=T_1}^{T_2-1} \subset X_2$ such that

$$\xi_t = x_t,\ t = T_1, \dots, T_2 - 1,\ |\xi_{T_2} - x_{T_2}| \leq 2\delta_1,$$

$$\eta_t = y_t,\ t = T_1, \dots, T_2 - 1,\ |\eta_{T_2} - y_{T_2}| \leq 2\delta_1$$

the following assertions hold.

1. If $M \geq 0$ and the pair $\{x_t\}_{t=T_1}^{T_2}$, $\{y_t\}_{t=T_1}^{T_2}$ is

$$(\{f_t\}_{t=T_1}^{T_2-1}, x_{T_1}, x_{T_2}, y_{T_1}, y_{T_2}, T_1, T_2, M)\text{-good},$$

then the pair $\{\xi_t\}_{t=T_1}^{T_2}$, $\{\eta_t\}_{t=T_1}^{T_2}$ is $(\{f_t\}_{t=T_1}^{T_2-1}, \xi_{T_1}, \xi_{T_2}, \eta_{T_1}, \eta_{T_2}, T_1, T_2, M + \delta_0)$-good.

2. If $M \geq 0$ and the pair $\{x_t\}_{t=T_1}^{T_2}$, $\{y_t\}_{t=T_1}^{T_2}$ is

$$(\{f_t\}_{t=T_1}^{T_2-1}, (x_{T_2}, y_{T_2})^-, T_1, T_2, M)\text{-good},$$

then the pair $\{\xi_t\}_{t=T_1}^{T_2}$, $\{\eta_t\}_{t=T_1}^{T_2}$ is $(\{f_t\}_{t=T_1}^{T_2-1}, (\xi_{T_2}, \eta_{T_2})^-, T_1, T_2, M+\delta_0)$-good.

By Theorem 8.5, there exist $\delta \in (0, \delta_1)$ and a natural number $L_{1,1} > 2l$ such that the following property holds:

(P24) for each integer $T \geq 2L_{1,1}$, each $f_t \in \mathcal{M}$, $t = 0, \dots, T-1$ satisfying

$$\|f_t - f\| \leq \delta,\ t = 0, \dots, T-1,$$

each (Ω_1)-program $\{x_t\}_{t=0}^T$ and each (Ω_2)-program $\{y_t\}_{t=0}^T$ satisfying

$$|x_T - x_f| \leq \delta,\ |y_T - y_f| \leq \delta$$

and for which at least one of the following conditions holds:

(i) $x_0 \in \bar{Y}_1(l)$, $y_0 \in \bar{Y}_2(l)$, the pair $\{x_t\}_{t=0}^T$, $\{y_t\}_{t=0}^T$ is

$$(\{f_t\}_{t=0}^{T-1}, x_0, x_T, y_0, y_T, 0, T, \delta)\text{-good};$$

(ii) the pair $\{x_t\}_{t=0}^T$, $\{y_t\}_{t=0}^T$ is $(\{f_t\}_{t=0}^{T-1}, (x_T, y_T)^-, 0, T, \delta)$-good we have for all $t = L_1, \dots, T$,

$$|x_t - x_f| \leq \delta_1,\ |y_t - y_f| \leq \delta_1.$$

In view of (8.293), there exists a natural number $L_{1,2}$ such that

$$|x_t^f - x_f| \leq \delta_1,\ |y_t^f - y_f| \leq \delta_1 \text{ for all integers } t \geq L_{1,2}. \tag{8.298}$$

Property (P24) and (8.292) imply that there exists a natural number $L_{1,3}$ such that for each $\xi_1 \in \cup\{\bar{Y}_1(l) : l = 1, 2, \dots\}$ and each $\xi_2 \in \cup\{\bar{Y}_2(l) : l = 1, 2, \dots\}$ for all integers $t \geq L_{1,3}$, we have

$$|x_t^{(\xi_1,\xi_2)} - x_f| \leq \delta_1,\ |y_t^{(\xi_1,\xi_2)} - y_f| \leq \delta_1. \tag{8.299}$$

Set

$$L_1 = 2(L_{1,1} + L_{1,2} + L_{1,3}). \tag{8.300}$$

Assume that an integer $T \geq L_1$, each $f_t \in \mathcal{M}$, $t = 0, \dots, T-1$ satisfy

$$\|f_t - f\| \leq \delta/T,\ t = 0, \dots, T-1, \tag{8.301}$$

$\{x_t\}_{t=0}^T$ is an (Ω_1)-program, $\{y_t\}_{t=0}^T$ is an (Ω_2)-program,

$$|x_T - x_f|,\ |y_T - y_f| \leq \delta \tag{8.302}$$

and at least one of conditions (i) and (ii) holds.

Property (P24), conditions (i), (ii), and (8.300)–(8.302) imply that for all $t = L_1, \dots, T$,

$$|x_t - x_f|,\ |y_t - y_f| \leq \delta_1. \tag{8.303}$$

In the case of condition (i) set

$$\xi_t = x_t,\ \eta_t = y_t,\ t = 0, \dots, T-1,\ \xi_T = x_T^{(x_0,y_0)},\ \eta_T = y_T^{(x_0,y_0)} \tag{8.304}$$

and in the case of condition (ii) set

$$\xi_t = x_t,\ \eta_t = y_t,\ t = 0, \dots, T-1,\ \xi_T = x_T^f,\ \eta_T = y_T^f. \tag{8.305}$$

It follows from conditions (i), (ii), (8.298), (8.299), and (8.303)–(8.305) that $\{\xi_t\}_{t=0}^T$ is an (Ω_1)-program and $\{\eta_t\}_{t=0}^T$ is an (Ω_2)-program. Property (P23), (8.298), (8.299), (8.301), and (8.303)–(8.305) imply that in the case of condition (i) the pair $\{\xi_t\}_{t=0}^T$, $\{\eta_t\}_{t=0}^T$ is

$$(\{f_t\}_{t=0}^{T-1}, x_0, x_T^{(x_0,y_0)}, y_0, y_T^{(x_0,y_0)}, 0, T, \delta + \delta_0)\text{-good}$$

and in the case of condition (ii) the pair $\{\xi_t\}_{t=0}^T$, $\{\eta_t\}_{t=0}^T$ is

$$(\{f_t\}_{t=0}^{T-1}, (x_T^f, y_T^f)^-, 0, T, \delta + \delta_0)\text{-good}.$$

In the case of condition (i) for all integers $t \geq 0$ set,

$$x_t^* = x_t^{(x_0,y_0)},\ y_t^* = y_t^{(x_0,y_0)} \tag{8.306}$$

and in the case of condition (ii) for all integers $t \geq 0$ set,

$$x_t^* = x_t^f,\ y_t^* = y_t^f. \tag{8.307}$$

By (8.301) and (8.304)–(8.307), in both cases

$$\begin{aligned}
-2\delta_0 + \sum_{t=0}^{T-1} f_t(\xi_t, \xi_{t+1}, \eta_t, \eta_{t+1}) &\leq \sum_{t=0}^{T-1} f_t(x_t^*, x_{t+1}^*, \eta_t, \eta_{t+1}) \\
\leq \sum_{t=0}^{T-1} f(x_t^*, x_{t+1}^*, \eta_t, \eta_{t+1}) + \delta &\leq \sum_{t=0}^{T-1} f(x_t^*, x_{t+1}^*, y_t^*, y_{t+1}^*) + \delta \\
\leq \sum_{t=0}^{T-1} f(\xi_t, \xi_{t+1}, y_t^*, y_{t+1}^*) + \delta &\leq \sum_{t=0}^{T-1} f_t(\xi_t, \xi_{t+1}, y_t^*, y_{t+1}^*)) + 2\delta \\
&\leq \sum_{t=0}^{T-1} f_t(\xi_t, \xi_{t+1}, \eta_t, \eta_{t+1}) + 2\delta_0.
\end{aligned} \tag{8.308}$$

Clearly, $\{2^{-1}(\xi_t + x_t^*)\}_{t=0}^T$ is an (Ω_1)-program and $\{2^{-1}(\eta_t + y_t^*)\}_{t=0}^T$ is an (Ω_2)-program. By (8.304)–(8.307),

$$\sum_{t=0}^{T-1} f(2^{-1}(\xi_t + x_t^*), 2^{-1}(\xi_{t+1} + x_{t+1}^*), y_t^*, y_{t+1}^*) \geq \sum_{t=0}^{T-1} f(x_t^*, x_{t+1}^*, y_t^*, y_{t+1}^*), \tag{8.309}$$

$$\sum_{t=0}^{T-1} f(x_t^*, x_{t+1,}^* 2^{-1}(\eta_t + y_t^*), 2^{-1}(\eta_{t+1} + y_{t+1}^*)) \le \sum_{t=0}^{T-1} f(x_t^*, x_{t+1}^*, y_t^*, y_{t+1}^*). \tag{8.310}$$

Set

$$E_1 = \{t \in \{0, \dots, T-1\} : \max\{|\xi_t - x_t^*|, |\xi_{t+1} - x_{t+1}^*| > \epsilon\}, \tag{8.311}$$

$$E_2 = \{t \in \{0, \dots, T-1\} : \max\{|\eta_t - y_t^*|, |\eta_{t+1} - y_{t+1}^*| > \epsilon\}. \tag{8.312}$$

In view of the convexity–concavity assumptions, for all $g \in \{f_0, \dots, f_{T-1}, f\}$ and all $t = 0, \dots, T-1$,

$$g(2^{-1}(\xi_t + x_t^*), 2^{-1}(\xi_{t+1} + x_{t+1}^*), y_t^*, y_{t+1}^*)$$

$$\le 2^{-1}g(\xi_t, \xi_{t+1}, y_t^*, y_{t+1}^*) + 2^{-1}g(x_t^*, x_{t+1}^*, y_t^*, y_{t+1}^*), \tag{8.313}$$

$$g(x_t^*, x_{t+1,}^* 2^{-1}(\eta_t + y_t^*), 2^{-1}(\eta_{t+1} + y_{t+1}^*))$$

$$\ge 2^{-1}g(x_t^*, x_{t+1,}^* y_t^*, y_{t+1}^*) + 2^{-1}g(x_t^*, x_{t+1}^*, \eta_t, \eta_{t+1}). \tag{8.314}$$

Property (P21), (8.301), and (8.311) imply that for every $t \in E_1$,

$$2^{-1}f_t(\xi_t, \xi_{t+1}, y_t^*, y_{t+1}^*) + 2^{-1}f_t(x_t^*, x_{t+1}^*, y_t^*, y_{t+1}^*)$$

$$-f_t(2^{-1}(\xi_t + x_t^*), 2^{-1}(\xi_{t+1} + x_{t+1}^*), y_t^*, y_{t+1}^*)$$

$$\ge 2^{-1}f(\xi_t, \xi_{t+1}, y_t^*, y_{t+1}^*) + 2^{-1}f(x_t^*, x_{t+1}^*, y_t^*, y_{t+1}^*)$$

$$- f(2^{-1}(\xi_t + x_t^*), 2^{-1}(\xi_{t+1} + x_{t+1}^*), y_t^*, y_{t+1}^*) - 3\delta \ge 4\delta_0 - 3\delta. \tag{8.315}$$

Property (P22), (8.301), and (8.312) imply that for every $t \in E_2$,

$$f_t(x_t^*, x_{t+1,}^* 2^{-1}(\eta_t + y_t^*), 2^{-1}(\eta_{t+1} + y_{t+1}^*))$$

$$-2^{-1}f_t(x_t^*, x_{t+1,}^* y_t^*, y_{t+1}^*) - 2^{-1}f_t(x_t^*, x_{t+1}^*, \eta_t, \eta_{t+1})$$

$$\ge f(x_t^*, x_{t+1,}^* 2^{-1}(\eta_t + y_t^*), 2^{-1}(\eta_{t+1} + y_{t+1}^*))$$

$$- 2^{-1}f(x_t^*, x_{t+1,}^* y_t^*, y_{t+1}^*) - 2^{-1}f(x_t^*, x_{t+1}^*, \eta_t, \eta_{t+1}) - 3\delta \ge 4\delta_0 - 3\delta. \tag{8.316}$$

By (8.301), (8.308), (8.309), and (8.313),

$$\sum_{t=0}^{T-1} f_t(x_t^*, x_{t+1,}^* y_t^*, y_{t+1}^*) \le \sum_{t=0}^{T-1} f(2^{-1}(\xi_t + x_t^*), 2^{-1}(\xi_{t+1} + x_{t+1}^*), y_t^*, y_{t+1}^*) + 2\delta$$

$$\le 2^{-1} \sum_{t=0}^{T-1} f_t(\xi_t, \xi_{t+1}, y_t^*, y_{t+1}^*)$$

$$+2^{-1} \sum_{t=0}^{T-1} f_t(x_t^*, x_{t+1}^*, y_t^*, y_{t+1}^*) - (4\delta_0 - 3\delta)\mathrm{Card}(E_1) + 2\delta$$

$$\leq 2^{-1}(\sum_{t=0}^{T-1} f_t(x_t^*, x_{t+1,}^* y_t^*, y_{t+1}^*) + 4\delta_0 + 4\delta)$$

$$+2^{-1}\sum_{t=0}^{T-1} f_t(x_t^*, x_{t+1,}^* y_t^*, y_{t+1}^*) - (4\delta_0 - 3\delta)\mathrm{Card}(E_1) + 2\delta,$$

$$0 \leq 4\delta + 2\delta_0 - (4\delta_0 - 3\delta)\mathrm{Card}(E_1) \leq 2\delta_0 - 3\delta_0\mathrm{Card}(E_1) + 4\delta$$

$$< 3\delta_0 - 3\delta_0\mathrm{Card}(E_1),$$

$$\mathrm{Card}(E_1) = 0, \ E_1 = \emptyset.$$

By (8.301), (8.308), (8.310), (8.314), and (8.316),

$$\sum_{t=0}^{T-1} f_t(x_t^*, x_{t+1,}^* y_t^*, y_{t+1}^*)$$

$$\geq \sum_{t=0}^{T-1} f_t(x_t^*, x_{t+1,}^* 2^{-1}(\eta_t + y_t^*), 2^{-1}(\eta_{t+1} + y_{t+1}^*)) + 2\delta$$

$$\geq 2^{-1}\sum_{t=0}^{T-1} f_t(x_t^*, x_{t+1,}^* y_t^*, y_{t+1}^*)$$

$$+2^{-1}\sum_{t=0}^{T-1} f_t(x_t^*, x_{t+1,}^* \eta_t, \eta_{t+1}) + (4\delta_0 - 3\delta)\mathrm{Card}(E_2) + 2\delta$$

$$\geq 2^{-1}\sum_{t=0}^{T-1} f_t(x_t^*, x_{t+1,}^* y_t^*, y_{t+1}^*) + 2^{-1}(\sum_{t=0}^{T-1} f_t(x_t^*, x_{t+1,}^* y_t^*, y_{t+1}^*) - 4\delta_0)$$

$$+(4\delta_0 - 3\delta)\mathrm{Card}(E_2),$$

$$0 \geq -2\delta_0 + 3\delta_0\mathrm{Card}(E_2), \ \mathrm{Card}(E_2) = 0, \ E_2 = \emptyset.$$

Since $E_1, E_2 = \emptyset$ it follows from (8.311) and (8.312) that for all $t = 0, \ldots, T$,

$$|\xi_t - x_t^*| \leq \epsilon, \ |\eta_t - y_t^*| \leq \epsilon. \tag{8.317}$$

In the case of condition (i) in view of (8.299), (8.303), (8.304), (8.306), and (8.317), for all $t = 0, \ldots, T-1$,

$$|x_t - x_t^{(x_0,y_0)}| \leq \epsilon, \ |y_t - y_t^{(x_0,y_0)}| \leq \epsilon,$$

$$|x_T - x_T^{(x_0,y_0)}| \leq |x_T - x_f| + |x_f - x_T^{(x_0,y_0)}| \leq 2\delta_1 < \epsilon,$$

$$|y_T - y_T^{(x_0,y_0)}| \leq |y_T - y_f| + |y_f - y_T^{(x_0,y_0)}| \leq 2\delta_1 < \epsilon.$$

In the case of condition (ii) in view of (8.298), (8.303)–(8.305), (8.307), and (8.317), for all $t = 0, \dots, T-1$,

$$|x_t - x_t^f| \le \epsilon,\ |y_t - y_t^f| \le \epsilon,$$

$$|x_T - x_T^f| \le |x_T - x_f| + |x_f - x_T^f| \le 2\delta_1 < \epsilon,$$

$$|y_T - y_T^f| \le |y_T - y_f| + |y_f - y_T^f| \le 2\delta_1 < \epsilon.$$

Lemma 8.35 is proved. □

Note that Lemma 8.35 can be applied to the triplet $(\bar{f}, \bar{\Omega}_1, \bar{\Omega}_2)$.

8.17 Proof of Theorem 8.31

By Lemma 8.35 applied to the triplet $(\bar{f}, \bar{\Omega}_1, \bar{\Omega}_2)$ there exist a natural number $L_1 > L$ and $\delta_0 \in (0, \epsilon)$ such that the following property holds:

(P25) for each integer $T \ge L_1$, each $f_t \in \mathcal{M}$, $t = 0, \dots, T-1$ satisfying

$$\|f_t - f\| \le \delta_0 T^{-1},\ t = 0, \dots, T-1,$$

each $(\bar{\Omega}_1)$-program $\{\xi_t\}_{t=0}^T$, each $(\bar{\Omega}_2)$-program $\{\eta_t\}_{t=0}^T$ satisfying

$$|\xi_T - x_f| \le \delta_0,\ |\eta_T - y_f| \le \delta_0$$

and such that the pair $\{\xi_t\}_{t=0}^T$, $\{\eta_t\}_{t=0}^T$ is $(\{\bar{f}_t\}_{t=0}^{T-1}, (\xi_T, \eta_T)^-, 0, T, \delta_0)$-good we have for all $t = 0, \dots, T$,

$$|\xi_t - \bar{x}_t^f| \le \epsilon,\ |\eta_t - \bar{y}_t^f| \le \epsilon.$$

By Theorem 8.5, there exist a natural number $L_2 > 2l$ and $\delta_1 \in (0, \delta_0)$ such that the following property holds: (P26) for each integer $T \ge 2L_2$, each $f_t \in \mathcal{M}$, $t = 0, \dots, T-1$ satisfying

$$\|f_t - f\| \le \delta_1,\ t = 0, \dots, T-1,$$

each (Ω_1)-program $\{x_t\}_{t=0}^T$ and each (Ω_2)-program $\{y_t\}_{t=0}^T$ for which

$$x_0 \in \bar{Y}_1(l),\ y_0 \in \bar{Y}_2(l)$$

and the pair $\{x_t\}_{t=0}^T$, $\{y_t\}_{t=0}^T$ is $(\{f_t\}_{t=0}^{T-1}, x_0, y_0, 0, T, \delta_1)$-good, for all $t = L_2, \dots, T - L_2$,

$$|x_t - x_f| \le \delta_0,\ |y_t - y_f| \le \delta_0.$$

Set

$$L_0 = 2L_1 + 2L_2 + 4,\ \delta = L_0^{-1}\delta_1. \tag{8.318}$$

Assume that an integer $T \ge L_0$, $f_t \in \mathcal{M}$, $t = 0, \dots, T-1$ satisfies

$$\|f_t - f\| \le \delta,\ t = 0, \dots, T-1, \tag{8.319}$$

$\{x_t\}_{t=0}^T$ is an (Ω_1)-program and $\{y_t\}_{t=0}^T$ is an (Ω_2)-program such that

$$x_0 \in \bar{Y}_1(l),\ y_0 \in \bar{Y}_2(l) \tag{8.320}$$

and the pair $\{x_t\}_{t=0}^T$, $\{y_t\}_{t=0}^T$ is $(\{f_t\}_{t=0}^{T-1}, x_0, y_0, 0, T, \delta)$-good. Together with property (P26) and (8.318)–(8.320) this implies that for all $t = L_2, \dots, T - L_2$,

$$|x_t - x_f| \le \delta_0,\ |y_t - y_f| \le \delta_0. \tag{8.321}$$

In view of (8.318),

$$L_2 + L_1 < T - L_2 - L_1. \tag{8.322}$$

By (8.321) and (8.322),

$$|x_{T-L_2-L_1} - x_f| \le \delta_0,\ |y_{T-L_2-L_1} - y_f| \le \delta_0. \tag{8.323}$$

It is clear that the pair of sequences $\{x_t\}_{t=T-L_2-L_1}^T$, $\{y_t\}_{t=T-L_2-L_1}^T$ is $(\{f_t\}_{t=T-L_2-L_1}^{T-1}, x_{T-L_2-L_1}, y_{T-L_2-L_1}, T - L_2 - L_1, T, \delta)$-good. Set

$$\xi_t = x_{T-t},\ \eta_t = y_{T-t},\ t = 0, \dots, L_1 + L_2. \tag{8.324}$$

It is easy to see that $\{\xi_t\}_{t=0}^{L_1+L_2}$ is an $(\bar{\Omega}_1)$-program and $\{\eta_t\}_{t=0}^{L_1+L_2}$ is an $(\bar{\Omega}_2)$-program. By (8.323) and (8.324),

$$|\xi_{L_2+L_1} - x_f| \le \delta_0,\ |\eta_{L_2+L_1} - y_f| \le \delta_0. \tag{8.325}$$

Proposition 8.30 implies that the pair of sequences $\{\xi_t\}_{t=0}^{L_1+L_2}$, $\{\eta_t\}_{t=0}^{L_1+L_2}$ is $(\{\bar{f}_{T-t-1}\}_{t=0}^{L_1+L_2-1}, (\xi_{L_2+L_1}, \eta_{L_2+L_1})^-, 0, L_2 + L_1, \delta)$-good. Combined with property (P25), (8.318), (8.319), and (8.325) this implies that for all $t = 0, \dots, L_1 + L_2$,

$$|\xi_t - \bar{x}_t^f| \le \epsilon,\ |\eta_t - \bar{y}_t^f| \le \epsilon. \tag{8.326}$$

In view of (8.324) and (8.326), for all $t = 0, \dots, L_1 + L_2$,

$$\epsilon \ge |\xi_t - \bar{x}_t^f| = |x_{T-t} - \bar{x}_t^f|,\ \epsilon \ge |\eta_t - \bar{y}_t^f| = |y_{T-t} - \bar{y}_t^f|.$$

Theorem 8.31 is proved.

8.18 Proof of Theorem 8.32

Theorem 8.32 follows from Proposition 8.30 and the next result.

Theorem 8.36. *Let $\epsilon \in (0, 1)$ and $L \ge 1$ be an integer. Then there exist a natural number $L_0 > L$ and $\delta > 0$ such that for each integer $T \ge L_0$, each $f_t \in \mathcal{M}$, $t = 0, \dots, T - 1$ satisfying*

$$\|f_t - f\| \le \delta,\ t = 0, \dots, T - 1,$$

each (Ω_1)-program $\{x_t\}_{t=0}^T$ and each (Ω_2)-program $\{y_t\}_{t=0}^T$ such that the pair $\{x_t\}_{t=0}^T$, $\{y_t\}_{t=0}^T$ is $(\{f_t\}_{t=0}^{T-1}, 0, T, \delta)$-good, for all $t = 0, \dots, L$,

$$|x_t - x_t^f| \le \epsilon,\ |y_t - y_t^f| \le \epsilon.$$

Proof. By Lemma 8.35, there exist a natural number $L_1 > L$ and $\delta_0 \in (0, \epsilon)$ such that the following property holds:

(P27) for each integer $T \geq L_1$, each $f_t \in \mathcal{M}$, $t = 0, \ldots, T-1$ satisfying

$$\|f_t - f\| \leq \delta_0 T^{-1},\ t = 0, \ldots, T-1,$$

each (Ω_1)-program $\{\xi_t\}_{t=0}^T$, each (Ω_2)-program $\{\eta_t\}_{t=0}^T$ satisfying

$$|\xi_T - x_f| \leq \delta_0,\ |\eta_T - y_f| \leq \delta_0$$

and such that the pair $\{\xi_t\}_{t=0}^T$, $\{\eta_t\}_{t=0}^T$ is $(\{f_t\}_{t=0}^{T-1}, (\xi_T, \eta_T)^-, 0, T, \delta_0)$-good we have for all $t = 0, \ldots, T$,

$$|\xi_t - x_t^f| \leq \epsilon,\ |\eta_t - y_t^f| \leq \epsilon.$$

By Theorem 8.5, there exist a natural number

$$L_2 > 2L_1,\ \delta_1 \in (0, \delta_0) \tag{8.327}$$

such that the following property holds:

(P28) for each integer $T \geq 2L_2$, each $f_t \in \mathcal{M}$, $t = 0, \ldots, T-1$ satisfying

$$\|f_t - f\| \leq \delta_1,\ t = 0, \ldots, T-1,$$

each (Ω_1)-program $\{x_t\}_{t=0}^T$ and each (Ω_2)-program $\{y_t\}_{t=0}^T$ for which the pair $\{x_t\}_{t=0}^T$, $\{y_t\}_{t=0}^T$ is $(\{f_t\}_{t=0}^{T-1}, 0, T, \delta_1)$-good, for all $t = L_2, \ldots, T - L_2$, we have

$$|x_t - x_f| \leq \delta_0,\ |y_t - y_f| \leq \delta_0. \tag{8.328}$$

Set

$$L_0 = 2L_1 + 2L_2 + 4,\ \delta = L_0^{-1}\delta_1. \tag{8.329}$$

Assume that an integer $T \geq L_0$, $f_t \in \mathcal{M}$, $t = 0, \ldots, T-1$ satisfies

$$\|f_t - f\| \leq \delta,\ t = 0, \ldots, T-1, \tag{8.330}$$

$\{x_t\}_{t=0}^T$ is an (Ω_1)-program and $\{y_t\}_{t=0}^T$ is an (Ω_2)-program such that the pair $\{x_t\}_{t=0}^T$, $\{y_t\}_{t=0}^T$ is $(\{f_t\}_{t=0}^{T-1}, 0, T, \delta)$-good. By (8.329), (8.330) and property (P28) for all $t = L_2, \ldots, T - L_2$ (3.238) holds. In view of (8.329),

$$L_2 + L_1 < T - L_2. \tag{8.331}$$

By (8.331) and (8.328),

$$|x_{L_2+L_1} - x_f| \leq \delta_0,\ |y_{L_2+L_1} - y_f| \leq \delta_0. \tag{8.332}$$

It is clear that the pair of sequences $\{x_t\}_{t=0}^{L_2+L_1}$, $\{y_t\}_{t=0}^{L_2+L_1}$ is

$$(\{f_t\}_{t=0}^{L_2-L_1-1}, (x_{L_2+L_1}, y_{L_2+L_1})^-, 0, L_2 + L_1, \delta)\text{-good}.$$

Combined with property (P27), (8.329), (8.330), and (8.332) this implies that for all $t = 0, \ldots, L_1 + L_2$,

$$|x_t - x_t^f| \leq \epsilon,\ |y_t - y_t^f| \leq \epsilon.$$

Theorem 8.36 is proved. □

8.19 Proof of Theorem 8.33

Theorem 8.33 follows from Proposition 8.30 and the next result.

Theorem 8.37. *Let $\epsilon \in (0,1)$ and $L, l \geq 1$ be integers. Then there exist a natural number $L_0 > L$ and $\delta > 0$ such that for each integer $T \geq L_0$, each $f_t \in \mathcal{M}$, $t = 0, \dots, T-1$ satisfying*

$$\|f_t - f\| \leq \delta, \; t = 0, \dots, T-1,$$

each (Ω_1)-program $\{x_t\}_{t=0}^T$ and each (Ω_2)-program $\{y_t\}_{t=0}^T$ such that

$$x_0 \in \bar{Y}_1(l), \; y_0 \in \bar{Y}_2(l), x_T \in Y_1(l), \; y_T \in Y_2(l)$$

and the pair $\{x_t\}_{t=0}^T$, $\{y_t\}_{t=0}^T$ is $(\{f_t\}_{t=0}^{T-1}, x_0, x_T, y_0, y_T, 0, T, \delta)$-good, for all $t = 0, \dots, L$,

$$|x_t - x_t^{(x_0,y_0)}| \leq \epsilon, \; |y_t - y_t^{(x_0,y_0)}| \leq \epsilon.$$

Proof. By Lemma 8.35, there exist a natural number $L_1 > L$ and $\delta_0 \in (0, \epsilon)$ such that the following property holds:

(P29) for each integer $T \geq L_1$, each $f_t \in \mathcal{M}$, $t = 0, \dots, T-1$ satisfying

$$\|f_t - f\| \leq \delta_0 T^{-1}, \; t = 0, \dots, T-1,$$

each (Ω_1)-program $\{x_t\}_{t=0}^T$, each (Ω_2)-program $\{y_t\}_{t=0}^T$ satisfying

$$x_0 \in \bar{Y}_1(l), \; y_0 \in \bar{Y}_2(l),$$

$$|x_T - x_f| \leq \delta_0, \; |y_T - y_f| \leq \delta_0$$

and such that the pair $\{x_t\}_{t=0}^T$, $\{y_t\}_{t=0}^T$ is $(\{f_t\}_{t=0}^{T-1}, (x_0, x_T, y_0, y_T, 0, T, \delta_0)$-good we have for all $t = 0, \dots, T$,

$$|x_t - x_t^{(x_0,y_0)}| \leq \epsilon, \; |y_t - y_t^{(x_0,y_0)}| \leq \epsilon.$$

By Theorem 8.5, there exist a natural number

$$L_2 > 2l, \; \delta_1 \in (0, \delta_0)$$

such that the following property holds:

(P30) for each integer $T \geq 2L_2$, each $f_t \in \mathcal{M}$, $t = 0, \dots, T-1$ satisfying

$$\|f_t - f\| \leq \delta_1, \; t = 0, \dots, T-1,$$

each (Ω_1)-program $\{x_t\}_{t=0}^T$ and each (Ω_2)-program $\{y_t\}_{t=0}^T$ for which

$$x_0 \in \bar{Y}_1(l), \; y_0 \in \bar{Y}_2(l), \; x_T \in Y_1(l), \; y_T \in Y_2(l)$$

and the pair $\{x_t\}_{t=0}^T$, $\{y_t\}_{t=0}^T$ is $(\{f_t\}_{t=0}^{T-1}, x_0, x_T, y_0, y_T, 0, T, \delta_1)$-good, for all $t = L_2, \dots, T - L_2$, we have

$$|x_t - x_f| \le \delta_0, \; |y_t - y_f| \le \delta_0.$$

Set

$$L_0 = 2L_1 + 2L_2 + 4, \; \delta = L_0^{-1}\delta_1. \tag{8.333}$$

Assume that an integer $T \ge L_0$, $f_t \in \mathcal{M}$, $t = 0, \dots, T - 1$ satisfies

$$\|f_t - f\| \le \delta, \; t = 0, \dots, T - 1, \tag{8.334}$$

$\{x_t\}_{t=0}^T$ is an (Ω_1)-program and $\{y_t\}_{t=0}^T$ is an (Ω_2)-program such that

$$x_0 \in \bar{Y}_1(l), \; y_0 \in \bar{Y}_2(l), \; x_T \in Y_1(l), \; y_T \in Y_2(l) \tag{8.335}$$

and the pair $\{x_t\}_{t=0}^T$, $\{y_t\}_{t=0}^T$ is $(\{f_t\}_{t=0}^{T-1}, x_0, x_T, y_0, y_T, 0, T, \delta)$-good. By (8.333)–(8.335) and property (P30),

$$|x_{L_2+L_1} - x_f| \le \delta_0, \; |y_{L_2+L_1} - y_f| \le \delta_0. \tag{8.336}$$

It is clear that the pair of sequences $\{x_t\}_{t=0}^{L_2+L_1}$, $\{y_t\}_{t=0}^{L_2+L_1}$ is

$$(\{f_t\}_{t=0}^{L_2+L_1-1}, x_0, x_{L_2+L_1}, y_0, y_{L_2+L_1}, 0, L_2 + L_1, \delta)\text{-good}.$$

Combined with property (P29) and (8.333)–(8.336) this implies that for all $t = 0, \dots, L_1 + L_2$,

$$|x_t - x_t^{(x_0,y_0)}| \le \epsilon, \; |y_t - y_t^{(x_0,y_0)}| \le \epsilon.$$

Theorem 8.37 is proved. □

Theorems 8.5 and 8.37 imply the following result.

Theorem 8.38. *Let $\epsilon \in (0, 1)$ and $L, l \ge 1$ be integers. Then there exist a natural number $L_0 > L$ and $\delta > 0$ such that for each integer $T \ge L_0$, each $f_t \in \mathcal{M}$, $t = 0, \dots, T - 1$ satisfying*

$$\|f_t - f\| \le \delta, \; t = 0, \dots, T - 1,$$

each (Ω_1)-program $\{x_t\}_{t=0}^T$ and each (Ω_2)-program $\{y_t\}_{t=0}^T$ such that

$$x_0 \in \bar{Y}_1(l), \; y_0 \in \bar{Y}_2(l)$$

and the pair $\{x_t\}_{t=0}^T$, $\{y_t\}_{t=0}^T$ is $(\{f_t\}_{t=0}^{T-1}, x_0, y_0, 0, T, \delta)$-good, for all $t = 0, \dots, L$,

$$|x_t - x_t^{(x_0,y_0)}| \le \epsilon, \; |y_t - y_t^{(x_0,y_0)}| \le \epsilon.$$

8.20 Optimal Solutions for Games with Discounting

We continue to assume that (A3) and (A4) hold (see Sect. 8.9). In this chapter we prove the following results.

Theorem 8.39. *Let $\epsilon \in (0,1)$ and $L, l \geq 1$ be integers. Then there exist a natural number L_0, $\delta > 0$ and $\lambda \in (0,1)$ such that for each integer $T \geq L_0$, each $f_t \in \mathcal{M}$, $t = 0, \dots, T-1$ satisfying*

$$\|f_t - f\| \leq \delta, \; t = 0, \dots, T-1,$$

each sequence $\{\alpha_t\}_{t=0}^{T-1} \subset (0,1]$ satisfying

$$\alpha_i \alpha_j^{-1} \geq \lambda \text{ for each } i, j \in \{0, \dots, T-1\} \text{ such that } |i-j| \leq L_0,$$

each (Ω_1)-program $\{x_t\}_{t=0}^T$ and each (Ω_2)-program $\{y_t\}_{t=0}^T$ such that

$$x_0 \in \bar{Y}_1(l), \; y_0 \in \bar{Y}_2(l)$$

and the pair $\{x_t\}_{t=0}^T$, $\{y_t\}_{t=0}^T$ is

$$(\{\alpha_t f_t\}_{t=0}^{T-1}, x_0, y_0, 0, T, \delta \min\{\alpha_t : \; t = 0, \dots, T-1\})\text{-good},$$

for all $t = 0, \dots, L$,

$$|x_{T-t} - \bar{x}_t^f| \leq \epsilon, \; |y_{T-t} - \bar{y}_t^f| \leq \epsilon.$$

Theorem 8.39 is proved in Sect. 8.21.

Theorems 8.19 and 8.39 and Proposition 8.30 imply the following result.

Theorem 8.40. *Let $\epsilon \in (0,1)$ and $L \geq 1$ be an integer. Then there exist a natural number L_0, $\delta > 0$ and $\lambda \in (0,1)$ such that for each integer $T \geq L_0$, each $f_t \in \mathcal{M}$, $t = 0, \dots, T-1$ satisfying*

$$\|f_t - f\| \leq \delta, \; t = 0, \dots, T-1,$$

each sequence $\{\alpha_t\}_{t=0}^{T-1} \subset (0,1]$ satisfying

$$\alpha_i \alpha_j^{-1} \geq \lambda \text{ for each } i, j \in \{0, \dots, T-1\} \text{ such that } |i-j| \leq L_0,$$

each (Ω_1)-program $\{x_t\}_{t=0}^T$ and each (Ω_2)-program $\{y_t\}_{t=0}^T$ such that the pair $\{x_t\}_{t=0}^T$, $\{y_t\}_{t=0}^T$ is $(\{\alpha_t f_t\}_{t=0}^{T-1}, 0, T, \delta \min\{\alpha_t : \; t = 0, \dots, T-1\})$-good, for all $t = 0, \dots, L$,

$$|x_t - x_t^f| \leq \epsilon, \; |y_t - y_t^f| \leq \epsilon,$$

$$|x_{T-t} - \bar{x}_t^f| \leq \epsilon, \; |y_{T-t} - \bar{y}_t^f| \leq \epsilon.$$

The next theorem is proved in Sect. 8.22.

Theorem 8.41. *Let $\epsilon \in (0, r_f)$ and $L, l \geq 1$ be integers. Then there exist a natural number L_0, $\delta > 0$ and $\lambda \in (0,1)$ such that for each integer $T \geq L_0$, each $f_t \in \mathcal{M}$, $t = 0, \dots, T-1$ satisfying*

$$\|f_t - f\| \leq \delta,\ t = 0, \dots, T-1,$$

each sequence $\{\alpha_t\}_{t=0}^{T-1} \subset (0,1]$ satisfying

$$\alpha_i \alpha_j^{-1} \geq \lambda \text{ for each } i, j \in \{0, \dots, T-1\} \text{ such that } |i-j| \leq L_0,$$

each (Ω_1)-program $\{x_t\}_{t=0}^T$ and each (Ω_2)-program $\{y_t\}_{t=0}^T$ such that

$$x_0 \in \bar{Y}_1(l),\ y_0 \in \bar{Y}_2(l), x_T \in Y_1(l),\ y_T \in Y_2(l)$$

and the pair $\{x_t\}_{t=0}^T$, $\{y_t\}_{t=0}^T$ is

$$(\{\alpha_t f_t\}_{t=0}^{T-1}, x_0, x_T, y_0, y_T, 0, T, \delta \min\{\alpha_t :\ t = 0, \dots, T-1\}) \text{ good},$$

for all $t = 0, \dots, L$,

$$|x_t - x_t^{(x_0, y_0)}| \leq \epsilon,\ |y_t - y_t^{(x_0, y_0)}| \leq \epsilon,$$

$$|x_{T-t} - \bar{x}_t^{(x_T, y_T)}| \leq \epsilon,\ |y_{T-t} - \bar{y}_t^{(x_T, y_T)}| \leq \epsilon.$$

Theorems 8.19 and 8.41 and Proposition 8.30 imply the following result.

Theorem 8.42. *Let $\epsilon \in (0, r_f)$ and $L, l \geq 1$ be integers. Then there exist a natural number L_0, $\delta > 0$ and $\lambda \in (0,1)$ such that for each integer $T \geq L_0$, each $f_t \in \mathcal{M}$, $t = 0, \dots, T-1$ satisfying*

$$\|f_t - f\| \leq \delta,\ t = 0, \dots, T-1,$$

each sequence $\{\alpha_t\}_{t=0}^{T-1} \subset (0,1]$ satisfying

$$\alpha_i \alpha_j^{-1} \geq \lambda \text{ for each } i, j \in \{0, \dots, T-1\} \text{ such that } |i-j| \leq L_0,$$

each (Ω_1)-program $\{x_t\}_{t=0}^T$ and each (Ω_2)-program $\{y_t\}_{t=0}^T$ such that

$$x_0 \in \bar{Y}_1(l),\ y_0 \in \bar{Y}_2(l)$$

and the pair $\{x_t\}_{t=0}^T$, $\{y_t\}_{t=0}^T$ is

$$(\{\alpha_t f_t\}_{t=0}^{T-1}, x_0, y_0, 0, T, \delta \min\{\alpha_t :\ t = 0, \dots, T-1\})\text{-good},$$

for all $t = 0, \dots, L$,

$$|x_t - x_t^{(x_0, y_0)}| \leq \epsilon,\ |y_t - y_t^{(x_0, y_0)}| \leq \epsilon.$$

8.21 Proof of Theorem 8.39

By Theorem 8.31, there exist a natural number $L_1 > L$ and $\delta_1 > 0$ such that the following property holds:

(P31) for each integer $T \geq L_1$, each $f_t \in \mathcal{M}$, $t = 0, \dots, T-1$ satisfying

$$\|f_t - f\| \leq \delta_1,\ t = 0, \dots, T-1,$$

each (Ω_1)-program $\{\xi_t\}_{t=0}^T$ and each (Ω_2)-program $\{\eta_t\}_{t=0}^T$ such that

$$\xi_0 \in \bar{Y}_1(l),\ \eta_0 \in \bar{Y}_2(l)$$

and the pair $\{\xi_t\}_{t=0}^T$, $\{\eta_t\}_{t=0}^T$ is $(\{f_t\}_{t=0}^{T-1}, \xi_0, \eta_0, 0, T, \delta_1)$-good, for all $t = 0, \dots, L$,

$$|\xi_{T-t} - \bar{x}_t^f| \leq \epsilon,\ |\eta_{T-t} - \bar{y}_t^f| \leq \epsilon.$$

By Theorem 8.19, there exist a natural number $L_2 > 2L$ and $\delta_2 \in (0, \delta_1)$, $\lambda_1 \in (0,1)$ such that the following property holds:

(P32) for each integer $T > 2L_2$, each $f_t \in \mathcal{M}$, $t = 0, \dots, T-1$ satisfying

$$\|f_t - f\| \leq \delta_2,\ t = 0, \dots, T-1,$$

each sequence $\{\alpha_t\}_{t=0}^{T-1} \subset (0,1]$ such that for each pair of integers $i, j \in [0, T-1]$ satisfying $|i-j| \leq L_2$,

$$\alpha_i \alpha_j^{-1} \geq \lambda_1,$$

each (Ω_1)-program $\{x_t\}_{t=0}^T$ and each (Ω_2)-program $\{y_t\}_{t=0}^T$ such that

$$x_0 \in \bar{Y}_1(l),\ y_0 \in \bar{Y}_2(l)$$

and the pair $\{x_t\}_{t=0}^T$, $\{y_t\}_{t=0}^T$ is $(\{\alpha_t f_t\}_{t=0}^{T-1}, x_0, y_0, 0, T, \delta \min\{\alpha_t : t = 0, \dots, T-1\})$-good we have for all $t = L_2, \dots, T-L_2$,

$$|x_t - x_f| \leq r_f/4,\ |y_t - y_f| \leq r_f/4.$$

Fix

$$\delta = \min\{\delta_1, \delta_2, 1\}/2, \tag{8.337}$$

a number λ such that

$$1 > \lambda > \lambda_1,\ (\|f\| + 1)(1-\lambda)\lambda^{-1} < \delta \tag{8.338}$$

and

$$L_0 = 4L_1 + 4L_2 + 4. \tag{8.339}$$

Assume that an integer $T \geq L_0$, $f_t \in \mathcal{M}$, $t = 0, \dots, T-1$ satisfies

$$\|f_t - f\| \leq \delta,\ t = 0, \dots, T-1, \tag{8.340}$$

a sequence $\{\alpha_t\}_{t=0}^{T-1} \subset (0,1]$ satisfies

$$\alpha_i \alpha_j^{-1} \geq \lambda \text{ for each } i, j \in \{0, \dots, T-1\} \text{ such that } |i-j| \leq L_0, \tag{8.341}$$

$\{x_t\}_{t=0}^T$ is an (Ω_1)-program and $\{y_t\}_{t=0}^T$ is an (Ω_2)-program such that

$$x_0 \in \bar{Y}_1(l),\ y_0 \in \bar{Y}_2(l) \tag{8.342}$$

and the pair $\{x_t\}_{t=0}^T$, $\{y_t\}_{t=0}^T$ is $(\{\alpha_t f_t\}_{t=0}^{T-1}, x_0, y_0, 0, T, \delta \min\{\alpha_t : t = 0, \dots, T-1\})$-good. Property (P32) and (8.337)–(8.342) imply that for all $t = L_2, \dots, T - L_2$,

$$|x_t - x_f| \le r_f/4,\ |y_t - y_f| \le r_f/4. \tag{8.343}$$

In view of (8.339) and (8.343),

$$T - (L_2 + L_1) > L_2 + L_1,$$

$$|x_{T-L_2-L_1} - x_f| \le r_f/4,\ |y_{T-L_2-L_1} - y_f| \le r_f/4. \tag{8.344}$$

It is clear that the pair of sequences $\{x_t\}_{t=T-L_2-L_1}^T$, $\{y_t\}_{t=T-L_2-L_1}^T$ is

$$(\{\alpha_t f_t\}_{t=T-L_2-L_1}^{T-1}, x_{T-L_2-L_1}, y_{T-L_2-L_1},$$

$$T - L_2 - L_1, T, \delta \min\{\alpha_t : t = 0, \dots, T-1\})\text{-good}.$$

This implies that the pair of sequences $\{x_t\}_{t=T-L_2-L_1}^T$, $\{y_t\}_{t=T-L_2-L_1}^T$ is

$$(\{\alpha_{T-L_2-L_1}^{-1} \alpha_t f_t\}_{t=T-L_2-L_1}^{T-1}, x_{T-L_2-L_1}, y_{T-L_2-L_1}, T - L_2 - L_1, T, \delta)\text{-good}.$$

By (8.337)–(8.341), for all integers $t \in [T - L_1 - L_2, T - 1]$,

$$\|f - \alpha_{T-L_2-L_1}^{-1} \alpha_t f_t\| \le \|f - f_t\| + \|f_t - \alpha_{T-L_2-L_1}^{-1} \alpha_t f_t\|$$

$$\le \delta + (\|f\| + 1)|\lambda - 1|\lambda^{-1} \le \delta_1.$$

Combined with property (P31) and (8.344) this implies that for all $t = 0, \dots, L$,

$$|x_{T-t} - \bar{x}_t^f| \le \epsilon,\ |y_{T-t} - \bar{y}_t^f| \le \epsilon.$$

Theorem 8.39 is proved. □

8.22 Proof of Theorem 8.41

Theorem 8.41 follows from Proposition 8.30 and the next result.

Theorem 8.43. *Let $\epsilon \in (0, r_f)$ and $L, l \ge 1$ be integers. Then there exist a natural number L_0, $\delta > 0$ and $\lambda \in (0, 1)$ such that for each integer $T \ge L_0$, each $f_t \in \mathcal{M}$, $t = 0, \dots, T-1$ satisfying*

$$\|f_t - f\| \le \delta,\ t = 0, \dots, T-1,$$

each sequence $\{\alpha_t\}_{t=0}^{T-1} \subset (0, 1]$ *satisfying*

$$\alpha_i \alpha_j^{-1} \geq \lambda \text{ for each } i, j \in \{0, \ldots, T-1\} \text{ such that } |i-j| \leq L_0,$$

each (Ω_1)*-program* $\{x_t\}_{t=0}^{T}$ *and each* (Ω_2)*-program* $\{y_t\}_{t=0}^{T}$ *such that*

$$x_0 \in \bar{Y}_1(l),\ y_0 \in \bar{Y}_2(l),\ x_T \in Y_1(l),\ y_T \in Y_2(l)$$

and the pair $\{x_t\}_{t=0}^{T}$, $\{y_t\}_{t=0}^{T}$ *is* $(\{\alpha_t f_t\}_{t=0}^{T-1}, x_0, x_T, y_0, y_T, 0, T, \delta \min\{\alpha_t : t = 0, \ldots, T-1\})$*-good, for all* $t = 0, \ldots, L$,

$$|x_t - x_t^{(x_0,y_0)}| \leq \epsilon,\ |y_t - y_t^{(x_0,y_0)}| \leq \epsilon.$$

Proof. By Theorem 8.37, there exist a natural number $L_1 > L$, $\delta_1 > 0$ such that the following property holds:

(P33) for each integer $T \geq L_1$, each $f_t \in \mathcal{M}$, $t = 0, \ldots, T-1$ satisfying

$$\|f_t - f\| \leq \delta_1,\ t = 0, \ldots, T-1,$$

each (Ω_1)-program $\{\xi_t\}_{t=0}^{T}$ and each (Ω_2)-program $\{\eta_t\}_{t=0}^{T}$ such that

$$\xi_0 \in \bar{Y}_1(l),\ \eta_0 \in \bar{Y}_2(l)$$

and the pair $\{\xi_t\}_{t=0}^{T}$, $\{\eta_t\}_{t=0}^{T}$ is $(\{f_t\}_{t=0}^{T-1}, \xi_0, \xi_T, \eta_0, \eta_T, 0, T, \delta_1)$-good, for all $t = 0, \ldots, L$, we have

$$|\xi_t - x_t^{(x_0,y_0)}| \leq \epsilon,\ |\eta_t - y_t^{(x_0,y_0)}| \leq \epsilon.$$

By Theorem 8.19, there exist a natural number $L_2 > 2L$, $\delta_2 \in (0, \delta_1)$ and $\lambda_1 \in (0, 1)$ such that the following property holds:

(P34) for each integer $T > 2L_2$, each $f_t \in \mathcal{M}$, $t = 0, \ldots, T-1$ satisfying

$$\|f_t - f\| \leq \delta_2,\ t = 0, \ldots, T-1,$$

each sequence $\{\alpha_t\}_{t=0}^{T-1} \subset (0, 1]$ such that for each pair of integers $i, j \in [0, T-1]$ satisfying $|i - j| \leq L_2$,

$$\alpha_i \alpha_j^{-1} \geq \lambda_1,$$

each (Ω_1)-program $\{x_t\}_{t=0}^{T}$ and each (Ω_2)-program $\{y_t\}_{t=0}^{T}$ such that

$$x_0 \in \bar{Y}_1(l),\ y_0 \in \bar{Y}_2(l),\ x_T \in Y_1(l),\ y_T \in Y_2(l)$$

and the pair $\{x_t\}_{t=0}^{T}$, $\{y_t\}_{t=0}^{T}$ is $(\{\alpha_t f_t\}_{t=0}^{T-1}, x_0, x_T, y_0, y_T, 0, T, \delta \min\{\alpha_t : t = 0, \ldots, T-1\})$-good we have for all $t = L_2, \ldots, T - L_2$,

$$|x_t - x_f| \leq r_f/4,\ |y_t - y_f| \leq r_f/4.$$

Fix

$$\delta = \min\{\delta_1, \delta_2, 1\}/2, \tag{8.345}$$

a number λ such that

$$1 > \lambda > \lambda_1, \ (\|f\| + 1)(1 - \lambda)\lambda^{-1} < \delta \tag{8.346}$$

and

$$L_0 = 4L_1 + 4L_2 + 4. \tag{8.347}$$

Assume that an integer $T \geq L_0$, $f_t \in \mathcal{M}$, $t = 0, \dots, T-1$ satisfies

$$\|f_t - f\| \leq \delta, \ t = 0, \dots, T-1, \tag{8.348}$$

a sequence $\{\alpha_t\}_{t=0}^{T-1} \subset (0, 1]$ satisfies

$$\alpha_i \alpha_j^{-1} \geq \lambda \text{ for each } i, j \in \{0, \dots, T-1\} \text{ such that } |i - j| \leq L_0, \tag{8.349}$$

$\{x_t\}_{t=0}^T$ is an (Ω_1)-program and $\{y_t\}_{t=0}^T$ is an (Ω_2)-program such that

$$x_0 \in \bar{Y}_1(l), \ y_0 \in \bar{Y}_2(l), \ x_T \in Y_1(l), \ y_T \in Y_2(l) \tag{8.350}$$

and the pair $\{x_t\}_{t=0}^T$, $\{y_t\}_{t=0}^T$ is $(\{\alpha_t f_t\}_{t=0}^{T-1}, x_0, x_T, y_0, y_T, 0, T, \delta \min\{\alpha_t : t = 0, \dots, T-1\})$-good. Combined with property (P34) and (8.345)–(8.347), (8.349), and (8.350) this implies that for all $t = L_2, \dots, T - L_2$,

$$|x_t - x_f| \leq r_f/4, \ |y_t - y_f| \leq r_f/4. \tag{8.351}$$

In view of (8.347) and (8.351),

$$L_2 + L_1 < T - L_2,$$

$$|x_{L_2+L_1} - x_f| \leq r_f/4, \ |y_{L_2+L_1} - y_f| \leq r_f/4. \tag{8.352}$$

It is clear that the pair of sequences $\{x_t\}_{t=0}^{L_2+L_1}$, $\{y_t\}_{t=0}^{L_2+L_1}$ is

$$(\{\alpha_t f_t\}_{t=0}^{L_2+L_1-1}, x_0, x_{L_2+L_1}, y_0, y_{L_2+L_1}, 0, L_2 + L_1,$$

$$\delta \min\{\alpha_t : \ t = 0, \dots, T-1\})\text{-good.}$$

This implies that the pair of sequences $\{x_t\}_{t=0}^{L_2+L_1}$, $\{y_t\}_{t=0}^{L_2+L_1}$ is

$$(\{\alpha_0^{-1}\alpha_t f_t\}_{t=0}^{L_2+L_1-1}, x_0, x_{L_2+L_1}, y_0, y_{L_2+L_1}, 0, L_2 + L_1, \delta)\text{-good.}$$

By (8.345)–(8.349), for all integers $t \in [0, L_1 + L_2 - 1]$,

$$\|f - \alpha_0^{-1}\alpha_t f_t\| \leq \|f - f_t\| + \|f_t - \alpha_0^{-1}\alpha_t f_t\|$$

$$\leq \delta + (\|f\| + 1)|\lambda - 1|\lambda^{-1} \leq \delta_1.$$

Combined with property (P33) and (8.350) this implies that for all $t = 0, \dots, L$,

$$|x_t - x_t^{(x_0, y_0)}| \leq \epsilon, \ |y_t - y_t^{(x_0, y_0)}| \leq \epsilon.$$

Theorem 8.43 is proved.

References

[1] Alvarez O, Bardi M (2007) Ergodic problems in differential games. In: Advances in dynamic game theory. Birkhäuser, Boston, pp 131–152

[2] Anderson BDO, Moore JB (1971) Linear optimal control. Prentice-Hall, Englewood Cliffs

[3] Arkin VI, Evstigneev IV (1987) Stochastic models of control and economic dynamics. Academic, London

[4] Aseev SM, Kryazhimskiy AV (2004) The Pontryagin maximum principle and transversality conditions for a class of optimal control problems with infinite time horizons. SIAM J Control Optim 43:1094–1119

[5] Aseev SM, Veliov VM (2012) Maximum principle for infinite-horizon optimal control problems with dominating discount. Dyn Continuous Discrete Impuls Syst Ser B 19:43–63

[6] Atsumi H (1965) Neoclassical growth and the efficient program of capital accumulation. Rev Econ Stud 32:127–136

[7] Aubin JP, Ekeland I (1984) Applied nonlinear analysis. Wiley Interscience, New York

[8] Aubry S, Le Daeron PY (1983) The discrete Frenkel-Kontorova model and its extensions I. Physica D 8:381–422

[9] Bachir M, Blot J (2015) Infinite dimensional infinite-horizon Pontryagin principles for discrete-time problems. Set-Valued Var Anal 23:43–54

[10] Bardi M (2009) On differential games with long-time-average cost. In: Advances in dynamic games and their applications. Birkhäuser, Boston, pp 3–18

[11] Baumeister J, Leitao A, Silva GN (2007) On the value function for nonautonomous optimal control problem with infinite horizon. Syst Control Lett 56:188–196

[12] Blot J (2009) Infinite-horizon Pontryagin principles without invertibility. J Nonlinear Convex Anal 10:177–189

A.J. Zaslavski, *Discrete-Time Optimal Control and Games on Large Intervals*, Springer Optimization and Its Applications 119,
DOI 10.1007/978-3-319-52932-5

[13] Blot J, Cartigny P (2000) Optimality in infinite-horizon variational problems under sign conditions. J Optim Theory Appl 106:411–419
[14] Blot J, Hayek N (2000) Sufficient conditions for infinite-horizon calculus of variations problems. ESAIM Control Optim Calc Var 5:279–292
[15] Blot J, Hayek N (2014) Infinite-horizon optimal control in the discrete-time framework. SpringerBriefs in optimization. Springer, New York
[16] Bright I (2012) A reduction of topological infinite-horizon optimization to periodic optimization in a class of compact 2-manifolds. J Math Anal Appl 394:84–101
[17] Brock WA (1970) On existence of weakly maximal programmes in a multi-sector economy. Rev Econ Stud 37:275–280
[18] Carlson DA (1990) The existence of catching-up optimal solutions for a class of infinite horizon optimal control problems with time delay. SIAM J Control Optim 28:402–422
[19] Carlson DA, Haurie A, Leizarowitz A (1991) Infinite horizon optimal control. Springer, Berlin
[20] Carlson DA, Jabrane A, Haurie A (1987) Existence of overtaking solutions to infinite dimensional control problems on unbounded time intervals. SIAM J Control Optim 25:1517–1541
[21] Cartigny P, Michel P (2003) On a sufficient transversality condition for infinite horizon optimal control problems. Autom J IFAC 39:1007–1010
[22] Coleman BD, Marcus M, Mizel VJ (1992) On the thermodynamics of periodic phases. Arch Ration Mech Anal 117:321–347
[23] Damm T, Grune L, Stieler M, Worthmann K (2014) An exponential turnpike theorem for dissipative discrete time optimal control problems. SIAM J Control Optim 52:1935–1957
[24] De Oliveira VA, Silva GN (2009) Optimality conditions for infinite horizon control problems with state constraints. Nonlinear Anal 71:1788–1795
[25] Evstigneev IV, Flam SD (1998) Rapid growth paths in multivalued dynamical systems generated by homogeneous convex stochastic operators. Set-Valued Anal 6:61–81
[26] Gaitsgory V, Grune L, Thatcher N (2015) Stabilization with discounted optimal control. Syst Control Lett 82:91–98
[27] Gaitsgory V, Rossomakhine S, Thatcher N (2012) Approximate solution of the HJB inequality related to the infinite horizon optimal control problem with discounting. Dyn Continuous Discrete Impuls Syst Ser B 19:65–92
[28] Gale D (1967) On optimal development in a multi-sector economy. Rev Econ Stud 34:1–18
[29] Ghosh MK, Mallikarjuna Rao KS (2005) Differential games with ergodic payoff. SIAM J Control Optim 43:2020–2035
[30] Guo X, Hernandez-Lerma O (2005) Zero-sum continuous-time Markov games with unbounded transition and discounted payoff rates. Bernoulli 11:1009–1029

[31] Hayek N (2011) Infinite horizon multiobjective optimal control problems in the discrete time case. Optimization 60:509–529
[32] Hernandez-Lerma O, Lasserre JB (2001) Zero-sum stochastic games in Borel spaces: average payoff criteria. SIAM J Control Optim 39:1520–1539
[33] Jasso-Fuentes H, Hernandez-Lerma O (2008) Characterizations of overtaking optimality for controlled diffusion processes. Appl Math Optim 57:349–369
[34] Khan MA, Zaslavski AJ (2010) On two classical turnpike results for the Robinson-Solow-Srinivisan (RSS) model. Adv Math Econ 13:47–97
[35] Khlopin DV (2013) Necessity of vanishing shadow price in infinite horizon control problems. J Dyn Control Syst 19:519–552
[36] Kolokoltsov V, Yang W (2012) The turnpike theorems for Markov games. Dyn Games Appl 2:294–312
[37] Leizarowitz A (1985) Infinite horizon autonomous systems with unbounded cost. Appl Math Opt 13:19–43
[38] Leizarowitz A (1986) Tracking nonperiodic trajectories with the overtaking criterion. Appl Math Opt 14:155–171
[39] Leizarowitz A, Mizel VJ (1989) One dimensional infinite horizon variational problems arising in continuum mechanics. Arch Ration Mech Anal 106:161–194
[40] Leizarowitz A, Zaslavski AJ (2005) On a class of infinite horizon optimal control problems with periodic cost functions. J Nonlinear Convex Anal 6:71–91
[41] Lykina V, Pickenhain S, Wagner M (2008) Different interpretations of the improper integral objective in an infinite horizon control problem. J Math Anal Appl 340:498–510
[42] Makarov VL, Rubinov AM (1977) Mathematical theory of economic dynamics and equilibria. Springer, New York
[43] Malinowska AB, Martins N, Torres DFM (2011) Transversality conditions for infinite horizon variational problems on time scales. Optim Lett 5:41–53
[44] Mammadov M (2014) Turnpike theorem for an infinite horizon optimal control problem with time delay. SIAM J Control Optim 52:420–438
[45] Marcus M, Zaslavski AJ (1999) On a class of second order variational problems with constraints. Israel J Math 111:1–28
[46] Marcus M, Zaslavski AJ (1999) The structure of extremals of a class of second order variational problems. Ann Inst H Poincaré Anal Non Linéaire 16:593–629
[47] Marcus M, Zaslavski AJ (2002) The structure and limiting behavior of locally optimal minimizers. Ann Inst H Poincaré Anal Non Linéaire 19:343–370
[48] McKenzie LW (1976) Turnpike theory. Econometrica 44:841–866
[49] Mordukhovich BS (1990) Minimax design for a class of distributed parameter systems. Autom Remote Control 50:1333–1340

[50] Mordukhovich BS (2011) Optimal control and feedback design of state-constrained parabolic systems in uncertainly conditions. Appl Anal 90:1075–1109

[51] Mordukhovich BS, Shvartsman I (2004) Optimization and feedback control of constrained parabolic systems under uncertain perturbations. In: Optimal control, stabilization and nonsmooth analysis. Lecture notes in control information science. Springer, Berlin, pp 121–132

[52] Moser J (1986) Minimal solutions of variational problems on a torus. Ann Inst H Poincaré Anal Non Linéaire 3:229–272

[53] Ocana Anaya E, Cartigny P, Loisel P (2009) Singular infinite horizon calculus of variations. Applications to fisheries management. J Nonlinear Convex Anal 10:157–176

[54] Pickenhain S, Lykina V, Wagner M (2008) On the lower semicontinuity of functionals involving Lebesgue or improper Riemann integrals in infinite horizon optimal control problems. Control Cybern 37:451–468

[55] Porretta A, Zuazua E (2013) Long time versus steady state optimal control. SIAM J Control Optim 51:4242–4273

[56] Prieto-Rumeau T, Hernandez-Lerma O (2005) Bias and overtaking equilibria for zero-sum continuous-time Markov games. Math Methods Oper Res 61:437–454

[57] Reich S, Zaslavski AJ (2014) Genericity in nonlinear analysis. Springer, New York

[58] Rockafellar RT (1970) Convex analysis. Princeton University Press, Princeton

[59] Rubinov AM (1984) Economic dynamics. J Sov Math 26:1975–2012

[60] Samuelson PA (1965) A catenary turnpike theorem involving consumption and the golden rule. Am Econ Rev 55:486–496

[61] Trelat E, Zuazua E (2015) The turnpike property in finite-dimensional nonlinear optimal control. J Differ Equ 218:81–114

[62] von Weizsacker CC (1965) Existence of optimal programs of accumulation for an infinite horizon. Rev Econ Stud 32:85–104

[63] Zaslavski AJ (1987) Ground states in Frenkel-Kontorova model. Math USSR Izv 29:323–354

[64] Zaslavski AJ (1995) Optimal programs on infinite horizon 1. SIAM J Control Optim 33:1643–1660

[65] Zaslavski AJ (1995) Optimal programs on infinite horizon 2. SIAM J Control Optim 33:1661–1686

[66] Zaslavski AJ (1998) Turnpike theorem for convex infinite dimensional discrete-time control systems. J Convex Anal 5:237–248

[67] Zaslavski AJ (1999) Turnpike property for dynamic discrete time zero-sum games. Abst Appl Anal 4:21–48

[68] Zaslavski AJ (2000) Turnpike theorem for nonautonomous infinite dimensional discrete-time control systems. Optimization 48:69–92

[69] Zaslavski AJ (2005) The turnpike property of discrete-time control problems arising in economic dynamics. Discrete Continuous Dyn Syst B 5:861–880

[70] Zaslavski AJ (2006) Turnpike properties in the calculus of variations and optimal control. Springer, New York

[71] Zaslavski AJ (2007) Turnpike results for a discrete-time optimal control systems arising in economic dynamics. Nonlinear Anal 67:2024–2049

[72] Zaslavski AJ (2009) Two turnpike results for a discrete-time optimal control systems. Nonlinear Anal 71:902–909

[73] Zaslavski AJ (2010) Stability of a turnpike phenomenon for a discrete-time optimal control systems. J Optim Theory Appl 145:597–612

[74] Zaslavski AJ (2010) Optimization on metric and normed spaces. Springer, New York

[75] Zaslavski AJ (2011) Turnpike properties of approximate solutions for discrete-time control systems. Commun Math Anal 11:36–45

[76] Zaslavski AJ (2011) Structure of approximate solutions for a class of optimal control systems. J Math Appl 34:1–14

[77] Zaslavski AJ (2011) Stability of a turnpike phenomenon for a class of optimal control systems in metric spaces. Numer Algebra Control Optim 1:245–260

[78] Zaslavski AJ (2011) The existence and structure of approximate solutions of dynamic discrete time zero-sum games. J Nonlinear Convex Anal 12:49–68

[79] Zaslavski AJ (2012) A generic turnpike result for a class of discrete-time optimal control systems. Dyn Continuous Discrete Impuls Syst Ser B 19:225–265

[80] Zaslavski AJ (2012) Existence and turnpike properties of solutions of dynamic discrete time zero-sum games. Commun Appl Anal 16:261–276

[81] Zaslavski AJ (2013) Nonconvex optimal control and variational problems. Springer optimization and its applications. Springer, New York

[82] Zaslavski AJ (2013) Two turnpike results for dynamic discrete time zero-sum games. J Nonlinear Convex Anal 14:687–717

[83] Zaslavski AJ (2014) Turnpike properties of approximate solutions of nonconcave discrete-time optimal control problems. J Convex Anal 21:681–701

[84] Zaslavski AJ (2014) Turnpike phenomenon and infinite horizon optimal control. Springer optimization and its applications. Springer, New York

[85] Zaslavski AJ (2014) Turnpike properties for nonconcave problems. Adv Math Econ 18:101–134

[86] Zaslavski AJ (2014) Structure of solutions of discrete time optimal control problems in the regions close to the endpoints. Set-Valued Var Anal 22:809–842

[87] Zaslavski AJ (2014) Turnpike theory for dynamic zero-sum games. In: Proceedings of the workshop "Variational and optimal control problems on unbounded domains", Haifa, 2012. Contemporary mathematics, vol. 619, pp 225–247

[88] Zaslavski AJ (2014) Turnpike properties of approximate solutions of dynamic discrete time zero-sum games. J Dyn Games 2014:299–330

[89] Zaslavski AJ (2014) Stability of the turnpike phenomenon in discrete-time optimal control problems. SpringerBriefs in optimization. Springer, New York

[90] Zaslavski AJ (2015) Convergence of solutions of optimal control problems with discounting on large intervals in the regions close to the endpoints. Set-Valued Var Anal 23:191–204

[91] Zaslavski AJ (2015) Structure of approximate solutions of discrete time optimal control Bolza problems on large intervals. Nonlinear Anal 123/124:23–55

[92] Zaslavski AJ (2015) Discrete time optimal control problems on large intervals. Adv Math Econ 19:91–135

[93] Zaslavski AJ (2015) Turnpike theory of continuous-time linear optimal control problems. Springer optimization and its applications. Springer, New York

[94] Zaslavski AJ (2016) Convergence of solutions of concave discrete optimal control problems in the regions close to the endpoints. Commun Appl Nonlinear Anal 23:1–10

[95] Zaslavski AJ (2016) Structure of solutions of optimal control problems on large intervals: a survey of recent results. Pure Appl Funct Anal 1:123–158

Index

A.J. Zaslavski, *Discrete-Time Optimal Control and Games on Large Intervals*, Springer Optimization and Its Applications 119,
DOI 10.1007/978-3-319-52932-5

The manufacturer's authorised representative in the EU is Springer Nature Customer Service Centre GmbH, Europaplatz 3, 69115 Heidelberg, Germany. If you have any concerns regarding our products, please contact ProductSafety@springernature.com

Printed and bound by CPI Group (UK) Ltd, Croydon, CR0 4YY
05/10/2026
02228131-0009